Measurements of Optical Fibers and Devices: Theory and Experiments

The Artech House Telecom Library

Telecommunications: An Interdisciplinary Text, Leonard Lewin, ed.
Telecommunications in the U.S.: Trends and Policies, Leonard Lewin, ed.
The ITU in a Changing World by George A. Codding, Jr. and Anthony M. Rutkowski
Integrated Services Digital Networks by Anthony M. Rutkowski
The ISDN Workshop: INTUG Proceedings, G. Russell Pipe, ed.
The Deregulation of International Telecommunications by Ronald S. Eward
The Competition for Markets in International Telecommunications by Ronald S. Eward
Machine Cryptography and Modern Cryptanalysis by Cipher A. Deavours and Louis Kruh
Teleconferencing Technology and Applications by Christine H. Olgren and Lorne A. Parker
The Executive Guide to Video Teleconferencing by Ronald J. Bohm and Lee B. Templeton
World-Traded Services: The Challenge for the Eighties by Raymond J. Krommenacker
Microcomputer Tools for Communication Engineering by S.T. Li, J.C. Logan, J.W. Rockway, and D.W.S. Tam
World Atlas of Satellites, Donald M. Jansky, ed.
Communication Satellites in the Geostationary Orbit by Donald M. Jansky and Michael C. Jeruchim
Communication Satellites: Power Politics in Space by Larry Martinez
New Directions in Satellite Communications: Challenges for North and South, Heather E. Hudson, ed.
Television Programming across National Boundaries: The EBU and OIRT Experience by Ernest Eugster
Evaluating Telecommunication Technology in Medicine by David Conrath, Earl Dunn, and Christopher Higgins
Measurements in Optical Fibers and Devices: Theory and Experiments by Giovanni Cancellieri and Umberto Ravaioli
Fiber Optics Communications, Henry F. Taylor, ed.
Advanced Mathematics for Practicing Engineers by Kurt Arbenz and Jean-Claude Martin
Mathematical Methods for Engineering Applications by Kurt Arbenz and Alfred Wohlhauser
Principles of Secure Communication Systems by Don J. Torrieri
Telecommunication Systems by Pierre-Gerard Fontolliet
Signal Theory and Processing by Frederic de Coulon
Techniques in Data Communications by Ralph Glasgal
Advances in Computer Communications and Networking, Wesley W. Chu, ed.
The Public Manager's Phone Book by Paul Daubitz
The Cable Television Technical Handbook by Bobby Harrell

Measurements of Optical Fibers and Devices: Theory and Experiments

Giovanni Cancellieri
Umberto Ravaioli

ARTECH HOUSE, INC.
610 Washington St.
Dedham, MA 02026

International Standard Book Number: 0-89006-133-5
Library of Congress Catalog Card Number: 83-072775

86 87 10 9 8 7 6 5 4 3 2

TABLE OF CONTENTS

1 **LIGHT PROPAGATION IN IDEAL MULTIMODE FIBERS**

1.1 Outline of the Electromagnetic Model **1**

1.1.1 General Considerations 2

1.1.2 Exact Electromagnetic Theory of Round Step-Index Fibers 5

1.1.3 Approximate Evaluation of the Total Number of Guided Modes 7

1.1.4 WKB Approximate Solution for Graded-Index Fibers ... 8

1.2 The Ray-Optics Model **11**

1.2.1 Light Acceptance in Step Index Fibers 12

1.2.2 Correspondence between Modes and Rays 14

1.2.3 Light Acceptance in Graded-Index Fibers 17

1.2.4 General Ray-Optics in Inhomogeneous Media 19

1.2.5 Constants of the Motion 22

1.2.6 Ray Tracing 24

1.2.7 Ray Domain 28

1.2.8 The Radiance Law 29

1.2.9 Ray Distribution 32

1.2.10 Domain (r_1, θ_1) 34

1.3 Intermodal Time Dispersion **37**

1.3.1 Time Dispersion in Step-Index Fibers 37

1.3.2 α-Profiles 42

1.3.3 WKB Approximation for α-Profiles 44

1.3.4 Effects of Leaky Modes 49

1.3.5 Ray-Optics Approach for α-Profiles 50

1.3.6 The *rms* Pulsewidth 55

1.3.7 Simplified Description of the Launching Condition 57
1.3.8 Effects of the Launching Condition 59
1.3.9 Effects of Profile Imperfections 62
1.3.10 Compensation Effect of Time Dispersion in Fiber Chains 68

1.4 Intramodal Time Dispersion 69
1.4.1 Group Delays in the Presence of Material Dispersion 70
1.4.2 Effects of the Finite Extension of the Source Spectral Width 72
1.4.3 Wavelength of Zero Material Dispersion 73
1.4.4 Total Time Dispersion 74
1.4.5 Optimum Index Profile in the Presence of Material Dispersion 76
1.4.6 Linear Profile Dispersion 78
1.4.7 Nonlinear Profile Dispersion 79

2 REAL FIBERS
2.1 Intrinsic Attenuation 85
2.1.1 Causes of Intrinsic Attenuation 86
2.1.2 Absorption Loss 87
2.1.3 Scattering Loss 91

2.2 Fiber Perturbations 93
2.2.1 Kinds of Fiber Perturbation 95
2.2.2 Outline of the Electromagnetic Approach to the Study of Fiber Perturbations 96
2.2.3 Ray-Optics Approach to Core Diameter Fluctuations 98
2.2.4 Ray-Optics Approach to Microbending. 104
2.2.5 Optical Power Flow Equation in the Electromagnetic Model 110
2.2.6 Optical Power Flow Equation in the Ray-Optics Model 113

2.2.7 Solutions of the Optical Power Flow Equation 115
2.2.8 The Time-Dependent Equation 118

2.3 Transmission Properties of Perturbed Fibers 119
2.3.1 Exact Solution of the Time-Independent Equation 120
2.3.2 Perturbative Solutions of the Time-Dependent Equation 123
2.3.3 Evolution of the Attenuation Coefficient while Approaching Steady-State 127
2.3.4 Evolution of the Baseband Response while Approaching Steady-State 129
2.3.5 Optimization of the Cabling Process 133

2.4 Effects of Joints 136
2.4.1 Splices and Connectors 136
2.4.2 Exact Electromagnetic Model of a Connector 137
2.4.3 Linear Model of a Connector 141
2.4.4 Matricial Approach to the Study of Cascaded Fibers 142
2.4.5 Power Loss of a Connector between Two Unperturbed Fibers 144
2.4.6 Power Loss of a Connector between Two Perturbed Fibers 148
2.4.7 Outline of Connector Time Dispersion Properties 150

3 TRANSDUCERS
3.1 Optical Sources 155
3.1.1 Collimated Beam Lasers 156
3.1.2 Top Emitting LEDs 159
3.1.3 Edge Emitting LEDs 163
3.1.4 Laser Diodes 165
3.1.5 Some Imperfections of the Laser Emission 168
3.1.6 Laser Structures 171

3.1.7 Spectral Properties of LEDs and Laser Diodes 173
3.1.8 Modulation Properties of LEDs and Laser Diodes 174

3.2 Photodetectors 176
3.2.1 Review of the Existing Photodetectors 178
3.2.2 Solar Cells and PIN Photodiodes 180
3.2.3 Avalanche Photodiodes (APD) 184
3.2.4 Some Considerations about the Noise of Photodiodes 186
3.2.5 Spectral Properties of Photodiodes 189
3.2.6 Modulation Properties of Photodiodes 191

4 TRANSMISSION SYSTEMS
4.1 Transmission Systems Based on Intensity Modulation 197
4.1.1 LED and Laser Drivers 198
4.1.2 Types of Front Ends 201
4.1.3 Analog Receivers 204
4.1.4 Digital Receivers 205
4.1.5 Equalization 207
4.1.6 Ultimate Limits of Transmission Capacity and Repeater Spacing for Intensity Modulation Systems 210

4.2 Coherent Transmission Systems 211
4.2.1 Advantages of Coherent Systems over Traditional Systems 213
4.2.2 Main Problems of Practical Coherent Transmissions 214

5 SCANNING MASUREMENTS
5.1 Scanning Measurements on Fibers 217
5.1.1 Defraction Limit 218
5.1.2 Optical Arrangements 219
5.1.3 Differential Mode Attenuation Measurements 223
5.1.4 Differential Mode Delay Measurements 224

5.1.5 Indirect Tests for Investigating Mode Coupling Effects 225
5.1.6 Advantages of Domain (r_1, θ_1) 227

5.2 Scanning Measurements on Transducers 231
5.2.1 Scanning Measurements on LEDs and Laser Diodes 231
5.2.2 Scanning Measurements on Photodiodes 234

6 MEASUREMENTS ON TRANSDUCERS
6.1 Integrated Measurements 240
6.1.1 Electric Characterization of LEDs and Laser Diodes 241
6.1.2 Measurement of the Electro-Optical Transduction and Its Baseband Response 244
6.1.3 Electric Characterization of Photodiodes 248
6.1.4 Measurements of the Responsivity and Its Baseband Response 250
6.1.5 Reliability Tests on Laser Diodes 252

6.2 Differential Measurements 253
6.2.1 Measurement of the Spectral Distribution of the Emitted Power in LEDs and Laser Diodes 254
6.2.2 Measurements of Near and Far Field Patterns in LEDs and Laser Diodes 257
6.2.3 Measurements of Spatial Relative Delays of Emission in LEDs 260
6.2.4 Theoretical Model of Radiation Confinement and Spatial Relative Delays of Emission 262
6.2.5 Combined Spectral and Geometrical Differential Measurements in LEDs and Laser Diodes 268
6.2.6 Tests on the Spectral and Geometrical Dependence of the Responsivity 271

7 INDEX PROFILE MEASUREMENTS
7.1 Fiber Profiling 277
7.1.1 Near Field Scanning Technique 278

7.1.2	Refracted Near Field Technique	282
7.1.3	Slice Interferometry	284
7.1.4	Reflection Method	286
7.1.5	Transverse Interferometry	287
7.1.6	Transverse Focusing Technique	290
7.1.7	Brief Review of Other Techniques	291
7.1.8	Comparative Discussion	292
7.2	**Preform Profiling**	**293**
7.2.1	Techniques Also Employed for Fiber Profiling	295
7.2.2	Ray Tracing Method	296
7.2.3	Comparison between the Fiber and Its Parent Preform	297
8	**GEOMETRICAL MEASUREMENTS**	
8.1	**Measurements on the Fiber Cross Section**	**303**
8.1.1	Direct Fiber End Face Inspection	305
8.1.2	Near Field Techniques	306
8.1.3	Outer Diameter Measurements	309
8.1.4	Measurements of the Ellipticity of the Fiber Outer Surface	312
8.1.5	Measurements of NA	313
8.2	**Fiber Length Measurements**	**314**
8.2.1	Fiber Length Measurement by Backscattering	315
8.2.2	Fiber Length Measurement by a Lock-in Amplifier	315
9	**ATTENUATION MEASUREMENTS**	
9.1	**Review of the Existing Techniques**	**320**
9.1.1	The Cut-Back Technique	321
9.1.2	The Backscattering Technique (OTDR)	326
9.1.3	Technological Measurements	329
9.1.4	Measurements of Total Loss by Lateral Scattering Detection	333
9.1.5	The Insertion Loss Measurement	334

9.1.6 Joint Loss Measurements ... 334
9.1.7 On-Line Tests during Fiber Fabrication ... 337

9.2 Discussion on Cutback ... 338
9.2.1 Beam Launchers and Mode Scramblers ... 338
9.2.2 Accuracy of Cutback Measurements ... 341
9.2.3 Spectral Loss Measurements ... 344
9.2.4 Measurements Performed by Varying the Launching Condition ... 348

9.3 Discussion on Backscattering ... 351
9.3.1 Launching Techniques ... 351
9.3.2 Ray Distribution of the Backscattered Power ... 354
9.3.3 Effects of Fluctuations in the Fiber Parameters ... 357
9.3.4 Measurements of Local Additional Losses ... 360
9.3.5 Analysis of All Noise Sources in Backscattering Measurements ... 362
9.3.6 Dynamic Range Limitations ... 365
9.3.7 Accuracy of Backscattering Measurements ... 368
9.3.8 Outline of Optical Frequency Domain Reflectometry (OFDR) ... 370

9.4 Differential Mode Attenuation Measurements ... 371
9.4.1 DMA Measurements of Step Index Fibers ... 372
9.4.2 DMA Measurements of Graded Index Fibers ... 373
9.4.3 Discussion of the Accuracy of DMA Measurements ... 376

10 TIME DISPERSION MEASUREMENTS
10.1 General Considerations ... 386
10.1.1 Extrapolation of the Experimental Results of Time Dispersion Measurements ... 386
10.1.2 Measurements in the Time Domain ... 388
10.1.3 Measurements in the Frequency Domain ... 392

10.1.4 Measurements of Intermodal Time Dispersion 395
10.1.5 Material Dispersion Measurements 397
10.1.6 Experimental Evaluation of Profile Dispersion 400

10.2 Discussion of the Measurements in the Time Domain 403
10.2.1 Main Causes of Uncertainty in Time Domain Measurements 404
10.2.2 Use of a Lock-in Amplifier to Improve the SNR 409
10.2.3 Use of Pick-Up Procedure to Reduce the Overall Uncertainty 410
10.2.4 Evaluation of the Effective Input Pulse 412

10.3 Discussion on the Measurements in the Frequency Domain 412
10.3.1 The Fiber Transfer Function 413
10.3.2 Main Causes of Uncertainty in Frequency Domain Measurements 415
10.3.3 A Method for Improving the Maximum Dynamic Range 417
10.3.4 Evaluation of the Phase Response via a Hilbert Transform 418
10.3.5 Direct Measurement of the Phase Response 419
10.3.6 Overall Accuracy of Frequency Domain Measurements 422

10.4 Differential Mode Delay Measurements 424
10.4.1 Purposes of a DMD Measurement 424
10.4.2 DMD Measurements on Step Index Fibers 429
10.4.3 DMD Measurements on Graded Index Fibers 432
10.4.4 Discussion of DMD Measurement Accuracy 432

11 INDIRECT TESTS ON MODE COUPLING EFFECTS

11.1 Tests Based on Amplitude Mode Power Distributions 439

11.1.1 Measurements of Amplitude Mode Power Distributions under a Selective Excitation 440

11.1.2 Measurement of Amplitude Mode Power Distributions at the Steady-State 442

11.2 Tests Based on Amplitude and Delay Mode Power Distributions 443

11.2.1 Theorectial Model for the Measurement of the Mode-Dependent Attenuation and the Coupling Coefficient 443

11.2.2 Measurement of Mode-Dependent Attenuation and Coupling Coefficient 446

11.2.3 Theoretical Model for an Indirect Evaluation of the Fiber Baseband Response in the Presence of Mode Coupling 448

11.2.4 Measurement of the Fiber Baseband Response through an Indirect Test, in the Presence of Mode Coupling 452

12 MEASUREMENTS OF MONOMODE FIBERS

12.1 Index Profile and Geometrical Measurements on Monomode Fibers 458

12.1.1 Index Profiling in Monomode Fibers 458

12.1.2 Cutoff Wavelength Measurement 460

12.1.3 Fundamental Mode Spot Size Measurement 464

12.1.4 Equivalent Step Index Profile Determination 465

12.1.5 Measurements of Core-Cladding Concentricity and Core Ellipticity in Monomode Fibers 466

12.1.6 Measurements of NA in Monomode Fibers 467

12.2 Transmission Measurements of Monomode Fibers 468

12.2.1 Attenuation Measurements in Monomode Fibers 468

12.2.2 Time Dispersion Measurements in Monomode Fibers 470

A1 OPTICAL INSTRUMENTS FREQUENTLY USED IN OPTICAL FIBER MEASUREMENTS

1.a Monochromators 475
1.b Light Modulators 480
1.c Microscope Objectives 483
1.d Optical Directional Couplers 484

A2 ELECTRIC INSTRUMENTS FREQUENTLY USED IN OPTICAL FIBER MEASUREMENTS

2.a Lock-in Amplifier and Light-Chopper 487
2.b Boxcar Integrator 489

A3 FIBER END PREPARATION

3.a Fiber Cleaving Technique 493
3.b Fiber Break Angle Inspection Procedures 495

Preface

This book is intended for graduate students, researchers, and design engineers who are interested in optical communications. It can be considered as a self-contained text, although several other books may be used in addition, to approach specific arguments in more detail. Measurements in fiber optics require a wide spectrum of experience, often in different fields: e.g., traditional optics, spectrometry, interferometry, analog and digital electronics, mechanics, *et cetera*. Nevertheless, we have tried to describe the main principles on which the various experimental setups are based, recalling only a few fundamental concepts.

Some analytical passages in the first two chapters are intended to give an idea of the approximations which sometimes are made during experimental investigation. It is not necessary for the reader to follow the proposed organization of arguments. Each kind of experimental measurement is presented independently, although some of them have many common aspects, which have been carefully stressed.

It is very difficult to fix precise rules for the execution of the measurements, considering also that some international committees are actually working on standardization. Yet, at present, it is possible to suggest methodologies which can help in many practical cases. A very important aspect one has to consider is the function which a given cable or device must perform in a certain transmission system. In this sense, the same experimental investigation may appear as very different, depending on the object under test and the function expected from it. For instance, attenuation measurements on optical fibers require a very high accuracy, especially when the test fiber is a very low loss one. Nevertheless, remarkable care must be devoted to considering the kind of optical source employed in the system to which the fiber will be assigned, since different results, well above the experimental uncertainty, may occur using different optical sources.

We wish to thank G. C. Corazza, President of the Fondazione G. Marconi, for the impetus given to the research in our fiber optic laboratory, where a great part of this book was conceived. We are also profoundly grateful to C. G. Someda for his continuous encouragement, advice, and experienced support to our work. We are indebted to M. Mezzetti and P. Fantini for their cooperation in obtaining some of the theoretical and experimental results reported here, and also to G. Falciasecca, V. Rizzoli, P. Bassi, and M. Zoboli for helpful discussions.

Symbols Used in the Manuscript

$\overline{A}$ for vectors representing parameters which have more than one scalar component from the physical point of view, and for complex phasors

$\underline{A}$ for matrices having one or more columns, representing an ensemble of scalar variables, from the mathematical point of view

$\hat{A}$ for unit vectors, and for differential operators

$\cdot$ scalar product between vectors

$\times$ vector product between vectors

CHAPTER 1

LIGHT PROPAGATION IN IDEAL MULTIMODE FIBERS

Two approaches lead to the description of the light propagation in a multimode optical fiber:

i) the electromagnetic model

ii) the ray-optics model

The electromagnetic model can be also applied when few modes propagate. The ray-optics model, on the other hand, gives acceptable results only when many modes (at least several hundred) propagate. The main features of the latter model are the simplicity and immediateness of the description, and the possibility of an easy experimental inspection to support the theory.

In the following, after a brief outline of the electromagnetic model, a ray-optics approach to light propagation is developed. This is justified because multimode fibers, usually employed in optical links, carry many hundreds of modes. Nevertheless, we will recall the electromagnetic theory at times, in this and following chapters. In fact, some limitations exist in the experimental setups which will be described, due to the electromagnetic nature of light.

1.1 OUTLINE OF THE ELECTROMAGNETIC MODEL

An ideal dielectric wave guide consists of an interface between two lossless dielectric media. Generally there is an inner medium, characterized by the refractive index n_1, and an outer medium, with refractive index n_2. For the light to propagate $n_1 > n_2$ must be satisfied. In fact, the propagation without loss is based on a total internal reflection of guided waves at the interface.

A general dielectric waveguide is schematically shown in Fig. 1.1. The definition of dielectric waveguide can also be extended to the case in which only the outer medium is homogeneous. In such a case only n_2 is constant, whereas n_1 is a function of a set of coordinates defined in the transverse plane (x, y in Fig. 1.1). Condition $n_1 > n_2$ must be verified at least in some points of the inner region. The definition of ideal dielectric waveguide entails that geometrical or physical property of the structure is independent of the axial coordinate z. In Chapter 2, we will introduce real optical waveguides as those which are lossy and are characterized by a perturbation along z.

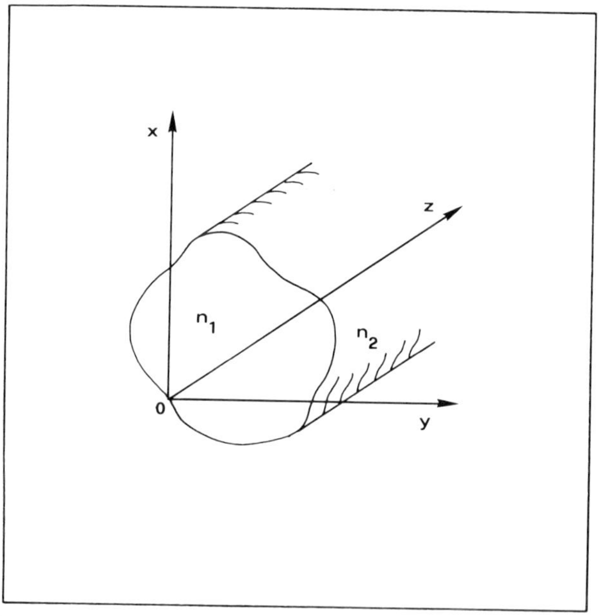

Fig. 1.1 - General dielectric waveguide.

1.1.1 General Considerations

The refractive index of an isotropic dielectric medium is defined as

$$n = \sqrt{\frac{\mu \epsilon}{\mu_o \epsilon_o}}$$

where μ, ϵ are the respective permeability and permittivity of the medium, and μ_o, ϵ_o are the permeability and permittivity of free space. Different values of n can be obtained by varying ϵ, keeping μ constant. Therefore,

$$\frac{n_1}{n_2} = \sqrt{\frac{\epsilon_1}{\epsilon_2}}$$

The dielectric optical waveguides employed for the transmission of information over long distances are usually made of $Si0_2$, which is properly doped in order to achieve the required index distribution in the inner medium. The most commonly used dopants are Ge, P, and

B. Such waveguides, made with round symmetry, are called round optical fibers. Nevertheless, the transverse section of real fibers turns out to be not perfectly round, showing an ellipticity of a few percent.

The inner medium of a round optical fiber is called the core, the outer medium the cladding. Two kinds of fibers are employed in communications:

i) step-index fibers, which are characterized by a homogeneous core whose uniform index is n_1

ii) graded-index fibers, with a gradual distribution $n_1(r)$ of the refractive index, where r represents the radial coordinate

The gradual distribution of the refractive index which characterizes the latter kind of fibers is intended for the equalization of the transit times of the guided modes, and therefore for the reduction of the time dispersion.

In the present section, we will recall briefly the electromagnetic theory of general propagation. Consider the electric field $\bar{\mathscr{E}}(r, \phi, z; t)$, and the magnetic field $\bar{\mathscr{H}}(r, \phi, z; t)$ of a monochromatic electromagnetic wave, where r, ϕ, z are cylindrical coordinates, and t denotes time dependence. These fields, in the absence of any sources, satisfy Maxwell's equations

$$\nabla \times \bar{\mathscr{E}}(t) = -\mu \frac{\partial \bar{\mathscr{H}}(t)}{\partial t}, \quad \nabla \times \bar{\mathscr{H}}(t) = \epsilon \frac{\partial \bar{\mathscr{E}}(t)}{\partial t} \tag{1.1}$$

Through a Fourier transform, letting ν_λ equal the optical frequency, such equations become

$$\nabla \times \bar{\mathscr{E}} = -i\, 2\pi \nu_\lambda \mu \bar{\mathscr{H}} \tag{1.2}$$

$$\nabla \times \bar{\mathscr{H}} = i\, 2\pi \nu_\lambda \epsilon \bar{\mathscr{E}}$$

where i is the imaginary unit, and

$$\bar{\mathscr{E}}(t) = \mathrm{Re}\,[\bar{\mathscr{E}} \exp(i\, 2\pi \nu_\lambda t)]$$

$$\bar{\mathscr{H}}(t) = \mathrm{Re}\,[\bar{\mathscr{H}} \exp(i\, 2\pi \nu_\lambda t)]$$

From equations (1.2) we can derive

$$\nabla \times (\nabla \times \bar{\mathscr{E}}) = \nabla(\nabla \cdot \bar{\mathscr{E}}) - \nabla^2 \bar{\mathscr{E}} = (2\pi \nu_\lambda)^2 \mu \epsilon \bar{\mathscr{E}}$$

Since $\nabla \cdot \bar{\mathscr{E}}$ is proportional to the electric charge density, and we have assumed that there are no sources, it follows that

$$\nabla^2 \overline{\mathscr{E}} = -(2\pi\nu_\lambda)^2 \mu \epsilon \overline{\mathscr{E}} \tag{1.3}$$

In a similar way

$$\nabla \times (\nabla \times \overline{\mathscr{H}}) = \nabla(\nabla \cdot \overline{\mathscr{H}}) - \nabla^2 \overline{\mathscr{H}} = (2\pi\nu_\lambda)^2 \mu \epsilon \overline{\mathscr{H}}$$

Since it is always true that $\nabla \cdot \overline{\mathscr{H}} = 0$, it follows that

$$\nabla^2 \overline{\mathscr{H}} = -(2\pi\nu_\lambda)^2 \mu \epsilon \overline{\mathscr{H}} \tag{1.4}$$

Equations (1.3) and (1.4) can be written in the form

$$\nabla^2 \overline{\mathscr{E}} + k^2 \overline{\mathscr{E}} = 0$$

$$\nabla^2 \overline{\mathscr{H}} + k^2 \overline{\mathscr{H}} = 0 \tag{1.5}$$

$$k^2 = (2\pi\nu_\lambda)^2 \mu \epsilon$$

which are called Helmholtz's equations. Maxwell's and Helmholtz's equations are the basis of any guided propagation. Considering z as the direction of propagation, it is justified to assume the trial solution

$$\overline{\mathscr{E}}(r, \phi, z) = \overline{E}(r, \phi) \exp(-i\beta z)$$

$$\overline{\mathscr{H}}(r, \phi, z) = \overline{H}(r, \phi) \exp(-i\beta z) \tag{1.6}$$

in which β represents a phase constant. Substituting (1.6) into (1.5):

$$\nabla_t^2 \overline{E} = -\chi^2 \overline{E}$$

$$\nabla_t^2 \overline{H} = -\chi^2 \overline{H}$$

$$\chi^2 = k^2 - \beta^2 \tag{1.7}$$

in which ∇_t^2 is a Laplacian operator in the transverse coordinates.

In a dielectric waveguide it is more convenient to express the wavenumber k as a function of the refractive index. Letting c equal the free space velocity of light, and λ the wavelength in a vacuum, the following relationships hold:

$$\nu_\lambda = \frac{c}{\lambda} = \frac{1}{\sqrt{\mu_o \epsilon_o}} \frac{1}{\lambda}$$

$$k = \frac{2\pi}{\lambda} n \tag{1.8}$$

Equations (1.7) must be written for both the inner and the outer medium. Letting

$$\chi_1^2 = k_1^2 - \beta^2$$

$$\chi_2^2 = \beta^2 - k_2^2 \qquad k_i^2 = (\frac{2\pi}{\lambda})^2 n_i^2 , \quad i = 1, 2 \tag{1.9}$$

it follows that

$$\nabla_t^2 \overline{E}_1 = -\chi_1^2 \overline{E}_1 \qquad \nabla_t^2 \overline{H}_1 = -\chi_1^2 \overline{H}_1 \tag{1.10}$$

$$\nabla_t^2 \overline{E}_2 = \chi_2^2 \overline{E}_2 \qquad \nabla_t^2 \overline{H}_2 = \chi_2^2 \overline{H}_2$$

Substituting (1.6) into (1.2), four relations are derived which give the transverse field components from the knowledge of the axial field components E_z H_z.[1] This makes it possible to consider only axial components in Equations (1.10), restricting ourselves to a scalar problem.

1.1.2 Exact Electromagnetic Theory of Round Step-Index Fibers

Assuming a set of cylindrical coordinates (r, θ, z), from (1.10) we have

$$\frac{1}{r}\frac{\partial}{\partial r}(r\frac{\partial \Psi_1}{\partial r}) + \frac{1}{r^2}\frac{\partial^2 \Psi_1}{\partial \phi^2} = -\chi_1^2 \Psi_1 \qquad \Psi_1 = E_{z1}, H_{z1}$$

$$\frac{1}{r}\frac{\partial}{\partial r}(r\frac{\partial \Psi_2}{\partial r}) + \frac{1}{r^2}\frac{\partial^2 \Psi_2}{\partial \phi^2} = \chi_2^2 \Psi_2 \qquad \Psi_2 = E_{z2}, H_{z2}$$

Through a separation of variables, letting

$$\psi_j(r, \phi) = R_j(r)\,\Phi_j(\phi) \qquad j = 1, 2$$

and substituting into the above general equations

$$\frac{d^2 R_1}{d r^2} + \frac{1}{r}\frac{d R_1}{d r} + (\chi_1^2 - \frac{\nu^2}{r^2}) R_1 = 0 , \quad \frac{d^2 \Phi_1}{d \phi^2} = -\nu^2 \Phi_1$$

$$\frac{d^2 R_2}{d r^2} + \frac{1}{r}\frac{d R_2}{d r} + (-\chi_2^2 - \frac{\nu^2}{r^2}) R_2 = 0 , \quad \frac{d^2 \Phi_2}{d \phi^2} = -\nu^2 \Phi_2 \tag{1.11}$$

in which ν^2 is a positive real constant. The general solution is

$$E_{z1}(r, \phi) = A_{1E} J_\nu(\chi_1 r) \exp(i\nu\phi)$$

$$H_{z1}(r, \phi) = A_{1H} J_\nu(\chi_1 r) \exp(i\nu\phi) \tag{1.12}$$

$$E_{z2}(r, \phi) = A_{2E} H_\nu^{(1)}(i\chi_2 r) \exp(i\nu\phi)$$

$$H_{z2}(r, \phi) = A_{2H} H_\nu^{(1)}(i\chi_2 r) \exp(i\nu\phi)$$

where J_ν is the ν-order Bessel function of the first kind, $H_\nu^{(1)}$ is the ν-order Hankel function of the first kind, and constants A_{1E}, A_{1H}, A_{2E}, A_{2H} which depend on the excitation condition, have a suitable physical dimension.

Next consider the continuity boundary conditions at the core-cladding interface. Such conditions regard the tangential field components. Letting a equal the core radius, we have the following relations

$$E_{z1}(a, \phi) = E_{z2}(a, \phi)$$

$$E_{\phi 1}(a, \phi) = E_{\phi 2}(a, \phi)$$

$$H_{z1}(a, \phi) = H_{z2}(a, \phi)$$

$$H_{\phi 1}(a, \phi) = H_{\phi 2}(a, \phi)$$

They lead to a set of four linear homogeneous equations in the integration constants A_{1E}, A_{1H}, A_{2E}, A_{2H}.[2]

Acceptable solutions are obtained by solving the characteristic equation of this set of linear homogeneous equations. So we are left with

$$\left[\frac{\epsilon_1}{\epsilon_2}\,\frac{a\,\chi_2^2}{\chi_1^2}\,\frac{J_\nu'(\chi_1 a)}{J_\nu(\chi_1 a)} + i\,\chi_2\,a\,\frac{H_\nu^{(1)\prime}(i\,\chi_2 a)}{H_\nu^{(1)}(i\,\chi_2 a)}\right]\left[\frac{a\,\chi_2^2}{\chi_1^2}\,\frac{J_\nu'(\chi_1 a)}{J_\nu(\chi_1 a)} + i\,\chi_2\,a\,\frac{H_\nu^{(1)\prime}(i\,\chi_2 a)}{H_\nu^{(1)}(i\,\chi_2 a)}\right] = \left[\nu\left(\frac{\epsilon_1}{\epsilon_2} - 1\right)\frac{\beta}{\chi_1^2}\,k_2\right]^2 \qquad (1.13)$$

in which the prime denotes differentiation of the functions with respect to their arguments. Permitted values of β are obtained as solutions of (1.13). They correspond to propagating modes and can be denoted by $\beta_{\nu\mu}$, where ν, called azimuthal mode number, depends on the order of the Bessel or Hankel function considered; whereas μ, called radial mode number, is due to the oscillating behavior of such functions.

Solutions (1.12) are based on the assumption that χ_1 and χ_2 are real quantities, i.e., χ_1^2 and χ_1^2 are real, positive quantities. From (1.9), in order to satisfy this requirement

$$k_2 \leq \beta \leq k_1 \tag{1.14}$$

Therefore, the permitted values $\beta_{\nu\mu}$ obtained from (1.13) belong to the range defined by (1.14).

A mode guided by an optical fiber generally has all its six field components, E_z, E_r, E_ϕ, H_z, H_r, H_ϕ, different from zero, except when $\nu = 0$. In fact, considering (1.12) and a well known property of Bessel and Hankel functions,[3] one obtains

$$\frac{\epsilon_1}{\epsilon_2} \; \frac{\chi_1}{\chi_2} \; \frac{J_1(\chi_1 a)}{J_0(\chi_1 a)} + i \; \frac{H_1^{(1)}(i\chi_2 a)}{H_0^{(1)}(i\chi_2 a)} = 0$$

$$\frac{\chi_1}{\chi_2} \; \frac{J_1(\chi_1 a)}{J_0(\chi_1 a)} + i \; \frac{H_1^{(1)}(i\chi_2 a)}{H_0^{(1)}(i\chi_2 a)} = 0$$

The first equation defines $TM_{0\mu}$ modes, which have $H_z = 0$. The second one defines $TE_{0\mu}$ modes, which have $E_z = 0$.[2]

It is possible to demostrate that for each pair (ν, μ) two solutions of (1.13) exist, which correspond to two modes, denoted $HE_{\nu\mu}$ and $EH_{\nu\mu}$[2] Moreover, considering the two polarization orientations of each mode, due to the possibility of a positive or negative $(\pm\nu)$ in (1.11), it follows that a fourfold degeneracy exists. In other words, four degenerate quasi-modes exhibit the same phase constant: $\beta_{\nu\mu}$. Note that the solution of the set of homogeneous equations in A_{1E}, A_{1H}, A_{2E}, A_{2H}, gives all these quantities as proportional to the excitation amplitude.

1.1.3 Approximate Evaluation of the Total Number of Guided Modes

Let us consider the following parameter

$$V^2 = (\chi_1^2 + \chi_2^2)\, a^2 = (n_1^2 - n_2^2)\left(\frac{2\pi}{\lambda}\, a\right)^2$$

We will demonstrate through very simple considerations that the total number of guided modes is proportional to V^2. The maximum light acceptance angle in air, θ_{aM}, is defined as

$$\theta_{aM} = \sin^{-1}\left(\sqrt{n_1^2 - n_2^2}\right)$$

This property will be demonstrated by the ray-optics model in the next paragraph. Considering that usually $(n_1 - n_2)/n_1 \ll 1$, θ_{aM} is a very small angle; therefore

$$\theta_{aM} \cong \sqrt{n_1^2 - n_2^2}$$

The maximum solid angle of light acceptance in air is then

$$\Omega_{aM} = \pi \theta_{aM}^2$$

The solid angle of a single electromagnetic radiation mode through an aperture having cross sectional area A is λ^2/A.[4] This is a linearly polarized mode, and thus can have two independent polarization orientations. Given the cross sectional area of the fiber core, $A = \pi a^2$, the total number of guided modes results

$$N = 2\frac{\Omega_{aM}}{\lambda^2} \pi a^2 = 2\pi^2 \frac{a^2}{\lambda^2}(n_1^2 - n_2^2) = \frac{V^2}{2}$$

This equation is approximate because it is based on a light acceptance angle derived from the ray-optics model, but the proportionality between N and V^2 is exact.

Performing an exact electromagnetic calculation, it is possible to show that for $V < 2.405$, a single mode propagates. This is the case for monomode fibers. In Table 1.1 some maximum permitted values of the core radius a *versus* $(n_1^2 - n_1^2)$ are given, to provide a monomode fiber.

Table 1.1 - Maximum core radius a *versus* the index difference $n_1^2 - n_2^2$, in order to have a monomode fiber.

$n_1^2 - n_2^2$	core radius a [μm]		
	$\lambda = 0.85\ \mu m$	$\lambda = 1.0\ \mu m$	$\lambda = 1.3 \mu m$
0.004	5.14	6.05	7.87
0.006	4.20	4.94	6.92
0.008	3.64	4.28	5.56
0.010	3.25	3.83	4.98
0.012	2.97	3.49	4.54
0.014	2.75	3.23	4.21
0.016	2.57	3.03	3.93
0.018	2.42	2.85	3.71
0.020	2.30	2.70	3.52

1.1.4 WKB Approximate Solution for Graded-Index Fibers

An exact electromagnetic theory of light propagation in graded-index optical fibers is not always available. For $n_1(r)$ as a function of the radial coordinate r in the fiber core, the first equation of (1.11) becomes

$$\frac{d^2 R_1}{d r^2} + \frac{1}{r} \frac{d R_1}{d r} + [k_1^2(r) - \beta^2 - \frac{\nu^2}{r^2}] R_1 = 0 \quad k_1(r) = \frac{2\pi}{\lambda} n_1(r) \tag{1.15}$$

Only when $n_1^2(r)$ has a parabolic form can this equation be solved in finite terms. However, by analogy with quantum mechanics, an approximate solution, called a WKB approximation, can be developed. Consider a trial solution consisting of a plane wave of the type

$$R_1(r) = R_o \exp[i U(r)]$$

in which R_o is a constant and $U(r)$ a function of r dependent on k_1. Assuming $n_1(r)$ varies slowly over wavelength distances, $U(r)$ can be expanded in power series of λ in the neighborhood of $\lambda = 0$. Since λ is proportional to $1/k_1$, such an expansion can be written as

$$U(r) = U_0(r) + \frac{1}{k_1} U_1(r) + \ldots$$

where $U_0(r), \ldots$ are functions of r independent of k_1.

Substituting into (1.15), as many differential equations as the number of terms in the expansion are obtained. Considering only the zero order term, we have the zero order WKB approximation. Considering the zero and first order terms, we have the first order WKB approximation, and so forth. Assuming λ/a is small, terms of higher order can be neglected. In the limit $\lambda/a \to 0$, only the zero order term, $U_0(r)$, remains. It is given by the following equation

$$-\left(\frac{d U_0}{d r}\right)^2 + [k_1^2(r) - \beta^2 - \frac{\nu^2}{r^2}] = 0$$

and hence

$$U_0(r) = \int_0^r [k_1^2(\zeta) - \beta^2 - \frac{\nu^2}{r^2}]^{1/2} d\zeta$$

Only if $U_0(r)$ assumes real values, $R_1(r)$ is characterized by an oscillating behavior, which is required for the propagation. Generally, there are two values of r, r_1 and r_2, between which $U_0(r)$ assumes real values, while, for $r < r_1$ and $r > r_2$, $R_1(r)$ exhibits an exponential decay. In Fig. 1.2 a graphical method which provides r_1 and r_2, for a given pair (β,ν) is shown Assuming $dn_1/dr \leq 0$, the behavior of $k_1^2(r)$ is of the type reported in dashed line. $k_1^2(r) - \nu^2/r^2$ is plotted in continuous line. For a given β^2, a couple of points r_1, r_2 are intercepted over the latter curve. They limit the region of field existence in the core section for the mode (β,ν).

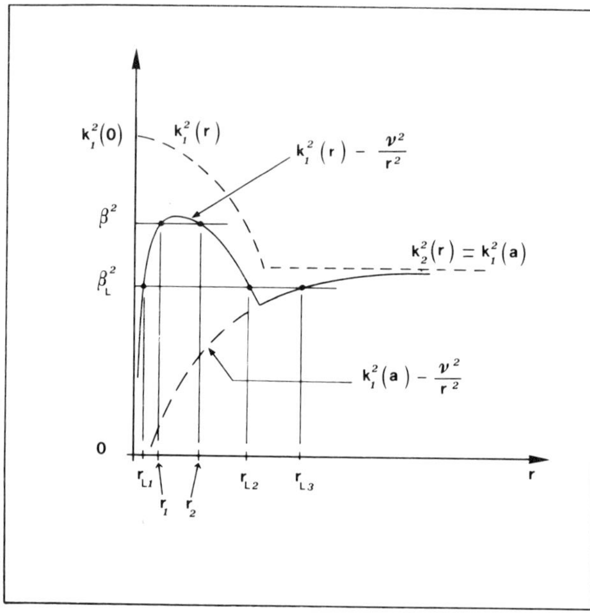

Fig. 1.2 - Graphical method for the determination of r_1 and r_2.

Between these two values of the radial coordinate, the field has an oscillating behavior. The number of oscillations must be an integer, and is given by

$$\mu = \frac{1}{\pi}\int_{r_1}^{r_2} [\, k_1^2(r) - \beta^2 - \frac{\nu^2}{r^2}\,]^{1/2}\, dr \tag{1.16}$$

Then each permitted value $\beta_{\nu\mu}$ can be related to a pair (ν,μ). By extending the function $k_1^2(r)$ beyond $r=a$, owing to the condition $k_2^2(r)=k_1^2(a)$, for a given $\nu \neq 0$, some permitted values of β_L, denoted β_L^2 in Fig. 1.2, also exist which lead to three interception points r_{L1}, r_{L2}, r_{L3}. Such modes have two regions of field existence: the first one inside the core, for $r_{L1} \leq r \leq r_{L2}$; the second one in the cladding, for $r \geq r_{L3}$. They exceed condition (1.14), therefore they cannot be defined rigorously guided modes. Developing a particular study of such modes,[5] it is possible to show that they are weakly attenuated. Nevertheless their influence may be appreciable over short distances from the optical sources, especially when the source is characterized by a wide angle of light emission and by a broad emitting surface. These modes are sometimes called "leaky," because they leak power or are sometimes called "tunnelling," by consequence of the analogy with quantum mechanics. They also exist in step-index fibers, and their number, both in step-and in graded-index fibers, increases with increasing ν.

Finally, for those values of β so small they exceed even the leaky guidance condition, there is a continuous spectrum of radiating modes. Their propagation is strongly attenuated, surviving in the core for a distance on the order of several centimeters. On the contrary, at the interface between the cladding and an outer medium (air or primary coating), cladding modes with very small value phase constants can survive for several meters from the optical source.

1.2 THE RAY-OPTICS MODEL

Ray-optics is based on the assumption $\lambda \rightarrow 0$. In this sense it coincides with the zero order WKB approximation. On the other hand, in ray-optics any variable can assume values which belong to a continuum. Adding suitable resonance conditions, like (1.16), to the ray-optics approach, a complete correspondence between the ray optics model and the electromagnetic model can be obtained. Nevertheless, when hundreds of modes propagate in a fiber, the quantization of the variables has negligible effects, just as classical mechanics replaces quantum mechanics when passing from microscopic to macroscopic dimensions.

The study of the light propagation in multimode fibers through the ray-optics model is founded on the definition of rays. A ray is characterized by a trajectory, which is the trace of the local direction of a plane wave. A ray trajectory depends on a certain initial condition, and for ideal fibers, it turns out to be periodic along the fiber axial coordinate. Owing to the round symmetry, it is not significant to distinguish rays whose initial conditions differ only in the azimuthal coordinate. Moreover, since the fiber length is always many times greater than the length of each ray period, even rays having the same trajectory, except for an axial shift, can be considered indistinguishable from each other. A family of indistinguishable rays is called ray congruence. A ray congruence in the ray-optics model corresponds to one mode in the electromagnetic model. In the following, except for an explicit advertisement, a certain ray will denote the whole ray congruence.

1.2.1 Light Acceptance in Step-Index Fibers

In Fig. 1.3 the longitudinal section of a step-index fiber is depicted in the neighborhood of its input end. A ray whose trajectory intersects the fiber axis is considered. Letting θ equal the propagation angle of the ray (the angle between the ray trajectory and the fiber axis), the correspondent angle in air is denoted by θ_a. Assuming the refractive index of air is equal to *1*, θ_a is related to θ through Snell's law

$$\sin \theta_a = n_1 \sin \theta \tag{1.17}$$

Letting $\alpha = \pi/2 - \theta$, the angle α characterizes the reflection of the ray at the core-cladding interface. In order to have total internal reflection,[1] it must be that

$$\alpha \geq \sin^{-1} \left(\frac{n_2}{n_1} \right)$$

Hence

$$\sin \theta \leq \sqrt{1 - \left(\frac{n_2}{n_1} \right)^2}$$

In accordance with (1.17), the maximum light acceptance angle in air, θ_{aM}, mentioned also in Sec. 1.1.3, can be written as

$$\sin \theta_{aM} = \sqrt{n_1^2 - n_2^2} \tag{1.18}$$

Such a quantity is called the *numerical aperture,* or simply *NA,* of the fiber. In Table 1.2 the maximum permitted values of θ and θ_a are reported, both in degrees and in radians, for some typical values of *NA* and the difference $n_1^2 - n_1^2$.

A ray whose trajectory does not intersect the fiber axis can be characterized by a greater value of the maximum light acceptance angle. A ray of this kind, having θ_a greater than θ_{aM} as defined by (1.18), is called a leaky ray, because it corresponds to a leaky mode, as will be better shown in the next section. For now, we will simply show that rays whose trajectories belong to outer regions in the core section may have greater *NA*'s. In Fig. 1.4 a transverse section of a fiber is depicted, showing the trace of a ray trajectory of this type. The angle between the local ray direction projected on the transverse plane and the radial direction is denoted by θ_ϕ. At the core-cladding interface, the total internal reflection takes place under the usual condition of angle α. But here this limit corresponds to

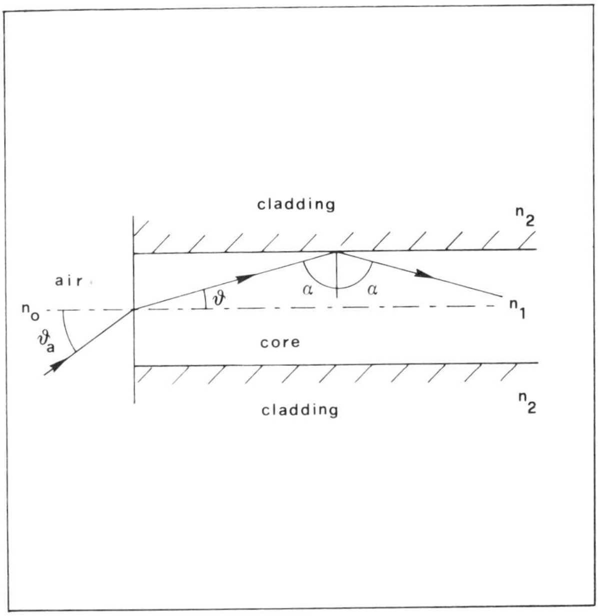

Fig. 1.3 - Longitudinal section of a step index fiber.

Table 1.2 - Maximum permitted values of θ and θ_a, assuming n_1 = 1.5. Only meridional rays have been considered.

NA	$n_1^2 - n_2^2$	θ_M deg	θ_M rad	θ_{aM} deg	θ_{aM} rad
0.10	0.010	3.82	0.067	5.74	0.100
0.12	0.014	4.59	0.080	6.89	0.120
0.14	0.020	5.36	0.093	8.05	0.140
0.16	0.026	6.12	0.107	9.21	0.160
0.18	0.032	6.89	0.120	10.37	0.181
0.20	0.040	7.66	0.134	11.54	0.201
0.22	0.048	8.43	0.147	12.71	0.222
0.24	0.058	9.21	0.161	13.89	0.242
0.26	0.068	9.98	0.174	15.07	0.263
0.28	0.078	10.76	0.188	16.26	0.284
0.30	0.090	11.54	0.201	17.46	0.305
0.32	0.102	12.32	0.215	18.66	0.326

$$\sin \theta \leq \frac{1}{\cos \psi_\phi} \sqrt{1 - \left(\frac{n_2}{n_1}\right)^2}$$

$$\sin \psi_\phi = \frac{r}{a} \sin \theta_\phi$$

and hence, from (1.17), it is

$$\sin \theta_{aM} = \sqrt{\frac{n_1^2 - n_2^2}{1 - (\frac{r}{a} \sin \theta_\phi)^2}} \tag{1.19}$$

A ray whose trajectory intersects the z-axis is called a meridional ray; a ray whose trajectory does not intersect the z-axis is called a skew ray. The quantity

$$\frac{\rho}{a} = \sin \psi_\phi$$

where ρ, the minimum distance of the ray trajectory from the fiber axis, is termed the eccentricity of the skew ray. This quantity is constant for a given ray. Equation (1.19) can be rewritten as

$$\sin \theta_{aM} = \sqrt{\frac{n_1^2 - n_2^2}{1 - (\frac{\rho}{a})^2}} \tag{1.20}$$

which allows one to conclude that the greater the eccentricity of the ray, the greater its acceptance angle. In other words, the greater the distance from the fiber axis, the greater the NA. Note that when equation (1.19) or (1.20) would give a $\sin \theta_{aM} > 1$, $\theta_{aM} = \pi/2$ must be assumed.

1.2.2 Correspondence between Modes and Rays

A plane wave whose motion is represented by a certain ray can be described as an electric field with the form

$$\bar{\mathscr{E}}(r, \phi, z) = \bar{\mathscr{E}}_o \exp(-i\, \bar{k} \cdot \bar{P}) \tag{1.21}$$

where $\bar{k}$, called the wave vector, has the same local direction as the ray, and modules $k_1 = (2\pi/\lambda)\, n_1$; $\bar{\mathscr{E}}_o$ is a constant vector; and the spatial position $\bar{P}$ in cylindrical coordinates is

$$\bar{P} = z\,\hat{z} + r\,\hat{r} + r\,\phi\,\hat{\phi}$$

in which $\hat{z}$, $\hat{r}$, $\hat{\phi}$ are unit vectors.

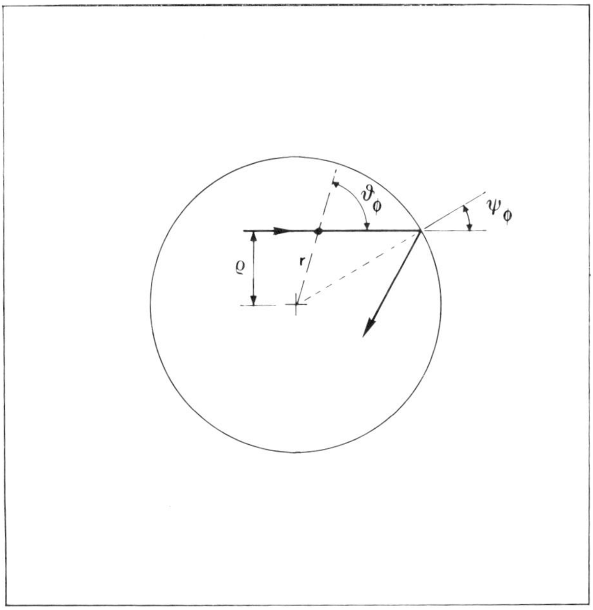

Fig. 1.4 - Core transverse section and trace of a skew ray trajectory.

The electromagnetic approach, developed in Sec. 1.1, is based on a solution of the following form

$$\overline{\mathscr{E}}(r, \phi, z,) = \overline{E}(r, \phi) \exp(-i \beta z) \tag{1.22}$$

Letting

$$\overline{k} = k_z \hat{z} + k_r \hat{r} + k_\phi \hat{\phi}$$

and comparing the exponentials of (1.21) and (1.22), we can see that the z-dependent part of $\overline{k} \cdot \overline{P}$ coincides with βz. Therefore,

$$\beta = k_z = k_1 \cos\theta \tag{1.23}$$

Equation (1.23) relates the phase constant β of electromagnetic model to the propagation angle θ of the ray optics model. Considering (1.14), the range of permitted values of θ is given by

$$n_2 \leq n_1 \cos\theta \leq n_1 \tag{1.24}$$

The right hand side of (1.24) entails that the minimum permitted value of θ is zero. Such a condition characterizes the axial ray. The left hand side of (1.24) coincides with the guidance condition for the light acceptance $\theta \leq \theta_M$.

Without the assumption of any resonance condition, the ray-optics model gives a continuum of θ values denoting guided rays, within boundaries (1.24). On the contrary, the electromagnetic model gives a finite number of permitted values for the variable β, within boundaries (1.14), which denote guided modes. Nevertheless the two couples of boundaries coincide. The ϕ-dependent part of $E(r, \phi)$ has the form $\exp(\pm i\nu\phi)$. Comparing with (1.21), we obtain

$$k_{\Phi} = \pm \frac{\nu}{r} \tag{1.25}$$

Finally, considering the modulus of k, it is possible to write

$$k_1^2 = k_z^2 + k_r^2 + k_\phi^2 = \beta^2 + k_r^2 + \frac{\nu^2}{r^2}$$

and hence

$$k_r^2 = k_1^2 - \beta^2 - \frac{\nu^2}{r^2}$$

Equation (1.25) represents an azimuthal resonance condition.

In order to obtain a complete correspondence between modes and rays, one has to add a radial resonance condition, which has the form

$$\mu = \frac{1}{\pi}\int_\rho^a k_r\, d r \tag{1.26}$$

where ρ, shown in Fig. 1.3, is the value of the radial coordinate at which k_r vanishes. The radial resonance condition (1.16) considered in the zero-order WKB approximate solution for graded-index fibers is an immediate extension of (1.26), with the only difference that $k_1^2(r)$ is assumed instead of k_1^2. This allows one to conclude that the zero-order WKB approximation is characterized by the same accuracy which characterizes the ray-optics model with the addition of suitable resonance conditions.

The wave vector $\bar{k}$ is depicted in Fig. 1.5. From this representation one can derive some relations between parameters of the ray-optics approach and parameters of the electromagnetic approach. For instance, it is

$$\nu^2 = k_1^2\, r^2 \sin^2\theta \sin^2\theta_\phi$$

$$\frac{\nu^2}{\beta^2} = r^2 \tan^2\theta \sin^2\theta_\phi \tag{1.27}$$

Moreover, a correspondence exists between $TM_{0\mu}$, $TE_{0\mu}$, modes and meridional rays: the first ones are characterized by $\nu = 0$, the latter by $\theta_{\phi} = 0, \pi$.

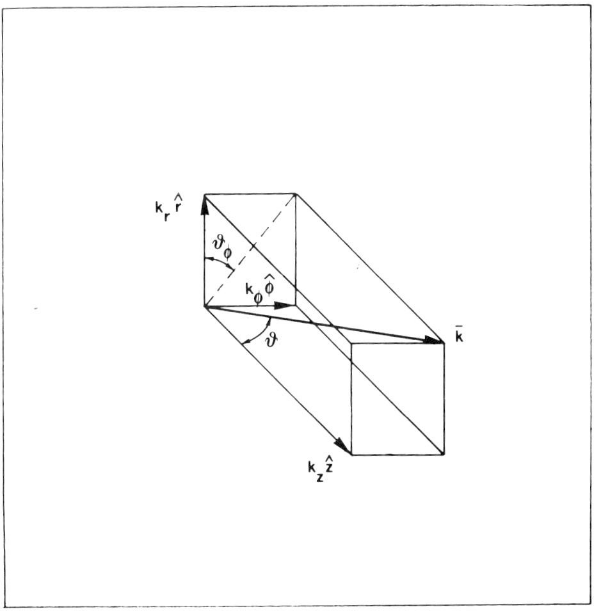

Fig. 1.5 - The propagation vector $\bar{k}$.

1.2.3 Light Acceptance in Graded-Index Fibers

For a better comprehension of the approximate solution based on the zero-order WKB method, reported in 1.1.4, we return briefly on this solution, with the aid of the ray-optics approach. The values r_1 and r_2 are those at which k_r vanishes, therefore they are called turning points. In other words, a ray trajectory exhibits an oscillating behavior between r_1 and r_2. For $\nu = 0$, only r_2 exists. From the electromagnetic point of view, the fields of such modes exist in a round region instead of an annular region. From the ray-optics point of view, an infinite number of meridional rays (one for each value of θ) fill the same region. For a given $\nu > 0$, the difference between r_2 and r_1 decreases with increasing β. When r_1 and r_2 coincide, the region of existence of the mode fields tends to a circumference, and the correspondent rays are called helical rays. In Fig. 1.6 the transverse projection of some ray trajectories are shown: m a meridional ray, s a skew ray, and h a helical ray.

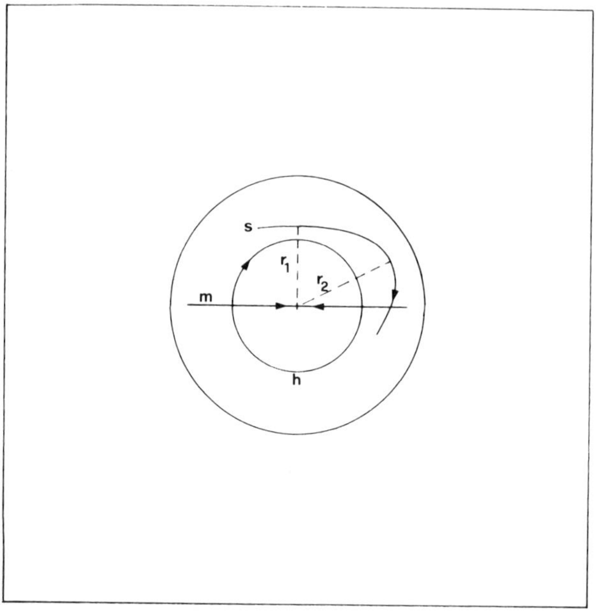

Fig. 1.6 - Transverse projections of several ray trajectories.

The light acceptance of a graded-index fiber, expressed through its *NA*, is a function of the radial coordinate and of the direction of the ray. Firstly we derive the expression of *NA* for meridional rays, which cannot be leaky. Then we will extend this expression to leaky skew rays. For $\nu = 0$, it is

$$k_r = \sqrt{k_1^2(r) - \beta^2} = \frac{2\pi}{\lambda}\sqrt{n_1^2(r) - n_1^2(r)\cos^2\theta}$$

Through the imposition of $k_r = 0$ when $r = a$, that is r_2, we obtain

$$n_1^2(a) - n_1^2(r)\cos^2\theta_M = 0$$

and hence

$$\sin^2\theta_M = \frac{n_1^2(r) - n_2^2}{n_1^2(r)}$$

Snell's law gives in this case

$$\sin\theta_{aM} = \sqrt{n_1^2(r) - n_2^2} \tag{1.28}$$

For $\nu \neq 0$, it is

$$k_r = \sqrt{k_1^2(r) - \beta^2 - \frac{\nu^2}{r^2}}$$

$$= \frac{2\pi}{\lambda}\sqrt{n_1^2(r) - n_1^2(r)\cos^2\theta - \frac{1}{r^2}n_1^2(r)\, r^2 \sin^2\theta \sin^2\theta_\phi}$$

Through the same imposition, we obtain

$$n_1^2(a) - n_1^2(r)\cos^2\theta_M - n_1^2(r)\sin^2\theta_M\left(\frac{r}{a}\sin\theta_\phi\right)^2 = 0$$

and hence

$$\sin^2\theta_M = \frac{n_1^2(r) - n_2^2}{n_1^2(r)\left[1 - \left(\frac{r}{a}\sin\theta_\phi\right)^2\right]}$$

Finally, from Snell's law

$$\sin\theta_{aM} = \sqrt{\frac{n_1^2(r) - n_2^2}{1 - \left(\frac{r}{a}\sin\theta_\phi\right)^2}} \tag{1.29}$$

Remark that in all the previous passages we have considered β^2 and ν^2 constant.

For graded-index fibers, as well as for step-index fibers, $r/a \sin\theta_\phi$ represents the eccentricity ρ/a of the ray. Therefore (1.29) can be rewritten as

$$\sin\theta_{aM} = \sqrt{\frac{n_1^2(r) - n_2^2}{1 - \left(\frac{\rho}{a}\right)^2}} \tag{1.30}$$

Also in this case, when it should be $\sin\theta_{aM} > 1$, $\theta_{aM} = \pi/2$ has to be assumed. Remark that (1.28), (1.29), (1.30) are the extensions of (1.18), (1.19), (1.20), with the only difference of assuming $n_1^2(r)$ instead of n_1^2. Those rays which exceed the *NA* defined by (1.28), but satisfy the *NA* expressed by (1.29), are called leaky rays. They correspond to the leaky modes introduced in 1.1.4.

1.2.4 General Ray-Optics in Inhomogeneous Media

Equations (1.2), using the definition of refractive index, can be written as

$$\nabla \times \bar{\mathscr{E}} = -i\, 2\pi\nu_\lambda n^2 \mu_0 \bar{\mathscr{H}}$$

$$\nabla \times \bar{\mathscr{H}} = i\, 2\pi\nu_\lambda n^2 \epsilon_0 \bar{\mathscr{E}}$$

Owing to the dualism between $\overline{\mathcal{E}}$ and $\overline{\mathcal{H}}$, in the following we can restrict ourselves only to $\overline{\mathcal{E}}$, without loss of generality. One can write

$$\nabla \times (\nabla \times \overline{\mathcal{E}}) = \nabla (\nabla \cdot \overline{\mathcal{E}}) - \nabla^2 \overline{\mathcal{E}} = (2 \pi \nu_\lambda^2)\, n^2 \mu_o \epsilon_o \overline{\mathcal{E}}$$

In inhomogeneous media, one must consider the electric displacement

$$\overline{D} = n^2 \epsilon_o \overline{\mathcal{E}}$$

whose divergence is proportional to the electric charge density. Since no sources are present, it is zero. Then, we are left with

$$n^2 \epsilon_o \nabla \cdot \overline{\mathcal{E}} + \epsilon_o \overline{\mathcal{E}} \cdot \nabla n^2 = 0$$

Through the substitution, instead of the first Helmholtz's equation (1.5), we have

$$\nabla^2 \overline{\mathcal{E}} + \nabla (\overline{\mathcal{E}} \cdot \frac{\nabla n^2}{n^2}) = - n^2 (2 \pi \nu_\lambda)^2 \mu_o \epsilon_o \overline{\mathcal{E}} \qquad (1.31)$$

Considering a set of rectangular coordinates, let us assume a trial solution of the form

$$\overline{\mathcal{E}} = \overline{E}(x, y, z) \exp[-i S(x, y, z)]$$

where $\overline{E}$ is a complex vector and S a scalar. Both are functions of all the spatial coordinates. Substituting into (1.31), we have

$$\begin{aligned} &[\nabla^2 \overline{E} - 2 i \nabla S \cdot \nabla \overline{E} - (\nabla S)^2 \overline{E} \\ &\quad + n^2 (2 \pi \nu_\lambda)^2 \mu_o \epsilon_o \overline{E}] \exp(-i S) \\ &\quad + \nabla [\overline{E} \exp(-i S) \cdot \frac{\nabla n^2}{n^2}] = 0 \end{aligned} \qquad (1.32)$$

in which $\nabla \overline{E}$ has tensorial meaning.

For a medium having *n(x,y,z)* slowly varying over wavelength distances, in the limit $\lambda \rightarrow 0$, it is

$$S = \overline{k} \cdot \overline{r}$$

as in a homogeneous medium. Since $|\overline{k}|$ is proportional to $1/\lambda$, also S can be assumed as proportional to $1/\lambda$. Therefore, when $\lambda \rightarrow 0$, in (1.32) there are terms of different magnitude. If we restrict ourselves only to terms of greater magnitude (on the order of $1/\lambda^2$), it remains

$$(\quad S)^2 = n^2 (2 \pi \nu_\lambda)^2 \mu_o \epsilon_o \qquad (1.33)$$

This equation is called the eikonal equation.

Let us indicate with $\overline{P}(s)$ a spatial position, where the variable s is defined along the ray trajectory, from a suitable starting point. The derivative $d\bar{P}/ds$ the represents a unit vector having the local direction of the ray. Therefore, from (1.33), it is possible to obtain

$$n \frac{d\bar{P}}{ds} = \frac{1}{2\pi\nu_\lambda\sqrt{\mu_o\epsilon_o}} \nabla S \qquad (1.34)$$

Taking the gradient of (1.33), we have

$$\nabla(\nabla S \cdot \nabla S) = (2\pi\nu_\lambda)^2 \mu_o \epsilon_o \nabla n^2$$

and thus

$$\nabla S \cdot \nabla(\nabla S) = (2\pi\nu_\lambda)^2 \mu_o \epsilon_o n \nabla n$$

From (1.34), it results

$$\frac{d\bar{P}}{ds} \cdot \nabla(\nabla S) = 2\pi\nu_\lambda\sqrt{\mu_o\epsilon_o}\nabla n$$

Since the following general property holds[1]

$$\frac{d\bar{P}}{ds} \cdot \nabla = \frac{d}{ds}$$

we have

$$\frac{d}{ds}(\nabla S) = 2\pi\nu_\lambda\sqrt{\mu_o\epsilon_o}\nabla n$$

and again from (1.34), we are left with

$$\frac{d}{ds}\left(n \frac{d\bar{P}}{ds}\right) = \nabla n \qquad (1.35)$$

which is called the ray equation.

Now we will demonstrate that Snell's law, which has been used in the previous sections, is only a particular case of the ray equation (1.35).

Let us consider a z axis perpendicular to the interface between two different media characterized by indices n_1 and n_2. Indicating with α_1 and α_2 the angles of the incident and of the refracted rays with respect to an x axis, parallel to the interface, it is possible to write for both media

$$\frac{d}{ds}\left(n_i \frac{dx}{ds}\right) = 0 \qquad \frac{dx}{ds} = \cos\alpha_i, \quad i = 1, 2$$

and then

$$n_1 \cos\alpha_1 = n_2 \cos\alpha_2$$

Indicating with θ_1 and θ_2 the angles of the incident and of the refracted rays with respect to the z axis, it is

$$\theta_i = \frac{\pi}{2} - \alpha_i , \; i = 1, 2$$

and hence

$$n_1 \sin \theta_1 = n_2 \sin \theta_2$$

1.2.5 Constants of the Motion

In a medium with round symmetry, the refractive index distribution satisfies the condition

$$n^2 (x, y, z) = n^2 (\sqrt{x^2 + y^2}, z)$$

therefore, letting $r^2 = x^2 + y^2$, it is

$$\frac{\partial n^2}{\partial x} = \frac{x}{r} \frac{\partial n^2}{\partial r}, \quad \frac{\partial n^2}{\partial y} = \frac{y}{r} \frac{\partial n^2}{\partial r}$$

Substituting into the x and y projections of (1.35), yields

$$x \frac{d}{ds} \left(n \frac{dy}{ds} \right) - y \frac{d}{ds} \left(n \frac{dx}{ds} \right) = 0$$

and hence

$$\frac{d}{ds} \left(n x \frac{dy}{ds} - n y \frac{dx}{ds} \right) = 0$$

This allows us to define a constant of the motion, that is a quantity which remains constant along a ray trajectory, as

$$h = n \left(x \frac{dy}{ds} - y \frac{dx}{ds} \right) \tag{1.36}$$

In Fig. 1.7 different coordinate sets are shown. We can distinguish:

i) rectangular coordinates x, y, z

ii) cylindrical coordinates r, ϕ, z

iii) intrinsic coordinates r, θ, θ_ϕ

The following relations, which will be useful in the remainder, can be easily derived from Fig. 1.7.

$$x = r \cos \phi \quad , \quad y = r \sin \phi$$

$$\frac{dz}{ds} = \cos \theta \, , \; r \frac{d\phi}{ds} = \sin \theta \sin \theta_\phi \tag{1.37}$$

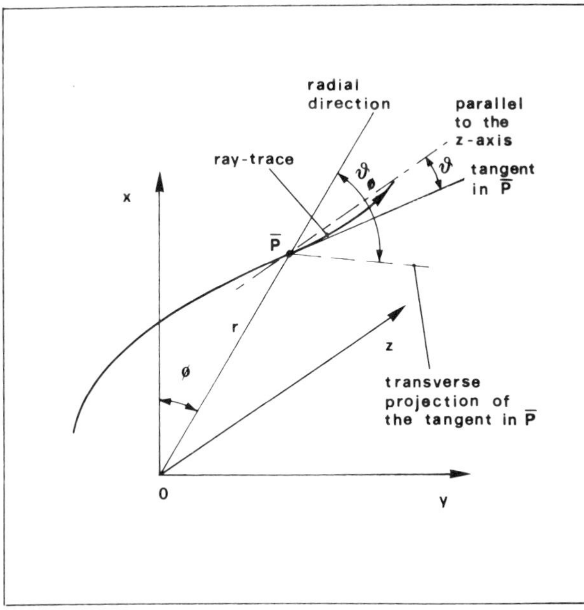

Fig. 1.7 - Different coordinate sets.

Expression (1.36) holds for a set of rectangular coordinates. When a set of cylindrical or intrinsic coordinates is used, from (1.37), we have

$$h = n r^2 \frac{d\phi}{ds} \tag{1.38}$$

$$h = nr \sin\theta \sin\theta_\phi \tag{1.39}$$

Let us consider a z invariant medium with round symmetry, like that of an ideal multimode fiber. For such a medium, it is

$$n^2(x, y, z) = n^2(\sqrt{x^2 + y^2})$$

Therefore, besides a first constant of the motion h, due to the round symmetry, a second constant of the motion can be obtained directly from the z projection of (1.35)

$$\frac{d}{ds}\left(n \frac{dz}{ds}\right) = 0$$

For rectangular or cylindrical coordinates, it is

$$E = n \frac{dz}{ds} \tag{1.40}$$

whereas, for intrinsic coordinates, it is

$$E = n \cos \theta \tag{1.41}$$

Comparing the results here obtained with those reported in the previous sections, and in particular (1.40) with (1.23), and (1.39) with the first of (1.27), we can conclude that the second constant of the motion E is proportional to β, and the first constant o the motion h is proportional to ν. In both cases, the proportionality factor is $2\pi/\lambda$. Dividing h by E, one obtains another constant, which for the three sets of coordinates results ordinately

$$\ell = \frac{h}{E} = x\frac{dy}{dz} - y\frac{dx}{dz} = r^2\frac{d\phi}{dz} = r \tan\theta \sin\theta_\phi \tag{1.42}$$

It coincides with the ratio ν/β, as shown by the second of (1.27). In Table 1.3 the different expressions of the constants of the motion for the three coordinate sets are resumed, and also their relations with some electromagnetic parameters are reported.

Table 1.3 - Different expressions of the constants of the motion.

	h	E	ℓ
retangular coordinates	$n\,(x\,dy/ds - y\,dx/ds)$	$n\,dz/ds$	$x\,dy/dz - y\,dx/dz$
cylindrical coordinates	$n\,r^2\,d\phi/ds$	$n \cos\theta$	$r^2\,d\phi/dz$
intrinsic coordinates	$nr \sin\theta \sin\theta_\phi$	$n \cos\theta$	$r \tan\theta \sin\theta_\phi$
relations with β, ν	$(\lambda/2\pi)\,\nu$	$(\lambda/2\pi)\,\beta$	ν/β

1.2.6 Ray Tracing

The study of the transmission properties of an optical fiber through the solution of the ray equation is based on a mathematical procedure called ray tracing. It consists of the determination of each ray trajectory, given a certain starting condition, which depends on the fiber excitation. Since all the properties of a ray depend only on a pair of constants of the motion, the domain in which rays can exist is a two-dimensional domain, where rays denoted by the same pair are indistinguishable. Nevertheless, in this doman, even when the fiber is uniformly excited, the density of rays is not uniform. Therefore, in order to weigh the contribution of each ray correctly, it is neccssary to know such a density, or ray distribution. This argument will be treated in Section 1.2.9.

A ray trajectory in a graded-index fiber is univocally defined by a suitable initial condition, which consists of the term (r_o, ϕ_o, z_o) as starting point, and the pair $(\theta_o, \theta_{\phi o})$ as starting direction. From this condition, one can obtain the two constants of the motion of the ray, as

$$h = n(r_o)\, r_o \sin\theta_o \sin\theta_{\phi o}$$

$$E = n(r_o) \cos\theta_o \tag{1.43}$$

The projection of the ray equation (1.35) on the radial direction is

$$\frac{d}{ds}\left(n \frac{dr}{ds}\right) - n r \left(\frac{d\phi}{ds}\right)^2 = \frac{dn}{dr}$$

therefore, one has

$$n \frac{dz}{ds} \frac{d}{dz}\left(n \frac{dz}{ds} \frac{dr}{dz}\right) - n^2 r \left(\frac{d\phi}{ds}\right)^2 = \frac{1}{2} \frac{dn^2}{dr}$$

From (1.38) and (1.40), we have

$$\frac{d^2 r}{dz^2} = \frac{h^2}{E^2} \frac{1}{r^3} + \frac{1}{2E^2} \frac{dn^2}{dr}$$

and multiplying by dr/dz, it follows

$$\frac{1}{2} \frac{d}{dz}\left[\left(\frac{dr}{dz}\right)^2\right] = \frac{h^2}{E^2 r^3} \frac{dr}{dz} + \frac{1}{2E^2} \frac{dn^2}{dr} \frac{dr}{dz}$$

Integration between r_1 and a certain r yields

$$\left(\frac{dr}{dz}\right)^2 = \frac{n^2(r)}{E^2} - \frac{h^2}{E^2 r^2} - \frac{n^2(r_1)\, r_1^2 - h^2}{E^2 r_1^2}$$

From (1.39) and (1.41), considering $r = r_1$, where it is $\theta_\phi = \pi/2$, one can write

$$n^2(r_1)\, r_1^2 - h^2 = E^2 r_1^2$$

and hence

$$\frac{dr}{dz} = \pm \sqrt{\frac{n^2}{E^2} - \frac{h^2}{E^2} \frac{1}{r^2} - 1} \tag{1.44}$$

The positive sign holds when $r(z)$ increases from r_1 to r_2, the negative sign when it decreases from r_2 to r_1.

From the knowledge of the starting condition (1.43), the solution of the differential equation (1.44) gives all the information in order to obtain the ray trajectory.[6] In fact, knowing *r(z)*, we have

$$\theta(z) = \cos^{-1}\left[\frac{E}{n(r(z))}\right]$$

$$\theta_\phi(z) = \sin^{-1}\left[\frac{h}{n[r(z)]\,r(z)\sin\theta(z)}\right]$$

Finally $\phi(z)$ can be obtained through the solution of the following differential equation

$$\frac{d\phi}{dz} = \frac{h}{E}\frac{1}{r^2(z)}$$

that is

$$\phi(z) = \phi_o + \frac{h}{E}\int_{z_o}^{z}\frac{dz}{r^2(z)}$$

Function *r(z)* is periodic along *z*, with period

$$Z(E,h) = 2\int_{r_1}^{r_2}\frac{dr}{\left(\frac{dr}{dz}\right)} = 2\int_{r_1}^{r_2}\frac{dr}{\sqrt{\frac{n^2}{E^2} - \frac{h^2}{E^2}\frac{1}{r^2} - 1}} \qquad (1.45)$$

and ranges between r_1 and r_2, according with the results of the previous sections. Also functions $\theta(z)$ and $\theta_\phi(z)$ are periodic with the same period, because they depend directly on *r(z)*. $\theta(z)$ ranges between a minimum value θ_2, reached when $r = r_2$, and a maximum value θ_1, reached when $r = r_1$. $\theta_\phi(z)$ also ranges between a minimum value $\theta_{\phi 1}$, reached when r moves from r_1 to r_2, and a maximum value $\theta_{\phi 2}$, reached when r moves from r_2 to r_1.

Meridional rays and helical rays exhibit particular behaviors. For meridional rays, $r_1 = 0$, $\theta_2 = 0$, $\theta_{\phi 1} = 0$, $\theta_{\phi 2} = \pi$. For helical rays, it is $r_1 = r_2$, $\theta_1 = \theta_2$, $\theta_{\phi 1} = \theta_{\phi 2} = \pi/2$. In Fig. 1.8, the behaviors of *r(z)*, $\theta(z)$, and $\theta_\phi(z)$ along a ray period are shown, for a skew ray, a meridional ray, and a helical ray. Assuming the origin of the z axis either at the beginning or at the middle of the period *Z(E,h)*, *r(z)* and $\theta(z)$ are even functions, whereas $\theta_\phi(z)$ is an odd function.[7]

Finally, function $\phi(z)$ has the following behavior

$$\phi(z) = vz + p(z)$$

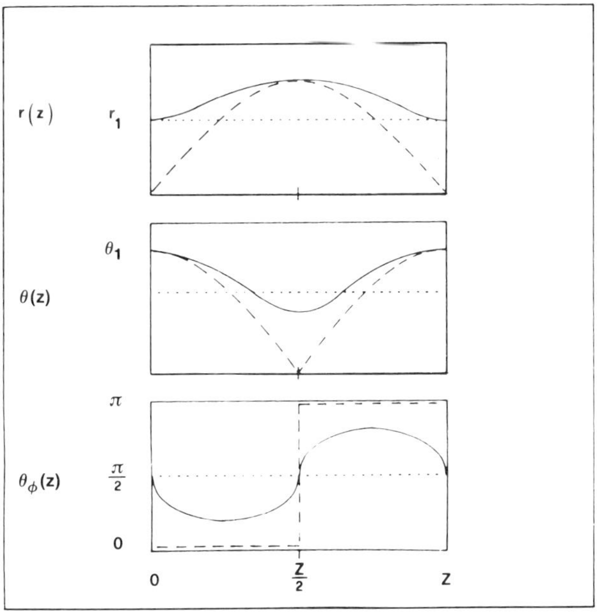

Fig. 1.8 - Behaviors of a skew ray (continuous line), a meridional ray (dashed line), and a helical ray (dotted line).

where *p(z)* is periodic with period

$$\Phi(E,h) = 2\,\frac{h}{E}\int_{r_1}^{r_2} \frac{dr}{r^2\left(\dfrac{dr}{dz}\right)}$$

$$= 2\,\frac{h}{E}\int_{r_1}^{r_2} \frac{dr}{r^2\sqrt{\dfrac{n^2}{E^2} - \dfrac{h^2}{E^2}\,\dfrac{1}{r^2} - 1}}$$

and υ is a constant dependent on E and h. Precisely, it results

$$\upsilon(E,h) = \frac{\Phi(E,h)}{Z(E,h)}$$

In other words, $\phi(z)$ has a pseudoperiodic behavior,[6] as shown in Fig. 1.9, where a skew ray, a meridional ray, and a helical ray are considered. For meridional rays it is $\Phi = \pi$, for helical rays it is $\Phi = 0$.

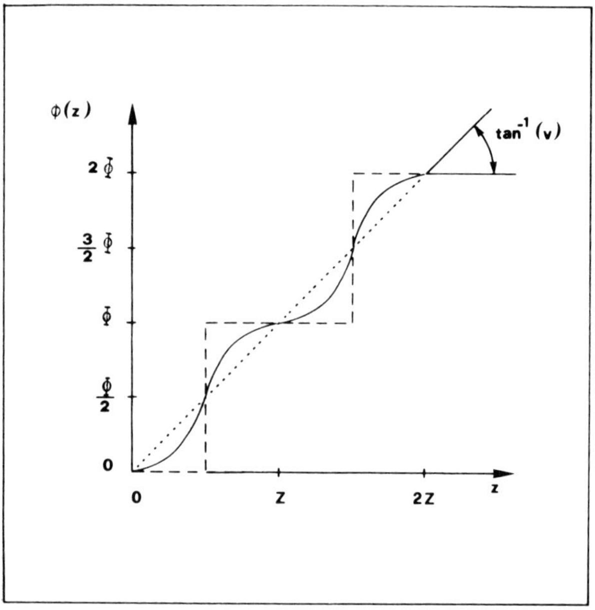

Fig. 1.9 - Pseudoperiodic behavior of ϕ (z) for a skew ray (continuous line), a meridonial ray (dashed line), and a helical ray (dotted) line.

1.2.7 Ray Domain

The domain of guided rays, in the plane *(E, h),* can be divided into two regions:

i) the region of strictly guided rays, denoted G

ii) the region of leaky rays, denoted L

Region G is bounded by the following curves

i) the meridional ray curve, denoted m

ii) the helical ray curve, denoted e

iii) the curve of strictly guidance condition, denoted g

Region L is bounded by the following curves:

i) curve g

ii) curve e

iii) the curve of leaky guidance condition, denoted ℓ

Curve m is the straight line

$$h = 0$$

Curve e is defined by the following parametric equations

$$h = \sqrt{-\frac{r^3}{2}\frac{dn^2}{dr}}$$

$$E = \sqrt{n^2(r) + \frac{r}{2}\frac{dn^2}{dr}} \qquad (1.46)$$

In fact, from the projection of the ray equation (1.35) on the radial direction

$$\frac{d}{ds}\left(n\frac{dr}{ds}\right) - nr\left(\frac{d\phi}{ds}\right)^2 = \frac{dn}{dr}$$

letting $dr/dz = 0$, one obtains the first of (1.46). From (1.44), letting again $dr/dz = 0$, one obtains the second of (1.46). Curve g is derived from the definition of NA without leaky rays (1.28). Substituting $n^2(r)$ to $n_1^2(r)$ and $n^2(a)$ to n_2^2, one can easily obtain

$$E = n(a)$$

Finally curve ℓ is derived from the definition of numerical aperture with leaky rays (1.29). With the same substitutions, independently of the shape of $n(r)$, it results

$$E = \sqrt{n^2(a) - \frac{h^2}{a^2}}$$

In Fig. 1.10, the two regions G and L are shown for a general index profile $n(r)$, $dn^2/dr \leq 0$ inside the fiber core. Note that the first of (1.46) entails that helical rays cannot exist when $dn^2/dr > 0$.

1.2.8 The Radiance Law

The primary characteristic of an optical source is its radiance, which is defined as density of optical power per unit area and unit solid angle. Any planar extended source has a radiance B which can be expressed, in the most general case, as a function of two coordinates which denote the point of the emitting planar surface and two coordinates which denote the direction of emission.

An elementary area is taken conventionally normal to the emission direction. Then, using a pair of cylindrical coordinates (r, ϕ) to denote the point and an intrinsic pair (θ, θ_ϕ) to denote the direction, such an elementary area can be written as

$$dS_n = \cos\theta \, r \, dr \, d\phi$$

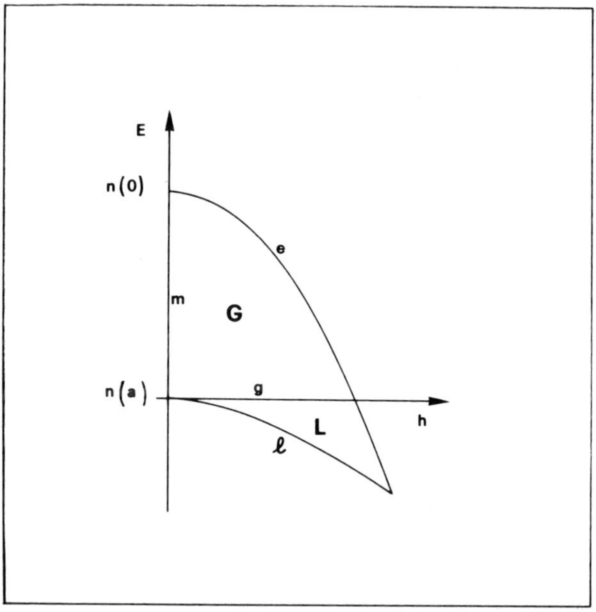

Fig. 1.10 - Regions of strictly guided (*G*) and leaky (*L*) rays in domain (*E, h*).

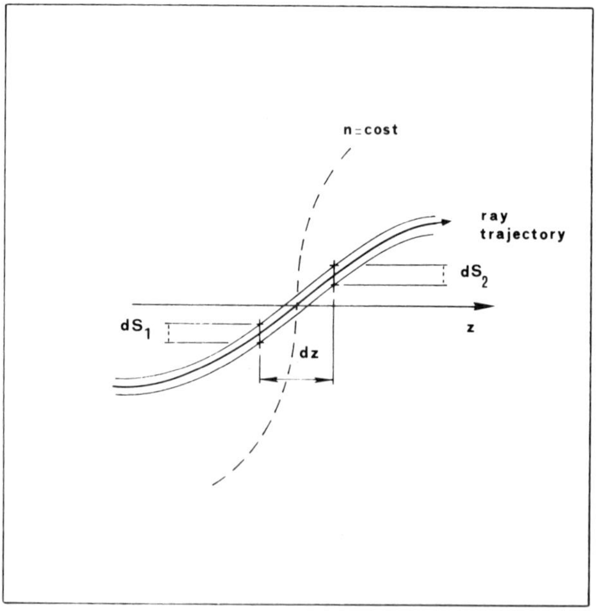

Fig. 1.11 - Ray picture for demonstrating the radiance law.

The elementary angle, with the same choice of coordinates, is

$$d\,\Omega = \sin\theta\, d\theta\, d\theta_\phi$$

Letting P the optical power, one can write

$$B(r, \phi, \theta, \theta_\phi) = \frac{d\,P}{d\,S_n\, d\,\Omega}$$

In a homogeneous lossless medium, the radiance propagates invariantly along each ray trajectory, which is a straight line with direction (θ, θ_ϕ), from the starting point (r, ϕ). In inhomogeneous lossless media, the ray trajectories are generally curved depending on the index distribution $n(\bar{P})$, where $\bar{P}$ denotes a spatial position. Yet, the quantity B/n^2 remains constant along each ray trajectory. The last property is called the radiance law.

In order to demonstrate this assert,[8] let us consider a surface on which n is constant. Assuming a z axis normal to this surface in a point 0, as shown in Fig. 1.11, we consider two elementary areas dS_1 and dS_2 along a ray trajectory, both normal to the z axis and a distance dz from each other. We let $d\Omega_1 = \sin\theta_1\, d\,\theta_1\, d\,\theta_{\phi 1}$ and $d\Omega_2 = \sin\theta_2\, d\,\theta_2\, d\,\theta_{\phi 2}$ the divergences of the optical power densities carried by the ray respectively across dS_1 and dS_2. Finally, we let n_1 and n_2, the refractive indices in dS_1 and dS_2, being $n_2 = n_1 + dn$. The radiances $B(n_1)$ and $B(n_2)$, respectively defined across dS_1 and dS_2, for continuity, can be related as

$$B(n_2) = B(n_1) + \frac{dB}{dn}\, dn$$

From the definition of radiance, considering that in a lossless medium dP remains constant, we have

$$\frac{B(n_2)}{B(n_1)} = \frac{d\,S_{n1}\, d\,\Omega_1}{d\,S_{n2}\, d\,\Omega_2} = \frac{\cos\theta_1\, d\,S_1 \sin\theta_1\, d\theta_1\, d\theta_{\phi 1}}{\cos\theta_2\, dS_2 \sin\theta_2\, d\theta_2\, d\theta_{\phi 2}}$$

One can assume dS_1 to be equal to dS_2, except for infinitesimal quantities of higher order, and a similar assumption can be made for $d\theta_{\phi 1}$ and $d\theta_{\phi 2}$. It remains

$$\frac{B(n_2)}{B(n_1)} = \frac{d\sin^2\theta_1}{d\sin^2\theta_2}$$

Through the differentiation of Snell's law, we have

$$\frac{B(n_2)}{B(n_1)} = \frac{n_2^2}{n_1^2}$$

Also a thermodynamical proof, which we will not report for simplicity, can be developed to demonstrate the radiance law.[8] The definition of radiance will be useful in the next section in order to obtain the distribution of optical power in the domain of guided rays. Another typical parameter which is sometimes used to describe optical sources is the irradiance I, which consists of the integrated radiance over a 2π solid angle around the direction normal to the emitting surface. So we have

$$I(r,\phi) = \frac{dP}{dS_n} = \int_0^{2\pi} d\,\theta_\phi \int_0^{\frac{\pi}{2}} B(r,\phi,\theta,\theta_\phi)\sin\theta\cos\theta\, d\,\theta$$

Both definitions of radiance and irradiance will be useful in Chapter 3 to classify the optical sources.

1.2.9 Ray Distribution

The total optical power launched into a fiber can be written as

$$P = \int_0^{2\pi} d\phi \int_0^{2\pi} d\theta_\phi \int_0^{a} r\, dr \int_0^{\theta_M} B(r,\phi,\theta,\theta_\phi)\sin\theta\cos\theta\, d\,\theta \qquad (1.47)$$

where $B(r, \phi, \theta, \theta_\phi)$ is the radiance distribution, due to the optical sources, at the input section of the fiber. The correspondent density of optical power in region (G or G + L), denoted $W(E, h)$, can be formally related to P through the following equation

$$P = \iint W(E,h)\, dE\, dh \qquad (1.48)$$

in which the double integral is extended to G, if we consider

$$\theta_M = \sin^{-1}\left[\sqrt{1 - \frac{n^2(a)}{n^2(r)}}\right]$$

or to the whole region ($G + L$), if we consider

$$\theta_M = \sin^{-1}\left\{\sqrt{\frac{1}{1-\left(\frac{r}{a}\sin\theta_\phi\right)^2}\left[1-\frac{n^2(a)}{n^2(r)}\right]}\right\}$$

Simple geometrical considerations yield

$$dr = \tan\theta\cos\theta_\phi\, dz$$

through the differentiation of (1.39) and of (1.41) squared, we have

$$dh = n(r)\, r\sin\theta\cos\theta_\phi\, d\,\theta_\phi$$

$$E\, dE = -n^2(r)\ \sin\theta\cos\theta\, d\theta$$

which give

$$d\,\theta_\phi\, r\, dr \sin\theta \cos\theta\, d\theta \;=\; -\,\frac{dE\, dh\, dz}{n^2(r)}$$

Comparing (1.47) and (1.48), in which the extremes of the E integration must be reversed to have a positive sign, for both G and $(G + L)$, it is

$$W(E, h) = 2\int_0^{2\pi} d\phi \int_0^{Z(E,h)} \frac{B(r(z), \phi, \theta(z), \theta_\phi(z))}{n^2(r(z))}\, dz =$$

$$= 4\int_0^{2\pi} d\phi \int_{r_1}^{r_2} \frac{B(r, \phi, \theta(r), \theta_\phi(r))}{n^2(r)\left(\dfrac{dr}{dz}\right)}\, dr \tag{1.49}$$

A uniform launching condition is characterized by the radiance in air

$$\frac{B(r, \phi, \theta, \theta_\phi)}{n^2(r)} = B_o$$

where B_o is a constant value. Even in this condition, the intrinsic distribution of rays in the ray domain is not uniform. In fact this distribution turns out to be

$$W_o(E, h) = 8\,\pi\, B_o \int_{r_1}^{r_2} \frac{dr}{\left(\dfrac{dr}{dz}\right)} = 4\,\pi\, B_o\, Z(E, h) \tag{1.50}$$

Expression (1.49) gives the ray distribution for the most general radiance distribution B. In such an expression the factor 2 takes into account the two possible orientations of each ray trajectory. Expression (1.50) gives the ray distribution for a uniform excitation in air, it is called also ray density in the domain of guided rays.

As expected, $W_o(E, h)$ is proportional to the axial length of the ray period. In fact the longer the period the greater the probability of finding one of the set of indistinguishable rays (E, h) on the transverse input section of the fiber. In fig. 1.12 a typical behavior of $Z(E, h)$ is reported against E, with h as a parameter. Considering B_o in (1.47), one obtains

$$P_o = 2\,\pi^2 B_o \int_0^a [\,n^2(r) - n^2(a)\,]\, r\, dr \tag{1.51}$$

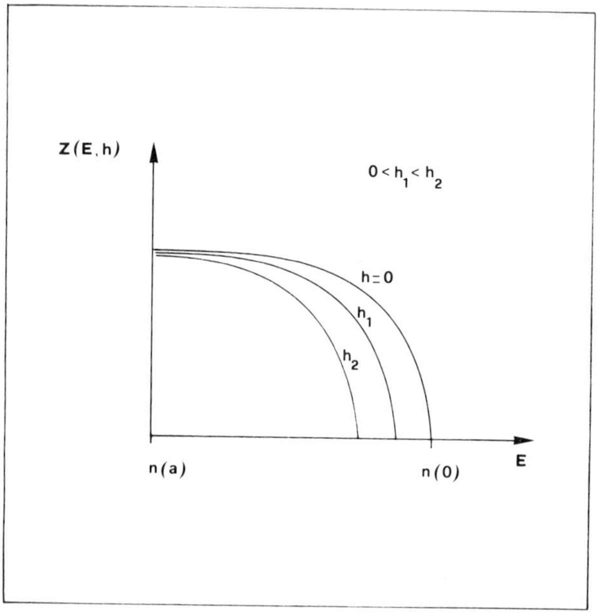

Fig. 1.12 - Typical behavior of $Z(E, h)$ vs. E, with h as a parameter.

when only region G is considered, and

$$P_o = 2\pi^2 B_o \int_0^a \frac{n^2(r) - n^2(a)}{\sqrt{1 - (\frac{r}{a})^2}} rdr \qquad (1.52)$$

when the whole region $(G + L)$ is considered. In such relations pedix o denotes that the total optical power P is launched using a uniform excitation in air.

1.2.10 Domain (r_1, θ_1)

The two constants of the motion E, h cannot be easily identified in order to scan the output section of the fiber, or to launch a given ray family into the fiber. Therefore, in order to test the propagation behavior of ray families experimentally, a more suitable pair of constants of the motion must be used. At this purpose, one can assume a new pair (r_1, θ_1) instead of the pair (E,h), where r_1 is the minimum distance of the ray trajectory from the fiber axis, and θ_1 the propagation angle when $r = r_1$, as it is shown in Fig. 1.8.[9]

Since the variable ϕ does not influence the determination of E and h, we will restrict ourselves to demonstrate that any starting point $(r_o,$

$\theta_o, \theta_{\phi o}$), which originates a ray (E,h), can be reproduced by a suitable pair (r_1, θ_1). We have

$$E = n(r_o)\cos\theta_o$$

$$h = n(r_o)\, r_o \sin\theta_o \sin\theta_{\phi o}$$

When $r_o = r_1$, $\theta_o = \theta_1$, it is $\theta_{\phi o} = \pi/2$ for skew rays, and $\theta_{\phi o} = 0, \pi$ for meridional rays. Therefore, one can write

$$E = n(r_1)\cos\theta_1 \qquad (1.53)$$
$$h = n(r_1)\, r_1 \sin\theta_1$$

the latter of these relations being true both for $\theta_{\phi o} = \pi/2$, and for $\theta_{\phi o} = 0, \pi$. For meridional rays, when it is anywhere $dn/dr \leq 0$, we have $r_1 = 0$ and hence $h = 0$. On the contrary, if in some r intervals it is $dn/dr > 0$, then a value of r_1 different from zero may exist, however it is $\theta_1 = 0$, and hence again $h = 0$. In other words, meridional rays, defined by the condition $h = 0$, when mapped in domain (r_1, θ_1), do not always correspond to $r_1 \neq 0$, while in domain (E, h) it was not possible.

From the transform relations (1.53), it is possible to derive boundaries and ray distribution in domain (r_1, θ_1). Also in this case, region G and region $(G + L)$ are bounded by the same curves, m, e, g, ℓ, introduced in 1.2.7 for domain (E, h). Here we have:

i) curve m which is given by the following expressions

$$r_1 = 0 \qquad \left(\frac{dn}{dr} \leq 0\right)$$

$$\theta_1 = 0 \qquad \left(\frac{dn}{dr} > 0\right)$$

ii) curve e which can be written as

$$\sin^2\theta_1 = -\frac{r_1}{2}\left.\frac{dn^2}{dr}\right|_{r = r_1}$$

iii) curve g in the form

$$\sin^2\theta_1 = 1 - \frac{n^2(a)}{n^2(r_1)}$$

iv) curve ℓ in the form

$$\sin^2\theta_1 = \frac{1}{1 - (r_1/a)^2}\left[1 - \frac{n^2(a)}{n^2(r_1)}\right]$$

The last expression holds when $n^2(a)/n^2(r_1) \geq (r_1/a)^2$, wheras it is $\sin^2\theta_1 = 1$ when $n^2(a)/n^2(r_1) < (r_1/a)^2$.

In Fig. 1.13, the two regions G and L are shown for a general index profile $n(r)$, having $dn/dr \leq 0$ anywhere. Remark that the region, bounded by curve e and by the lines $\theta_1 = 0, r_1 = \text{a}$, represents a domain (r_2, θ_2) which can be considered as the conjugate of domain (r_1, θ_1).

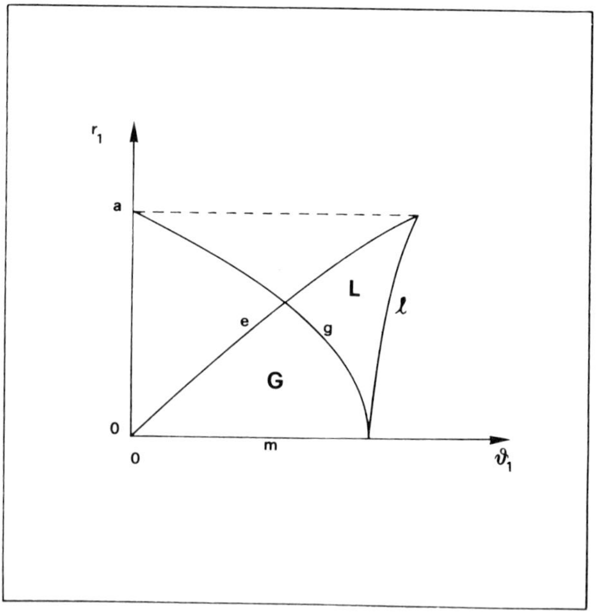

Fig. 1.13 - Regions of strictly guided (G) and leaky (L) rays in domain (r_1 , θ_1).

As for the ray distribution, simple geometrical considerations and the differentiation of equations (1.53),[9] yield

$$d\theta_\phi = \frac{\tan\theta_1}{r\tan\theta\cos\theta_\phi}\, dr_1$$

$$\sin\theta\cos\theta\, d\theta = \frac{n^2(r_1)}{n^2(r)}\sin\theta_1\cos\theta_1\, d\theta_1$$

$$dr = \tan\theta\cos\theta_\phi\, dz$$

Therefore, comparing the expression of the total guided power

$$P = \iint W(r_1, \theta_1)\, dr_1\, d\theta_1 \tag{1.54}$$

with (1.47), one obtains

$$W(r_1, \theta_1) = 4\, n^2(r_1) \sin^2\theta_1 \int_0^{2\pi} d\phi \int_{r_1}^{r_2} \frac{B(r, \phi, \theta(r), \theta_\phi(r))}{n^2(r)\left(\frac{dr}{dz}\right)} dr \tag{1.55}$$

For uniform excitation, we have

$$W_o(r_1, \theta_1) = 4\,\pi\, B_o\, n^2(r_1) \sin^2\theta_1\, Z(r_1, \theta_1) \tag{1.56}$$

in which it is

$$Z(r_1, \theta_1) = 2\int_{r_1}^{r_2} \frac{dr}{\left(\frac{dr}{dz}\right)}$$

$$\left(\frac{dr}{dz}\right)^2 = \frac{n^2(r)}{n^2(r_1)\cos^2\theta_1} - \left(\frac{r_1}{r}\tan\theta_1\right)^2 - 1$$

1.3 INTERMODAL TIME DISPERSION

Time dispersion in multimode fibers is due to the spread in the group delays for the guided modes. Moreover, each mode exhibits a group delay which depends on the wavelength of the optical power. Therefore if the optical source is characterized by a spectrum of finite width $\Delta\lambda$, an intramodal time dispersion is present, in addition to an intermodal time dispersion, and it increases with increasing $\Delta\lambda$. Usually, when a very accurate compensation of the intermodal time dispersion is performed by a suitable index profile, even a very small width $\Delta\lambda$ can produce remarkable effect on the overall time dispersion. On the other hand, when the index profile is not optimized, even for very large values of $\Delta\lambda$, the effects of intramodal time dispersion are negligible.

The influence of λ on the group delays is called material dispersion. Since it depends on the dopants employed, in order to develop a general theory of the compensation which can be performed by an optimized index profile, in the present paragraph we will consider the refractive index $n(r)$ as independent of the wavelength, ignoring material dispersion. In the next subsection we will also take into account material dispersion.

1.3.1 Time Dispersion in Step-Index Fibers

Time dispersion effects in step-index fibers can be derived through simple geometrical considerations. A ray characterized by a propagation angle θ exhibits a group delay per unit distance coincident with the inverse of the axial component of its group velocity. It can be written

$$\tau(\theta) = \frac{n_1}{c} \frac{1}{\cos\theta} \tag{1.57}$$

where n_1 is the core refractive index, and c is the free space velocity of light. Since θ is usually very small, $\cos\theta$ can be expanded in power series, and, neglecting terms of order higher than two, it results

$$\tau(\theta) = \frac{n_1}{c}\left(1 + \frac{\theta^2}{2}\right) \tag{1.58}$$

where n_1/c is the delay which characterizes the axial ray, $(n_1/c)(\theta^2/2)$ is the relative delay distribution with respect to this axial ray.

In step-index fibers θ depends only on the constant of the motion E. Precisely, since it is $E = n_1\cos\theta$, from (1.57) we have

$$\tau(E) = \frac{n_1^2}{c} \cdot \frac{1}{E} \tag{1.59}$$

It is, however, more convenient in this case to consider the propagation angle θ as a variable denoting the modes or rays.

A phasor which represents the effects of a unit optical power carried by a ray denoted by a certain angle θ, at a distance z from the optical source, can be written as

$$\bar{u}(\theta) = \exp[-i\,\omega\,\tau(\theta)\,z]$$

where ω is the modulation circular frequency, and i is the imaginary unit. We assume that the optical source is unaffected by any relative spatial delay of emission. All rays having the same θ can be considered as indistinguishable. Thus, in order to take into account the whole spread, letting $B(r, \phi, \theta, \theta_\phi)$ the radiance distribution at the input fiber section, we consider a phasor $\bar{H}(\omega,z)$ which represents the base band response of the fiber at the distance z. It can be written as

$$\bar{H}(\omega, z) = \frac{\int_0^{2\pi} d\phi \int_0^{2\pi} d\theta_\phi \int_0^{a} r\,dr \int_0^{\theta_M} B(r, \phi, \theta, \theta_\phi)\,\bar{u}(\theta) \sin\theta\cos\theta\,d\theta}{\int_0^{2\pi} d\phi \int_0^{2\pi} d\theta_\phi \int_0^{a} r\,dr \int_0^{\theta_M} B(r, \phi, \theta, \theta_\phi) \sin\theta\cos\theta\,d\theta}$$

For a uniform excitation, given by

$$B(r, \phi, \theta, \theta_\phi) = n_1^2 B_o$$

it follows

$$\bar{H}(\omega, z) = \frac{\int_0^{\theta_M} \bar{u}(\theta) \sin\theta \cos\theta \, d\theta}{\int_0^{\theta_M} \sin\theta \cos\theta \, d\theta} \tag{1.60}$$

Considering that $\sin\theta_M \cong \theta_M$, and making use also of Euler's formula, we have

$$|\bar{H}(\omega, z)| = \frac{\left|\sin\left(\omega \frac{n_1}{c} \frac{\theta_M^2}{4} z\right)\right|}{\omega \frac{n_1}{c} \frac{\theta_M^2}{4} z}$$

$$arg\,[\bar{H}(\omega, z)] = -\,\omega \frac{n_1}{c}\left(1 + \frac{\theta_M^2}{4}\right) z \tag{1.61}$$

Such expressions hold for strictly guided rays; i.e., for a numerical aperture given by (1.18). When leaky rays are considered too, instead of (1.60) we have

$$\bar{H}(\omega, z) = \frac{\int_0^{2\pi} d\theta_\phi \int_0^a r dr \int_0^{\theta_M(r,\theta_\phi)} \bar{u}(\underline{\ell}) \sin\theta \cos\theta \, d\theta}{\int_0^{2\pi} d\theta_\phi \int_0^a r dr \int_0^{\theta_M(r,\theta_\phi)} \sin\theta \cos\theta \, d\theta}$$

in which the numerical aperture, given by (1.19), is a function of both r and θ_ϕ. Here the assumption $\sin\theta_M \cong \theta_M$ is not always justified. In fact, for certain r and θ_ϕ, it can be $\sin\theta_M = 1$. Moreover, instead of (1.58), one must consider (1.57). The integral cannot be expressed in finite terms, however it can be seen that the presence of leaky rays produces and increase in time dispersion; i.e., a reduction in the bandwidth.

From (1.61) one can conclude that the baseband response of a step-index fiber, under a uniform excitation, neglecting leaky rays, suffers only an amplitude distortion. In fact arg $[\bar{H}]$ is proportional to ω. The 3 dB bandwidth B_ω can be obtained as a function of the distance z, from the following relation

$$\frac{\left|\sin\left(B_\omega \frac{n_1}{c} \frac{\theta_M^2}{4} z\right)\right|}{B_\omega \frac{n_1}{c} \frac{\theta_M^2}{4} z} = \frac{1}{2}$$

Since this expression is function of the product $B_\omega z$, it is possible to state that $B_\omega(z)$ decreases as $1/z$. This is a general property, true not only for step-index fibers, but also for graded-index fibers, and is due to the multimodal propagation. The amplitude of the baseband response $|H(\omega, z)|$ is then usually reported against the product ωz. It exhibits a series of lobes spaced spaced by zeros, as shown in Fig. 1.14, for typical values of n_1 and θ_M.

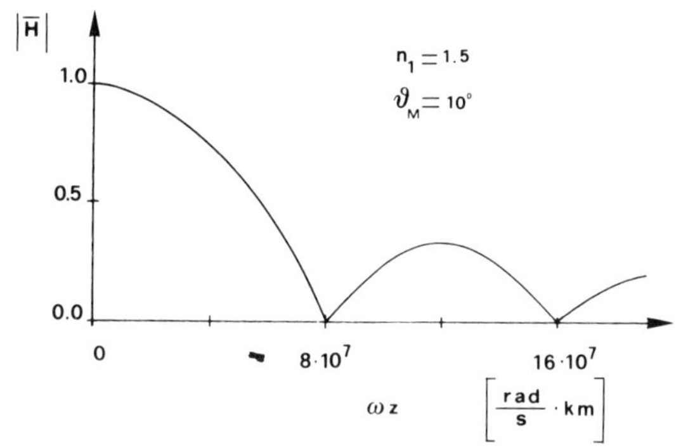

Fig. 1.14 - Amplitude baseband response of a step index fiber, under a uniform excitation and neglecting leaky rays.

An alternative way for the description of time dispersion is that based on the impulse response of the fiber. Baseband response and impulse response are a Fourier transform pair. The impulse response of a step-index fiber can calculated through the following general expression

$$h(t, z) = \frac{\int_0^{2\pi} d\phi \int_0^{2\pi} d\theta_\phi \int_0^a r\,dr \int_0^{\theta_M} B(r, \phi, \theta, \theta_\phi)\,\delta[t - \tau(\theta) z] \sin\theta \cos\theta\, d\theta}{\int_0^{2\pi} d\phi \int_0^{2\pi} d\theta_\phi \int_0^a r\,dr \int_0^{\theta_M} B(r, \phi, \theta, \theta_\phi) \sin\theta \cos\theta\, d\theta}$$

in which δ represents a Dirac pulse function. In this case, the calculation can be performed considering guided rays only, or guided and leaky rays. Neglecting leaky rays, under a uniform excitation, it is

$$h(t,z) = constant \qquad \frac{n_1}{c} z \leq t \leq \frac{n_1}{c}\left(1 + \frac{\theta_M^2}{2}\right) z$$

$$h(t,z) = 0 \qquad t < \frac{n_1}{c} z \quad \frac{n_1}{c}\left(1 + \frac{\theta_M^2}{2}\right) z < t \tag{1.62}$$

Also in this case, the assumption $\sin \theta_M \cong \theta_M$ has been considered. Also taking into account leaky rays, it is not possible to write $h(t, z)$ in finite terms. Nevertheless the effect of leaky rays can be seen as a tail which must be added to the impulse respone (1.62).

In Table 1.4 pulsewidths $(n_1/c)(\theta_M^2/2)$ and 3 dB bandwidths $B_\omega/2\pi$, expressed in modulation frequencies, are reported for comparison, with reference to a unit distance, for different values of θ_M, assuming $n_1 = 1.5$. Such results show how the fiber numerical aperture NA must be chosen as a trade-off between two opposite requirements:

i) increasing of light acceptance, which requires a high value of NA

ii) reducing time dispersion, which requires a low value of NA

Table 1.4 - Pulsewidth and 3 dB bandwidth for step-index fibers having different numerical apertures, with $n_1 = 1.5$. Leaky rays have been neglected.

θ_M [deg]	Pulsewidth [ns/km]	3 dB Bandwidth [MHz · km]
4	12	50
6	27	22
8	49	12
10	76	8
12	110	6
14	149	4.5
16	195	3.5

1.3.2 α-Profiles

A widely investigated class of index profiles depending on a single parameter, called α-profiles, is given by

$$n_1^2(r) = n_1^2(0)\left[1 - 2\Delta\left(\frac{r}{a}\right)^\alpha\right] \tag{1.63}$$

where Δ imposes the index difference between core and cladding, a is the core radius, and parameter α controls the profile shape.[10] The main properties which make α-profiles of particular practical interest will be stressed in the following. In Fig. 1.15 some α-profile index distributions are reported. Practical values of α range between 1 and ∞. Note that for $\alpha = \infty$ we have the step-index profile. Now we will particularize for α-profiles all the expressions of the boundary curves which contain domain (E, h) or domain (r_1, θ_1). First of all, from (1.28) and (1.29), NA turns out to be

$$\sin\theta_{aM} = n_1(0)\sqrt{2\Delta\left[1 - \left(\frac{r}{a}\right)^\alpha\right]} \qquad \text{(guided rays)} \tag{1.64}$$

$$\sin\theta_{aM} = n_1(0)\sqrt{\frac{2\Delta\left[1 - \left(\frac{r}{a}\right)^\alpha\right]}{1 - \left(\frac{r}{a}\sin\theta_\phi\right)^2}} \qquad \text{(guided and leaky rays)} \tag{1.65}$$

The helical ray curve e in domain (E, h), form (1.46), can be obtained as

$$h = n_1(0)(\Delta\alpha)^{1/2}\, r\left(\frac{r}{a}\right)^{\alpha/2}$$

$$E = n_1(0)\sqrt{1 - (\alpha + 2)\Delta\left(\frac{r}{a}\right)^\alpha}$$

whereas in domain (r_1, θ_1) it can be written explicity as

$$\sin^2\theta_1 = n_1^2(0)\,\Delta\alpha\left(\frac{r_1}{a}\right)^\alpha$$

Curves m, g, ℓ do not assume particular forms for α-profiles in domain (E,h). Nevertheless in domain (r_1, θ_1), curves g and ℓ can be obtained respectively as

$$\sin^2\theta_1 = \frac{2\Delta\left[1 - \left(\frac{r_1}{a}\right)^\alpha\right]}{1 - 2\Delta\left(\frac{r_1}{a}\right)^\alpha},$$

$$\sin^2 \theta_1 = \frac{1}{1 - (r_2/a)^2} \; \frac{2\Delta \left[1 - \left(\frac{r_1}{a}\right)^{\alpha}\right]}{1 - 2\Delta \left(\frac{r_1}{a}\right)^{\alpha}}$$

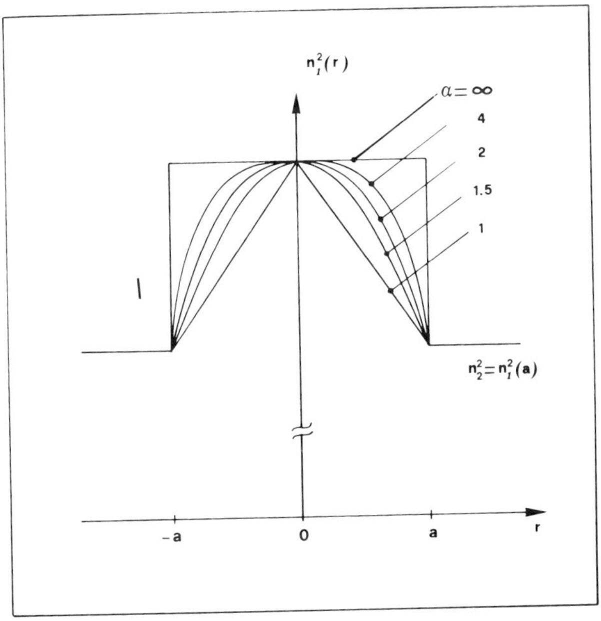

Fig. 1.15 - Some α-profiles.

Fig. 1.16 shows the shape of some regions G and L, for various α.[9] In particular, for $\alpha = 2$ (parabolic profile), curve e and curve ℓ turn out to be practically straight lines. When $\alpha < 2$, for any r_1, the maximum $\theta_1 (r_1)$ of leaky rays is always smaller than the maximum θ_1 for meridional rays $(\theta_M)_m$. When $\alpha > 2$, the maximum $\theta_1 (r_1)$ for leaky rays is alway greater than $(\theta_M)_m$. When $\alpha = \infty$ (step-index profile), a value r_{1s} exists such that, for $r_1 > r_{1s}$, the maximum $\theta_1 (r_1)$ is $\pi/2$. It is then possible to conclude that when $\alpha < 2$, and $\alpha \cong 2$, the assumption $\sin \theta_1$ is justified, but when $\alpha \gg 2$, it must be carefully verified. In particular, for large values of α, it fails when $r_1 \cong r_{1s}$. r_{1s} can be obtained by

$$n_1 (r_{1s}) \, r_{1s} = n_1 (a) \, a$$

and hence

$$\frac{r_{1s}}{a} - 2\Delta \left(\frac{r_{1s}}{a}\right)^{\alpha + 1} = 1 - 2\Delta$$

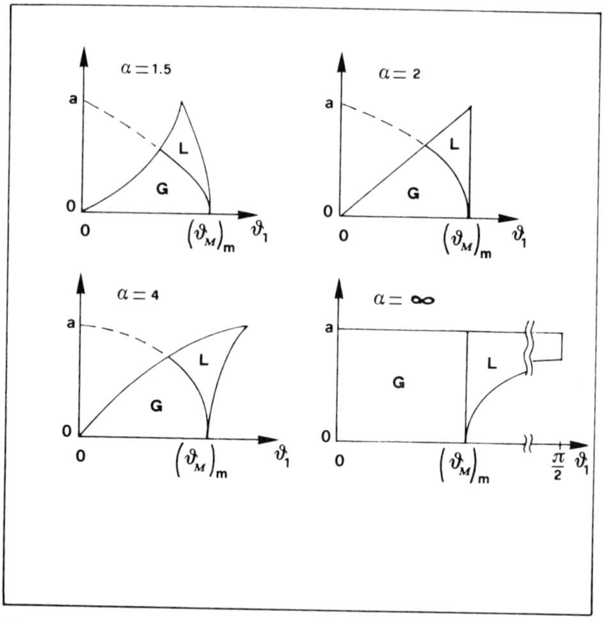

Fig. 1.16 - *G* and *L* regions in domain (r_1, θ_1) for different α.

One of the most interesting properties of an α-profile is that the group delays of the modes or rays can be expressed as a function of only one variable (β in the electromagnetic model, E in the ray-optics model). The next three sections will be devoted to discussing such a property, in order to evaluate time dispersion in α-profile fibers. Another interesting property of this profile is that it accurately reproduces practical graded-index distributions.

1.3.3 WKB Approximation for α-Profiles

The total number of guided modes, for any graded-index fiber, through the WKB approach, is given by

$$M = 4\int_0^{\nu_{max}} d\nu \int_0^{\mu_{max}(\nu)} d\mu$$

where factor 4 takes into account the fourfold degeneracy of each pair (ν, μ). From (1.16), considering that $\mu_{max}(\nu)$ is reached when β approaches its minimum permitted value k_2, we can write[10]

$$M = \frac{4}{\pi}\int_0^{\nu_{max}} d\nu \int_{r_1[\nu]}^{r_2[\nu]} \sqrt{k_1^2(r) - k_2^2 - \frac{\nu^2}{r^2}}\, dr$$

Changing the order of integration leads to

$$M = \frac{4}{\pi} \int_0^a dr \int_0^{r\sqrt{k_1^2(r) - k_2^2}} \sqrt{k_1^2(r) - k_2^2 - \frac{\nu^2}{r^2}}\, d\nu$$

and hence

$$M = \int_0^a \sqrt{k_1^2(r) - k_2^2}\, r\, dr = \frac{2\pi}{\lambda} \int_0^a \sqrt{n_1^2(r) - n_2^2}\, r\, dr \qquad (1.66)$$

The analogy between (1.51) and (1.66) confirms that when the discrete spectrum of the modes is replaced by a continuum of rays, the WKB approximation gives the same results of the ray-optics model. Considering (1.66), and pointing out that the maximum permitted value of β is reached, for a certain μ, when $\nu = 0$, the number of modes $m(\beta)$, whose phase constant is greater than a given β, [10] is

$$m(\beta) = \int_0^{r_2(0)} \sqrt{k_1^2(r) - \beta^2}\, r\, dr$$

For a α-profile we have

$$m(\beta) = \frac{\alpha}{\alpha + 2} a^2 \Delta k_0^2 \left(\frac{k_0^2 - \beta^2}{2 \Delta k_0^2} \right)^{\frac{2+\alpha}{\alpha}} \qquad (1.67)$$

$$k_o = \frac{2\pi}{\lambda} n_1(0)$$

and

$$M = \frac{\alpha}{\alpha + 2} a^2 \Delta k_0^2$$

The group velocity of a mode denoted by the phase constant β is [1]

$$v_g = 2\pi \frac{\partial \nu_\lambda}{\partial \beta}$$

where ν_λ is the optical frequency. The group delay per unit distance is then

$$\tau = \frac{1}{v_g} = \frac{1}{2\pi} \frac{\partial \beta}{\partial \nu_\lambda}$$

Considering that $\lambda = c/\nu_\lambda$, we have

$$k_o = \frac{2\pi}{c} \nu_\lambda n_1(0)$$

and hence

$$\tau = \frac{n_1(0)}{c} \frac{\partial \beta}{\partial k_o} \tag{1.68}$$

For a general graded-index profile, β is a function of ν and μ, therefore τ is also a function of ν and μ. For an α-profile, on the other hand, equation (1.67) holds. Inversion of (1.67) leads to a function $\beta(m, k_o)$, which substituted into (1.68), gives a function $\tau(m, k_o)$. Finally, using (1.67) again, the m dependence can be suppressed, and it results $\tau(\beta, k_o)$, in which k_o is a constant dependent only on the fiber characteristics.

In this way we have demonstrated that, for any α-profile, τ is a function of β only. The rather complex expression of $\tau(\beta)$ can be simplified by replacing β with a new variable ϵ, defined as

$$\epsilon = \frac{1}{2}\left(1 - \frac{\beta^2}{k_0^2}\right) \tag{1.69}$$

which ranges from 0, when $\beta = k_o$, to Δ, when $\beta = k_2$. From (1.67), we have the following relation between m and ϵ

$$\frac{m}{M} = \left(\frac{\epsilon}{\Delta}\right)^{\frac{2+\alpha}{\alpha}} \tag{1.70}$$

Considering also the condition $\Delta \ll 1$, $\tau(\beta) = \tau(\epsilon)$ can be expanded in power series of ϵ. Neglecting terms of order greater than two in this expansion, we finally obtain

$$\tau(\epsilon) = \frac{n_1(0)}{c}\left(1 + \frac{\alpha - 2}{\alpha + 2}\epsilon + \frac{3\alpha - 2}{\alpha + 2}\frac{\epsilon^2}{2}\right) \tag{1.71}$$

Two cases must be taken into account:

i) $\alpha \cong 2$

ii) $\dfrac{|\alpha - 2|}{\Delta} \gg 1$

In the first case, both the α-dependent terms in (1.71) have non-negligible effects, and the shape of curve $\tau(\epsilon)$ is parabolic. In the other case, the first ϵ-dependent term is predominant, and curve $\tau(\epsilon)$ exhibits a linear behavior. In Fig. 1.17 some typical behaviors of $\tau(\epsilon)$ are shown. When $\alpha > 2$, the profile is called undercompensated, when $\alpha < 2 - 2\Delta$, it is called overcompensated. These definitions arise from the purpose of a compensation with respect to the fundamental mode ($\epsilon = 0$). When $2 - 2\Delta \leq \alpha \leq 2$, curve $\tau(\epsilon)$ exhibits values either smaller or greater than $n_1(0)/c$, therefore the profile is neither undercompensated nor overcompensated.

We consider now the particular case $\alpha = \infty$ (step-index fibers). From (1.69), the relation between the propagation angle θ and the variable ϵ is

$$\epsilon = \frac{1}{2} \sin^2 \theta \cong \frac{\theta^2}{2}$$

Substituting into (1.71), for $\alpha = \infty$, expression (1.58) can be obtained, confirming the validity of the ray-optics approach developed for step-index fibers in 1.3.1.

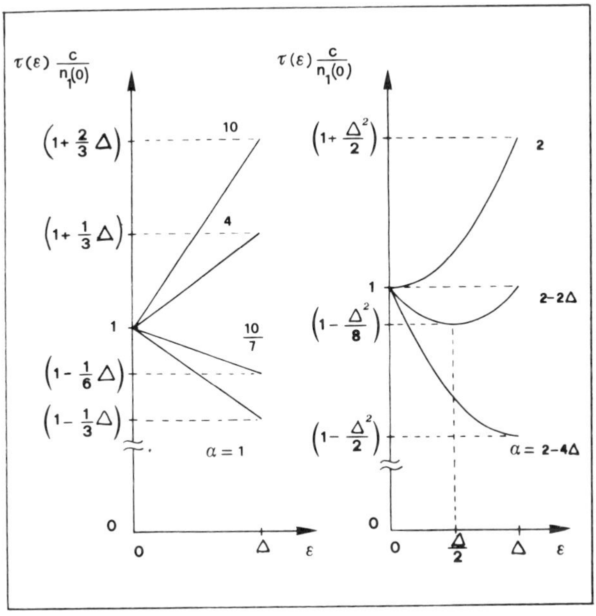

Fig. 1.17 - Typical behavior of $\tau(\varepsilon)$, for various α.

In order to evaluate the impulse response of an α-profile fiber, we can assume the transit time of the fundamental mode to be the time reference, and thus

$$t = \left[\tau(\epsilon) - \frac{n_1(0)}{c}\right] z$$

$$= \frac{n_1(0)}{c} \left(\frac{\alpha - 2}{\alpha + 2} \epsilon + \frac{3\alpha - 2}{\alpha + 2} \frac{\epsilon^2}{2} \right) z$$

Such a relation stresses that there is a correspondence between the time and the variable ϵ. Since m modes are characterized by a given value of ϵ, through (1.70), there is a correspondence between the time and the variable m. Under a uniform excitation of all the modes, the impulse response of the fiber can be written as

$$h(t, z) = \frac{1}{M} \frac{dm}{dt} \tag{1.72}$$

Developing all the calculations, one can easily obtain the following results.[10] When $\alpha \cong 2$, it is

$$h(t, z) = constant, \quad 0 \le t \le \frac{n_1(0)}{c} \frac{\Delta^2}{2} z$$

$$h(t, z) = 0 \qquad t < 0$$

$$\frac{n_1(0)}{c} \frac{\Delta^2}{2} z < t \tag{1.73}$$

When $|\alpha - 2| / \Delta \gg 1$, $h(t, z)$ is proportional to $|t|^{2/\alpha}$, for $t_1 \le t \le 0$, if $\alpha < 2$, and for $0 \le t \le t_2$, if $\alpha > 2$, being

$$t_1 = t_2 = \frac{n_1(0)}{c} \frac{\alpha - 2}{\alpha + 2} \Delta z$$

and it is $h(t, z) = 0$ elsewhere.

To summarize, the following properties can be stated:

i) α-profiles are characterized by mode distributions and group delays which depend on a single variable (m or ϵ)

ii) time dispersion of α-profile fibers can be evaluated giving the impulse response in analytical closed from, provided that $\Delta \ll 1$

iii) for any α, time dispersion increases linearly with the distance z from the optical source

iv) when α is close to 2, time dispersion increases as Δ^2, on the contrary when $|\alpha - 2|/\Delta \gg 1$, time dispersion increases as Δ

v) the optimum value of α, which minimizes the temporal width of transit times is $\alpha_{ow} = 2 - 2\,\Delta$, and the minimum in such a temporal width of transit times per unit distance turns out to be $(n_1(0)/c)\,(\Delta^2/8)$

vi) such an optimum value of α is very critical, in fact a 1% deviation of α leads to an almost doubling of the time dispersion

Since material dispersion has been ignored in this model, one can expect that by taking material dispersion into account another optimum α_{oW} occurs. In the next paragraph we will show how it depends on the particular dopants employed in the profile manufacturing process, but, however, it is not far from $\alpha = 2$.

Before concluding the present section, we will give an outline about the study of time dispersion based on the baseband response, instead of that based on the impulse response. For a uniform excitation, we have here

$$\overline{H}(\omega, z) = \frac{1}{M} \int_0^M \overline{u}(m)\, dm$$

where

$$\overline{u}(m) = \exp[-i\,\omega\,\tau(m)\,z]$$

in which, from (1.70) and (1.71), τ can be expressed as a function of m, as

$$\tau(m) = \frac{n_1(0)}{c} \tag{1.74}$$

$$\left[1 + \frac{\alpha - 2}{\alpha - 2}\,\Delta \left(\frac{m}{M}\right)^{\frac{\alpha}{\alpha+2}} + \frac{3\alpha - 2}{\alpha + 2}\,\frac{\Delta^2}{2}\left(\frac{m}{M}\right)^{\frac{2\alpha}{\alpha+2}}\right]$$

1.3.4 Effects of Leaky Modes

All the results of the previous section have been carried out considering strictly guided modes only. If leaky modes are taken into account, different results are obtained. For an undercompensated profile, a tail takes place after the impulse response of strictly guided modes. For an overcompensated profile, a similar tail precedes the impulse response of strictly guided modes. Such conclusions can be easily derived from Fig. 1.17, considering that leaky modes are characterized by $\epsilon > \Delta$.

A particular behavior occurs when $2 - 4\Delta \leq \alpha \leq 2 - 2\Delta$. Here, time delays of leaky modes can be smaller than those of strictly guided modes. So, in order to obtain the optimum valuc of α in the presence of leaky modes, the maximum permitted value of ϵ for such modes ϵ_{Lmax}, must be considered. From Fig. 1.2, it is possible to obtain the maximum ν, ν_{Lmax}, which corresponds to the minimum β_L^2 given by

$$\frac{d}{dr}\left[k_1^2(r) - \frac{\nu_{Lmax}^2}{r^2}\right]\Bigg|_{r=a} = 0$$

that is

$$\nu_{Lmax}^2 = \alpha k_0^2 \Delta a^2$$

For this value of ν, one obtains $r_{3L} = a$. The general relation between ϵ and r_{3L} can be derived from

$$k_1^2(a) - \beta_L^2 - \frac{\nu^2}{r_{3L}^2} = 0$$

resulting in

$$r_{3L}^2 = \frac{\nu^2}{2 k_0^2 (\epsilon_L - \Delta)}$$

Finally, replacing a^2 with this expression of r_{3L}^2, in the above expression of ν_{Lmax}^2, one obtains

$$\epsilon_{Lmax} = \Delta\left(1 + \frac{\alpha}{2}\right)$$

We can conclude that the maximum permitted value of ϵ_L increases with increasing α, therefore the tails caused by leaky modes in undercompensated fibers are longer than in overcompensated ones. In particular, when $\alpha \cong 2$ it is $\epsilon_{max} \cong 2\,\Delta$. It is possible to demonstrate[11] that the optimum value of α in order to minimize the overall temporal width of transit times in the presence of leaky modes results in $2 - 4\Delta$. Such an overall temporal width coincides with that obtained with strictly guided modes for the same value of α. Note that, since leaky modes suffer an excess attenuation with respect to strictly guided modes, the above optimum value is of interest only for very short distances from the optical source.

1.3.5 Ray-Optics Approach for α-Profiles

Considering a domain (ϵ, h) instead of (E, h) for α-profiles, the expressions of some boundary curves can be simplified. From (1.69) and keeping in mind the relation between β and E, we have

$$\epsilon = \frac{1}{2}\left[1 - \frac{E^2}{n_1^2(0)}\right] \tag{1.75}$$

Boundary curves in domain (ϵ, h) are as follows:

i) curve m of meridional rays, $h = 0$ again

ii) cure e of helical rays, from (1.3.2) and (1.75)

$$\epsilon = \frac{\alpha + 2}{2} \Delta \left[\frac{1}{n_1(0)\sqrt{\Delta \alpha}} \frac{h}{a} \right]^{\frac{2\alpha}{2+\alpha}} \tag{1.76}$$

iii) curve g of the strictly guidance condition

$$\epsilon = \Delta \tag{1.77}$$

iv) curve ℓ of the leaky guidance condition

$$\epsilon = \Delta + \frac{1}{2 n_1^2(0)} \frac{h^2}{a^2} \tag{1.78}$$

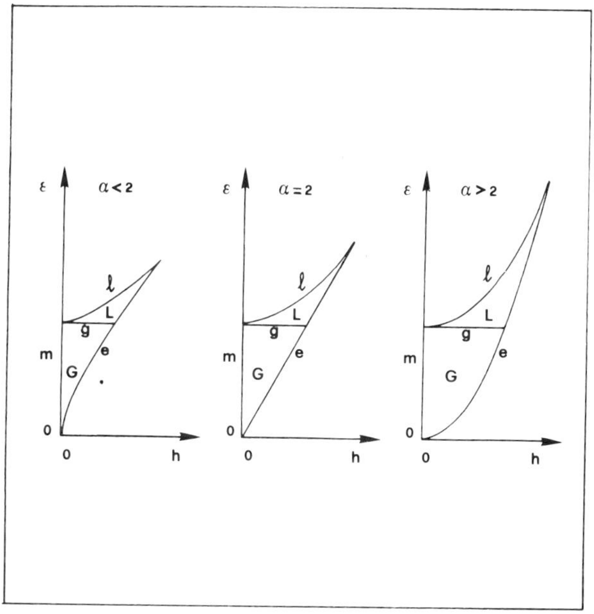

Fig. 1.18 - Domain (ε, h) for different classes of the profile parameter α.

The last curve is independent of the shape of $n(r)$, as in domain (E,h). In Fig. 1.18 some regions G and L are shown for different classes of α. Comparing Fig. 1.18 with Fig. 1.16, a correspondence between θ_1 and ϵ, as well as between r_1 and h, can be observed. For studying time dispersion in α-profiles, domain (ϵ, h) is a better convenience than domain (r_1, θ_1), because τ depends on ϵ only. We will demonstrate this property again, through a ray-optics approach. Taking the average time delay over a ray period as time delay per unit distance, we have

$$\tau(\epsilon, h) = \frac{1}{Z(\epsilon, h)} \int_0^{Z(\epsilon, h)} \frac{dz}{\upsilon(z) \cos\theta(z)}$$

where $\upsilon(z)$ is the local group velocity along the ray trajectory, whose direction diverges from the z axis by $\theta(z)$. Thus it is that

$$\upsilon(z) = \frac{c}{n(r(z))} = \frac{c}{n^2(r(z))} \quad \frac{E}{\cos\theta(z)}$$

and hence

$$\tau(\epsilon, h) = \frac{1}{c\, n_1(0)\sqrt{1 - 2\epsilon}} \quad \frac{1}{Z(\epsilon, h)} \int_0^{Z(\epsilon, h)} n^2(r(z))\, dz$$

in which (1.75) has also been used.

Changing the variable of integration, we are left with

$$\tau(\epsilon, h) = \frac{1}{c\, n_1(0)\sqrt{1 - 2\epsilon}} \quad \frac{\displaystyle\int_{r_1}^{r_2} \frac{n^2(r)\, dr}{\left(\frac{dr}{dz}\right)}}{\displaystyle\int_{r_1}^{r_2} \frac{dr}{\left(\frac{dr}{dz}\right)}} \tag{1.79}$$

where dr/dz is a function of r, dependent on ϵ, h. Precisely, from (1.44) it is

$$\left(\frac{dr}{dz}\right)^2 = \frac{2\left[\epsilon - \Delta\left(\frac{r}{a}\right)^\alpha\right]}{1 - 2\epsilon} - \frac{1}{n_1^2(0)(1 - 2\epsilon)} \quad \frac{h^2}{r^2}$$

$\tau(\epsilon, h)$ can be obtained in finite terms enucleating $Z(\epsilon, h)$ from the integral at the numerator. Letting $\epsilon \ll 1$ expression (1.71) is obtained again.[12] Now we will show that only for α-profiled τ is function of the single variable E. Let us assume a general profile

$$n_1^2(r) = n_1^2(0)\,[1 - f(r)]$$

where $f(r)$ is an unknown function. Letting

$$E^2 = n_1^2(0)\,(1 - B) \tag{1.80}$$

where B is another variable, we can write

$$\frac{dr}{dz} = \pm \frac{p(E, h, r)}{rE}$$

$$p(E, h, r) = \sqrt{n_1^2(0)[B - f(r)]r^2 - h^2} \tag{1.81}$$

Considering again the average time delay over a ray period as time delay per unit distance,

$$\tau(E,h)\sqrt{1-B}\int_{r_1}^{r_2}\frac{r\,dr}{p(E,h,r)}=\frac{n_1(0)}{c}\int_{r_1}^{r_2}\frac{[1-f(r)]r\,dr}{p(E,h,r)}$$

and hence

$$\left[1-\frac{c\,\tau(E,h)\sqrt{1-B}}{n_1(0)}\right]\int_{r_1}^{r_2}\frac{r\,dr}{p(E,h,r)}=\int_{r_1}^{r_2}\frac{f(r)\,r\,dr}{p(E,h,r)} \quad (1.82)$$

Since it is that

$$p(E,h,r_1)=p(E,h,r_2)=0$$

we have

$$\int_{r_1}^{r_2}\left(\frac{dp}{dr}\right)dr=0$$

so that, from the second of (1.81), we obtain

$$\int_{r_1}^{r_2}\left[B-f(r)-\frac{r}{2}\frac{df}{dr}\right]\frac{r\,dr}{p(E,h,r)}=0$$

and thus

$$B\int_{r_1}^{r_2}\frac{r\,dr}{p(E,h,r)}=\int_{r_1}^{r_2}\left[1+\frac{r}{2f(r)}\frac{df}{dr}\right]\frac{f(r)\,r\,dr}{p(E,h,r)} \quad (1.83)$$

Dividing (1.82) by (1.83), we are left with

$$\frac{n_1(0)-c\,\tau(E,h)\sqrt{1-B}}{n_1(0)\,B}=\frac{\displaystyle\int_{r_1}^{r_2}\frac{f(r)\,r\,dr}{p(E,h,r)}}{\displaystyle\int_{r_1}^{r_2}\left[1+\frac{r}{2f(r)}\frac{df}{dr}\right]\frac{f(r)\,r\,dr}{p(E,h,r)}}$$

In order to have $\tau(E,h)$ a function of B only, that is to say of E only, the h dependence of $p(E,h,r)$ must vanish. This entails the following condition

$$1+\frac{r}{2f}\frac{df}{dr}=C_o$$

where C_o is a constant. The general solution of this differential equation is

$$f(r)=f(a)\left(\frac{r}{a}\right)^{2(C_o-1)}$$

which, letting $f(a) = 2\Delta$ and $2(C_0 - 1) = \alpha$, gives the usual form of an α-profile. Finally, $B = 2\epsilon$.

Two ways can be followed in order to study the fiber time dispersion:

i) a differential approach, based on an equation of the type (1.72), which leads to the impulse response $h(t, z)$

ii) an integral approach, based on the evaluation of a phasor $\overline{H}(\omega, z)$, which gives amplitude and argument of the baseband response

In order to obtain results which can be compared with those derived from the WKB approac, we consider a uniform excitation of all the rays. From the differential of (1.75), we have

$$d\epsilon = -\frac{E\,dE}{n_1^2(0)}$$

Therefore the ray distribution in domain (ϵ, h), under a uniform excitation, from (1.50), is given by

$$W_o(\epsilon, h) = 4\pi B_o \frac{n_1(0)}{\sqrt{1 - 2\epsilon}} Z(\epsilon, h) \tag{1.84}$$

which must be integrated between $\epsilon = 0$ and $\epsilon = \Delta$.

The amount of rays whose ϵ is smaller than a given ϵ can be written as

$$m(\epsilon) = \int_0^\epsilon d\epsilon \int_0^{h_e(\epsilon)} W_o(\epsilon, \mathrm{h})\, dh$$

where $h_e(\epsilon)$ represents the helical ray value of constant h for a given ϵ, and can be obtained inverting (1.76). The total amount of rays then results in

$$M = \int_0^\Delta d\epsilon \int_0^{h_e(\epsilon)} W_o(\epsilon, \mathrm{h})\, dh$$

Note that these two quantities correspond to $m(\beta)$ and M of the WKB approach. The impulse response $h(t, z)$ of the fiber can be obtained by applying (1.72).

The baseband response, on the other hand, can be obtained from the phasor

$$\overline{H}(\omega, \mathrm{z}) = \frac{1}{M} \int_0^\Delta d\epsilon \int_0^{h_e(\epsilon)} W_o(\epsilon, h)\, \overline{u}(\epsilon)\, dh$$

$$\overline{u}(\epsilon) = \exp[-i\omega\tau(\epsilon) z] \tag{1.85}$$

It is possible to express in finite terms only h-integration, [12] which results in

$$\int_{0}^{h_e(\epsilon)} W_o(\epsilon, h)\, dh = 2\,\pi^2 B_o a^2 \quad \left(\frac{\epsilon}{\Delta}\right)^{2/\alpha}$$

In all the previous expressions we have restricted our attention to strictly guided rays only. In order to extend these result to the whole domain $(G+L)$, we must consider this expression before the integration over ϵ. Inverting (1.78), it results

$$W_o(\epsilon) = \int_{0}^{h_e(\epsilon)} W_o(\epsilon, h)\, dh \quad (\epsilon \leq \Delta)$$

$$\int_{h_L(\epsilon)}^{h_e(\epsilon)} W_o(\epsilon, h)\, dh \quad (\epsilon > \Delta)$$

where $h_L(\epsilon)$ represents the leaky guidance condition in h for a given ϵ. Inverting (1.78), it results in

$$h_L(\epsilon) = n_1(0)\, a \sqrt{2(\epsilon - \Delta)}$$

1.3.6 The rms Pulsewidth

A very simple way to evaluate time dispersion in multimode fibers is based on the study of the *rms* pulsewidth. It allows one to avoid all the calculations required for the determination of the impulse response. Moreover, it takes into account the real influence of tails in impulse response, which sometime are very long but carry a very sm^ll amount of power.

The general expression of the *rms* pulsewidth is

$$\sigma(z) = \left[\frac{M_2}{M_0} - \left(\frac{M_1}{M_0}\right)^2\right]^{1/2} \tag{1.87}$$

M_0, M_1, M_2 are moments given by

$$M_n = \int_{-\infty}^{+\infty} t^n\, h(t, z)\, dt, \quad n = 0, 1, 2 \tag{1.88}$$

where $h(t, z)$ is the impulse response of the fiber. Alternatively, they are given by

$$M_n = i^n \frac{\partial^n}{\partial \omega^n} \overline{H}(\omega, z)\Big|_{\omega = 0}, \quad n = 0, 1, 2 \tag{1.89}$$

where $\overline{H}(\omega,z)$ is the phasor of the baseband response of the fiber. One can observe that the ω differentiation of $\overline{u}(E,h) = \exp|-i\omega\tau(E,h)z|$ in (1.89) can be performed easily, even when $\tau(2E,h)$ exhibits are not simple expressions. Therefore equation (1.89) is usually preferred to derive $\sigma(z)$.

For an α-profile fiber, under a uniform excitation, and neglecting leaky rays, we can write

$$M_0 = \frac{1}{M}\int_0^{\Delta} d\epsilon \int_0^{h_e(\epsilon)} W_o(\epsilon, h)\, dh$$

$$M_1 = \frac{1}{M}\int_0^{\Delta} \tau(\epsilon)\, d\epsilon \int_0^{h_e(\epsilon)} W_o(\epsilon, h)\, dh$$

$$M_2 = \frac{1}{M}\int_0^{\Delta} \tau^2(\epsilon)\, d\epsilon \int_0^{h_e(\epsilon)} W_o(\epsilon, h)\, dh$$

As for the first expression, observing that the double integral in M_0 represents the total amount of rays guided by the fiber, i.e., M itself, it is $M_0 = 1$. Therefore M_1 and M_2 represent the mean values of the time delay and of the square of the time delay respectively.

Taking into account integral (1.86), after some calculations, it results

$$\sigma(z) = \sigma_o(\alpha) z \tag{1.90}$$

where the *rms* pulsewidth per unit distance $\sigma_o(\alpha)$ turns out to be

$$\sigma_o(\alpha) = \frac{n_1(0)}{c} \frac{\alpha\,\Delta}{[(\alpha+2)(3\alpha+2)]^{1/2}} (A + B\,\Delta + C\,\Delta^2)^{1/2} \tag{1.91}$$

for

$$A = \left(\frac{\alpha - 2}{2\alpha + 2}\right)^2$$

$$B = \frac{(\alpha - 2)(3\alpha - 2)}{(4\alpha + 2)(\alpha + 1)}$$

$$C = \frac{(3\alpha - 2)^2}{(5\alpha + 2)(3\alpha + 2)}$$

Differentiating in α, the optimum value of the profile parameter, α_{orms}, which minimizes the *rms* pulsewidth, is obtained as

$$\alpha_{orms} = 2 - \frac{12}{5}\Delta \tag{1.92}$$

The correspondent minimum in σ_o turns out to be

$$\sigma_{om} \cong \frac{n_1(0)}{c}\, 2.87 \cdot 10^{-2}\, \Delta^2 \tag{1.93}$$

This minimum in time dispersion is very critical. A variation of a few percent in α, around α_{orms}, leads to an increase in the time dispersion of more than one order of magnitude.

In Table 1.5 the overall temporal width of transit times and the *rms* pulsewidth per unit distance are reported for different values of α, taking $\Delta = 0.01$ and $n_l(0) = 1.5$. In this Table, the values of the 3 dB bandwidth, which exhibit an optimum shifted further are also reported.

Table 1.5 - Time dispersion for α-profile fibers, with $\Delta = 0.01$, and $n_o = 1.5$. Leaky modes have been neglected.

α	Overall Temporal Width [ns/km]	*rms* Pulsewidth [ps/km]	3 dB Bandwidth [GHz · km]
1.8	2.63	548	0.350
1.9	1.28	234	0.820
1.960	0.260	51	6
1.965	0.198	36	14
1.970	0.144	23	28
1.975	0.110	14	17
1.980	0.064	18	11
1.985	0.096	30	6
1.990	0.139	44	4
1.995	0.190	58	3
2.000	0.250	73	2.5
2.1	1.22	369	0.490
2.2	2.38	657	0.275
2.5	5.55	1460	0.124
3.0	10.00	2650	0.068
4.0	16.65	4523	0.040
∞	50.00	14000	0.012

1.3.7 Simplified Description of the Launching Condition

In the previous sections, the time dispersion properties of α-profile fibers have been investigated assuming a uniform excitation of all the modes or rays. Now we will show that a nonuniform excitation can heavily influence the shape of the baseband response of the fiber, especially when a near-optimum profile is considered. Note that such effects of the launching condition are sometimes of lesser importance in real fibers. In fact, the perturbation suffered by the fiber induces a power exchange among the guided modes. Therefore, for very long fibers, the effects of the launching condition are practically negligible. Nevertheless, in order to study near-optimum α-profiles for ideal fibers, the launching condition must be carefully considered.

For $\alpha \cong 2$, under uniform excitation, from (1.85) and (1.86), the result is

$$\overline{H}(\omega, z) = \frac{\int_0^{\Delta} \exp[-i\,\omega\,\tau(\epsilon)\,z\,]\,\epsilon\, d\epsilon}{\int_0^{\Delta} \epsilon\, d\epsilon} \qquad (1.94)$$

A nonuniform excitation can be taken into account introducing into this expression an optical power distribution $\overline{W}_i(\epsilon, \omega)$, in which $|\overline{W}_i(\epsilon, \omega)|$ accounts for a nonuniform amplitude in the excitation, while arg $|\overline{W}_i(\epsilon, \omega)|$ accounts for a nonuniform phase in the excitation. Thus we can write

$$\overline{H}(\omega, z) = \frac{\int_0^{\Delta} \overline{W}_i(\epsilon, \omega) \exp[-i\,\omega\,\tau(\epsilon)\,z]\,\epsilon\, d\epsilon}{\int_0^{\Delta} \overline{W}_i(\epsilon, 0)\,\epsilon\, d\epsilon} \qquad (1.95)$$

Note that this model is not completely general, which is different from the one which assumes the most general expression of the radiance $B(r, \phi, \theta, \theta_\phi)$, and leads to the optical power density in the ray domain given by (1.49).

Nevertheless, the model based on the optical power distribution $W_i(\epsilon, \omega)$ can be considered acceptable when the optical source (LED or *laser diode*) is characterized by an axis of the emission pattern which coincides with the fiber axis. In this case, in fact, if the direction of emission can be assumed as approximately parallel to the fiber axis, the displacement r from the central point of emission (intersection of the fiber axis with the emitting surface) turns out to be proportional to $\sqrt{\epsilon}$. This property can be easily demonstrated considering the relation

$$\beta = \frac{2\pi}{\lambda}\, n_1(r) \cos\theta$$

which for $\theta \cong 0$, and from (1.69), gives

$$n_1^2(1 - 2\epsilon) \cong n_1^2(r)$$

and hence

$$\epsilon \cong \Delta \left(\frac{r}{a} \right)^{\alpha} \tag{1.96}$$

Finally, in the limit $\alpha \cong 2$, we are left with

$$r \cong a \left(\frac{\epsilon}{\Delta} \right)^{1/2} \tag{1.97}$$

The optical sources usually employed for transmission systems can be divided into two classes:

i) top emitting devices (e.g., Burrus-type LEDs)

ii) edge emitting devices (e.g., ultra-high radiance LEDs, to laser diodes)

As will be shown in Capter 6, a model based on a transmission line can be used to study the propagation of a modulating signal along the emitting surface. This modulating signal is a voltage whose amplitude and delay distributions along the emitting surface are approximately the same which characterize the optical power distribution $\overline{W}_i(\epsilon, \omega)$. We have a radial transmission line in a plane parallel to the junction for top emitting devices, and a linear transmission line on the intersection of the junction plane with the emitting side wall for edge emitting devices.

In both cases, however, from (1.97), one obtains an optical power distribution $\overline{W}_i(\epsilon, \omega)$, whose amplitude and phase can be expanded in power series of ϵ. To help the discussion of the effects of the launching condition on the fiber baseband responde,[13] the curves obtained can be fitted by the aid of a function which depends on a single free parameter. A suitable curve for taking into account the nonuniform amplitude has the following gaussian form

$$| \overline{W}_i (\epsilon, \omega) | = \exp \left[- \frac{\epsilon^2}{2A} \right] \tag{1.98}$$

in which parameter A allows one to simulate a large family of optical sources. Next, we can account for the nonuniform phase through a parabolic argument, writing

$$\arg [\overline{W}_i (\epsilon, \omega)] = \exp (\pm i \, \omega \, B \, \epsilon^2) \tag{1.99}$$

where B is a real positive quantity independent of ω. The positive sign holds for top emitting devices, where as the negative sign holds for edge emitting devices.

In the following section, after a brief discussion of the baseband response under a uniform excitation through (1.94), we will study separately the two effects of nonuniform excitation. Then, (1.95) and (1.98) will be used for the effect of a nonuniform amplitude, while (1.95) and (1.99) for the effect of a nonuniform phase.

1.3.8 Effects of the Launching Condition

Substituting (1.71) into (1.94), one can obtain the baseband response of a quasi-parabolic fiber under a uniform excitation in analytical form.[13]

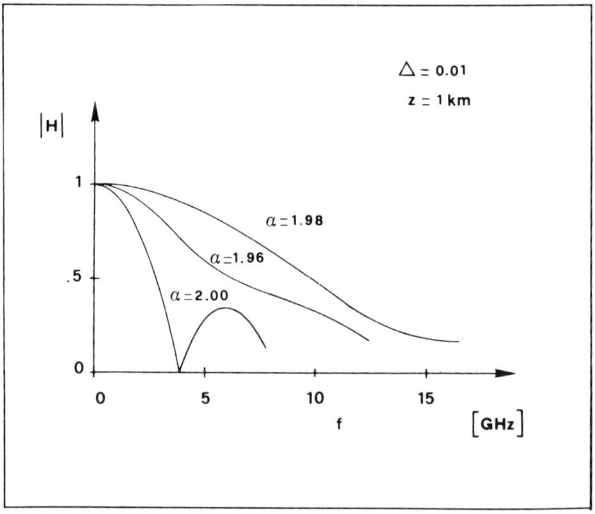

Fig. 1.19 - Amplitude baseband responses of near-optimum α-profile fibers under a uniform excitation [from (13)].

Nevertheless, it is rather complicated. So we restrict ourselves, in Fig. 1.19, to showing the amplitudes of the baseband response of three near optimum fibers, for $\Delta = 0.01$, over a reference distance of 1 km. From Fig. 1.19 it is possible to understand the very critical role played by α for the width and for shape of the fiber baseband response. The following properties can be deduced: when $\alpha \geq 2 - 2\,\Delta$, $|\overline{H}(\omega, z)|$ exhibits a series of lobes separated by minima. Only for $\alpha = 2$ can these minima assume zero values. Note that the baseband response for $\alpha = 2$, can be obtained also through a Fourier transform of (1.73). When $\alpha < 2 - 2\Delta$, however, $|\overline{H}(\omega, z)|$ exhibits different concavities separated by points of inflection.

A nonuniform amplitude in the excitation, such as that given by (1.98), leads to a modification of the above curves. Some examples of these modifications show in Fig. 1.20. A nonuniform $|\overline{W}_i(\epsilon, \omega)|$ induces a vertical shift of the curves, causing a variation in the 3 dB bandwidth. Nevertheless, the positions of minima or points of inflection remain approximately the same. There are different effects for different near optimum fibers.[13] When $\alpha = 2$ and $\alpha = 2 - 2\Delta$, the 3 dB bandwidth increases as the spatial extension of the optical source A is reduced. This behavior characterizes also non-near optimum fibers. An exception occurs when $\alpha = 2 - 4\Delta$, here we have a value A (from Fig. 1.20, $A = 0.15 \cdot 10^{-5}$), for which the 3 dB bandwidth turns out to be equal to that obtained for a uniform excitation ($A = \infty$). If A is smaller than $0.15 \cdot 10^{-5}$, it decreases under the value of a uniformly excited fiber with increasing A, till a minimum is reached for $A = 0.125 \cdot 10^{-4}$. A further increase in A entails an increase in the 3 dB bandwidth until the value of a uniformly excited fiber is reached.

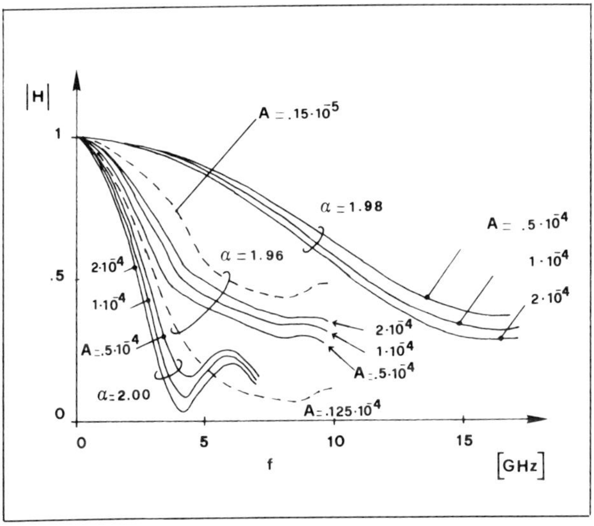

Fig. 1.20 - Amplitude baseband responses of near-optimum α-profile fibers under a nonuniform amplitude excitation [from (13)].

Considering now a nonuniform phase in the excitation, we restrict ourselves to the negative sign in (1.99) only, since only for edge emitting devices the intrinsic limitation in the baseband response due to the optical source can be considered fairly well above the limitation due to a nonuniform distribution of spatial delays of emission. We have to refer to a given length z of the transmission link, in fact the spread in the modal delay distribution along the fiber increases with z, whereas the distribution of spatial delays of emission depends on the optical source only.[13] Assuming a typical value $z = 10$ km, the curves reported in Fig. 1.21 can be obtained. The choice of the negative sign in (1.99) entails an improvement in the baseband response for overcompensated fibers, whereas there is a worsening for undercompensated fibers. In particular, when $\alpha = 2 - 2\Delta$, a very strong dependence on B can be observed. Finally, these results become more evident as z decreases.

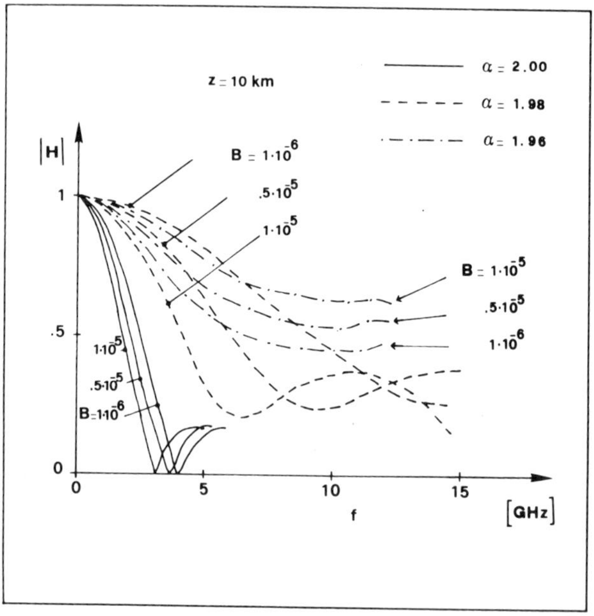

Fig. 1.21 - Amplitude baseband responses of near-optimum α-profile fibers under a nonuniform phase excitation [from (13)].

1.3.9 Effects of Profile Imperfections

Among the fiber fabrication processes, the chemical vapor deposition (CVD) technique is one of the most used. In this technique,[14] a preform is produced depositing layers of pure or doped SiO_2 inside a tube of mineral quartz glass. The tube, which is usually horizontally mounted, rotates on its own axis, ensuring an axially symmetric deposition. Heating is provided over a fraction of its length, to achieve a homogeneous chemical reaction among tetrachloride $SiCl_4$ and chlorides of the particular dopants which are emplyed (e.g., Ge, P, B). They are vaporized and mixed with oxygen. A temperature of 1400°C or more is required for the reactants to fuse in glassy layers. The amount of dopants is designed to achieve a suitable index profile. The tube is then collapsed to a solid preform by means of a higher temperature, of about 2000°C, and afterwards it is drawn into a fiber.

Keeping in mind the fabrication steps of the CVD technique outlined above, the main imperfections which may affect the index profile of real fibers can be understood. They are:

i) the presence of an axial dip in the refractive index distribution, due to outdiffusion of dopants, because of the higher temperature adopted when the preform is collapsed

ii) the presence of a staircaselike profile, due to the finite number of layers, each characterized by a different dopant concentration

In order to take into account such perturbations of the index profile, simple modifications in a α-profile can be considered. Nevertheless, the group delays per unit distance turn out to be functions of two independent variables. For instance, one could employ the expression of the group delay per unit distance in domain (E, h), for general graded-index profile $n_1(r)$ which results in

$$\tau(E,h) = \frac{1}{c}\frac{1}{E}\frac{\displaystyle\int_{r_1(E,h)}^{r_2(E,h)} \frac{n_1^2(r)\,dr}{\left(\frac{dr}{dz}\right)}}{\displaystyle\int_{r_1(E,h)}^{r_2(E,h)} \frac{dr}{\left(\frac{dr}{dz}\right)}}$$

$$\frac{dr}{dz} = \sqrt{\frac{n_1^2(r)}{E^2} - \frac{h^2}{E^2}\frac{1}{r^2} - 1} \tag{1.100}$$

Alternatively, domain (r_1, θ_1) can be assumed.

Keeping in mind the relations between variables E, h, and variables r_1, θ_1, developed in 1.2.10, from (1.100) it is possible to obtain

$$\tau(r_1,\theta_1) = \frac{1}{c}\frac{1}{n_1(r_1)\cos\theta_1}\frac{\displaystyle\int_{r_1}^{r_2(r_1,\theta_1)} \frac{n_1^2(r)\,dr}{\left(\frac{dr}{dz}\right)}}{\displaystyle\int_{r_1}^{r_2(r_1,\theta_1)} \frac{dr}{\left(\frac{dr}{dz}\right)}}$$

$$\frac{dr}{dz} = \sqrt{\frac{n_1^2(r)}{n_1^2(r_1)\cos\theta_1} - \left(\frac{r_1}{r}\tan\theta_1\right)^2 - 1}$$

The main advantage in the use of the latter domain is that it allows one to relate directly the spatial extension of the profile perturbation of its influence on the distribution of relative group delays per unit distance. In other words, reporting τ as a function of θ_1 with r_l as a parameter gives an immediate evaluation of the effects of the profile perturbation on the time dispersion properties of the fiber.

Now, an axial dip can be taken into account assuming a perturbed α-profile of the type[15]

$$n_l(r) = n_o \sqrt{1 - 2\Delta \left(\frac{r}{a}\right)^{\alpha}} \left[1 - A \exp\left(-\frac{1}{B^2}\left(\frac{r}{a}\right)^2\right)\right] \quad (1.101)$$

where A accounts for the depth of the dip, and B for its width. On the other hand, the shape of a staircaselike profile, being smoothed by dopant diffusion, can be assumed as a sinusoidal perturbation superimposed on a certain α-profile. So, we have[16]

$$n_l(r) = n_o \sqrt{1 - 2\Delta \left(\frac{r}{a}\right)^{\alpha} + \frac{d}{N}\Delta \sin\left[2\pi N \left(\frac{r}{a}\right)^{\alpha}\right]} \quad (1.102)$$

where N is the step number of the staircase; that is, the number of disposition layers of the CVD technique, and d is a coefficient accounting for the amount of diffusion.

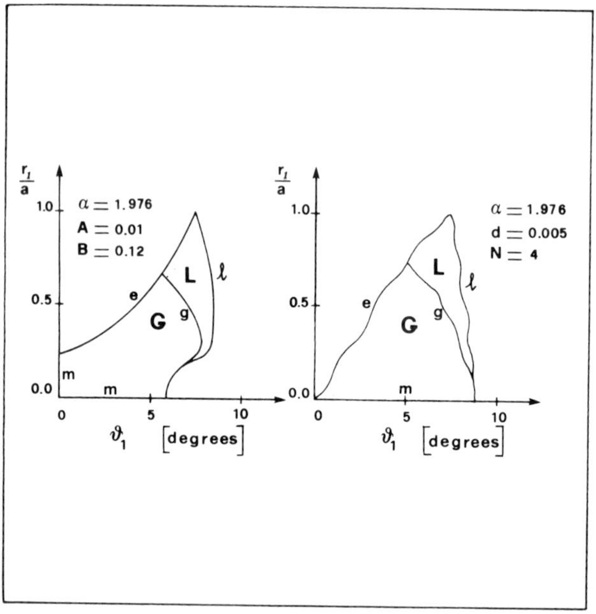

Fig. 1.22 - Modifications in the shape of domain (r_1, θ_1) due to index perturbations [from (17)].

In order to have $n_1(r) \geq n_o\sqrt{1 - 2\Delta}$, anywhere, the condition $A \leq \Delta$ must be fulfilled in (1.101). To avoid the oscillating part of (1.102) giving rise to positive values of dn^2/dr, the conditon $d \leq 1/\pi$ must be added. In the present section we will investigate separately the effects of the two profile perturbations. n_o = 1.5 and Δ = *0.01* will be always considered.[17] Numerical computations are needed.

Fig. 1.22 modifications in the shape of domain (r_1, θ_1) are reported. The optimum value for the unperturbed profile α = 1.976 has been assumed, whereas A = 0.01, B = 0.12, N = 4, and d = 0.005 have been considered as typical parameters of the two perturbations. For the axial dip, the whole region where $dn^2/dr > 0$ is a region characterized by the presence of meridional rays (identified by condition θ_1 = 0, which entails h = 0), having $r_1 \neq 0$.

In Fig. 1.23 the distribution of relative group delays per unit distance, $\tau(r_1, \theta_1) - n_o/c$, is reported for the unperturbed profile. The dashed line h refers to the relative group delays of helical rays. Each continuous line is denoted by a certain value of r_1/a as a parameter. Fig. 1.24 shows two distributions of relative group delays, both obtained in the case of an axial dip with A = 0.01 and B = 0.12. The first, in continuous lines (except for h), refers to α = 1.976 and shows how the axial dip can produce negative relative delays for rays having $r_1/a \leq 0.2$, which are influenced by the profile depression. The second, in dashed and dotted lines (except for h), refers to α = 1.85, that is the value of α which minimizes the *rms* pulsewidth in the presence of a dip characterized by the given values of A and B.

Note that in both cases, rays having $r_1/a > 0.3$ are not influenced at all by the profile depression. The effects of the dip in overcompensated fibers are lower than in undercompensated fibers, because the latter exhibit relative delays having opposite sign with respect to those produced by the dip. Parameter A is not very influential on the fiber time dispersion. On the contrary, parameter B is of greater importance. The higher the value of B, the larger the difference between α_{orms} and the value of α which minimizes the *rms* pulsewidth in the presence of the dip. Comparing delay scales in Fig. 1.23 and in Fig. 1.24, we can conclude that the dip entails remarkable increase in the time dispersion. For instance, under a uniform excitation and considering only strictly guided rays, when α = 1.976 in the absence of the dip, σ_o = 0.01 ns/km, whereas when α = 1.85 in the presence of the profile perturbation, although the *rms* pulsewidth is minimized, it is σ_o = 0.6 ns/km.

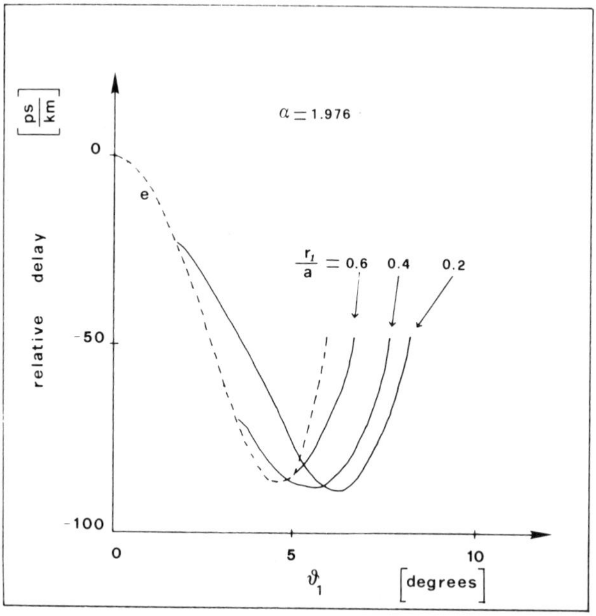

Fig. 1.23 - Relative group delays per unit distance *vs.* θ_1 with r_1/a as a parameter for an unperturbed optimum α-profile [from (17)].

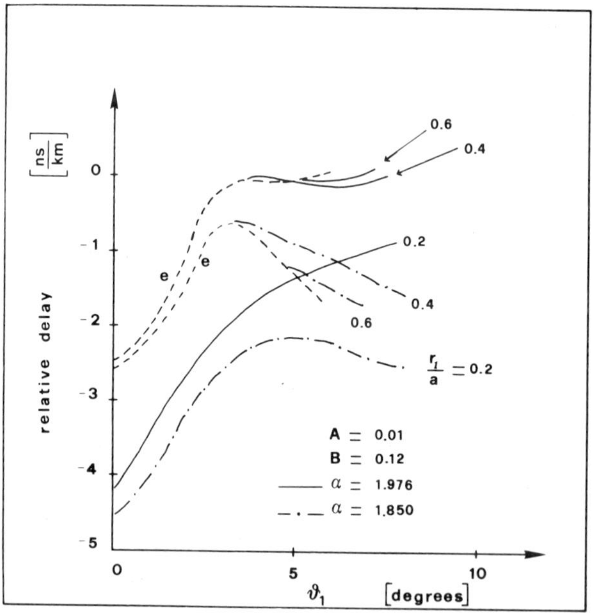

Fig. 1.24 - Relative group delays per unit distance *vs.* θ_1 with r_1/a as a parameter for profiles affected by an axial dip [from (17)].

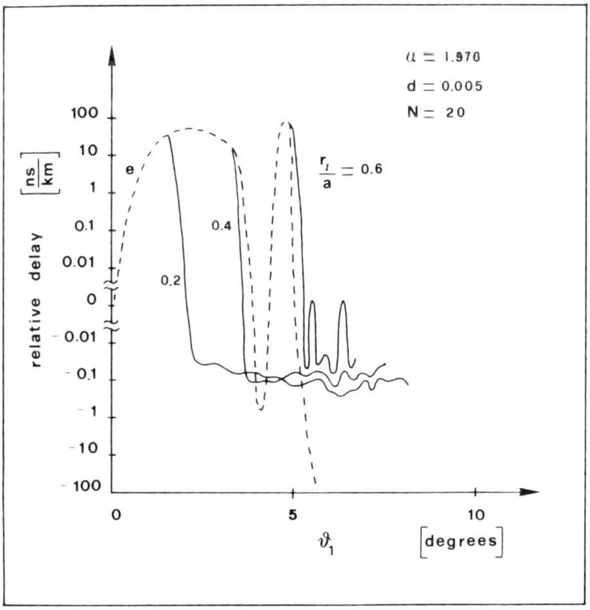

Fig. 1.25 - Relative group delays per unit distance *vs.* θ_1 with r_1/a as a parameter, for a profile affected by a sinusoidal perturbation[from (17)].

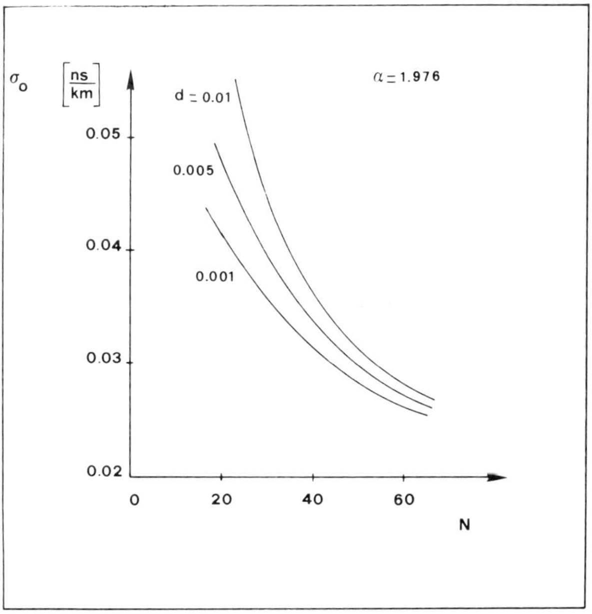

Fig. 1.26 - Effects due to the number *N* of deposition layers.

A sinusoidal perturbation, as it is given by (1.102) influences the distribution of relative group delays in a strong way. Large fluctuations of relative group delays are shown in the example of Fig. 1.25, where α = 1.976, N = 20, and d = 0.005 have been assumed. Rays having a small difference between r_1 and r_2, i.e., close to the helical ray condition, are much more influenced by this type of perturbation. On the contrary, rays whose r_1 and r_2 differ considerably, traveling inside a wide region of the core section along their period $Z(r_1, \theta_1)$, suffer only an average effect of the perturbation, so that their group delays exhibit smaller oscillations.

For a given d, the overall effect of this profile perturbation is softened if N increases. Fig. 1.26 reports the *rms* pulsewidth per unit distance, under a uniform excitation of strictly guided rays *versus* N, for different values of d as a parameter. When N approaches a value of about 70, the effects of the sinusoidal perturbation can be practically neglected, even for values of d that are not small. This limit condition depends on the number of rays which are considered in the numerical computations of the moments M_0, M_1, M_2, to obtain σ_0. The maximum precision which can be requested is reached when this number of rays is close to the number of nondegenerate modes guided by the fiber, which can be calculated from (1.66). Assuming $\alpha \cong 2$, $\lambda = 1$ μm, a = 50 μm, one obtains about 1000 modes on the whole. Eliminating the fourfold degeneracy, we are left with about 250 rays, which are distributed in the region G of domain (r_1, θ_1) according to (1.65).

As for the evaluation of the group delay $\tau(r_1, \theta_1)$, all the results reported in the present section have been derived performing a numerical integration of both integrals in its expression. A Gauss-Chebicef quadrature formula,[13] with 30 steps, is sufficient for the precision requested.[9]

1.3.10 Compensation Effect of Time Dispersion in the Fiber Chain

When overcompensated and undercompensated fibers are installed in a transmission link, an intrinsic compensation in the overall time dispersion takes place. Considering α-profile fibers, without errors in the joints, it is possible to demonstrate[18] that the overall time dispersion of a link remains very low, provided that the average value of the various coefficients α is close to the optimum, even when remarkable differences characterize them. A simplified analysis of this problem can be carried out for near-optimum fibers, in the limit $2/\alpha \cong 1$, for which the difference in the optical power distribution, under a uniform excitation, can be neglected.

In this case, assuming two fibers have equal lengths, the condition for minimizing time dispersion results in

$$\int_0^{\Delta}[\tau_1(\epsilon)+\tau_2(\epsilon)]^2\epsilon\, d\epsilon \int_0^{\Delta}\epsilon\, d\epsilon = \left(\int_0^{\Delta}[\tau_1(\epsilon)+\tau_2(\epsilon)]\epsilon\, d\epsilon\right)^2$$

where $\tau_1(\epsilon)$ and $\tau_2(\epsilon)$ are the group delay distributions of the two fibers. Such a condition has been derived letting (1.187) equal to zero for $\tau(\epsilon)=\tau_1(\epsilon)+\tau_2(\epsilon)$. Now, replacing $\tau_1(\epsilon)$and$\tau_2(\epsilon)$ with (1.71), where α_1 is assumed for the first fiber and α_2 for the second fiber, one obtains

$$\frac{1}{12}(\alpha_1\alpha_2-4)^2+\frac{\Delta}{5}(3\alpha_1\alpha_2+2\alpha_1+2\alpha_2-4)(\alpha_1\alpha_2-4)$$

$$+\frac{\Delta^2}{8}(3\alpha_1\alpha_2+2\alpha_1+2\alpha_2-4)^2=0$$

which gives a relation between α_1 and α_2.

The above equation, at the first order in Δ gives $\alpha_1\alpha_2\cong 4$ but at the second order would give inaccurate results.

In principle, in the limit of such an approximation, even for very long fiber chains, the order in which the fibers are installed is unimportant. Yet, when wider variations in α are considered, and the optical power distribution depends on α, the order of installation may influence the overall time dispersion.[19] Finally, taking into account possible errors in the joints, even the compensation in a couple of fibers may give much smaller effects than in the absence of joint errors. The evaluation of the time dispersion effects due to an imperfect joint between two equal fibers will be discussed in Section 2.4.

1.4 INTRAMODAL TIME DISPERSION

The refractive index of a fiber depends on the wavelength λ of the light which propagates in the fiber itself. This dependence is caused by the dipole moments of the molecules distribution in the dielectric medium, and is responsible for both the material dispersion and the intrinsic attenuation due to absorption. Thus, resonance optical wavelengths occur, which depend either on the host material, commonly $Si0_2$, or on the dopants used for the profile synthesis. Nevertheless, for any choice of dopant, the behavior of the squared index n_2 can be well-fitted by Sellmeir's law[20]

$$n^2 = 1+\sum_{i=1}^{N}\frac{a_i\lambda^2}{\lambda^2-b_i} \tag{1.103}$$

where a_i and b_i are constants. Usually $N = 3$ is enough for the precision requested.

In the present paragraph we will describe the effects of such a λ dependence. Each mode suffers a time dispersion, owing to the differences in the values of the refractive index that any spectral component of the light encounters during the propagation. For this reason, such an effect is called intramodal time dispersion. It becomes more severe with increasing spectral width of the optical source. Moreover, even when this spectral width tends to zero, equation (1.103) must be taken into account to obtain an index profile which minimizes intermodal time dispersion. In fact, as will be shown in the following, the derivative $dn/d\lambda$ is involved in the calculation of the group delay per unit distance τ, which characterizes each mode or ray.

1.4.1 Group Delays in the Presence of Material Dispersion

Recalling the definition of group delay per unit distance, introduced in 1.3.3, we have

$$\tau = \frac{1}{2\pi}\frac{\partial\beta}{\partial\nu_\lambda} = \frac{1}{2\pi}\frac{\partial\beta}{\partial k_o}\frac{\partial k_o}{\partial\lambda}\frac{\partial\lambda}{\partial\nu_\lambda}$$

Since

$$\frac{\partial k_o}{\partial\lambda} = n_1(0)\frac{d}{d\lambda}\left(\frac{2\pi}{\lambda}\right) + \frac{2\pi}{\lambda}\frac{d}{d\lambda}\left[n_1(0)\right]$$

$$= -n_1(0)\frac{2\pi}{\lambda^2} + \frac{2\pi}{\lambda}\frac{d}{d\lambda}\left[n_1(0)\right]$$

and

$$\frac{\partial\lambda}{\partial\nu_\lambda} = -\frac{\lambda^2}{c}$$

we are left with

$$\tau(\lambda) = \frac{n_1(0)}{c}\left[1 - \frac{\lambda}{n_1(0)}\frac{d}{d\lambda}n_1(0)\right]\frac{\partial\beta}{\partial k_o} \tag{1.104}$$

Comparing with (1.68), it can be seen that when the λ dependence of the refractive index is taken into account, the expression of the group delay per unit distance exhibits a corrective term. In other words, an equivalent index

$$N_1(0) = n_1(0)\left[1 - \frac{\lambda}{n_1(0)}\frac{d}{d\lambda}\; n_1(0)\right]$$

called group index, must be considered instead of $n_1(0)$

In particular, for the axial ray (or fundamental mode), we have $\partial\beta/\partial k_o = 1$. Since this ray travels in a region in which the index is constant, it is possible to conclude that the local group velocity along a general ray trajectory results in

$$v(z, \lambda) = \frac{c}{n_1(z)\left[1 - \frac{\lambda}{n_1(z)}\frac{d}{d\lambda}\; n_1(z)\right]}$$

instead of $v(z) = c/n_1(z)$. Therefore, all the calculations reported in the previous paragraph, for the determination of the group delay of a ray, can be corrected taking into account the λ dependence of n_1 in this way. In particular, from (1.100), one obtains the following expression of the group delay per unit distance of a general ray denoted by the constants of the motion E, h

$$\tau(E, h, \lambda) = \frac{1}{cE}\;\frac{\displaystyle\int_{r_1(E,h)}^{r_2(E,h)} \frac{n_1(r,\lambda)\,[\,n_1(r,\lambda) - \lambda\,\frac{\partial n_1}{\partial\lambda}\,]\,dr}{\sqrt{n_1^2(r,\lambda)/E^2 - h^2/(Er)^2 - 1}}}{\displaystyle\int_{r_1(E,h)}^{r_2(E,h)} \frac{dr}{\sqrt{n_1^2(r,\lambda)/E^2 - h^2/(Er)^2 - 1}}} \tag{1.105}$$

The quantity $(\lambda/n)\,dn/d\lambda$, for a general index n, can be experimentally measured, in order to perform the correction. Considering quasi-monochromatic optical sources, i.e., sources having spectral width $\Delta\lambda \to 0$, the intermodal time dispersion, which is the only cause of signal distortion of the fiber in this case, must be evaluated through equation (1.105) as well. Therefore, since $(\lambda/n)\;dn/d\lambda$ depends on the particular dopants employed, for a given class of index profiles, the determination of the optimum leads one to different results depending on the dopants themselves. Considering α-profiles, the optimum values α_{ow} or α_{orms}, defined in 1.3.3 and 1.3.6, turn out to be different from those which have been obtained ignoring material dispersion. However, they remain not far from $\alpha = 2$.[9] Generally, numerical calculations are required to evaluate the group delay distribution and therefore to obtain the optimum. Under suitable conditions, analytical results can be derived. We will return to this in 1.4.5.

1.4.2 Effects of the Finite Extension of the Source Spectral Width

From (1.104), the intramodal time disperson due to the finite extension $\Delta\lambda$ of the source spectral width can be defined as

$$\Delta \tau = \frac{d\tau}{d\lambda}\Delta\lambda \tag{1.106}$$

This quantity gives, as a first approximation, the spread in the time delay per unit distance caused by the presence of different spectral components in the optical power carried by a certain mode. From the following expression of the time delay per unit distance

$$\tau = \frac{1}{2\pi}\frac{\partial\beta}{\partial\lambda}\frac{\partial\lambda}{\partial\nu_\lambda} = -\frac{1}{2\pi}\frac{\lambda^2}{c}\frac{\partial\beta}{\partial\lambda}$$

we obtain

$$\frac{d\tau}{d\lambda} = -\frac{1}{2\pi c}\left(\lambda^2\frac{\partial^2\beta}{\partial\lambda^2} + 2\lambda\frac{\partial\beta}{\partial\lambda}\right) \tag{1.107}$$

This parameter is properly called material dispersion.

The exact evaluation of the partial derivatives in (1.107) is possible only for step-index and parabolic-index fibers. Nevertheless, the following considerations can be carried out for the most general index distribution. From (1.107) two distinct effects can be distinguished:

i) the true material dispersion, due to the λ dependence of the core refractive index n_1

ii) additional effect, called waveguide dispersion, which arises from the wavelength dependence of the equation (1.13) that gives the permitted values of β

It is possible to demonstrate that the fundamental mode of a step-index fiber is characterized by an overall derivative $(d\tau/d\lambda)_{m+w}$, in which m denotes material dispersion and w waveguide dispersion,[21] whose expression is

$$\left(\frac{d\tau}{d\lambda}\right)_{m+w} = \frac{2\pi}{c}\frac{1}{\lambda^2\beta^3}\left[\lambda^2\beta^2\left[\left(\frac{\lambda}{2\pi}\right)^2\chi_1\frac{d^2\chi_1}{d\lambda^2} - n_1\frac{d^2 n_1}{d\lambda^2}\right]\right.$$

$$+\ \chi_1^2\left(n_1 - \lambda\frac{dn_1}{d\lambda}\right)^2$$

$$+\ \lambda\frac{d\chi_1}{d\lambda}\left[2n_1\chi_1\left(n_1 - \lambda\frac{dn_1}{d\lambda}\right)\right.$$

$$+ 2\left(\frac{\lambda}{2\pi}\right)^2 \beta^2 \chi_1 + n_1^2 \lambda \frac{d\chi_1}{d\lambda} \Big] \Big]$$

where χ_1^2 is given from (1.9).

The true material dispersion can be obtained letting χ_1 and its derivatives equal zero. Therefore, $\beta = 2\pi n_1/\lambda$, and thus

$$\left(\frac{d\tau}{d\lambda}\right)_m = - \frac{\lambda}{c} \frac{d^2 n_1}{d\lambda^2}$$

We can conclude that:

i) material dispersion and waveguide dispersion are not additive

ii) waveguide dispersion can be evaluated only through the electromagnetic model of the propagation

iii) when many modes propagate, i.e., when the ray-optics model is suitable for describing the propagation, waveguide dispersion can be neglected

iv) for step-index fibers with many modes in propagation, even material dispersion is usually negligible

v) for monomode fibers both material and waveguide dispersion must be taken into account *

1.4.3 Wavelength of Zero Material Dispersion

From Sellmeier's law (1.103), a wavelength λ_m may exist for which $d^2n/d\lambda^2 = 0$, provided that $N \geq 2$. At this wavelength, which depends on the nature and on the amount of the dopants employed, material dispersion vanishes. For undoped SiO_2, $\lambda_m \cong 1.27\,\mu m$ and for the usual doped SiO_2 $\lambda_m \cong 1.3\ \mu m$.[22] Besides, zero material dispersion can be reached, employing glasses not based on SiO_2, even at higher wavelengths,[23] where attenuation decreases, as will be shown in Section 2.1. For instance, with fluoride glasses, it is $\lambda_m \cong 1.7\ \mu m$, and moreover material dispersion remains small even up to $\lambda = 4\ \mu m$.[24]

The approximate evaluation of intramodal dispersion through (1.106) is based on a series expansion of $\beta(\nu_\lambda)$ truncated at the second order term. Considering also the third order term, we have

$$\beta(\nu_\lambda) \cong \beta(\nu_{\lambda o}) + \beta'(\nu_{\lambda o})\,(\nu_\lambda - \nu_{\lambda o})$$
$$+ \frac{1}{2}\beta''(\nu_{\lambda o})\,(\nu_\lambda - \nu_{\lambda o})^2$$
$$+ \frac{1}{6}\beta'''(\nu_{\lambda o})\,(\nu_\lambda - \nu_{\lambda o})^3$$

*See notes pg. 497

where the derivatives β', β'', β''', can be easily calculated. Letting $\lambda_o = c/\nu_{\lambda o}$ the central wavelength of emission of the optical source, if $d^2 n_1/d\lambda^2 \,|_{\lambda=\lambda_o} = 0$, then only the effects of β''' $(\nu_{\lambda o})$ remain.

Neglecting terms involved in waveguide dispersion, we are left with

$$\beta'''(\nu_{\lambda o}) = \frac{2\pi}{c^3}\lambda_o^5 \left.\frac{d^3 n_1}{d\lambda^3}\right|_{\lambda=\lambda_o}$$

Therefore the third derivative of $n_1(\lambda)$ is present, at $\lambda_o = \lambda_m$, in this residual dispersion. Usually, it turns out to be of few ps/km, so in practice it affects only monomode fibers.[25]

1.4.4 Total Time Dispersion

In this section we will evaluate the total time dispersion of a multimode fiber through the *rms* pulsewidth of its impulse response. The optical guided power can be considered as a function of the time t, the wavelength λ, and the distance z.

$$P(t, z, \lambda) = \sum_m P_m(z, \lambda)\,\delta(t - \tau_m(\lambda) z) \tag{1.108}$$

where m denotes the mode dependence, and δ is a Dirac pulse function. $P_m(z, \lambda)$ defines a certain spectral component of the optical power guided by the mth mode.

The total optical guided power can be written as

$$P(t, z) = \int_0^\infty P(t, z, \lambda)\, d\lambda$$

Letting $S(\lambda)$ the spectral distribution of the optical source, its mean wavelength and *rms* pulsewidth are respectively

$$\lambda_o = \frac{\int_0^\infty S(\lambda)\,\lambda\, d\lambda}{\int_0^\infty S(\lambda)\, d\lambda}$$

$$\sigma_s = \frac{\int_0^\infty (\lambda - \lambda_o)\, S(\lambda)\, d\lambda}{\int_0^\infty S(\lambda)\, d\lambda} \tag{1.109}$$

The *rms* pulsewidth is given again by (1.87) and (1.88). From (1.108) we have

$$M_n = z^n \sum_m \int_0^\infty P_m(z, \lambda)\,\tau_m^n(\lambda)\, d\lambda \tag{1.110}$$

Note that if the spectral attenuation of the fiber can be considered practically uniform within the interval in which $S(\lambda) \neq 0$, it is

$$P_m(z, \lambda) = k_m S(\lambda) \exp(-\gamma z)$$

where γ is the intrinsic attenuation coefficient, and k_m is a constant accounting for the excitation of the mth mode. Normalizing as

$$\sum_m k_m = 1$$

we have

$$M_0(z) = \exp(-\gamma z) \int_0^\infty S(\lambda)\, d\lambda$$

Substitution of (1.110) into (1.87) yields[26]

$$\sigma^2(z) \cong \sigma^2_{inter}(z) + \sigma^2_{intra}(z) \tag{1.111}$$

where

$$\sigma^2_{inter}(z) = z^2 \Big[\left(< \tau^2(\lambda_o) > - < \tau(\lambda_o) >^2 \right)$$

$$+ \frac{\sigma_s^2}{\lambda_o^2} \left(< \lambda_o^2 \tau''(\lambda_o)\, \tau(\lambda_o) > - < \lambda_o^2 \tau''(\lambda_0) > < \tau(\lambda_o) > \right)\Big]$$

$$\sigma^2_{intra}(z) = z^2 \frac{\sigma_s^2}{\lambda_o^2} < [\lambda_o \tau'(\lambda_o)]^2 >$$

Primes ′ and ″ denote the first and second derivatives with respect to λ, and symbol $< >$ represents an average value over the mode distribution. For instance,

$$< \tau(\lambda_o) > = \frac{\sum_m \tau_m(\lambda_o) \int_0^\infty P_m(z, \lambda)\, d\lambda}{M_0}$$

The approximation in (1.11) consists of neglecting terms of order greater than two in σ_s/λ_o. for LED sources σ_s/λ_o ranges between 10^{-2} and 10^{-1} whereas for laser sources it is typically smaller than $10.^{-3}$

From (1.111) we can conclude that the total $\sigma^2(z)$ is proportional to z^2, and consists of two additive terms:

i) the first is due to intermodal time dispersion, with a main contribution of a λ independent index profile (considered at $\lambda = \lambda_o$), added to a corrective term, proportional to σ_s^2/λ_o^2, in which the second derivative of $\tau_m(\lambda)$ is involved:

ii) the second, due to intramodal time dispersion, is proportional to σ_s^2/λ_o^2 and depends only on the first derivative of $\tau_m(\lambda)$.

Keeping in mind that in the nth derivative of $\tau_m(\lambda)$ the $(n+1)$the derivative of $n_1 1(\lambda)$ is present, the corrective term in σ^2_{inter} does not vanish completely when $d^2n_1/d\lambda^2 = 0$, therefore it takes into account the third-order dispersion introduced in 1.4.3.
If waveguide dispersion is neglected, σ^2_{intra} vanishes when $d^2n_1/d\lambda_2 = 0$. Some authors have studied the dispersion properties of fiber-glass in order to minimize both intermodal and intramodal time dispersion. Two ways can be followed:

i) assuming that the index profile does not change in shape varying λ, but changes only in its absolute value (linear profile dispersion)

ii) assuming that the index profile changes both in shape and in absolute value

1.4.5 Optimum Index Profile in the Presence of Material Dispersion

We will demonstrate here that α-profile fibers exhibit group delays $\tau_m(\lambda)$ depending on a single mode variable, as well as in the absence of material dispersion. Let us assume the following general profile distribution

$$n_1^2(r, \lambda) = n_1^2(0)[\, 1 - f(r, \lambda)\,] \tag{1.112}$$

The group delay per unit distance of a ray denoted by E, h, from (1.105), can be written as

$$\tau(E, h, \lambda) = \frac{1}{c\sqrt{1 - B}} \frac{\int_{r_1}^{r_2} n_1(r, \lambda)\left(1 - \frac{\lambda}{n_1}\frac{\partial n_1}{\partial \lambda}\right) dr}{\int_{r_1}^{r_2} p(E, h, r, \lambda)\, dr} \tag{1.113}$$

where B is given by (1.80) and we have let

$$p(E, h, r, \lambda) = \sqrt{n_1^2(0)[\, B - f(r, \lambda)\,]r^2 - h^2}$$

Considering the group index defined in 1.4.1, one can introduce the following parameter[27]

$$q(r, \lambda) = \frac{n_1(0)}{N_1(0)} \frac{\lambda}{f(r, \lambda)} \frac{\partial f}{\partial \lambda} \tag{1.114}$$

and hence, substituting (1.112) into (1.113), this relation is derived

$$\left[1 - \frac{c\,\tau(E,h,\lambda)\sqrt{1-B}}{N_1(0)}\right]\int_{r_1}^{r_2}\frac{r\,dr}{p(E,h,r,\lambda)}$$

$$= \int_{r_1}^{r_2}\frac{\left[1 - \frac{q(r,\lambda)}{2}\right]f(r,\lambda)\,r\,dr}{p(E,h,r,\lambda)} \tag{1.115}$$

which represents the extension of (1.82). Equation (1.83) remains unchanged except for the substitution of $f(r)$ and $p(E, h, r)$ with $f(r, \lambda)$ and $p(E, h, r, \lambda)$ respectively. So we have

$$B\int_{r_1}^{r_2}\frac{r\,dr}{p(E,h,r,\lambda)} = \int_{r_1}^{r_2}\left[1 + \frac{r}{2f(r,\lambda)}\ \frac{\partial f}{\partial f}\right]\frac{f(r,\lambda)\,r\,dr}{p(E,h,r,\lambda)} \tag{1.116}$$

Dividing (1.115) and (1.116), one finally obtains

$$\frac{N_1(0) - c\,\tau(E,h,\lambda)\sqrt{1-B}}{N_1(0)\,B} = \frac{\displaystyle\int_{r_1}^{r_2}\frac{\left[1 - \frac{q(r,\lambda)}{2}\right]f(r,\lambda)\,r\,dr}{p(E,h,r,\lambda)}}{\displaystyle\int_{r_1}^{r_2}\left[1 + \frac{r}{2f(r,\lambda)}\ \frac{\partial f}{\partial r}\right]\frac{f(r,\lambda)\,r\,dr}{p(E,h,r,\lambda)}}$$

In order to have τ as a function of B only, that is of E only, the h-dependence of $p(E, h, r, \lambda)$ must vanish. This entails the following condition [27], called Marcatili's condition

$$1 + \frac{r}{2f(r,\lambda)}\ \frac{\partial f}{\partial r} = C_o(\lambda)\left[1 - \frac{q(r,\lambda)}{2}\right] \tag{1.117}$$

where $C_o(\lambda)$ is a constant depending on the wavelength λ only. Now we assume that q is constant with respect to r. This means that $\lambda/n\ dn/d\lambda$ depends on the dopant employed but not on its concentration. In fact, $f(r, \lambda)$ in this case is a product of two functions, the first dependent on r only, the second on λ only. This is the case of an α-profile with linear profile dispersion, i.e.,

$$f(r,\lambda) - 2\,\Delta(\lambda)\left(\frac{r}{a}\right)^{\alpha}$$

From (1.117)

$$\alpha = 2\left[C_o(\lambda)\left(1 - \frac{q(\lambda)}{2}\right) - 1\right]$$

Note that the λ-dependences of $C_o(\lambda)$ and of $q(\lambda)$ must be perfectly complementary to have α effectively independent of λ. Finally, from (1.113), when a linear profile dispersion is present, we have

$$q(\lambda) = \frac{n_1(0)}{N_1(0)} \frac{\lambda}{\Delta} \frac{d\Delta}{d\lambda}$$

In conclusion, we have shown that α-profile fibers exhibit group delays which depend on a single mode variable, even in the presence of material dispersion, provided that a linear profile dispersion is present. Marcatili's condition (1.117) is a more general condition which a profile function $f(r, \lambda)$ must satisfy in order to have group delays depending on a single mode variable. The optimization of this class of profile functions is easier than the optimization of profile functions leading to group delays which depend on two mode variables. Such an optimization will be discussed separately for linear and non-linear profile dispersion in the next two sections.

1.4.6 Linear Profile Dispersion

Curves $q(\lambda)$ can be obtained experimentally for different glasses.[28] From this knowledge it is possible to determine the value $\alpha_{orms}(\lambda)$ which minimizes the *rms* intermodal pulsewidth under the assumption of a linear profile dispersion. The group delay per unit distance $\tau(\epsilon, \lambda)$, for α-profiles, considering also the presence of $q(\lambda)$, can be written as[26]

$$\tau(\epsilon, \lambda) = \frac{N_1(0)}{c}$$

$$\left[1 + \frac{\alpha - 2 + 2q(\lambda)}{\alpha + 2}\epsilon + \frac{3\alpha - 2 + 4q(\lambda)}{\alpha + 2}\frac{\epsilon^2}{2}\right] \qquad (1.118)$$

Comparing with (1.71), the role of $q(\lambda)$ can be emphasized. From (1.118), considering a uniform excitation of all the modes, and following procedure very similar to that reported in 1.3.6, it is possible to obtain easily an expression like (1.91) for the *rms* pulsewidth.[26] The optimum α then results.

$$\alpha_{orms}(\lambda) = 2 - 2q(\lambda) - \frac{[4 - 2q(\lambda)][3 - 2q(\lambda)]}{5 - 4q(\lambda)}\Delta$$

Note that when $q(\lambda) \rightarrow 0$ this expression leads to the optimum value (1.92) in the absence of material dispersion.

Since $q(\lambda)$ may be an order of magnitude greater than Δ, sometimes $\alpha_{orms}(\lambda)$ deviates significantly from the optimum value without material dispersion. In Fig. 1.27, $\alpha_{orms}(\lambda)$ is reported for some glasses based on SiO_2.[28] If two or more dopants are used, it is possible to obtain the optimum at different wavelengths, or over an extended interval around a fixed wavelength. For instance, one can assume a profile function of the type[29]

$$f(r, \lambda) = \sum_{i=1}^{N} 2\Delta_i(\lambda)\left(\frac{r}{a}\right)^{\alpha_i}, \quad \sum_{i=1}^{N} \Delta_i = \Delta$$

in which each power law is obtained from a different dopant. This function satisfies Marcatili's condition (1.117) with $q(r, \lambda)$ dependent on λ only, if

$$\alpha_i = 2\left[C_o(\lambda)\left(1 - \frac{1}{2}\frac{n_1(0)}{N_1(0)}\frac{\lambda}{\Delta_i}\frac{d\Delta_1}{d\Delta}\right) - 1\right], i = 1, 2, \ldots N$$

which represents a relation between each α_i and the correspondent $\Delta_i(\lambda)$.

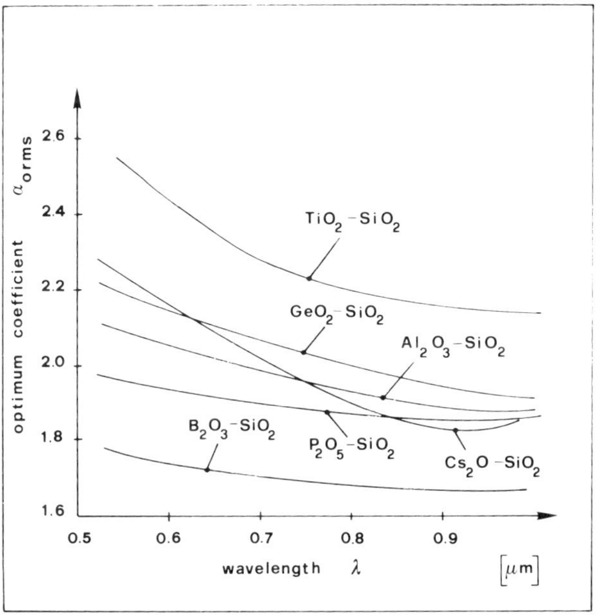

Fig. 1.27 - Optimum value α_{orms} *vs.* λ for different types of glasses [from (28)].

1.4.7 Nonlinear Profile Dispersion

When nonlinear profile dispersion is to be considered, and it is the general case, $q(r, \lambda)$ is a function of both r and λ, therefore, in order to satisfy condition (1.117), the quantity $(r/2f)\,\partial f / \partial r$ must be a function of both r and λ. In this case an α-profile can be defined only at a fixed wavelength.

The optimization, under condition (1.117), can be performed considering τ as a function of variable B only. From the passages reported in 1.4.5, this function can be written in explicit form[27] as

$$\tau(B, \lambda) = \frac{N_1(0)}{c} \; \frac{1 - B / C_o(\lambda)}{\sqrt{1 - B}}$$

The optimization over an extended wavelength interval requires the imposition of suitable conditions both on the r- and on the λ-dependence of the profile function, leading to remarkable technological problems.[30],[31] In reference (27) and (31), some modified α-profile functions have been proposed for optimization over an extended wavelength interval, under condition (1.117), but their practical use appears doubtful.

An alternative way for attaining the profile optimization consists of ignoring condition (1.117), considering group delays dependent on two mode variables, and minimizing the *rms* pulsewidth at different single wavelengths via numerical computations.[9], [32] Experimental data on profile dispersion, for different dopants, can be drawn from suitable curves which give $n_1\, \partial n_1/\partial\lambda$ *versus* n_1^2, at some fixed wavelengths.[32]

In the case of linear profile dispersion, we have

$$f(r, \lambda) = 2\,\Delta(\lambda)\, g\left(\frac{r}{a}\right)$$

where g is a general function of r/a. So

$$\frac{\partial}{\partial \lambda}\left[n_1^2(r, \lambda)\right] = \left[1 - 2\,\Delta(\lambda)\, g\left(\frac{r}{a}\right)\right] \frac{d}{d\lambda}\left[n_1^2(0)\right]$$

$$- 2\, n_1^2(0)\, g\left(\frac{r}{a}\right) \frac{d\Delta}{d\lambda}$$

and hence

$$n_1 \frac{\partial n_1}{\partial \lambda} = \frac{1}{2} \frac{\partial}{\partial \lambda}\left[n_1^2(r, \lambda)\right] = A\, n_1^2 + B$$

where

$$A = \frac{1}{2}\left[\frac{1}{n_1^2(0)} \frac{d}{d\lambda}\left[n_1^2(0)\right] + \frac{1}{\Delta} \frac{d\Delta}{d\lambda}\right]$$

$$B = -\frac{1}{2}\, n_1^2(0)\, \frac{1}{\Delta} \frac{d\Delta}{d\lambda}$$

are constant at a given wavelength. On the other hand, in the case of nonlinear profile dispersion, we have only

$$\frac{\partial}{\partial \lambda}[n_1^2(r, \lambda) = [1 - f(r, \lambda)] \frac{d}{d \lambda} [n_1^2(0)] - n_1^2(0) \frac{\partial f}{\partial \lambda}$$

which can never be considered as a linear function of $n_1^2 (r, \lambda)$. From these observations, when and only when experimental data give at any wavelength $n_1 \, \partial n_1/\partial\lambda$ as a linear function of n_1^2, a linear profile dispersion can be assumed.

REFERENCES

1. Ramo, S., Whinnery J. R., and Van Duzer T. *Fields and Waves in Communications Electronics,* New York: Wiley & Sons (1965).
2. Marcuse, D. *Light Transmission Optics* New York: Van Nostrand Reynolds (1972).
3. Abramowitz, M., and Stegun, I. A. *Handbook of Mathematical Functions,* New York: Dover Pub. (1968).
4. Toraldo di Francia, G. "Degrees of Freedom of Image," *N. Opt. Soc. Am.*59, pp. 799-80 (1969).
5. Snyder, A. W. "Leaky-Ray Theory of Optical Waveguides of Circular Cross Section," *Appl. Phys.,* 4, pp. 273-290 (1974).
6. Eve, M. "Rays and Time Dispersion in Multimode Graded Core Fibers," *Opt. & Quant. Electron.,* 8, pp. 285-293 (1976).
7. Cancellieri, G. and Fantini, P. "Numerical Calculations of Ray Optical Paths and Relative Delays in Graded-Index Optical Fibers," *Alta Freq.,* 48, pp. 414-419 (1979).
8. Di Vita, P. and Vannucci, R. "The Radiance Law in Radiation Transfer Processes," *Appl. Phys.,* 7, pp. 249-255 (1975).
9. Cancellieri, G. "New Ray Optics Approach to the Propagation in Graded-Index Optical Fibers," *Opt. Comm.,* 29, pp. 287-292 (1979).
10. Gloge, D. "Multimode Theory of Graded-Core Fibers," *Bell Syst. Tech. J.,* 52, pp. 1563-1578 (1973).
11. Adams, M. J., Payne, D. N. and Salden, F. M. E. "Mode Transit Times in Near-Parabolic Index Optical Fibers," *Electron. Lett.,* 11, pp. 389-391 (1975).

12. Ankiewicz, A. and Pask, C. "Geometric Optics Approach to Light Acceptance and Propagation in Graded-Index Fibers," *Opt. & Quant. Electron.,* 9, pp. 87-109 (1977).
13. Cancellieri, G., and Ravaioli, U. "Baseband Response of Multimode Optical Fibers under Different Excitation Conditions," *J. Opt. Comm.,* 3, pp. 13-18 (1982).
14. Küppers D., and Koenings, J. "Preform Fabrication by Deposition of Thousands of Layers with the Aid of Plasma Activated CVD," 2nd E.C.O.C., Paris, pp. 49-54 (1976).
15. Cancellieri, G., Fantini, P., and Mezzetti, M. "Effects of a Central Dip on the Time Dispersion Characteristics of Graded-Index Optical Fibers," IEEE/AP.S Int. Simp., Quebec, B.2/AP-9, (1980).
16. Unger H. G. *Planar Optical Waveguides and Fibres,* Oxford: Clarendon Press (1977).
17. Cancellieri, G., and Ravaioli, U., "Time Dispersion in Optical Fibers with Perturbed Profiles," *Alta Freq.,* 52, pp. 18-23 (1983).
18. Galeati, P., Rizzoli, V., and Someda, C. G. "Bandwidth Behaviour of Tandem Connected Graded-Index Multimode Fibers," *J. Opt. Comm.,* 2, pp. 54-58 (1981).
19. Eve, M., "Statistical Model for the Prediction of the Bandwidth of an Optical Route," *Electron. Lett.,* 13, pp. 315-316 (1977).
20. Born, M., and Wolf, E. *Principles of Optics,* Edinburgh: Pergamon Press (1964).
21. Marcuse, D. "Interdependence of Waveguide and Material Dispersion," *Appl. Opt.,* 18, pp. 2930-2932 (1979).
22. Wemple, S. H. "Material Dispersion in Optical Fibers," *Appl. Opt.,* 18, pp. 31-35 (1979).
23. Nassau, K. "The Material Dispersion Zero in Infrared Optical Waveguide Material," *Bell Syst. Tech. J.,* 60, pp. 327-336 (1981).
24. Jeunhomme, L., Poignant, H., and Monerie, M. "Material Dispersion Evaluation in a Fluoride Glass," *Electron. Lett.,* 17, pp. 808-809 (1981).
25. Miyagi, M., and Nishida, S. "Pulse Spreading in a Single Mode Fiber due to Third Order Dispersion," *Appl. Opt.,* 18, pp. 678-682 (1979).

26. Olshansky, R., and Keck, D. B. "Pulse Broadening in Graded-Index Optical Fibers," 15, pp. 483-491 (1976).
27. Marcatili, E. A. J. "Modal Dispersion in Optical Fibers with Arbitrary Numerical Aperture and Profile Dispersion," *Bell Syst. Tech. J.,* 56, pp. 49-63 (1977).
28. Presby, H. M., and Kaminow, I. P. "Binary Silica Optical Fibers: Refractive Index and Profile Dispersion Measurements," *Appl. Opt.,* 15, pp. 3029-3036 (1976).
29. Olshansky, R. "Multiple α Index Profiles," *Appl. Opt.,* 18, pp. 683-689 (1979).
30. Arnaud, J. A. "Optimum Profiles for Dispersive Multimode Fibers," *Opt. & Quant. Electron.,* 9, pp. 111-119 (1977).
31. Geckeler, S. "Compensation of Profile Dispersion in Graded-Index Optical Fibers," *Electron. Lett.,* 15, pp. 682-683 (1979).
32. Arnaud, J. A., and Fleming, J. W. "Pulse Broadening in Multimode Optical Fibers with Large $\Delta n/n$: Numerical Results," *Electron. Lett.,* 12, pp. 1167-169 (1976).

CHAPTER 2

REAL FIBERS

In the previous chapter we have presented the theory of light propagation in ideal fibers, i.e., lossless fibers whose geometrical and physical properties are perfectly axially invariant. In the present chapter, we will take into account the effects of intrinsic loss and of perturbations affecting real fibers. Intrinsic loss leads to an optical power decay along the fiber length. Fiber perturbations cause an exchange of power, leading to variations in the optical power distribution among the modes or rays along the fiber length. This power exchange is usually called mode coupling.

Also leaky and even radiating modes, as well as guided modes, are involved in mode coupling. Therefore, fiber perturbations cause an excess attenuation. In this sense we call the loss of an unperturbed fiber intrinsic. Another cause of excess loss, which must be taken into account in the design of optical transmission systems is due to fiber joints. Perturbations and joints cause not only an excess attenuation, but also a variation in the time dispersion properties of the optical link.

2.1 INTRINSIC ATTENUATION

The optical power decay in multimode fibers could not be rigorously expressed by an exponential law of the type

$$P(z) = P(0)\exp(-\gamma z) \tag{2.1}$$

where γ is the intrinsic power attenuation coefficient. In fact γ depends on the particular mode or ray considered. Nevertheless, equation (2.1) is widely used, being experimentally verified with fairly good approximation, when the fiber is unperturbed. From the theoretical point of view, it is possible to consider equation (2.1) as averaged over the modes or rays in propagation in an unperturbed fiber. Nevertheless, the mode-dependence of γ is of critical importance in order to study mode coupling effects in perturbed fibers, as will be shown in the following.

Coefficient γ in (2.1) is usually measured in km^{-1}. More conveniently, a power attenuation per km, A, is considered, letting

$$A = \frac{1}{z}\, 10 \log_{10}\left[\frac{P(0)}{P(z)}\right]\left[\frac{dB}{km}\right]$$

So, substituting into (2.1) we have the relation

$$A \cong 4.34\, \gamma \left[\frac{dB}{km}\right] \tag{2.2}$$

2.1.1 Causes of Intrinsic Attenuation

Two are the main causes of the fiber intrinsic attenuation:

i) absorption loss

ii) scattering loss

Absorption loss is due to the conversion of optical power into heat. It is a wavelength selective process, because the absorption occurs when the photon energy is approximately equal to the gap between two energy levels of the dielectric waveguide material. Scattering loss is mainly due to the granular structure which this amorphous material exhibits at a microscopic dimension. Also inhomogenities, like small gas bubbles trapped in the glass, may be sometimes another remarkable source of scattering loss.

Absorption entails directly the loss of the optical power, which is converted into heat. On the contrary, scattering produces deviations in the direction of propagation of the optical power. Some of this deviated power is given to radiating modes, and becomes lost for the transmission along the waveguide. The scattered light, which is still guided and can propagate, forward and backward, may appear to cause mode coupling, as well as external perturbations do. However, the improvements of today's low-loss fibers make scattering-induced mode coupling practically negligible.

In order to support this assertion, we will estimate the fraction F of the scattered power which is guided forward in a step-index fiber. Assuming $\theta_M = \pi/20$, and considering only strictly guided rays, if the scattering is isotropic, we have

$$F = \frac{\int_0^{\theta_M} \sin\theta\, d\theta}{2\int_0^{\pi/2} \sin\theta\, d\theta} = \frac{\theta_M^2}{4} \cong 6\cdot 10^{-3}$$

The power transitions among guided rays can be taken into account through the following transport equation[1]

$$\frac{\partial W}{\partial z} = -[2\int_0^{\pi/2} K(\theta, \theta') \sin\theta' \, d\theta' + \gamma_a] W(\theta, z) + \int_0^{\theta_M} K(\theta', \theta) W(\theta', z) \sin\theta' \, d\theta'$$

in which $W(\theta, z)$ represents the optical power distribution at the fiber section z, $K(\theta, \theta')$ is the probability density for a ray θ to be converted into a ray θ', finally γ_a is the attenuation coefficient due to absorption. The first integral represents the power lost because of scattering by ray θ, the second one represents the power received by ray θ from the other guided rays.

If an isotropic scattering is assumed and the scattering centers are considered as uniformly distributed in the fiber core, being θ_M being rather small, $K(\theta, \theta')$ results in practical independence from θ and θ'. Therefore we have

$$\frac{\partial W}{\partial z} = -(\gamma_s + \gamma_a) W(\theta, z) + K[\int_0^{\theta_M} W(\theta', z) \sin\theta' \, d\theta' - W(\theta, z) \int_0^{\theta_M} \sin\theta' \, d\theta']$$

where γ_s is the attenuation coefficient due to scattering and results in

$$\gamma_s = K[\int_0^{\theta_M} \sin\theta' \, d\theta' + 2\int_{\theta_M}^{\pi/2} \sin\theta' \, d\theta']$$

In this way, all the backward scattered power is taken as lost for the propagation. The negative term within square brackets in the expression of $\partial W/\partial z$ represents the power exchange among the rays and turns out to be at least F times smaller than the term $\gamma_s W(\theta, z)$, which represents the power loss. Therefore $W(\theta, z)$ can be considered as practically independent of z, except for the attenuation factor exp $[-(\gamma_s + \gamma_a)z]$, as assumed for real uperturbed fibers.

The approximations made in this model, and particularly the assumption of a step-index fiber and isotropic scattering, do not affect the generality of the conclusions, provided that the numerical aperture of the fiber is not too high.

2.1.2 Absorption Loss

The transitions of electrons and molecules between different energy states cause light absorption which occurs at the peak wavelength

$$\lambda_p = \frac{hc}{E_2 - E_1}$$

where $E_2 - E_1$ is the energy gap between the two states, h is Planck's constant, and c is the free space velocity of light. The electronic transitions involve mainly shorter wavelengths, typically in the ultraviolet region. The molecular vibrational transitions, on the contrary, occur mainly at longer wavelengths, typically in the infrared region.

The absorption curve around a peak wavelength λ_p should have a Lorentzian shape, as it is for the resonance curve of a harmonic oscillator. Far from λ_p, the curve should decrease exponentially.[(2)] Nevertheless, owing to the superimposition of the effects of many absorption peaks or tails, the above behavior is difficult to observe in practice. Moreover, when a very strong absorption peak is present, its overtones and even the combinations of fundamental peaks and overtones having different origins can be experimentally measured. Therefore the prediction of the spectral absorption loss is difficult, also when an accurate chemical analysis of the material has been performed. Even when the fiber has been produced, it is not so easy to recognize overtones and combinations. In fact, sometimes the central wavelengths of secondary peaks may shift owing to the tails of primary peaks.

For example, Table 2.1 reports the assignment of overtones and combinations of the OH spectrum for silica (SiO_2), on the basis of experimental data.[(3)] The intensities of the peaks are measured in

Table 2.1 - Experimental results for positions and intensities of OH absorption bands in fused silica [from (3)].

Wavelength [nm]	Intensity [dB/km]	Identification	
1370	2900	$2\nu_{\lambda o}$	1st overtone
1230	150	$2\nu_{\lambda o} + \nu_{\lambda 1}$	
1125	3.4	$2\nu_{\lambda o} + 2\nu_{\lambda 1}$	
1030	0.4	$2\nu_{\lambda o} + 3\nu_{\lambda 1}$	
950	72	$3\nu_{\lambda o}$	2nd overtone
880	6.6	$3\nu_{\lambda o} + \nu_{\lambda 1}$	
825	0.8	$3\nu_{\lambda o} + 2\nu_{\lambda 1}$	
775	0.1	$3\nu_{\lambda o} + 3\nu_{\lambda 1}$	
725	6.4	$4\nu_{\lambda o}$	3rd overtone
685	0.9	$4\nu_{\lambda o} + \nu_{\lambda 1}$	
585	0.5	$5\nu_{\lambda o}$	4th overtone

dB/km. $\nu_{\lambda 0}$ is the fundamental frequency of the OH vibration (corresponding to $\lambda_0 \cong 2730$ nm), and $\nu_{\lambda 1}$ the fundamental vibration frequency of the bond Si-O (corresponding to $\lambda_1 \cong 8 \mu m$). The content of water is very critical for fiber absorption loss. For instance, a concentration of 1 ppm of water leads to an absorption loss of about 1.25 dB/km in correspondence with the second overtone of the OH vibrational frequency ($\lambda \cong 950$ nm).

Ion contaminations of transition metals cause other absorption peaks. Even very small traces of these impurities in the glass are enough to give rise to very high absorption loss. Table 2.2 shows the concentrations of the main transition metals, which lead to an absorption loss of about 1 dB/km and the wavelength of the corresponding peak of absorption.(4) These kinds of impurities mainly affect fibers whose glass is melted in platinum crucibles, or fibers pulled by the double crucible technique.(5)

The peaks of absorption due to metallic ions are practically absent in today's low-loss fibers, and even the absorption loss due to the OH content has been greatly reduced. Fig. 2.1 reports the spectral attenuation per km, $A(\lambda)$, of two fibers. The upper curve refers to a classical fiber characterized by some peaks due to the absorption bands of Table 2.1. The lower curve refers to a low loss fiber of recent fabrication, with only a very small peak corresponding to the first overtone of the OH vibrational frequency. The content of water of such a fiber can be estimated to be a few ppb.

The upper curve in Fig. 2.1 shows three minima in the absorption loss, approximately around the wavelengths: 850 nm, 1200 nm, and 1500 nm. They are called ordinately: the first, the second, and the

Table 2.2 - Concentration of ions of transition metals which lead to an absorption loss of about 1 dB/km [from (4)].

Ion	Wavelength [μm]	Concentration [ppb in weight]
Cu	0.800	0.45
Fe	1.100	0.40
Ni	0.650	0.20
V	0.475	0.90
Cr	0.675	0.40
Mn	0.500	0.90

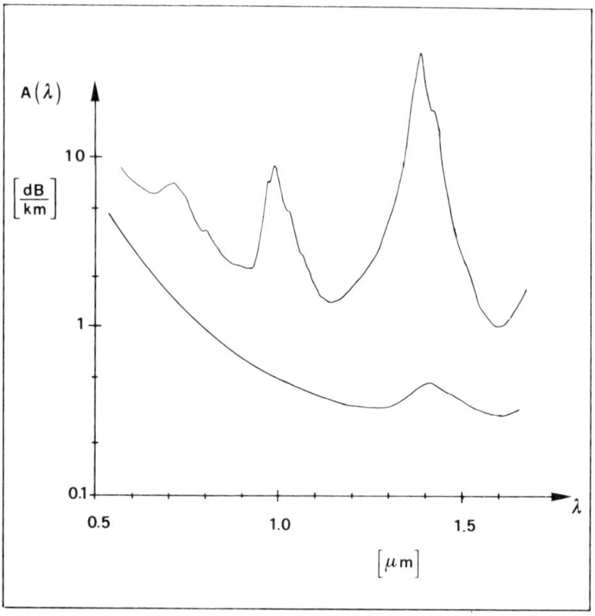

Fig. 2.1 - Spectral attenuation of typical fibers: the upper curve represents a classical behavior, the lower curve is characterized by very low loss.

third window of the spectral attenuation. The technology of transmitting and receiving transducers has been developed starting from the first window. The second and the third windows are more interesting for transmission systems, because of their lower attenuation, both for absorption and scattering. In fact the peaks of absorption arise from the more uniform curve of scattering loss, which, as will be shown in the next section, decreases with increasing wavelength. This decreasing behavior is fairly evident in the lower curve of Fig. 2.1, for which the scattering loss is predominant.

At longer wavelengths, a further reduction of the absorption loss can be obtained. This is possible by using particular dopants for the profile synthesis. The trapping of OH radicals at the sites of such dopants leads to a shift of the fundamental OH vibration from $\lambda \cong 2.73$ μm to wavelengths higher than 3 μm. This technique makes it possible to obtain a minimum attenuation of 0.23 dB/km at $\lambda \cong 1.62$ μm, employing GeO_2 and P_2O_5 as dopants.[6] More drastic reductions can be attained at higher wavelengths when other types of glass are used. For example, a GeO_2 glass exhibits the fundamental vibration of the bond Ge-O around $\lambda = 10$ μm. However, the technology of such fibers has not been developed to the point where ultra-low attenuation is available. At present, the minima in the spectral loss, which have been measured around $\lambda = 2\ \mu m$, are nearly 10 dB/km.[7]

In Fig. 2.2 a comparison is reported between the theoretical predictions of spectral loss for germanium oxide glass fibers and for fused silica fibers.[7] The small peak at $\lambda \cong 2.5\ \mu m$, which characterizes the first curve, represents the effect of the fundamental OH vibration, assuming a water content of 1 ppb.

2.1.3 Scattering Loss

The glass which constitutes the basic material for fiber fabrication has a disordered structure, and its density is not microscopically uniform but fluctuates around an average value. These density variations entail fluctuations in the value of the refractive index, whose extension is typically much smaller than a wavelength distance. So, the main type of scattering which occurs is Rayleigh scattering. The differential cross section of this scattering depends on light polarization. Considering a linearly polarized wave, in a

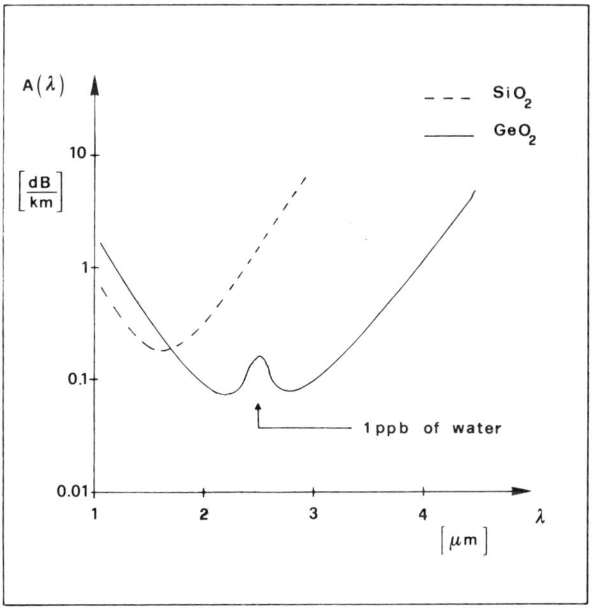

Fig. 2.2 - Comparison between the theoretical predictions of spectral loss for germanium oxide glass fibers and for fused silica fibers [from (7)].

plane which contains the line of polarization we have the differential cross section

$$\sigma_{11}(\psi) = \sigma_o \cos^2\psi$$

where σ_o is a constant and ψ represents the angle between the directions of the incident and of the scattered ray. On the other hand, in a plane orthogonal to the line of polarization, we have a uniform differential cross section, i.e.,

$$\sigma_{\perp}(\psi) = \sigma_o$$

In the general case, we are with unpolarized light, therefore

$$\sigma(\psi) = \sigma_o (1 + \cos^2\psi) \tag{2.3}$$

In order to relate this differential cross section to the probability density of the ray transition $K(\theta, \theta')$, introduced in 2.1.1, we consider an incident ray with propagation angle θ and a scattered ray with propagation angle θ'. It is easy to demonstrate that

$$\cos\psi = \cos\theta \cos\theta' + \sin\theta \sin\theta' \cos\phi$$

where the angle ϕ is defined in Fig. 2.3. Finally

$$K(\theta, \theta') = C_o \int_0^{\pi} \sigma(\psi)\, d\phi \tag{2.4}$$

where C_o is a normalizing constant, in order to have

$$2\int_0^{\pi/2} \sin\theta'\, d\theta' \int_0^{\pi/2} K(\theta, \theta') \sin\theta\, d\theta = 1 \tag{2.5}$$

Substituting (2.3) into (2.4), and considering also (2.5), we have from reference (1)

$$K(\theta, \theta') = \frac{3}{16} \quad \sigma_o [3(\cos^2\theta \cos^2\theta' + 1) - (\cos^2\theta \cos^2\theta')]$$

Since both θ and θ' are smaller than θ_M, which is usually a very small angle, $K(\theta, \theta')$, as a first approximation, can be taken as independent of θ and θ', as assumed in 2.1.1.

Developing an electromagnetic model, based on an equivalent short electric dipole, it is possible to obtain the expression of σ_o. Integrating in θ over the angle $\pi - \theta_M \cong \pi$, the scattering attenuation coefficient γ_s is obtained, which results[(8)] in

$$\gamma_s = \frac{8\pi^3}{3} \; n_1^8 p^2 K T_e \beta \; \frac{1}{\lambda^4} \tag{2.6}$$

where n_1 is the core refractive index, p is the photoelastic coefficient, K is Boltzmann's constant, T_e is the temperature at which the glass can reach a state of thermal equilibrium (for SiO_2 about 1900°K),

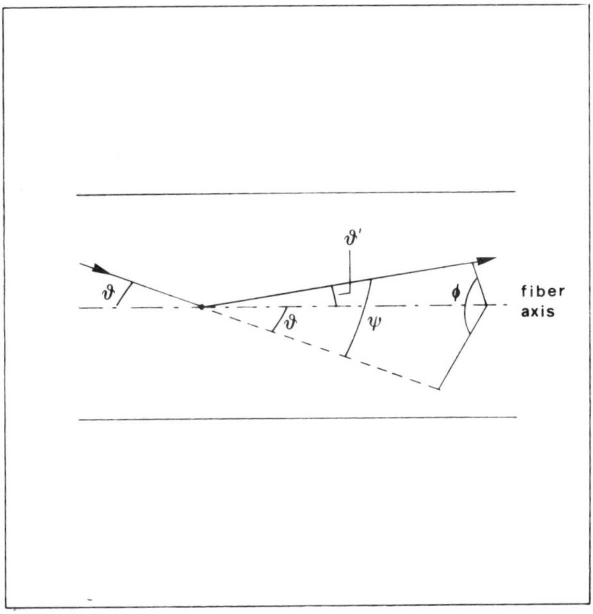

Fig. 2.3 - Definition of angles in a scattering process: ϕ belongs to the fiber cross section.

finally β is the isothermal compressibility.

The proportionality with $1/\lambda^4$, given by (2.6), is the most characteristic property of this process. For graded-index fibers, a much more complex geometrical model must be taken into account. Nevertheless, the proportionality with $1/\lambda^4$ remains, because it is present in the expression of the differential cross section of the scattering centers.

When centers of greater size are present, i.e., comparable with the dimension of the wavelength, other types of scattering occur, which usually exhibit a practically uniform dependence on λ. This is the case of microfractures or gas bubbles trapped in the glass. In the low-loss fibers recently produced, only Rayleigh scattering is practically present. This justifies the use of higher wavelengths for transmission. The minimum limit in the attenuation per km due to Rayleigh scattering, A_s, from experimental data, can be assumed to be about 0.2 dB/km in the neighborhood of $\lambda = 1.5\ \mu$m. This should lead to $A_s = 0.1$ dB/km just at $\lambda \cong 1.8\ \mu$m.

2.2 FIBER PERTURBATIONS

As fiber perturbations we will consider here all the geometrical imperfections of real fibers, particularly the variations along the fiber axis of the geometrical properties which for ideal fibers have been assumed to be perfectly axial-invariant. These variations can have a deterministic or a statistical nature. The study of their influence on the transmission properties of the fiber is fairly different in the two cases.

When a deterministic model of the fiber perturbation is assumed, the fiber is not characterized by axially uniform guided modes or rays, but they depend in a very complex way on the particular fiber section considered. On the contrary, when a statistical model of the fiber perturbation is assumed, the optical power distribution, among the modes or rays of an equivalent ideal fiber, evolves along the fiber length z with statistical properties, which can be derived from the statistics of the perturbation itself. If the analysis is restricted to a set of macroscopic properties, by following the statistical approach it is possible to consider an esemble of fibers of the same type and affected by the same kind of perturbation whose properties can be statistically investigated.

The deterministic approach, however, is useful for calculating the power transistions among modes or rays which are locally defined along the fiber length. This deterministic approach, as well as the general theory of the propagation in ideal multimode fibers, can be developed either through an electromagnetic model, or through ray optics. Only after this step, the statistical properties of the fiber perturbation can be taken into account. If the fiber perturbation is considered as a stationary random process, the optical power distribution among the modes or rays locally defined tends asymptotically to a steady-state condition, except for an attenuation factor. Such an asymptotical distribution is usually independent of the launching condition, whereas for an unperturbed fiber the launched optical power distribution does not change in shape at any distance.

In the present paragraph, after a brief outline of the electromagnetic model, we develop a ray-optics approach to the propagation in a perturbed fiber. Then, we will show that both the electromagnetic model and the ray-optics model, although with different assumptions in the two cases, lead to an equation called *the optical power flow equation,* which describes the evolution of the optical power distribution along the fiber length. In the next paragraph we will study the solution of this equation, first in the time independent case, which holds when the optical power is not modulated, and then in the time dependent case, which holds when it is modulated by an external signal.

2.2.1 Kinds of Fiber Perturbation

The usual processes of fiber fabrication give rise to two main kinds of perturbations:

i) core diameter fluctuations

ii) axial bending

Core diameter fluctuations usually have correlation lengths on the order of some meters. On the other hand, axial bending, especially when induced during the cabling process, can have a very short correlation length, in which case it is called microbending.

In order to develop a rigorous analysis, the index variation due to the elastic strain imposed on the fiber by microbending should be taken into account.[9] Nevertheless, since the bending radius is always well above the usual values of core diameter, it is possible to neglect this index variation and consider a transmission medium characterized by circular symmetry around a curved axis.

Core diameter fluctuations can be taken into account through the perturbed index profile

$$\widetilde{n}_1\,(r,z) = n_1\left\{r\,[1 - f(z)]\right\} \tag{2.7}$$

where $f(z)$ represents the fluctuations. Here we have circular symmetry around a straight axis, this justifies the dependence on the radial coordinate r. But axial bending requires the use of two transverse coordinates, e.g., x and y, and the following perturbed index profile can be taken into account

$$\widetilde{n}_1\,(x,y,z) = n_1\quad[x + f_x(z),\, y + f_y(z)] \tag{2.8}$$

where $f_x(z)$ and $f_y(z)$ represents the axial displacements along two orthogonal axes, respectively denoted x and y. In (2.7) $n_1(r)$ and in (2.8) $n_1(x, y)$ denote a reference unperturbed profile.

Note that $f(z)$ in (2.7) is a pure number, which can be either positive or negative, usually of few percent. On the other hand $f_x(z)$ and $f_y(z)$ in (2.8) are spatial lengths. Large bending (e.g., with a bending radius on the order of many centimeters) has no appreciable effects on the propagation because it produces very slow variations in the index gradient along any ray trajectory, with respect to the unperturbed case. Therefore the power transitions among the rays can be considered as successions of quasi-stationary states of propagation in an unperturbed fiber. Microbending, however, largely affects the propagation. Nevertheless, it consists of axial displacements which are usually small with respect to the core radius a. Then, for practical purposes, the absolute values of $f_x(z)$ and $f_y(z)$ can be assumed smaller than $a/10$.

Equation (2.7) represents a continuous variation in the shape of the index profile, while equation (2.8) expresses a rigid displacement of a given index profile. In Fig. 2.4 the two models are compared. The first can be considered along any radial direction, the second has been depicted in the xz plane. For the first model, all the statistical properties of the perturbation are present in $f(z)$, while for the second model the cross-correlation between $f_x(z)$ and $f_y(z)$ also must be considered.

2.2.2 Outline of the Electromagnetic Approach to the Study of Fiber Perturbations

There are many possible descriptions of the power exchange among the modes, which are locally defined in a perturbed waveguide. Marcuse[10] has considered two classes of modes:

i) modes of an ideal waveguide, e.g., that defined by $n_1(r)$ or $n_1(x, y)$ in (2.7) and (2.8)

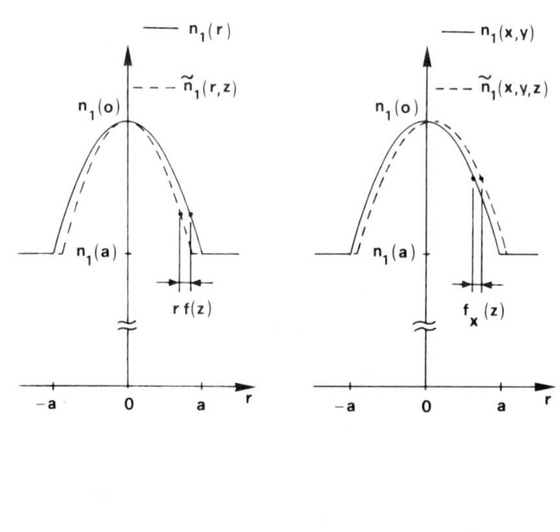

Fig. 2.4 - Effects of $f(z)$ in equation (2.1), and of $f_x(z)$ in equation (2.2) (from [16]).

ii) local normal modes, which are defined for a particular z, considering a hypothetical waveguide, whose axial direction is constant but whose transverse section is that of the actual waveguide

In Fig. 2.5 the difference between these two definitions is shown.

The above definitions are not restricted only to the two main kinds of fiber perturbations introduced in the previous section. The power distribution among the modes of the ideal waveguide or the local normal modes changes along z. Nevertheless, by expanding the electromagnetic field in terms of such modes, coupling coefficients can be obtained which represent the transitions of optical power from one mode to another. A proper superimposition of modes of the ideal waveguide, or of local normal modes, with z dependent coefficients, satisfies Maxwell's equations and the boundary conditions for any z.

Each type of expansion has some advantages for a certain kind of perturbation. Usually, inhomogeneities in the core region, like scattering centers, are better described through an expansion in terms of

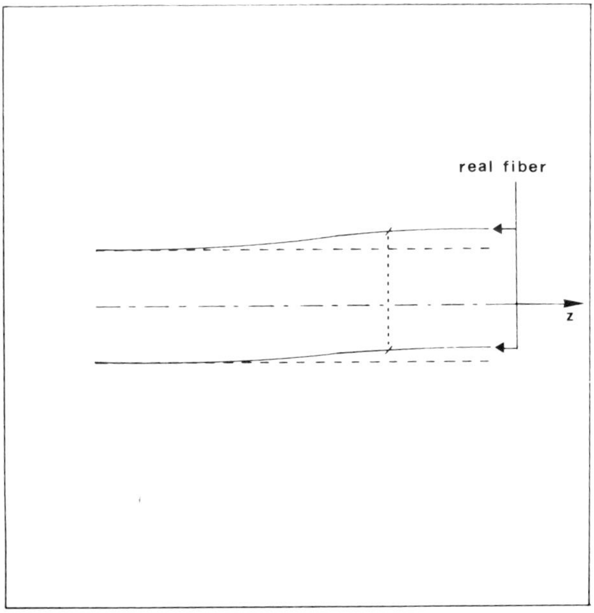

Fig. 2.5 - Ideal waveguide (dashed line) and a section in which local normal modes are defined (dotted line).

modes of the ideal waveguide. On the other hand, when geometrical perturbations such as core diameter fluctuations or microbending are present, an expansion in terms of local normal modes appears preferable.[10]

Developing this type of expansion, and considering the average power carried by each mode, Marcuse has obtained a set of coupled power equations, having the following form [10]

$$\frac{d\,W_m}{dz} = -\gamma_m W_m + \sum_{n=1}^{M} h_{mn}(W_n - W_m),\ m = 1, 2, \ldots M \qquad (2.9)$$

where γ_m is the power attenuation coefficient of the mth mode, $W_m(z)$ is the optical power carried by this mode, and h_{mn} is a power transition coefficient between the nth and the mth mode. In (2.9) a general M mode fiber is considered. All the geometrical and statistical properties of the perturbation are contained in h_{mn}. It is $h_{nm} = h_{mn}$.

The coefficients h_{mn} can be obtained in analytical form only when step-index or parabolic-index fibers are considered. In fact, the exact expressions of the electromagnetic fields and of the propagation constants of the local normal modes are available for such types of multimode fibers only, as was shown in 1.1. Such expressions are necessary for calculating the expansions which lead to (2.9). In the general case, however, h_{mn} can be written as

$$h_{mn} = K_{mn} \int_{-\infty}^{+\infty} R(\zeta) \exp[-i(\beta_n - \beta_m)\zeta]\, d\zeta \qquad (2.10)$$

where K_{mn} takes into account the behavior of the fields of the two modes, β_n and β_m are their phase constants, finally $R(\zeta)$ is the autocorrelation function of the perturbation.

2.2.3 Ray-Optics Approach to Core Diameter Fluctuations

When core diameter fluctuations affect the fiber, since the perturbed medium maintains circular symmetry, the constant of the motion h is not involved in the process of power transfer among the rays. In other words, rays having different h do not exchange any power between each other but since the medium is not z invariant, the constant of the motion E is involved in this process. Then, we can consider a ray as defined by a fixed value of h, and a value of E depending on z. The purpose of the present approach is to relate function $E(z)$ to the perturbation $f(z)$, so that the statistical properties of the latter can be easily transferred to $E(z)$ itself.

Substituting (2.7) into the z projection of the ray equation (1.35), we have[11]

$$\frac{d}{dz}\left(\tilde{n}_1 \frac{dz}{ds}\right) = -\frac{dn_1}{dr'}\, r\, \frac{df}{dz}$$

where $r' = r[1 - f(z)]$ is the independent variable which describes the index profile in the unperturbed case. Letting θ be the local ray propagation angle, it results in $dz/ds = \cos\theta$ and hence

$$\cos\theta \frac{d}{dz}(\tilde{n}_1 \cos\theta) = -\frac{dn_1}{dr'}\, r\, \frac{df}{dz} \tag{2.11}$$

$E(z)$ can be generalized as

$$E(z) = \tilde{n}_1 \cos\theta \tag{2.12}$$

Considering a small perturbation, i.e., $f(z) \ll 1$, r' can be taken as practically coincident with r, thus

$$n_1(r') \cong n_1(r)$$

$$\frac{dn_1}{dr'} \cong \frac{dn_1}{dr}$$

Equation (2.11) can therefore be rewritten as

$$\tilde{n}_1 \cos\theta \frac{d}{dz}(\tilde{n}_1 \cos\theta) \cong -n_1 \frac{dn_1}{dr}\, r\, \frac{df}{dz}$$

From (2.12), finally it is

$$\frac{dE}{dz} = g(E, h, z)\, \frac{df}{dz} \tag{2.13}$$

$$g(E, h, z) = -\frac{r}{2E} \frac{dn_1^2}{dr}$$

The function $g(E, h, z)$ depends on E, h, z, because the radial coordinate r is covered by the ray in a way which depends on E and h, with an approximately oscillating behavior along z. In other words, $g(E, h, z)$ is characterized by a rapidly varying part, which follows the oscillating behavior of the ray trajectory, and by a more slowly varying part, which follows the variation of $E(z)$, due to the perturbation $f(z)$. In the remainder, instead of considering rays which belong to any fiber section and characterized by E as a function of z, we will consider rays which belong to a certain local fiber section, in which they have a constant E. The power transitions among these rays will be derived from (2.13), giving a description very similar to that based on expansions in terms of local normal modes proposed in the electromagnetic approach.

In this model, the optical power distribution W in domain (E,h), which has been introduced in 1.2.9 for unperturbed fibers, will result in a function of z. It can be obtained through a general transport equation of the type

$$W(E, h, z) = \iint_G p(E, h; E_o, h_o; z)\, W(E_o, h_o, 0)\, dE_o\, dh_o \tag{2.14}$$

where $p(E, h; E_o, h_o; z)$ represents the transition probability density between a ray (E_o, h_o) and a ray (E, h) over the distance z. Finally $\iint_G$ denotes a double integration in the region G of guided rays. A factor, which in function p should take into account the intrinsic attenuation of the optical guided power, has been ignored. The two following conditions can be assumed about p

$$\lim_{z \to 0} p(E, h; E_o, h_o; z) = \delta(E - E_o)\, \delta(h - h_o) \tag{2.15}$$

$$p(E, h; E_o, h_o; z) = \iint_G p(E, h; E_1, h_1; z - z_1)\, p(E_1, h_1; E_o, h_o; z_1)\, dE_1\, dh_1 \tag{2.16}$$

where δ denotes a Dirac pulse function. The latter condition is true for any z_1 between 0 and z, and is usually known as the Chapman-Kolmogorof theorem.

The evolution of p, under conditions (2.15) and (2.16), can be studied through the standard Fokker-Planck theory,[12] and this will be done in the next section for microbending. Here, it is possible to derive the statistical properties of $E(z)$ directly from those of $f(z)$. As an application example, we consider an α-profile $n_1(r)$ given by (1.63), and a stationary normal stochastic process $f(z)$,[13] whose correlation function is

$$R_f(\zeta) = A \exp(-\beta \zeta)$$

where A and β are constants. $1/\beta$ is dimensionally a spatial length. Since $g(E, h, z)$, as $r(z)$, is locally periodic along z with period $Z(E, h)$, it can be expanded in Fourier series, leading to

$$g(E, h, z) = \sum_{n=0}^{\infty} g_n(E, h) \cos\left[2\pi n \frac{z}{Z(E, h)}\right]$$

where

$$g_o(E, h) = \frac{2}{Z(E, h)} \int_0^{\frac{Z(E,h)}{2}} g(E, h, z)\, dz$$

$$g_n(E,h) = \frac{4}{Z(E,h)} \int_0^{\frac{Z(E,h)}{2}} g(E,h,z) \cos\left[2\pi n \frac{z}{Z(E,h)}\right] dz$$

$n = 1, 2, \ldots$ (2.17)

As introduced in 2.2.1, the correlation length $1/\beta$ for this type of perturbation is on the order of several meters, whereas the ray oscillation period ranges typically between 10^{-3} and 10^{-2} m. Therefore $1/\beta \gg Z(E, h)$, so it is possible to consider the zero-order term only, in the expansion of $g(E, h, z)$, i.e., its value averaged over a ray oscillation period. Substituting the first of (2.17) into the second of (2.13), when n_1 is an α-profile, we obtain [11]

$$g_o(E,h) = \frac{n_1^2(0)}{E} \frac{\alpha}{\alpha+2}\left[1 - \frac{E^2}{n_1^2(0)}\right]$$

Finally, from the first of (2.13), it is

$$\frac{dE}{dz} = \frac{n_1^2(0)}{E} \frac{\alpha}{\alpha+2}\left[1 - \frac{E^2}{n_1^2(0)}\right]\frac{df}{dz}$$

Now, E can be replaced by variable ϵ, which has been defined by (1.75), so that

$$\frac{d\epsilon}{dz} = -\epsilon K(\alpha) \frac{df}{dz}$$

$$K(\alpha) = \frac{2\alpha}{\alpha+2}$$

Through an integration and a power expansion, truncated at the first order term, we are left with

$$\epsilon(z) - \epsilon_o = -\epsilon_o K(\alpha)\,[f(z) - f(0)]$$

where we have let $\epsilon(0) = \epsilon_o$. This equation allows one to relate the correlation function of $\epsilon(z)$ and that of $f(z)$. Precisely,

$$R_\epsilon(\zeta) = \epsilon_o^2 K^2(\alpha)\, R_f(\zeta) = \epsilon_o^2 K^2(\alpha)\, A \exp(-\beta\zeta) \qquad (2.18)$$

The probability density that a ray (ϵ_o, h) is converted into a ray (ϵ, h) after a distance z, coincides with the probability density of the event $\epsilon(z)$ conditioned by the event ϵ_o. Without loss of generality, $f(0) = 0$ can be assumed, so that this conditioned probability density results:[13]

$$p(\epsilon, \epsilon_o, z) = \frac{1}{\sqrt{2\pi\sigma(z)}} \exp\left[-\frac{(\epsilon-\epsilon_o)^2}{2\sigma^2(z)}\right]$$

$$\sigma(z) = \epsilon_o K(\alpha)\sqrt{A\,[1 - \exp(-2\beta z)]} \qquad (2.19)$$

The *rms* deviation $\sigma(z)$, initially zero, tends asymptotically to the value $\epsilon_o K(\alpha)\sqrt{A}$. Equation (2.19) can be assumed to be the transition probability density in the sense of equation (2.14), only observing that, since h is a constant, and since variable ϵ replaces variable E, this equation must be rewritten as

$$W(\epsilon,h,z) = \int_{\epsilon_e(h)}^{\Delta} p(\epsilon, \epsilon_o, z)\, W(\epsilon_o, h, 0)\, d\epsilon_o \tag{2.20}$$

where $\epsilon_e(h)$ represents the helical ray value of ϵ for a given h, obtained by equation (1.76).

Equation (2.19) is not rigorously correct, in fact it has been derived considering an unbounded domain of guided rays. However, G is a bounded region. For $\epsilon > \Delta$ we have transitions towards leaky or radiating rays, while for $\epsilon < \epsilon_e(h)$ transitions cannot occur, in fact no rays exist in this region. Therefore, an absorbing wall must be considered at $\epsilon = \Delta$, and a reflecting wall at $\epsilon = \epsilon_e(h)$. The effects of this correction, taking the usual values of A and therefore of $\sigma(z)$, are non-negligible only for rays which belong to very narrow strips close to $\epsilon_e(h)$ and Δ. This however entails an excess attenuation, which occurs during the reaching of the asymptotical configuration characterized by the condition $z \rightarrow \infty$, but well approximated for $z \gg 1/\beta$.[11]

In order to discuss the variations in the ray distribution produced by this kind of fiber perturbation, the asympotic values of the normalized *rms* deviation

$$\lim_{z \to \infty} \frac{\sigma(z)}{\Delta} = \frac{\epsilon_o K(\alpha)}{\Delta} \sqrt{A}$$

are reported *versus* the normalized initial condition ϵ_o/Δ in Fig. 2.6, for various α and $A = 10^{-4}$ (a very small perturbation), $A = 10^{-2}$ (a small perturbation). As expected, the asympotic deviation increases for larger values of A. Nevertheless, the parameter α also has an important role.

Although in this case we have directly derived the expression of the transition probability density through a statistical approach, the function $p(\epsilon, \epsilon_o, z)$ satisfies the following differential equation

$$\frac{\partial p}{\partial z} = -\gamma p - \frac{\partial}{\partial \epsilon}(M_1 p) + \frac{1}{2}\frac{\partial^2}{\partial \epsilon^2}(M_2 p) \tag{2.21}$$

where

$$M_1 = - \frac{\beta (\epsilon - \epsilon_o)}{A \epsilon_o^2 K^2 (\alpha)}$$

$$M_2 = 2A \beta \epsilon_o^2 K^2 (\alpha)$$

and in which γ takes into account the intrinsic attenuation of the fiber, which has been ignored till now. This equation can be considered as a Fokker-Planck type equation, characterized by the scattering moments M_1 and M_2.

Following the standard Fokker-Plank theory, it is possible to show that a simpler expression of p, which satisfies a heat-diffusion type equation, can be obtained. Nevertheless the last model appears only roughly approximated. In fact, equation (2.21) tends to a heat-diffusion type equation if z is so small it justifies the following truncated expansion

$$\exp(-2\beta z) \cong 1 - 2\beta z$$

This approximation is introduced just by the standard Fokker-Planck theory, in which, after a z integration, one has to consider the limit $z \to 0$,[14] but it is not critical when a microscopic correlation is considered.

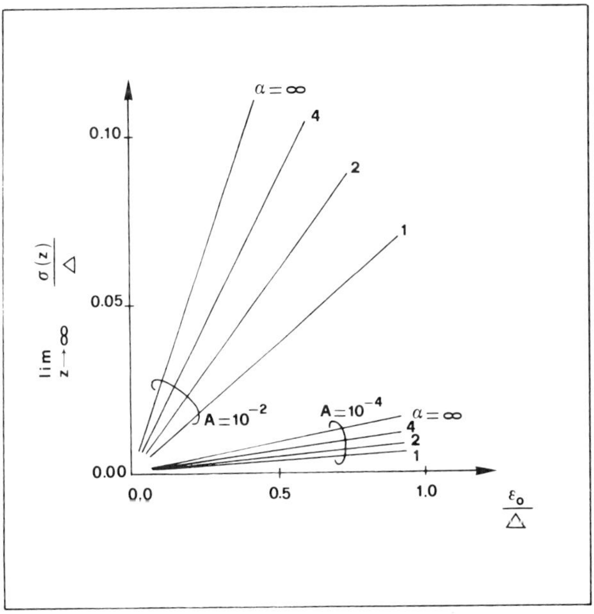

Fig. 2.6 - Asymptotic normalized *rms* deviation *vs.* ε_o/Δ [from (11)].

In conclusion, core diameter fluctuations do not entail a considerable power exchange among the rays. The asymptotical power distribution usually differs only in little part from that imposed by the launching condition. An excess attenuation is the main effect of such a perturbation. Also from an electromagnetic model similar conclusions can be drawn.[15]

2.2.4 Ray-Optics Approach to Microbending

When microbending affects the fiber, we have not a medium with circular symmetry. Therefore both the constants of the motion are involved in the process of power transfer among the rays. Substituting (2.8) into the z projection of the ray equation (1.35),[16] we have

$$\frac{dE}{dz} = \frac{1}{2E}\left(\frac{\partial \tilde{n}_1^2}{\partial x}\frac{df_x}{dx} + \frac{\partial \tilde{n}_1^2}{\partial y}\frac{df_y}{dz}\right) \tag{2.22}$$

As regards the other constant, instead of h, it is convenient to employ

$$\ell = x\,\frac{dy}{dz} - y\,\frac{dx}{dz}$$

Now, considering a general perturbed profile $\tilde{n}_1$ from the x and y projection of the ray equation (1.35), we have

$$\frac{d}{ds}\left(y\,\tilde{n}_1\,\frac{dx}{ds}\right) = y\frac{\partial \tilde{n}_1}{\partial x} + \tilde{n}_1\,\frac{dx}{ds}\,\frac{dy}{ds}$$

$$\frac{d}{ds}\left(x\,\tilde{n}_1\,\frac{dy}{ds}\right) = x\frac{\partial \tilde{n}_1}{\partial y} + \tilde{n}_1\,\frac{dx}{ds}\,\frac{dy}{ds}$$

Subtracting these equations, we are left with

$$\frac{dz}{ds}\,\frac{d}{dz}\left[n_1\frac{dz}{ds}\left(x\,\frac{dy}{dz} - y\,\frac{dx}{dz}\right)\right] = x\,\frac{\partial \tilde{n}_1}{\partial y} - y\,\frac{\partial \tilde{n}_1}{\partial x}$$

which yields

$$\frac{d\ell}{dz} = -\,\frac{\ell}{E}\,\frac{dE}{dz} + \frac{1}{2E^2}\left(x\,\frac{\partial n_1^2}{\partial y} - y\,\frac{\partial n_1^2}{\partial x}\right) \tag{2.23}$$

Considering a small perturbation, i.e., $f_x/x,\ f_y/y \ll 1$, as for core diameter fluctuations, it is possible to assume

$$\frac{\partial \tilde{n}_1^2}{\partial x} \cong \frac{\partial n_1^2}{\partial x} \tag{2.24}$$

$$\frac{\partial \tilde{n}_1^2}{\partial y} \cong \frac{\partial n_1^2}{\partial y}$$

Moreover, owing to the circular symmetry of n_1, we have

$$x \frac{\partial n_1^2}{\partial y} - y \frac{\partial n_1^2}{\partial x} = 0$$

In conclusion, (2.22) and (2.23) can be approximately rewritten as

$$\frac{dE}{dz} \cong \frac{1}{2E} \left(\frac{\partial n_1^2}{\partial x} \frac{df_x}{dz} + \frac{\partial n_1^2}{\partial y} \frac{df_y}{dz} \right)$$

$$\frac{d\ell}{dz} \cong - \frac{\ell}{2E^2} \left(\frac{\partial n_1^2}{\partial x} \frac{df_x}{dz} + \frac{\partial n_1^2}{\partial y} \frac{df_y}{dz} \right) \tag{2.25}$$

while the following differential equation holds between E and ℓ

$$\frac{dE}{d\ell} = - \frac{E}{\ell} \tag{2.26}$$

The last equation defines a set of curves in the domain (E,ℓ). Recalling the relation among E, ℓ, and h, which holds for unperturbed fibers,

$$h = E\ell$$

We can conclude that on these curves h remains constant. In other words, in this order of approximation, the medium exhibits a circular symmetry again. This is a consequence of the assumptions (2.24), which lead to $dh/dz \cong 0$. On the other hand, assuming

$$\frac{\partial \tilde{n}_1^2}{\partial x} \cong \frac{\partial n_1^2}{\partial x} + f_x \frac{\partial^2 n_1^2}{\partial x^2}$$

$$\frac{\partial \tilde{n}_1^2}{\partial y} \cong \frac{\partial n_1^2}{\partial y} + f_y \frac{\partial^2 n_1^2}{\partial y^2}$$

one would have

$$\frac{dh}{dz} \cong \frac{1}{2E} \left[x f_y \frac{\partial^2 n_1^2}{\partial y^2} - y f_x \frac{\partial^2 n_1^2}{\partial x^2} \right]$$

This expression is certainly a better approximation, but since it depends on f_x and f_y, instead of df_x/dz and df_y/dz, as the first of (2.25), it would lead to a more complicated statistical analysis.

So, for the sake of simplicity, we restrict ourselves to (2.24), keeping in mind that the discussion will be somehow approximated. Then, depending on the choice of the pair of constants of the motion, the evolution of the optical power distribution in the domain of guided rays can be described through a function $W(E, h, z)$ or through a function $W'(E, \ell, z)$. Following the approach, a one-dimension Fokker-Planck theory can which leads to this equation

$$\frac{\partial W}{\partial z} = - \gamma W - \frac{\partial}{\partial E} (M \quad\quad (M_2 W) \tag{2.27}$$

From the standard Fokker-Planck theory,[14]

$$M_1 = \lim_{\Delta z \to 0} \frac{1}{\Delta z} \int_0^{\Delta z} < \frac{dE}{dz} > dz$$

$$M_2 = \lim_{\Delta z \to 0} \frac{1}{\Delta z} \int_0^{\Delta z} dz' \int_0^{\Delta z} < \frac{dE}{dz}(z') \frac{dE}{dz}(z'') > dz'' \tag{2.28}$$

Note that (2.21) and (2.27), in spite of the similar form, have quite different meanings. The first one describes the evolution of a transition probability density between an initial and a final condition, respectively ϵ_0 and ϵ. The other one describes the evolution of an optical power distribution. A proper initial condition must be added to equation (2.27), to take into account the optical power distribution at the fiber launching section, whereas the model based on equation (2.21) requires an initial condition only as $W(\epsilon_o, h, 0)$ in (2.20). In the following, we will calculate M_1 and M_2 for a microbending perturbation characterized by microscopic correlation. Next, we will return briefly to the case of core diameter fluctuations to make a comparison between these two types of perturbation, especially when they have a sinusoidal behavior with a spatial period of the same order of the ray oscillation period. As known, this can occur in practice only for microbending.

As a consequence of the assumption of small perturbations, $r(z)$ can be considered as locally periodic along z, with period $Z(E, h)$. Moreover, so the azimuthal angle $\phi(z)$ can be considered as locally pseudo-periodic it can be written

$$\phi(z) = vz + p(z)$$

where v and $p(z)$ have been introduced in 1.2.6. Therefore

$$\frac{\partial n_1^2}{\partial x} = \frac{x}{r} \frac{d n_1^2}{dr} = \cos \phi(z) \frac{d n_1^2}{dr} =$$

$$= \left[\cos v \cos p(z) - \sin vz \sin p(z) \right] \frac{d n_1^2}{dr}$$

The functions cos $p(z)\, dn_1^2/dr$ and sin $p(z)\, dn_1^2/dr$ can be expanded in Fourier series and by simple trigonometric considerations so we are left with

$$\frac{\partial n_1^2}{\partial x} = \frac{1}{2} \sum_{n=0}^{\infty} [(F_n^p + F_n^d) \cos (2 \pi \Omega_n^+ z)$$

$$+ (F_n^p - F_n^d) \cos (2 \pi \Omega_n^- z)]$$

where $F_n^p(E, h)$ and $F_n^d(E, h)$ are suitable coefficients, finally it is

$$\Omega_n^{\pm} = \frac{v}{2\pi} \pm \frac{n}{Z} \tag{2.29}$$

In a similar way, except for the substitution of cos $\phi(z)$ with sin $\phi(z)$, we have

$$\frac{\partial n_1^2}{\partial y} = \frac{1}{2} \sum_{n=0}^{\infty} \Big[(F_n^p + F_n^d) \sin (2 \pi \Omega_n^+ z)$$

$$+ (F_n^p - F_n^d) \sin (2 \pi \Omega_n^- z) \Big]$$

Considering df_x/dz and df_y/dz as stationary random processes, with zero croscorrelation, the correlation functions involved in the calculation of M_2 are

$$R_x(\zeta) = \left< \frac{df_x}{dz}(z) \frac{df_x}{dz}(z+\zeta) \right>$$

$$R_y(\zeta) = \left< \frac{df_y}{dz}(z) \frac{df_y}{dz}(z+\zeta) \right>$$

$$R_{xy}(\zeta) = \left< \frac{df_x}{dz}(z) \frac{df_y}{dz}(z+\zeta) \right> = 0$$

Owing to the stationarity,

$$\left< \frac{df_x}{dz} \right> = \left< \frac{df_y}{dz} \right> = 0$$

and hence $M_1 = 0$. From the second of (2.28), the first of (2.25) and the above expressions of $R_x(\zeta)$ and $R_y(\zeta)$, we are left with

$$M_2 = \frac{1}{4E^2} \lim_{\Delta z \to 0} \frac{1}{\Delta z} \int_0^{\Delta z} dz' \int_0^{\Delta z} dz'' \Big[\frac{\partial n_1^2}{\partial x}(z') \frac{\partial n_1^2}{\partial x}(z'') R_x(z'-z'')$$

$$+ \frac{\partial n_1^2}{\partial y}(z') \frac{\partial n_1^2}{\partial y}(z'') R_y(z'-z'') \Big] \tag{2.30}$$

This expression is completely general. A particular case, characterized by a very simple formalism, is that of a microscopic symmetric correlation, for which we have

$$R_x (z' - z'') = R_y (z' - z'') = A\,\delta\,(z' - z'')$$

where δ denotes a Dirac pulse function, whereas A takes into account the amplitude of a perturbation. $1/A$ is dimensionally a spatial length. In this case, we obtain

$$M_2 = \frac{A}{4E^2}\;\frac{1}{Z(E,h)}\int_0^{Z(E,h)}\left(\frac{dn_1^2}{dr}\right)^2 dz \tag{2.31}$$

On the other hand, in the most general case, substituting the expressions obtained for $\partial n_1^2/\partial x$ and $\partial n_1^2/\partial y$ into (2.30), the following rather complex expression is obtained

$$M_2 = \frac{1}{16}\,\frac{1}{E^2}\sum_{n=0}^{\infty} F_n^2(E,h)\left[S(\Omega_n^+) + S(\Omega_n^-)\right] \tag{2.32}$$

where

$$F_n^2(E,h) = (F_n^p)^2 + (F_n^d)^2$$

and $S(\Omega)$ denote the power spectrum of the perturbation.

Before discussion of (2.32), we will apply the standard Fokker-Planck theory to the case of core diameter fluctuations. From (2.13), expanding in Fourier series $g(E, h, z)$, an equation of the type (2.27) is obtained, in which

$$M_1 = \lim_{\Delta z \to 0} \frac{1}{\Delta z}\int_0^{\Delta z} g(E,h,z) < \frac{df}{dz} > dz = 0$$

$$M_2 = \lim_{\Delta z \to 0} \frac{1}{\Delta z}\int_0^{\Delta z} dz' \int_0^{\Delta z} dz''\, g(E,h,z')\, g(E,h,z'')$$

$$< \frac{df}{dz}(z')\,\frac{df}{dz}(z'') >$$

$$= \frac{1}{2}\sum_{n=0}^{\infty} g_n(E,h)\int_{-\infty}^{+\infty} R(\zeta)\cos\left[2\pi n\,\frac{\zeta}{Z(E,h)}\right] d\zeta$$

where the coefficients $g_n(E, h)$ are given by (2.17) and $R(\zeta)$ represents the correlation function of the process df/dz. Introducing its power spectrum S, it is

$$M_2 = \frac{1}{2}\sum_{n=0}^{\infty} g_n(E,h)\, S\left[\frac{n}{Z(E,h)}\right] \tag{2.33}$$

This expression, as well as (2.32), has been derived without any assumption on the shape of the index profile and on the correlation function of the perturbation. Now we will consider a sinusoidal perturbation, which is characterized by a sharply peaked power spectrum around a circular spatial frequency Ω_c. Note that this definition has significance only if it is given locally. In fact, if the perturbation were rigorously sinusoidal, it should have an infinite correlation length, while the above expressions have been derived under the condition $R_x(\Delta z) = R_y(\Delta z) = R(\Delta z) \cong 0$. From (2.32) and (2.33), we can conclude that a sinusoidal perturbation gives rise to a very high loss when its spatial frequency Ω_c coincides with a given set of spatial frequencies.

When core diameter fluctuations affect the fiber, these spatial frequencies are the inverses of the ray oscillation periods, and their integer multiples. When microbending affects the fiber, there are all the spatial frequencies given by (2.29). Since Ω_n^- are generally negative, they can be reported on the possitive side of the spectrum, obtaining a distribution for each ray, as shown in Fig. 2.7. This distribution is determined by $Z(E, h)$ and $v(E, h)$. Since $v(E, h)$ gives

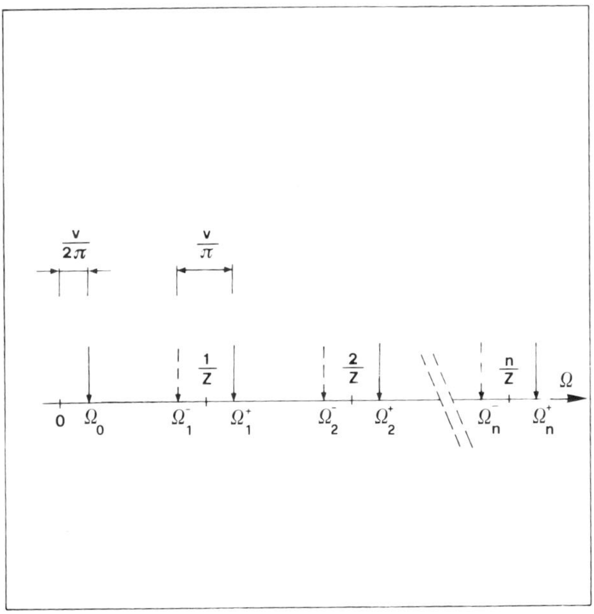

Fig. 2.7 - Distribution of the spatial frequencies which characterize the microbending loss for a given ray (from [16]).

a measure of the skewness of the ray, it can be observed that meridional rays suffer core diameter fluctuations and microbending in the same way. On the contrary, skew rays offer a wider spectrum of spatial frequencies in the neighborhood of the points $n/[Z(E,h)]$, in the case of microbending. From these considerations, it is possible to state that, in practical cases, a microbending perturbation with oscillating behavior and spatial period between 10^{-3} and 10^{-2} m can strongly affect the propagation, giving rise to a remarkable excess loss.

2.2.5 Optical Power Flow Equation in the Electromagnetic Model

In equation (2.9), we can consider that, as a first approximation, only adjacent modes exchange power, therefore one can write

$$\frac{\partial W_m}{\partial z} = -\gamma_m W_m + h_{m+1}(W_{m+1} - W_m) - h_m(W_m - W_{m-1}), \quad m = 1, 2, \ldots, M \tag{2.34}$$

where only M coefficients h_m, instead of M^2 coefficients h_{mn}, have been considered. The equations (2.34), when M is very large, can be transformed into

$$\frac{\partial W_m}{\partial z} = -\gamma_m W_m + h_{m+1}\frac{\partial W_{m+1}}{\partial m} - h_m \frac{\partial W_m}{\partial m}, \quad m = 1, 2, \ldots, M$$

in which we have considered the approximations

$$W_{m+1} - W_m \cong \frac{\partial W_{m+1}}{\partial m}\Delta m$$

$$W_m - W_{m-1} \cong \frac{\partial W_m}{\partial m}\Delta m$$

with $\Delta m = 1$. Rigorously, the W_m should be continuous functions of m, while m is an integer number. Nevertheless, formally this can be done, changing the dimension of h_m.

Repeating this approximation for the last terms, a single differential equation can be written, taking m as a continuous variable,

$$\frac{\partial W(m)}{\partial z} = -\gamma(m)\, W(m) + \frac{\partial}{\partial m}\left[h(m)\frac{\partial W(m)}{\partial m}\right] \tag{2.35}$$

Two main problems must be solved in order to apply (2.35) in practice:

i) finding a single variable m which denotes the modes

ii) determining the mode coupling function $h(m)$ from the knowledge of the kind of fiber perturbation and its statistical properties.

First of all, the choice of the independent variable m depends on the class of index profiles considered. Then, the choice of m must be done so that modes having the same m can be considered as indistinguishable for the determination of the coefficients h_{mn} through the electromagnetic model introduced in 2.2.2. In other words, the fiber perturbation must give rise to changes in m only, and not in any other mode variable. Olshansky,[17] considering the class of α-profiles, has assumed variable $m(\beta)$, introduced in 1.3.3, which satisfies the above requirements, for the usual fiber perturbation.

As regards the second problem, since an exact electromagnetic model of the propagation in ideal fibers exists only for $\alpha = \infty$ (step-index profile) and $\alpha = 2$ (parabolic index profile), in these two cases only, it is possible to obtain $h(m)$ from the exact expression of h_{mn}. For example, from Marcuse's expressions,[10] when microbending affects the fiber, with a perturbation of the type (2.2), we have

$$h_{\infty}(m) = \frac{1}{8}\left[\frac{2\pi}{\lambda}\,\frac{n_1(0)}{a}\right]^2 S(\Delta\beta)\,,\ \alpha = \infty$$

$$h_2(m) = \frac{1}{8}\left[\frac{2\pi}{\lambda}\,\frac{n_1(0)}{a}\right]^2 \left(\frac{m}{M_2}\right)^{1/2} S(\Delta\beta)\,,\ \alpha = 2 \qquad (2.36)$$

where M_2 is the total number of modes, defined by (1.66), when $\alpha = 2$, S is the power spectrum of d^2f_x/dz^2 and d^2f_y/dz^2, finally the argument $\Delta\beta$ represents the difference $(\beta_m - \beta_{m-1})$. A difference in the phase constants of the modes plays a role similar to that of Ω_n^+, Ω_n^- in the ray optics approach.

Olshansky has assumed for $h_\alpha(m)$ the following general expression,[17] which satisfies (2.36) for particular cases,

$$h_\alpha(m) = \frac{1}{8}\left[\frac{2\pi}{\lambda}\,\frac{n_1(0)}{a}\right]^2 \left(\frac{m}{M_\alpha}\right)^{\frac{2}{\alpha+2}} S(\Delta\beta) \qquad (2.37)$$

This assumption is somehow restrictive. Furthermore, mode coupling occurs rigorously just between adjacent modes only for parabolic index fibers. [18] Nevertheless, the use of (2.37) leads to a simple analytical form for (2.35), which can be easily solved in finite terms. So, though this analysis has to be considered as roughly approximated, it gives quantitative results which can be immediately understood.

From experimental data, the power spectrum of d^2f_x/dz^2 and d^2f_y/dz^2 (which mean curvature of the perturbed fiber), can be approximated well by

$$S(\Delta\beta) = A\,(D\,\Delta\beta)^{-2p} \tag{2.38}$$

where A takes into account, as in the ray-optics approach, the amplitude of the perturbation, D is a measure of its correlation length, and p is a constant, whose value is typically 0, 1, or 2, depending on the nature of the external stress, and on the fiber mechanical properties.

At this point, another consideration must be done. When $\alpha = \infty$ or $\alpha = 2$, for any integer n, there is a group of $2(n + 1)$ quasi-degenerate modes, i.e., modes having β very close to one another. So, it is possible to assume that this property holds also for all the other values of α. Then equation (2.34), which gives the single optical power W_m of the mth mode, can be replaced by a new equation giving the optical power W_n averaged over the nth group of $2(n + 1)$ quasi-degenerate modes. The total number of these groups of modes $N(\alpha)$ is related to $M(\alpha)$ by

$$M(\alpha) = \sum_{n=1}^{N(\alpha)} 2\,(n+1) \cong N^2(\alpha)$$

thus we have $m/M = (n/N)^2$. The new equation which substitutes (2.34) is[17]

$$\begin{aligned} 2\,(n+1)\,\frac{\partial W_n}{\partial z} &= -\,2\,(n+1)\,\gamma_n W_n + 2\,(n+1)\,h_{n+1}(W_{n+1} - W_n) \\ &\quad - 2n\,h_n\,(W_n - W_{n-1}), \\ &\qquad n = 1, 2, \ldots, N \end{aligned}$$

Finally, equation (2.35) must be replaced by

$$\frac{\partial W(n)}{\partial z} = -\,\gamma(n)\,W(n) + \frac{1}{n}\,\frac{\partial}{\partial n}\left[n\,h(n)\,\frac{\partial W(n)}{\partial n}\right] \tag{2.39}$$

where $W(n)$, $\gamma(n)$, and $h(n)$ are, respectively, the optical power, the attenuation coefficient, and the coupling coefficient, averaged over the nth group of quasi-degenerate modes. Inverting (1.67), and considering the substitution of m/M with n/N, we have

$$\beta(n) = \frac{2\pi}{\lambda}\,n_1(0)\left[1 - 2\Delta\left(\frac{n}{N}\right)^{\frac{2\alpha}{\alpha+2}}\right]^{1/2} \tag{2.40}$$

from which $\Delta\beta$ can be obtained.

In order to solve (2.39) in finite terms as a Bessel-type equation, following reference (17), the coupling coefficient $h_\alpha(n)$ can be taken as

$$h_\alpha(n) = h_o \left(\frac{n}{N}\right)^{-2q} \tag{2.41}$$

where h_o and q are suitable constants. Substituting (2.41) into (2.39), and introducing a normalized mode group number $x_g = n/N$, which ranges between 0 and 1, we have the optical power flow equation

$$\frac{\partial W}{\partial z} = -\gamma W + \frac{h_o}{N^2} \frac{1}{x_g} \frac{\partial}{\partial x_g} \left(x_g^{1-2q} \frac{\partial W}{\partial x_g} \right) \tag{2.42}$$

where γ is in general a function of x_g, and W is a function of x_g and z.

In the particular case of microbending, comparing (2.42) with (2.37) and (2.38), in which $\Delta\beta$ is obtained from (2.40), the following relation[17] between parameters p and q holds

$$q = \frac{p(\alpha - 2) - 2}{\alpha + 2} \tag{2.43}$$

2.2.6 Optical Power Flow Equation in the Ray-Optics Model

The optical power flow equation in the ray-optics model is a Fokker-Planck type equation like (2.27), in which M_1 and M_2 can be derived either directly from statistical considerations,[19] or from the standard Fokker-Planck theory.[12] When a single variable is not enough for describing the process of power exchange among the rays, it is possible to use a two-dimension Fokker-Planck equation.[20] The ray optics approach is not limited by considering coupling effects only between adjacent modes.[21] Nevertheless it suffers other simplifying assumptions: for instance, we have a perturbation so small it justifies approximation (2.24) or the paraxial ray approximation,[12],[20],[21] which consists of neglecting the value of Δ with respect to the unity. In conclusion, both possible approaches to the study of perturbed fibers appear approximated, nevertheless they allow one to obtain some useful insight into the practical effects of mode coupling. In the present section, we will write equation (2.27) for a particular case of microbending, in order to give an example of this equation in explicit form.

Considering microbending with zero mean value of the derivative of the axis displacement and microscopic correlation, we obtain $M_1 = 0$, while M_2 is given by (2.31), which can be calculated in finite terms only for a parabolic index profile. In this case,[16] we have

$$M_2(E) = A \frac{\Delta}{a^2} n_1^2(0) \left[\frac{n_1^2(0)}{E^2} - 1 \right]$$

Substituting into (2.21), and considering the variable ϵ defined by (1.75), instead of E, we are left with

$$\frac{\partial W}{\partial z} = - \left[\gamma - 3A \frac{\Delta}{a^2} \frac{1}{(1 - 2\epsilon)^2} \right] W + \frac{2 - \epsilon}{1 - 2\epsilon} A \frac{\Delta}{a^2} \frac{\partial W}{\partial \epsilon}$$

$$+ A \frac{\Delta}{a^2} \epsilon \frac{\partial^2 W}{\partial \epsilon^2}$$

This differential equation cannot be solved in finite terms, but in the para-axial ray approximation ($\epsilon \ll 1$), it leads to

$$\frac{\partial W}{\partial z} \cong - \gamma W + A \frac{\Delta}{a^2} \left(\epsilon \frac{\partial^2 W}{\partial \epsilon^2} + 2 \frac{\partial W}{\partial \epsilon} + 3 W \right) \tag{2.44}$$

which exhibits a Bessel-type second member, very similar to that of (2.42). Note that the assumption of γ independent of ϵ, implicitly made in 2.2.4, can now be removed without loss of generality, in order to attain a complete analogy between (2.42) and (2.44).

As for the differences between the electromagnetic approach and the ray-optics approach, the general equation (2.35) can be seen as a Fokker-Planck type equation with $M_1 = 0$, independently of the variable m or ϵ employed, provided that the quantity $M_2 = 2h(m)$ is a constant. This is the case of a heat-diffusion type equation, which often represents a too rough approximation. Only for step-index fibers, using the propagation angle θ as independent variable, and considering just meriodional rays, this type of equation appears suitable for the description of microbending effects in perturbed fibers.[19],[22]

In conclusion, the optical power flow equation, independent of the approach used and the variable u assumed for denoting modes or rays, has the following general form

$$\frac{\partial W}{\partial z} = - \gamma W + \hat{D} W \tag{2.45}$$

where $\hat{D}$ is a linear second-order differential operator with respect to u. For step-index fibers, a diffused expression of $\hat{D}$ is $A\ (\partial^2/\partial\theta^2)$, where A takes into account the amplitude of the fiber perturbation. For graded-index fibers, through the electromagnetic approach we have a Bessel-type operator, whereas the ray-optics approach involves a more complex expression. Nevertheless, since when $\alpha = 2$ in the para-axial ray approximation $\hat{D}$ becomes a Bessel-type operator, we can assume that this property holds approximately also for other values of α near 2.

Though the last assumption can appear somehow restrictive, it is justified by the importance to have a proper operator $\hat{D}$ for a closed form solution of (2.45). Therefore, in the following, in order to derive time-independent and also time-dependent solutions, we will consider two cases only:

i) the harmonic operator $\hat{D} = A(\partial^2/\partial\theta^2)$, suitable for step index fibers only, neglecting skew rays, but with the advantage of a simpler formalism

ii) the Bessel-type operator obtained by Olshansky $\hat{D} = (h^\circ/N^2)(1/x_g)\,\partial/\partial x_g\,[\,x_g^{1-2q}(\partial W/\partial x_g)]$, which holds for general graded-index fibers, neglecting mode coupling between non-adjacent modes

All the properties which are derived for these two operators can be considered of general validity, except for the particular analytical expression of the solution.

Another reason for the choice of these two operators is that they are self-adjoint. As known,[23] when an operator is self-adjoint, its solutions Q_n are orthogonal functions. This property will be very useful, in order to solve the optical power flow equation. Such orthogonality can be expressed through the following conditions in the two cases

$$\int_0^{\theta_M} Q_n(\theta)\,Q_m(\theta)\,d\theta = 0, \qquad m \neq n, \qquad (\textit{step-index fibers})$$

$$\int_0^{1} Q_n(x_g)\,Q_m(x_g)\,x_g\,dx_g = 0, \quad m \neq n, \quad (\textit{graded-index fibers})$$

The integrating factor $x_g\,dx_g$ arises from the mode degeneracy, while for step-index fibers with only meridional rays, this mode degeneracy is ignored, and the integrating factor remains simply $d\theta$.

2.2.7 Solutions of the Optical Power Flow Equation

Two kinds of solutions can be obtained for equation (2.45):

i) the unbounded solution

ii) the bounded solution

The unbounded solution is based on the assumption that the intrinsic attenuation coefficient γ is an increasing function of the mode variable u. Therefore, modes or rays having large values of u are strongly attenuated and their contribution can be neglected at the steady-state. Then, the maximum value u_m, at which lim $z \to \infty$ $W(u, z)$ can be considered as practically different from zero, is well under the maximum permitted value u_M which characterizes modes or rays in the unperturbed fiber.[24] On the contrary, for the bounded solution, γ is constant when $0 \leq u \leq u_M$, whereas $\gamma \to \infty$ when $u > u_M$. This leads to the boundary condition

$$W(u_M, z) = 0 \tag{2.46}$$

which holds for any z. For step-index fibers the unbounded solution appears justified, especially when two different glasses, or glass and silicone, make up the core-cladding interface. In fact, the reflection at this interface may produce a diffusion, and a consequent loss for radiation which increases with increasing propagation angle. On the contrary, for graded-index fibers, where there are no reflecting surfaces because the ray trajectory changes its direction continuously, and also for step-index fibers with a smoothed transition between core and cladding, as it is when a single doped glass is used, the bounded solution is to be preferred.

In order to develop an example of an unbounded solution for step-index fibers, we can consider a θ-dependent intrinsic attenuation coefficient of the type

$$\gamma(\theta) = \gamma_o + b\,\theta^2$$

where γ_o and b are constants. Equation (2.45), when $\widehat{D}$ assumes the harmonic form, then becomes

$$\frac{\partial W}{\partial z} = -(\gamma_o + b\,\theta^2)\,W + A\,\frac{\partial^2 W}{\partial \theta^2} \tag{2.47}$$

Considering the trial function

$$W(\theta, z) = Q(\theta)\exp(-\lambda z)$$

where $Q(\theta)$ is a function of the only θ, and λ is a constant whose dimension is the inverse of distance, we have

$$A\,\frac{d^2 Q}{d\theta^2} = -(\lambda - \gamma_o - b\,\theta^2)\,Q$$

This equation is satisfied by Hermite-Gaussian polynomials. The general solution of (2.47) can be written as

$$W(\theta, z) = \sum_n c_n\,Q_n(\theta)\exp(-\lambda_n z)$$

in which constants c_n are imposed by the launching condition, and

$$\lambda_n = \gamma_o + (2n + 1)\sqrt{Ab}$$

$$Q_n(\theta) = H_n(\theta)\exp\left(-\frac{\sqrt{b}\,\theta^2}{2\sqrt{A}}\right),\ n=0,1,2\ldots$$

where $H_n(\theta)$ represents the nth Hermite polynomial. The constant λ_n has the meaning of attenuation coefficient of the nth configuration $Q_n(\theta)$, which is a sort of permitted optical power distribution, called nth statistical configuration of the perturbed fiber.

Very far from the launching section $z = 0$, i.e., when $z \rightarrow \infty$, a single statistical configuration dominates, precisely the one which is characterized by the least attenuation coefficient. Since λ_n increases as n increases, this fundamental statistical configuration is that denoted by $n = 0$. When practically only the fundamental statistical configuration influences the optical power distribution, the steady-state condition is reached. This occurs after a distance which becomes shorter and shorter with increasing A and b. At the steady-state, the optical power distribution exhibits a gaussian shape with *rms* width proportional to $(A/b)^{1/4}$. Therefore, the fiber numerical aperture depends on z, and at the steady-state it should be practically independent of the maximum permitted value θ_M which characterizes the unperturbed fiber. The last property has not been confirmed at all by experimental measurements. In practice, NA depends effectively on z, but, even at the steady-state, it maintains a dependence on the value of θ_M, which is imposed by the index difference between core and cladding. This experimental result confirms that an unbounded solution, even for step-index fibers, suffers some restrictive assumptions, and suggests that it is advisable to develop a bounded solution.

In this case, we will study the general equation (2.45), and then we will particularize the results for harmonic and Bessel-type operators. We assume, also in this case, a trial function having the form

$$W(u,z) = Q(u)\exp(-\lambda z)$$

Thus we have

$$\hat{D}Q = -(\lambda - \gamma)\,Q \qquad (2.48)$$

where we must keep in mind that γ is a constant. The boundary condition (2.46), which leads to $Q(u_M) = 0$, makes this one an eigenvalue equation, where Q_n are the eigenfunctions, and

$$\Gamma_n = \gamma - \lambda_n,\ n = 1, 2, \ldots$$

are the eigenvalues. The general solution of (2.48) can be written as

$$W(u,z) = \sum_n c_n \, Q_n(u) \exp(-\lambda_n z)$$

where constants c_n are imposed by the launching condition.

Also in this case, ordering λ_n with increasing n, we obtain the steady-state solution $Q_1(u) \exp(-\lambda_1 z)$. As for reaching the steady-state, it occurs after a distance which becomes shorter and shorter as the amplitude of the fiber perturbation increases. On the other hand, $Q_1(u)$ turns out to be independent of the amplitude of the fiber perturbation, but depends only on u_M. In this sense, the bounded solution is in better agreement with experimental results.

2.2.8 The Time-Dependent Equation

A time dependent equation is necessary when the optical power distribution W is modulated by an external signal; therefore, it is a function also of the time t. The group delay per unit distance $\tau(u)$, which characterizes the mode or ray denoted by u, must be taken into account in the optical flow equation, which results

$$\frac{\partial W}{\partial z} + \tau(u) \frac{\partial W}{\partial t} = -\gamma W + \hat{D} W$$

Letting $\overline{W}(u, z, \omega)$ the Fourier transform of $W(u, z, t)$, where ω represents the modulation circular frequency, this equation becomes

$$\frac{\partial \overline{W}}{\partial z} = -[\gamma + i\,\omega\tau(u)]\,\overline{W} + \hat{D}\,\overline{W} \tag{2.49}$$

where i denotes the imaginary unit.

In such a time dependent problem, we are interested mainly in the changes of the time dispersion properties which characterize the signal propagation, with respect to the case of an unperturbed fiber. Here, we will show only that, at the steady state, the *rms* pulsewidth increases as the square root of z, instead of z, as for unperturbed fibers. The steady-state solution of (2.49) can be written in the following general form[25]

$$\overline{W}(u,z,\omega) = Q(u,\omega) \exp[-\lambda(\omega) z] \exp[-i\omega T_r(u,\omega)] \exp[-i\omega\,\tau_\infty(\omega) z] \tag{2.50}$$

in which $Q(u, \omega)$ and $T_r(u, \omega)$ are independent of z, owing to the definition of steady-state, $\lambda(\omega)$ represents a ω dependent attenuation coefficient, finally $\tau_\infty(\omega)$ denotes the time delay per unit distance independent of u, i.e., common to all the modes or rays. From (1.87) and (1.89), letting

$$\overline{H}(\omega, z) = \frac{\int_G \overline{W}(u, z, \omega,)\, du}{\int_G \overline{W}(u, z, 0,)\, du}$$

we have a very complicated expression for the *rms* pulsewidth $\sigma(z)$. Nevertheless, assuming that $\lambda(\omega)$, $\tau_\infty(\omega)$, $Q(u, \omega)$, $T_r(u, \omega)$ are slowly varying in the neighborhood of $\omega = 0$, i.e.,

$$\left.\frac{d\lambda}{d\omega}\right|_{\omega=0} = \left.\frac{d\tau_\infty}{d\omega}\right|_{\omega=0} = \left.\frac{\partial Q}{\partial\omega}\right|_{\omega=0} = \left.\frac{\partial T_r}{\partial\omega}\right|_{\omega=0} = 0 \tag{2.51}$$

we are left with[25]

$$\sigma(z) = \sqrt{s_0 + s_2 z} \tag{2.52}$$

where

$$s_0 = \left[\int_G Q(u,0)\, du \int_G \left[Q(u,0)\, T_r(u,0) - \left.\frac{\partial^2 Q}{\partial\omega^2}\right|_{\omega=0} \right] du - \left[\int_G Q(u,0)\, T_r(u,0)\, du \right]^2 \right] \Big/ \left[\int_G Q(u,0)\, du \right]^2$$

$$s_2 = \left.\frac{d^2\lambda}{d\omega^2}\right|_{\omega=0}$$

From (2.52), we can conclude that for large distances, at which $s_2 z \gg s_0$, $\sigma(z)$ increases as $\sqrt{z}$. This is an advantage of perturbed fibers with respect to unperturbed ones. It is balanced by an increase in the attenuation, which has been pointed out also considering the time independent equation. To summarize, mode coupling due to fiber perturbations increases the attenuation and reduces the time dispersion. This property is certainly verified for long distances, where the steady-state is practically reached. For intermediate distances the launching condition has appreciable effect on the transmission properties of a real fiber, leading to different situations, which must be studied with more attention. This will be done in the next section.

2.3 TRANSMISSION PROPERTIES OF PERTURBED FIBERS

A real, perturbed fiber is always in a condition which can be considered as intermediate between the unperturbed situation, i.e., without mode coupling, and the steady-state situation. Moreover, the latter condition is a statistical concept, and its properties must be assumed as averaged over an ensemble of perturbed fibers with similar perturbations. In order to extrapolate the results of experimental measurements over different fiber lengths, it is then useful to develop a statistical theory of the propagation in perturbed fibers. Nevertheless, such an extrapolation, especially for quantitative results, must be taken with some care. In fact, the solution of the optical power flow equation is strongly dependent on the index profile, on the kind, amplitude and statistics of the perturbation. Furthermore, the stationarity of such a perturbation should be verified in practice.

The qualitative results offered by the statistical theory are of some help, especially when critical situations must be avoided in the installation of transmission systems. For instance, a great deal of attention must be payed to the estimate of the excess attenuation due to the perturbation. Yet, it would be rather unsafe to rely on the increase of the bandwidth. In other words, the uncertainty on the index profile shape and on the perturbation parameters makes it rather difficult to take practical advantages from mode coupling effects.

On the other hand, the transmission system whose repeater spacing is limited by the bandwidth of the transmission medium can be an advantage for mode coupling, because its baseband response turns out to be less critical than that of an unperturbed medium with respect to index profile and launching condition. This makes it possible to compensate for this baseband response effectively, by the use of suitable equalizer circuits, while this is very difficult for unperturbed fibers.

2.3.1 Exact Solution of the Time Independent Equation

At first, the case of a harmonic operator $\widehat{D}$ will be studied, since it needs a simpler formalism. Then we will consider a Bessel-type operator. Assuming

$$Q_n(\theta) = \sqrt{\frac{2}{\theta_M}} \cos\left(n \, \frac{\pi}{2} \, \frac{\theta}{\theta_M}\right) \quad (n \; odd)$$

allows one to impose the orthonormality condition

$$\int_0^{\theta_M} Q_n(\theta)\, Q_m(\theta)\, d\theta = \delta_{nm}, \quad n, m = 1, 3, \ldots$$

where δ_{nm} denotes Kronecker's symbol. The general solution of (2.45) is then

$$W(\theta, z) = \sum_n c_n\, Q_n(\theta) \exp(-\lambda_{on} z) \tag{2.53}$$

where pedix o means that we deal with the ω independent part of λ_n, which can be written as

$$\lambda_{on} = \gamma + A\left(\frac{n\,\pi}{2\,\theta_M}\right)^2 \quad (n\ odd) \tag{2.54}$$

Finally, letting $W_o(\theta)$ be the launched power distribution, we have

$$c_n = \sqrt{\frac{2}{\theta_M}}\int_0^{\theta_M} W_o(\theta)\cos\left(n\frac{\pi}{2}\frac{\theta}{\theta_M}\right)d\theta \qquad (n\ odd)$$

Since the present model neglects the contribution of skew rays, only launching conditions consisting of a proper power distribution in the domain of meridional rays can be considered.

In Fig. 2.8 the first three statistical configurations are reported. Only the fundamental one is always positive between $\theta = 0$ and $\theta = \theta_M$. It does not mean, however, that we can consider negative values of the optical power distribution $W(\theta, z)$, which would have no physical meaning. In fact, among the coefficients c_n for a given real launching condition, c_1 is greater than the other ones, to avoid the presence of negative values in $W(\theta, 0)$. So, because of the factors $\exp(-\lambda_{on}\, z)$, the fundamental statistical configuration will be more dominant with increasing z.

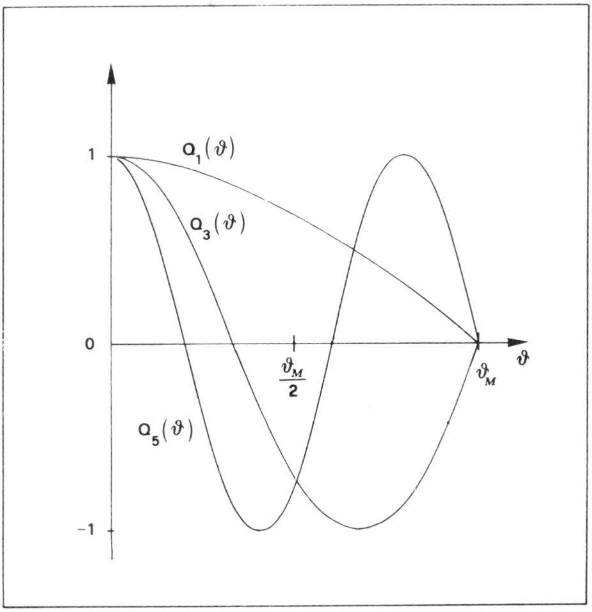

Fig. 2.8 - First three statistical configurations.

The total optical guided power at the fiber section z is

$$P(z) = \int_0^{\theta_M} W(\theta, z)\, d\theta$$

Substituting this expression into the general equation (2.45), with $\hat{D} = A(\partial^2/\partial\theta^2)$, through an integration in θ, we obtain

$$\frac{dP}{dz} = -\gamma P + A\left(\left.\frac{\partial W}{\partial \theta}\right|_{\theta=\theta_M} - \left.\frac{\partial W}{\partial \theta}\right|_{\theta=0}\right)$$

Any statistical configuration Q_n is characterized by the property $dQ_n/d\theta|_{\theta=0} = 0$, which leads to

$$\frac{dP}{dz} = -\gamma P + A\left.\frac{\partial W}{\partial \theta}\right|_{\theta=\theta_M} - \tag{2.55}$$

From this equation,[26] the term $-A\,(\partial W/\partial\theta|_{\theta=\theta_M})$ can be considered as the fraction of guided power which, at the fiber section z, is lost owing to the perturbation. When we are with a real launching condition, this is always a positive quantity. In fact it is $W(\theta_M, z) = 0$, and, for $\theta \lesssim \theta_M$, we have $W(\theta, z) > 0$.

It is possible to observe that, for a given θ_o, between 0 and θ_M, the quantity $-A\,(\partial W/\partial\theta|_{\theta=\theta_o})$ represents the optical power which is transferred from rays having $\theta \le \theta_o$ to rays having $\theta > \theta_o$. At the steady-state, since $-A\,(\partial W/\partial\theta|_{\theta=\theta_o})$ increases with increasing θ_o, a transfer towards the outer rays dominates the power exchange. On the contrary, for other statistical configurations, $-A\,(\partial W/\partial\theta|_{\theta=\theta_o})$ may have either positive or negative values. Then, the optical power is transferred sometimes from the inner to the outer rays, sometimes from the outer to the inner rays.

Considering now a Bessel-type operator, the assumption

$$Q_n(x_g) = \frac{(1+q)^{1/2}}{|J_{\nu+1}[Z_n(\nu)]|}\, x_g^q J_\nu[Z_n(\nu)\, x_g^{1+q}], \quad \nu = -\frac{q}{1+q}$$

where J_ν represents the ν-order Bessel function, and $Z_n(\nu)$ is its nth zero, allows one to impose the orthonormality condition[17]

$$\int_0^1 Q_n(x_g)\, Q_m(x_g)\, 2x_g\, dx_g = \delta_{nm}, \quad n, m = 1, 2\ldots$$

in which $2x_g\, dx_g$ takes into account the degeneracy. The general solution of equation (2.45) is then

$$W(x_g, z) = \sum_n c_n Q_n(x_g) \exp(-\lambda_{on} z)$$

where λ_{on} is given by

$$\lambda_{on} = \gamma + \frac{h_o}{N^2}(1+q)^2 Z_n^2(\nu) \tag{2.56}$$

Finally, letting $W_o(x_g)$ be the launched power distribution, we have

$$c_n = \frac{(1+q)^{1/2}}{|J_{\nu+1}[Z_n(\nu)]|}\int_0^1 W_o(x_g)\, x_g^q J_\nu[Z_n(\nu)x_g^{1+q}]\, 2x_g\, dx_g$$

Also this model, assuming a single variable x_g, allows one to consider only some particular launching conditions, and precisely those which exhibit a uniform excitation of each group of quasi-degenerate modes.

The total optical guided power at the fiber section z is

$$P(z) = \int_0^1 W(x_g)\, 2x_g\, dx_g$$

The property $(\partial W/\partial\theta|_{\theta=0}) = 0$ is now replaced by

$$x_g^{1-2q}\left.\frac{\partial W}{\partial x_g}\right|_{x_g=0} = 0$$

and all the considerations made about the quantity $-A(\partial W/\partial\theta|_{\theta=\theta_0})$, for a harmonic operator, can be repeated as well about the quantity $(-h_o/N^2)\,(x_g^{1-2q})\,(\partial W/\partial W/\partial x_g|_{x_g=x_{go}})$ where x_{go} is a given value of x_g between 0 and 1. As for the shape of the first statistical configurations $Q_n(x_g)$, we have a behavior very similar to that of $Q_n(\theta)$ reported in Fig. 2.8, except for the presence of q as a parameter, which influences the slope of each curve but not its zero.

2.3.2 Perturbative Solutions of the Time-Dependent Equation

An exact solution of (2.49) does not exist. Nevertheless, a perturbative solution can be obtained assuming the time independent exact solution as unperturbed case. First of all, when a step-index fiber is considered, we have

$$\tau(\theta) = \frac{n_1}{c}\left(1 + \frac{\theta^2}{2}\right) \tag{2.57}$$

For graded-index fibers, assuming an α-profile without material dispersion for simplicity, from (1.74) and keeping in mind that $x_g^2 = m/M$, we have

$$\tau(x_g) = \frac{n_1(0)}{c}\left(1 + \frac{\alpha-2}{\alpha+2}\Delta\, x_g^{\frac{2\alpha}{\alpha+2}} + \frac{3\alpha-2}{\alpha+2}\,\frac{\Delta^2}{2}x_g^{\frac{4\alpha}{\alpha+2}}\right)$$

When a harmonic operator $\widehat{D}$ is considered, through a second order perturbative method,[26] the general expression of $\overline{W}(\theta,z,\omega)$ results

$$\overline{W}(\theta, z, \omega,) = \sum_n c_n \overline{Q}_n(\theta, \omega) \exp[-\lambda_n(\omega) z] \exp(-i\omega \tau_{\infty n} z)$$

2.58)

Each statistical configuration $\overline{Q}_n(\theta,\omega)$ is characterized by an ω-dependent attenuation coefficient

$$\lambda_n(\omega) = \lambda_{on} - \omega^2 \sum_{\substack{r \\ n\neq r}} \frac{|a_{nr}|^2}{\lambda_{on} - \lambda_{or}} \quad (n\ odd) \tag{2.59}$$

and by a time delay per unit distance common to all the rays

$$\tau_{\infty n} = a_{nn} \quad (n\ odd) \tag{2.60}$$

It is

$$a_{nm} = \int_0^{\theta_M} \tau(\theta) Q_n^{(o)}(\theta) Q_m^{(o)}(\theta) d\theta, \quad n, m = 1, 3 \ldots \tag{2.61}$$

in which $Q_n^{(0)}(\theta)$ and $Q_m^{(0)}(\theta)$ and denotes two statistical configurations of the unperturbed equation, which were labeled simply $Q_n(\theta)$ and $Q_m(\theta)$ in 2.3.1. Finally, $\overline{Q}_n(\theta, \omega)$ is given by

$$\overline{Q}_n(\theta, \omega) = Q_n^{(o)} - \omega^2 \sum_{\substack{s \\ n\neq s}} \frac{Q_s^{(o)}}{\lambda_{on} - \lambda_{os}}$$

$$\sum_{\substack{r \\ n\neq r}} \frac{(a_{rs} - a_{nn}\delta_{rs}) a_{nr}}{\lambda_{on} - \lambda_{or}}$$

$$+ i\omega \sum_{\substack{s \\ n\neq s}} \frac{a_{ns} Q_s^{(o)}}{\lambda_{on} - \lambda_{os}} \quad (n\ odd) \tag{2.62}$$

Substituting (2.57) into (2.61), we have

$$a_{nm} = \frac{n_1}{c}\left(\frac{2\theta_M}{\pi}\right)^2 \left[\frac{\cos(m+n)\frac{\pi}{2}}{(m+n)^2} + \frac{\cos(m-n)\frac{\pi}{2}}{(m-n)^2}\right], \quad n \neq m \tag{2.63}$$

$$a_{nn} = \frac{n_1}{c}\left[1 + \left(\frac{\theta_M}{\pi}\right)^2 \left(\frac{\pi^2}{6} - \frac{1}{n^2}\right)\right]$$

Coefficients c_n in equation (2.58) are imposed as usual by the launching condition.

It is possible to demonstrate[26] that the summation which multiplies

ω^2 in (2.59) may be either positive or negative. For $n = 1$, it is always negative, that is $\lambda_1(\omega)$ always increases with increasing ω, but for $n > 1$, $\lambda_n(\omega)$ may decrease with increasing ω. Therefore, when ω is greater than a given ω_x, the fundamental statistical configuration $\overline{Q}_1$ is not predominant, as it is in the time-independent case. However, for very large values of z, ω_x turns out to be outside the useful region of the fiber baseband response, so only those values of $\omega < \omega_x$ for which the first statistical configuration dominants are of interest.

At the steady-state, therefore, for practical purposes, we can write simply

$$\overline{W}(\theta, z, \omega) = \overline{Q}_1(\theta, \omega) \exp[-\lambda_1(\omega) z] \exp(-i\omega \tau_{\infty 1} z) \tag{2.64}$$

Comparing with (2.50), it is

$$Q \exp(-i\omega T_r) = \overline{Q}_1, \; \lambda = \lambda_1, \; \tau_\infty = \tau_{\infty 1}$$

From (2.59), we have

$$\left.\frac{d\lambda}{d\omega}\right|_{\omega=0} = 0$$

Finally, from (2.62) and (2.63), one can obtain explicit expressions for Q and T_r of (2.50), which lead to

$$\left.\frac{\partial Q}{\partial \omega}\right|_{\omega=0} = \left.\frac{\partial T_r}{\partial \omega}\right|_{\omega=0} = 0$$

Thus, the assumptions in (2.51), which are at the basis of the asymptotical decay of the fiber bandwidth as $1/\sqrt{z}$, are consistent with the present second-order perturbative solution of the time-dependent equation. The asymptotical behavior of the fiber baseband response will be investigated in detail in 2.3.4.

The amplitude $Q = |\overline{Q}_1|$ and the relative delay $T_r = 1/\omega \arg[\overline{Q}_1]$ of the fundamental statistical configuration are reported in Fig. 2.9. In order to discuss these two families of curves, we must consider briefly the expression of $\tau_{\infty 1}$, which turns out to be

$$\tau_{\infty 1} = \frac{n_1}{c}\left[1 + \left(\frac{\theta_M}{\pi}\right)^2\left(\frac{\pi^2}{6} - 1\right)\right] \cong \frac{n_1}{c}\left(1 + \frac{\theta_M^2}{16}\right)$$

Since n_1/c represents the group delay per unit distance common to all the rays of the unperturbed fiber, comparing the above expression with (2.57), it is possible to conclude that rays characterized approximately by $\theta = \theta_M/4$ exhibit the same group delay, with and

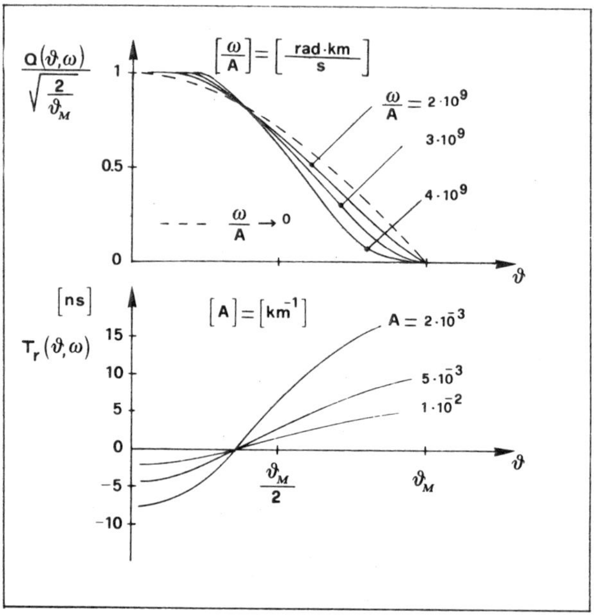

Fig. 2.9 - Amplitude and relative delay distributions computed from (2.64) (steady-state), with $n_1 = 1.5$ and $\theta_M = \pi/20$ [from (26)].

without the perturbation. In fact, Fig. 2.9 allows one to see that rays which are in the neighborhood of $\theta = \theta_M/4$ have also zero relative delays $T_r(\theta, \omega)$ at the steady-state.

This can be justified observing that inner rays give power to intermediate rays, but this power outflow, which is dominant owing to

$$-\left.\frac{\partial Q}{\partial \theta}\right|_{\theta=\theta_o} > 0\,, \quad 0 \leq \theta_o \leq \theta_M$$

is balanced in part by an opposite flow, which is responsible for the negative relative delays $T_r(\theta, \omega)$ for small θ. At $\theta \cong \theta_M/4$, transit times equilibrate.

Finally, we observe that the steady-state distribution (2.64) takes place only after a distance z much greater than a value z_m given by

$$\left(\tau_{\infty 1} - \frac{n_1}{c}\right) z_m = T_r\,(0,\omega)$$

Substituting the expression previously obtained, we have

$$z_m \cong \frac{1}{\lambda_{o3} - \lambda_{01}}$$

as expected.

Some of the results obtained for the harmonic operator can be easily extended to the case of a Bessel-type operator. Expression (2.58) is replaced by

$$W(x_g,z,\omega) = \sum_n c_n \overline{Q}_n(x_g, \omega) \exp[-\lambda_n(\omega) z] \exp(-i\omega\tau_{\infty n} z) \qquad (2.65)$$

(2.59) and (2.60) hold again, but the general expression of a_{nm} is now

$$a_{nm} = \int_0^1 \tau(x_g) \, Q_n^{(o)}(x_g) \, Q_m^{(o)}(x_g) \, 2x_g \, dx_g \,, \quad n,m = 1,2,\ldots \qquad (2.66)$$

$\overline{Q}_n(x_g, \omega)$ are given by (2.62), but it is not possible to obtain an explicit expression of Q and T_r, since (2.66) cannot be calculated in finite terms. Nevertheless, results obtained through numerical computations are very similar to those previously reported for the harmonic operator.

2.3.3 Evolution of the Attenuation Coefficient while Approaching Steady-State

During the reaching of the steady-state it is not possible to define an attenuation coefficient in a rigorous way, because the optical power decay is not exponential. Considering for simplicity the time independent equation, in the general case characterized by the mode variable u, we can write

$$W(u,z) = \sum_n c_n Q_n(u) \exp(-\lambda_{on} z)$$

The total guided power is

$$P(z) = \sum_n c_n \exp(-\lambda_{on} z) \int_0^{u_M} Q_n(u) \, du \qquad (2.67)$$

Nevertheless, an average attenuation coefficient $\lambda_a(z)$[26] can be defined through the following expression

$$\frac{P(z+\Delta z)}{P(z)} = \exp[-\lambda_a(z)\Delta z] \qquad (2.68)$$

When $z \to \infty$, it is $\lambda_a(z) \to \lambda_{o1}$. For shorter z, $\lambda_a(z)$ may be either greater than λ_{o1} or smaller than λ_{o1}. In fact the coefficients of the summation (2.67) may be positive or negative, depending on the launching condition. However, $\lambda_a(z)$ is always greater than the intrinsic attenuation coefficient γ. In order to give an example of the evaluation of $\lambda_a(z)$ while approaching steady-state, let us consider the case of a harmonic operator. The launching condition can be simulated by the function

$$W_o(\theta) = 1\ (0 \leq \theta \leq h),\ 0\ (h < \theta \leq \theta_M) \tag{2.69}$$

where parameter h takes into account the divergence of the source beam. When $h = 0$, only the axial ray is excited, while $h = \theta_M$ represents a uniform excitation of all the rays.

From $W_o(\theta)$ the coefficients c_n are obtained as functions of h. Through (2.67) and (2.68), provided that Δz is a short distance with respect to any distance $1/\lambda_{on}$ of interest, we are left with

$$\lambda_a(z) = \frac{\sum_n \lambda_{on} \frac{1}{n^2} \sin\left(n \frac{\pi}{2}\right) \sin\left(n \frac{\pi}{2} \frac{h}{\theta_M}\right) \exp(-\lambda_{on} z)}{\sum_n \frac{1}{n^2} \sin\left(n \frac{\pi}{2}\right) \sin\left(n \frac{\pi}{2} \frac{h}{\theta_M}\right) \exp(-\lambda_{on} z)} \quad (n\ odd)$$

In Fig. 2.10 $\lambda_a(z)$ is reported for three different values of A, and four different values of h. It has been assumed that $\gamma = 1\text{km}^{-1}$. It is possible to observe that the steady-state is reached after a longer distance if A is small. When the launching condition produces a function $W_o(\theta)$ similar to the steady-state power distribution $Q_1(\theta)$, i.e., for h between $\theta_M/2$ and $(3/4)\,\theta_M$, the steady-state is reached in the fastest way.

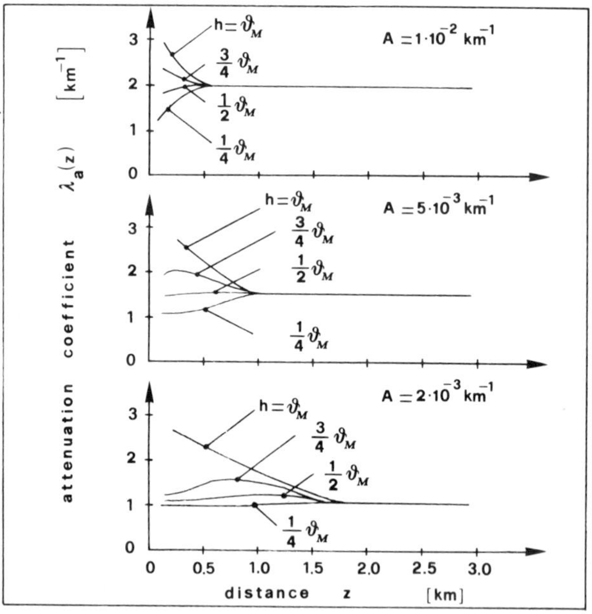

Fig. 2.10 - Evolution of the attenuation coefficient during the reaching of the steady state [from (26)].

For greater h, $\lambda_a(z)$ decreases with increasing z, since the optical power lost by rays having $\theta \cong \theta_M$ becomes immediatcly radiated power. On the contrary, for smaller h, $\lambda_a(z)$ increases, in fact the optical power lost by rays having small θ is transferred before to intermediate rays, and only later becomes radiated power. The overall behavior is not always monotonic: especially for h between $\theta_M/2$ and $(3/4)\ \theta_M$, a smoothed oscillation may be present.

Considering now a Bessel-type operator, we can observe that a launching condition expressed by (2.69), except for the substitution of θ with x_g, can take into account an excitation parallel to the fiber axis, whose spatial extension on the fiber input end face is measured by the parameter h, as discussed in 1.3.7. Then, as for the evolution of $\lambda_a(z)$, defined again through (2.68), similar considerations to those reported for the harmonic operator can be reported as well.

2.3.4 Evolution of the Baseband Response while Approaching Steady-State

The baseband response of the fiber, while reaching the steady-state, in the general case, can be written as

$$\overline{H}(\omega, z) = \left[\sum_n c_n \exp(-\lambda_{on} z) \exp(-i\omega\tau_{\infty n} z) \exp(-E_n^{(2)} \omega^2 z) \int_0^{u_M} [Q_n^{(o)} - \omega^2 Q_n^{(2)} + i\,\omega\, Q_n^{(1)}]\, du \right] \Big/ \sum_n c_n \exp(-\lambda_{on} z) \int_0^{u_M} Q_n^{(o)}\, du$$

where

$$E_n^{(2)} = \sum_{\substack{r \\ n \neq r}} \frac{|a_{nr}|^2}{\lambda_{or} - \lambda_{on}}$$

$$Q_n^{(1)} = \sum_{\substack{s \\ n \neq s}} \frac{|a_{ns}|^2}{\lambda_{on} - \lambda_{os}} Q_s$$

$$Q_n^{(2)} = \sum_{\substack{s \\ n \neq s}} \frac{Q_s}{\lambda_{on} - \lambda_{os}} \sum_{\substack{r \\ n \neq r}} \frac{(a_{rs} - a_{nn}\delta_{rs})\, a_{nr}}{\lambda_{on} - \lambda_{or}}$$

Except for very short distances from the optical source, only the first term between square brackets turns out to be predominant. So, considering also that $|E_1^{(2)}| \gg |E_2^{(2)}|, |E_3^{(2)}|, \ldots,$ we can rewrite the expression of the baseband response taking in separate terms the fundamental configuration and the other ones, i.e.

$$\bar{H}(\omega,z) \cong \frac{c_1 \exp(-\lambda_{o1} z) \exp(-i\omega\tau_{\infty 1} z) \exp(-E_n^{(2)} \omega^2 z) \int_0^{u_M} Q_1^{(o)}\, du}{\sum_n c_n \exp(-\lambda_{on} z) \int_0^{u_M} Q_n^{(o)}\, du}$$

$$+ \frac{\sum_{\substack{n \\ n>1}} c_n \exp(-\lambda_{on} z) \exp(-i\omega\tau_{\infty n} z) \exp(-E_n^{(2)} \omega^2 z) \int_0^{u_M} Q_n^{(o)}\, du}{\sum_n c_n \exp(-\lambda_{on} z) \int_0^{u_M} Q_n^{(o)}\, du} \tag{2.70}$$

Over short distances, both terms on the right hand side of (2.70) have effect on the baseband response, but the factors $\exp(-E_n^{(2)} \omega^2 z)$, for any n, are negligible. Therefore the bandwidth decay takes place as $1/z$, because of the spread in the phasors $\exp(-i\omega\, \tau_{\infty n} z)$. On the contrary, when the distance increases, the first term becomes more predominant, till the steady-state is reached. The factor $\exp(-i\omega\, \tau_{\infty 1} z)$ does not introduce any distortion, whereas $\exp(-E_1^{(2)} \omega^2 z)$, turns out to be responsible for the bandwidth decay, which takes place as $1/\sqrt{z}$.

The asymptotical expression of the fiber baseband response results

$$\lim_{z \to \infty} \bar{H}(\omega, z) = \exp(-i\omega\tau_{\infty 1} z) \exp(-E_1^{(2)} \omega^2 z) \tag{2.71}$$

which does not depend on the launching condition, but only on the index profile and on the fiber perturbation. These properties influence the value of $E_1^{(2)}$. However, the most formidable consequence of expression (2.71) is that its gaussian form, without any phase distortion, is completely general. Then, it can be easily equalized in the receiver circuit of a transmission system. On this particular feature, we will return in Chapter 4.

Till now, material dispersion has been ignored. Yet, this could lead to considerable errors when (2.71) is used as the baseband response of the fiber over very long distances. In fact, when a bandwidth much more extended than that of the unperturbed fiber is achieved, owing to the properties of the steady-state, the effects of material dispersion, characterized by a bandwidth decay always of the type 1/z, can be not negligible and even predominant. In conclusion, three asymptotes are present in the most general case in the bandwidth decay of a fiber, as shown in Fig. 2.11. Their slopes are ordinately $1/z$, $1/\sqrt{z}$, $1/z$. Their intersections depend on all the four properties which characterize the time dispersion of the real fiber considered, namely:

i) index profile

ii) launching condition

iii) fiber perturbation

iv) material dispersion

The index profile influences all three asymptotes. On the contrary, the launching condition is responsible only for the first one, the fiber perturbation of the second one, and material dispersion mainly of the third one. Fig. 2.11 shows in a qualitative way such effects. The arrows denote how the asymptotes shift with increasing values of the parameters. Such parameters are ordinately: h, which takes into account how the launched power distribution is confined between 0 and u_M; A, which represents the amplitude of the fiber perturbation; $\Delta\lambda_\nu$, which is the spectral width of the optical source.

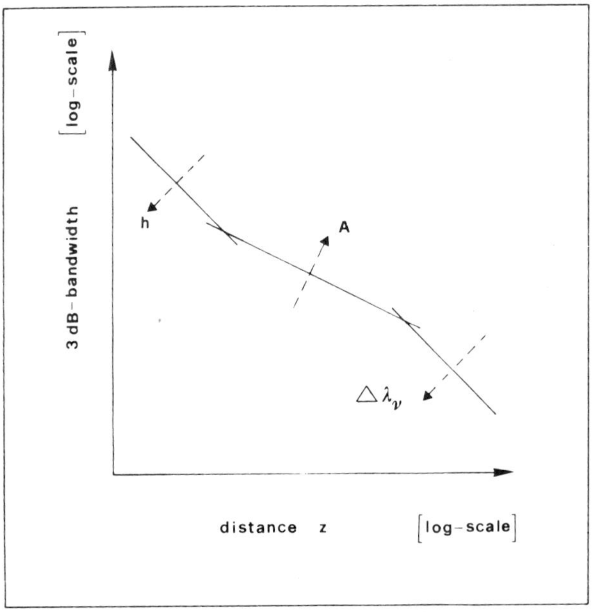

Fig. 2.11 - Effects over the 3 dB bandwidth of the launching condition (h) the amplitude of the fiber perturbation (A), and material dispersion ($\Delta\ \lambda_\nu$) [from (26)].

Considering also the attenuation properties of the fiber, the following observations can be made. An increase in h entails an increase not only of the time dispersion, but also of the attenuation, that is to say the smaller the value of h the better the performance of the transmission system, provided that the launched optical power is high enough. On the other hand, an increase in A entails a reduction of the time dispersion, but an increase of the attenuation. So, if one were able to control the value of A, he could take this value as a trade-off between two opposite requirements, depending on the system to be realized. Finally, since $\Delta\lambda_\nu$ does not entail any variation in the attenuation properties of the fiber, the smaller the value of $\Delta\lambda_\nu$ the better the performance of the transmission system.

For the case of a harmonic operator, expression (2.70) can be obtained in finite terms.[26] In Fig. 2.12 a typical result of this calculation is reported. It consists of the 3dB bandwidth *versus* the distance z, with A and h as parameters. For a step-index fiber, the intersection between the second and the third asymptotes occurs only for very large values of z, even when a LED source is employed, which is usually characterized by a large value of $\Delta\lambda_\nu$.

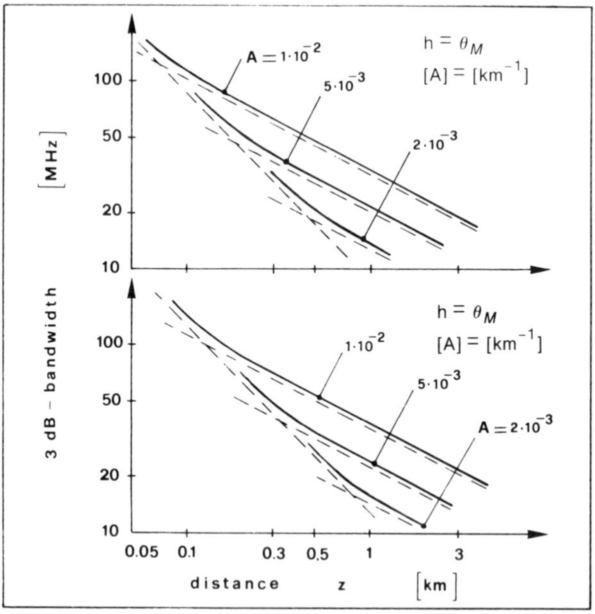

Fig. 2.12 - 3 dB bandwidth *vs.* z, with A and h as parameters, assuming $n_1 = 1.5$ and $\theta_M = \pi/20$ [from (26)].

As for the Bessel-type operator, in the simplifying assumption of a mode-independent coupling coefficient h_o, i.e., when $q = 0$ in (2.41), it is possible to obtain, through a numerical computation,[25] the following approximate expression for $E_1^{(2)}$

$$E_1^{(2)} = a(\alpha) \exp\left[-b(\alpha) \frac{h_o}{N^2}\right] \tag{2.72}$$

where $a(\alpha)$ and $b(\alpha)$ are positive quantities dependent on α. It is then confirmed that the steady-state bandwidth increases with increasing h_o, which is proportional to the amplitude of the fiber perturbation but depends also on the index gradient through the profile parameter α.*

2.3.5 Optimization of the Cabling Process

The excess attenuation at the steady-state over a distance z can be expressed through the quantity

$$G(z) = [\lambda_{o1}(h_o, \alpha) - \gamma]z$$

where h_o is proportional to the amplitude of the fiber perturbation and α is the profile parameter. Keeping in mind expression (2.52), which holds at the steady-state, the *rms* pulsewidth at the steady-state $\sigma_{ss}(z)$, for a very large distance z, can be written as

$$\sigma_{ss}(z) \cong \sqrt{s_2(h_o, \alpha)z}$$

$$s_2(h_o, \alpha) = \left.\frac{d\lambda_1}{d\omega}\right|_{\omega=0} = 2E_1^{(2)}$$

whereas for the uperturbed fiber we have

$$\sigma(z) = \sigma_o(\alpha)z$$

in which $\sigma_o(\alpha)$ has been reported in 1.3.6. Therefore, the reduction of time dispersion over a given distance z, due to the perturbation, can be taken into account through the quantity

$$R(z) = \frac{\sigma_{ss}(z)}{\sigma(z)} = \frac{1}{\sigma_o(\alpha)}\left[\frac{s_2(h_o, \alpha)}{z}\right]^{1/2}$$

The quality of a cabling process can be evaluated through the parameter [17]

$$V(h_o, \alpha) = R^2 G = \frac{s_2(h_o, \alpha)}{\sigma_0^2(\alpha)}\left[\lambda_{o1}(h_o, \alpha) - \gamma\right] \tag{2.73}$$

which is independent of z and must be as small as possible. V cannot be directly measured, but might be estimated from the experimental

*See notes pg. 497

evaluation of h_o and α. For example, considering a mode-independent coupling coefficient, from (2.56), we have

$$\lambda_{01}(h_o, \alpha) = \gamma + \frac{h_o}{N^2} Z_1^2(0) \tag{2.74}$$

and, from (2.72), it is possible to write

$$s_2(h_o, \alpha) = 2\,a(\alpha) \exp\left[-b(\alpha)\frac{h_o}{N^2}\right] \tag{2.75}$$

Substituting (2.74) and (2.75) into (2.73), we are left with

$$V(h_o, \alpha) = 2\,Z_1^2(0)\,\frac{a(\alpha)}{\sigma_o^2(\alpha)} \exp\left[-b(\alpha)\frac{h_o}{N^2}\right]\frac{h_o}{N^2}$$

Rigorously, N^2 should be function of α. Nevertheless, considering near-parabolic fibers, this dependence can be neglected as a first approximation, and in the above expression of V, only the explicit dependence on α can be taken into account. So, a family of curves is obtained, as reported in Fig. 2.13. All these curves have a maximum for

$$\frac{h_o}{N^2} = \frac{1}{b(\alpha)}$$

which may be reached in practice, and should be possibly avoided. The very high value of V obtained for $\alpha = 1.98$ is due to $\sigma_o^2(\alpha)$ which, in this situation, turns out to be very small. This means that a fiber with a very small time dispersion in the unperturbed case could be highly degraded when the steady-state is reached.

Now, we will consider that the cable manufacturer can control in some way the amplitude of the fiber perturbation introduced during the cabling process, though a reduction in this amplitude usually entails an increase in the cost of the production. So, from the knowledges of h_o/N^2 and α, it is possible to estimate whether the point which represents the behavior of the cabled fibers is on the left or on the right hand side of a maximum in a given curve $V(h_o/N^2, \alpha)$.[27] The properties of the fiber can be improved, reducing h_o in the first case, and paradoxically increasing h_o in the second case.

It is however of great importance to consider the transmission properties required for the fiber, depending on the transmission system to be realized. In fact, all the maxima in Fig. 2.13 occur for $h_o/N^2 \geq 0.2\,\text{km}^{-1}$, so, from (2.74), the correspondent excess attenuation coefficient $\lambda_{01} - \gamma$ is greater than 1 km^{-1}. Only very special fibers nowadays are likely to be requested with so high an excess attenuation coefficient, even if they exhibit a very small time dispersion at the steady-state. In other words, instead of a z-independent parameter $V(h_o/N^2, \alpha)$, two z-dependent parameters, $G(z)$ and $R(z)$, must be optimized, taking into account also the expected repeater spacing of the transmission system in which the fiber is to be employed. In Chapter 4 we will show that two limitations in the length of the repeater spacing exist, which are due to the attenuation and the bandwidth of the fiber respectively. Only for systems whose bandwidth limitation is very strong, an increase in h_o could be of interest.

Before concluding the present section, we will give typical values of the parameters taking into account the amplitude of the fiber perturbation, respectively A and h_o/N^2, for the harmonic and the Bessel-type operator. From (2.54), assuming an excess attenuation coefficient at the steady-state which ranges between 0.1 and 1 km^{-1}, being typically $\theta_M \cong \pi/20$, we have $10^{-3} \leq A \leq 10^{-2}$ km^{-1}. From (2.56), considering the same values for the excess attenuation coefficient at the steady-state, since $(1+q)^2 Z_l^2(\nu)$ ranges typically between 5 and 50, we have $2.10^{-3} \leq h_o/N^2 \leq 0.2$ km^{-1}. In particular, the higher values occur for q = 0.

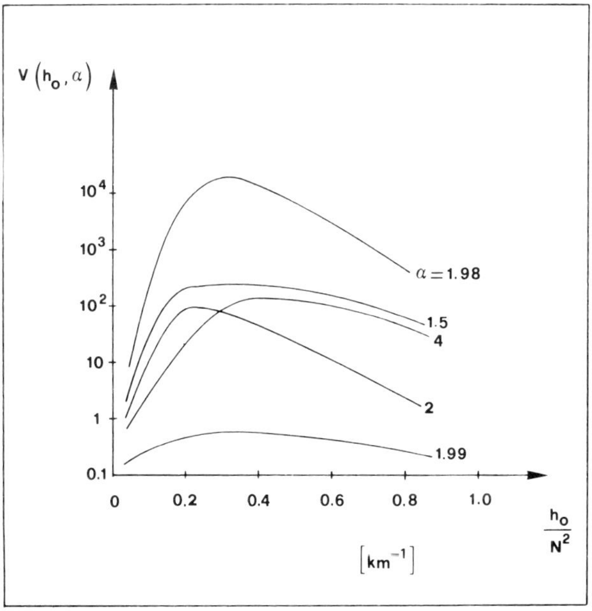

Fig. 2.13 - $V(h_o, \alpha)$ vs. h_o/N^2, for various α, with n_o = 1.5 and Δ = 0.01 [from (27)].

2.4 EFFECTS OF JOINTS

A joint between two fibers causes a variation in the optical power distribution among the guided modes or rays. Its main effect is the introduction of a power loss. The baseband response of the two cascaded fibers is also influenced by the joint. Finally, a joint, introducing a deterministic power exchange among the modes or rays, can alter the reaching of the steady-state in the sequence of two perturbed fibers. Rigorously, after each joint, a new process for the reaching of the steady-state should be considered.

Nowdays, most of the optical sources and detectors are supplied with fiber pigtails, so that only joints between fibers are to be made in a link. Two kinds of joints are usually considered:

i) demountable joints, or connectors

ii) permanent joints, or splices

Demountable joints are to be preferred at the transducers or at the terminations of a link, and permanent joints along the link, between two optical cables. In this sense, a splice is usually performed between two fibers having equal nominal parameters. Splices characterized by power losses as small as 0.1 dB may be realized on multimode fibers at present. Moreover, the mode conversion introduced is small, therefore their overall effect on the light propagation along a sequence of cascaded fibers is not dramatic.

Connectors, however, may cause a power loss up to 1 dB, and usually introduce a remarkable mode conversion. A particular phenomenon, which is related to the coherence properties of the optical source employed, and can be studied only through the electromagnetic theory of the light propagation, takes place in a connector. It consists of a nonlinear behavior of the connector with respect to the transmitted optical power, which leads to a degradation of the transmitted signal, called modal noise.

2.4.1 Splices and Connectors

A splice is realized by sealing the two fibers by fusion or by an epoxy resin. The alignment of the fibers is usually obtained by means of a capillary tube or of a plane surface with a *V* groove.[28] These techniques can be extended also to the case of several fibers (typically all the fibers supported by an optical cable) to be quickly joined.[29],[30]

The fiber alignment in conectors must be maintained through only a mechanical equilibrium. Because of the tolerances in this condition, repeated connections and disconnections show that typically, for a mean power loss of 0.8 dB, the *rms* deviation may be as large as 0.4 dB. The installation of connectors characterized by good performances may require many operations, and sometimes there is no convenience if they are manually adjusted and not self-aligning. In fact they need a remote monitoring of the transmitted power in order to obtain the maximum level of signal, optimizing a considerable number of mechanical movements. A partial improvement can be obained employing an adjustable connector with two rotatable eccentric sleeves containing the fibers,[(31)] where only the reciprocal rotation of the sleeves is required.

Fiber joints, including splices, but especially connectors, may be affected by:

i) differences between the parameters of the two fibers

ii) defects in the fiber end preparation

iii) intrinsic imperfections of the joint

Effects due to differences in diameters, numerical apertures, index profiles, and due to the ellipticity of the two cores cannot be reduced, even in adjustable joints. The ellipticity of the claddings and the eccentricity of the cores affect the joint when the lateral cladding surfaces are taken as a reference for the alignment.

The fiber and preparation is extremely important. A suitable cut has to be performed: it must be orthogonal and without surface imperfections or fractures, especially on the core section (see Appendix 3). Moreover, the fiber end face must be carefully cleaned to avoid the presence of dust or residual parts of the protective coating. Poor end preparations may entail power losses of some dB, even for splices.

The main intrinsic imperfections of the joint are:

i) axial separation

ii) lateral displacement (offset)

iii) angular misalignment (tilt)

iv) Fresnel reflection at the end faces

The lateral displacement turns out to be the worst imperfection, as it will be shown in the next section. As for the Fresnel reflection at the end faces, it can be practically avoided by the use of a suitable index matching oil.

2.4.2 Exact Electromagnetic Model of a Connector

Only the intrinsic imperfections of a connector will be taken into account here. Considering two M mode fibers with equal parameters, we will take into account the electric field only, but this is not a restrictive condition, because a perfectly dual description can be developed also for the magnetic field. Letting $E_m^{(1)}$ and $E_{m'}^{(2)}$ be the complex vectors representing the electric field of a given mode in the two fibers, the total electric fields in the two fibers can be written as

$$\overline{E}^{(1)}(t) = \sum_{m=1}^{M} A_m \exp(-i\, 2\, \pi\, \nu_\lambda\, t)\, \overline{E}_m^{(1)}$$
$$\overline{E}^{(2)}(t) = \sum_{m'=1}^{M} A_{m'} \exp(-i\, 2\, \pi\, \nu_\lambda\, t)\, \overline{E}_{m'}^{(2)} \tag{2.76}$$

where ν_λ is the optical frequency, A_m and $A_{m'}$ are complex coefficients which account for the excitation condition of modes m and m'.

In each fiber, the electric field $\overline{E}_m^{(i)}$ and the correspondent magnetic field $\overline{H}_m^{(i)}$ satisfy the orthonormality conditions

$$\int_S \overline{E}_m^{(i)} \times H_n^{(i)*} \cdot \hat{z}\, dS = \begin{matrix} 0 & (m \neq n) \\ 1 & (m = n) \end{matrix}, \; i=1,\, 2$$

where * denotes a complex conjugation, $\hat{z}$ is a unit vector along the fiber axis, finally S is the cross section of the fiber. Considering the total guided power in the two fibers, immediately before and after the connector, as a superimposition of the contributions from the different modes, it is

$$P^{(1)} = P_o \sum_{m=1}^{M} c_m$$
$$P^{(2)} = P_o \sum_{m'=1}^{M} c_m' \tag{2.77}$$

where P_o is a normalizing factor, and it is

$$c_m = |A_m|^2$$
$$c_{m'} = |A_{m'}|^2 \tag{2.78}$$

The determination of a formal relation between A_m and A_m' restricting ourselves to the case of lateral displacement only, is straightforward, and gives[32]

$$A_{m}' = \sum_{m=1}^{M} I_{mm'} A_{m}$$

$$I_{mm'} = \int_{S_c} \overline{E}_{m}^{(1)} \times \overline{H}_{m'}^{(2)*} \cdot \hat{z}\, dS$$

in which S_c is the cross section common to the two fibers. Thus, the expression of $c_{m'}$ as a function of c_m is not linear, in fact, from (2.78), it results in

$$c_{m'} = \left| \sum_{m=1}^{M} I_{mm'} \sqrt{c_m} \right|^2$$

A more general approach, not restricted to lateral displacement only, would give the same conclusion.

The rigorous electromagnetic approach to the study of a fiber connector shows the nonlinear behavior in power of the connector itself. So, a noticeable difficulty arises in the description of the signal degradation along an optical link. In particular, it is not possible to generalize the study of a single joint to that of a sequence of cascaded joints. Nevertheless, some conditions exist under which the electromagnetic model leads to a behavior approximately linear in power. Nonlinear effects may, however, be so large as to cause nonlinear distortion in the transmitted signal, degrading the performance of the optical link.

These effects are related to the coherence properties of the optical source.[33] In order to discuss this point, we will consider the excitation at the input section of the first fiber, whose length is z_l. Using the excitation coefficients $a_m \exp(i\,\phi_m)$, the first part of (2.76) must be replaced by

$$\overline{E}^{(1)}(t) = \sum_{m=1}^{M} a_m \exp(i\,\phi_m)$$

$$\exp[-i\,2\,\pi\,(\nu_\lambda - \nu_{\lambda o})\,\tau_m\, z_1]\exp(-i\,2\,\pi\,\nu_\lambda\, t)\,\overline{E}_m^{(1)}$$

where τ_m is the group delay per unit distance of the mth mode, and $\nu_{\lambda o}$ is an arbitrary central frequency. Note that $a_m = |A_m|$ is the amplitude, whereas ϕ_m denotes the phase at $\nu_\lambda = \nu_{\lambda o}$. Finally, unperturbed fibers have been assumed for the sake of simplicity.

The power loss of the connector, i.e., the ratio between total guided powers immediately before and after the connector, can be written as

$$L_o = \frac{\sum_{m=1}^{M} a_m^2}{\sum_{m'=1}^{M} |A_{m'}|^2}$$

Developing the calculations, we are left with

$$\frac{1}{L_o} = \frac{1}{\sum_{m=1}^{M} a_m^2} \sum_{m=1}^{M} \sum_{n=1}^{M} a_m a_n \cos [\, \phi_{mn} - 2\pi (\nu_\lambda - \nu_{\lambda o}) \tau_{mn} z_1 \,] F_{mn}$$

where

$$\phi_{mn} = \phi_m - \phi_n$$

$$\tau_{mn} = \tau_m - \tau_n$$

$$F_{mn} = \sum_{k=1}^{M} I_{mk} I_{nk}$$

The coefficients F_{mn} describe the power transfer between the first and the second fiber. Precisely, F_{mm} gives the fraction of the power guided by the mth mode of the first fiber, which is coupled with the second fiber. But F_{mn} $(m \neq n)$ represents an interference between the powers of the mth and of the nth modes of the first fiber, which is coupled with the second fiber. Large values of F_{mn} $(m \neq n)$ entail strong nonlinear distortion. Since this cause of signal degradation arises from the modal nature of the light propagation, it is called modal noise.

The coefficients I_{mk} depend on the optical frequency, precisely, on the difference $(\nu_\lambda - \nu_{\lambda o})$, i.e., on the spectral width of the optical source. A perfectly coherent source, which is characterized by the only optical frequency $\nu_{\lambda o}$, represents the worst case from the point of view of the nonlinearity.[33] This descreases with descreasing the coherence of the optical source, to the point where the behavior of the connector can be well approximated by a model linear in power. Moreover, the fiber length z_1, and also the differences in the group delays per unit distance, tend to destroy the coherence properties of the optical source.

Also the modulation bandwidth B of the transmitted signal influences the behavior of connectors. In fact, the spectral distribution of modal noise is always filtered in the receiver, over a modulation bandwidth which is strictly that of the useful signal. When the bandwidth (expressed in optical frequencies) of the optical source is large with respect to the modulation bandwidth of the signal, the connectors can be considered as approximately linear in power.[33]

In order to give an idea of the last condition, we consider a typical laser diode as an optical source whose central wavelength is $\lambda_o = 850$ nm, and has a spectral width $\Delta\lambda = 2$ nm. Its spectral bandwidth, expressed in optical frequencies, must be written as

$$\Delta \nu_\lambda = \frac{c}{\lambda_o - \frac{\Delta\lambda}{2}} - \frac{c}{\lambda_o + \frac{\Delta\lambda}{2}} \cong \nu_{\lambda o}\frac{\Delta\lambda}{\lambda_o}$$

where c is the free space velocity of light, and $\nu_{\lambda o} = c/\lambda_o$ is the central frequency of emission. It results $\Delta\nu_\lambda \cong 10^{12}$ Hz. Therefore, for such a laser, if $B < 10^9$ Hz, the above condition can be considered as verified. Using mode-locked lasers, the previous limit must be reduced by at least an order of magnitude.

2.4.3 Linear Model of a Connector

Recalling the expansion (2.77) for the optical guided power in the two fibers, if the connector is considered to produce a simple power exchange among the modes, it is possible to relate $c_{m'}$ and c_m through a linear equation of the type

$$c_{m'} = \sum_{m=1}^{M} K_{mm'}\, c_m$$

where $K_{mm'}$ are suitable coefficients. Therefore, from (2.77),

$$P^{(2)} = P_o \sum_{m'=1}^{M} \sum_{m=1}^{M} K_{mm'}\, c_m$$

The power loss of the connector is then

$$L_o = \frac{P^{(1)}}{P^{(2)}} = \frac{\sum_{m=1}^{M} c_m}{\sum_{m'=1}^{M} \sum_{m=1}^{M} K_{mm'}\, c_m}$$

Since the coefficents c_m do not vanish, one can conclude that L_o depends on the optical power distribution in the first fiber.

As regards the effect of a connector on the baseband response of the link of two cascaded fibers, letting z_1 and z_2 be their lengths, it is possible to write

$$\overline{H}^{(1+2)}(\omega, z_1 + z_2)$$

$$= \frac{\sum_{m'=1}^{M} \exp(-i\,\omega\,\tau_{m'} z_2) \sum_{m=1}^{M} K_{mm'}\, c_m \exp(-i\,\omega\,\tau_m z_1)}{\sum_{m'=1}^{M} \sum_{m=1}^{M} K_{mm'}\, c_m}$$

where ω is the modulation circular frequency. The baseband response of the link over the whole distance $z_1 + z_2$ without any connector (or with a perfect connector), can be written as

$$\overline{H}^{(1)}(\omega, z_1 + z_2) = \overline{H}^{(2)}(\omega, z_1 + z_2)$$

$$= \sum_{m=1}^{M} c_m \exp\left[-i\,\omega\,\tau_m (z_1 + z_2)\right] \Big/ \sum_{m=1}^{M} c_m$$

Comparing with the above expression of $\overline{H}^{(1+2)}(\omega, z_1 + z_2)$, the same baseband response is obtained only when

$$K_{mm'} = \delta_{mm'}$$

where $\delta_{mm'}$ denotes Kronecker's symbol.

The influence of the connector on the baseband response depends on the optical power distribution in the first fiber, as well as its power loss L_o. Note that, in the previous evaluation of the power decay and of the baseband response, only ideal fibers have been considered. In the next section, we will extend this model to the case of real fibers, and even to a sequence of many cascaded fibers. This is possible because a linear behavior of connectors, and more generally of joints, is assumed.

2.4.4 Matricial Approach to the Study of Cascaded Fibers

Taking into account an attenuation coefficient γ, a factor $\exp(-\gamma z_i)$, for $i = 1,2$, must be added between the input and the output powers guided by two unperturbed fibers. Letting $Q_m^{(i)}$ the optical power of the mth mode at the output of the ith fiber, it is related to the input optical power $W_m^{(i)}$ of the same mode by

$$Q_m^{(i)} = \exp(-\gamma z_i)\, W_m^{(i)}, \; i = 1,2$$

The power exchange at the connector can be expressed in matricial form, that is

$$\underline{W}^{(2)} = \underline{K}\, \underline{Q}^{(1)}$$

where $\underline{W}^{(2)}$ is a column-vector containing the $W_{m'}^{(2)}$, $\underline{Q}^{(1)}$ is a column-vector containing the $Q_m^{(1)}$, and finally, $\underline{K}$ is the $M \times M$ matrix of coefficients $k_{mm'}$. Considering the power decay along the two fibers, we have

$$\underline{Q}^{(2)} = \exp(-\gamma z_2)\, \underline{K} \exp(-\gamma z_1)\, \underline{W}^{(1)} = \exp(-\gamma z)\, \underline{K\ W^{(1)}}$$

where we have let $z = z_1 + z_2$.

Extending this model, n cascaded unperturbed fibers separated by $n - 1$ connectors (or more generally joints), from the point of view of the optical power decay, can be studied through the equation

$$\underline{Q}^{(n)} = \exp(-\gamma z)\, \underline{K}^{(n-1)}\, \underline{K}^{(n-2)} \ldots \underline{K}^{(1)}\, \underline{W}^{(1)}, \qquad (2.79)$$

where z is the overall length of the link, and $\underline{K}^{(1)}, \underline{K}^{(2)}, \ldots \underline{K}^{(n-1)}$ are the matrices of transmission coefficients of the connectors.

We will take into account now the time dispersion properties of connectors. Letting $\underline{Q}^{(i)}(\omega)$ be a column-vector containing the phasors which represent the modulated output powers of the modes of the ith fiber, it can be written

$$\underline{Q}^{(i)}(\omega) = \underline{D}(\omega, z_i)\, \underline{W}^{(i)}$$

where $\underline{D}(\omega, z_i)$ is a diagonal matrix of the type

$$\underline{D}(\omega, z_i) = \text{diag}\,[\exp(-i\omega\tau_1 z_i), \exp(-i\omega\tau_2 z_i), \ldots, \exp(-i\omega\tau_M z_i)]$$

In this way, the n cascaded fibers previously considered, from the point of view of the baseband response, can be studied through the equation

$$\underline{Q}^{(n)}(\omega) = \underline{D}(\omega, z_n)\, \underline{K}^{(n-1)}\, \underline{D}(\omega, z_{n-1}) \ldots \underline{K}^{(1)}\, \underline{D}(\omega, z_1)\, \underline{W}^{(1)} \qquad (2.80)$$

Precisely the baseband response of this optical link can be calculated as the ratio between

$$\sum_{m=l}^{M} Q_m^{[n]}(\omega) \qquad \text{and} \qquad \sum_{m=l}^{M} Q_m^{[n]}(0)$$

To summarize, equations (2.79) and (2.80) allow one to evaluate attenuation and time dispersion properties of a sequence of unperturbed cascaded fibers having equal single transmission properties. It is possible to consider also the case of perturbed fibers. In fact, equation (2.9) can be rewritten in matricial form as

$$\frac{d}{dz}(\underline{W}) = \underline{h}\, \underline{W} \qquad (2.81)$$

where $\underline{W}$ is a column-vector containing the powers W_m and $\underline{h}$ is a symmetric matrix containing the coupling coefficients h_{mn}, and, on its main diagonal, the intrinsic attenuation coefficients. This matrix $\underline{h}$ can be diagonalized, determining its eigenvalues and eigenvectors. Such a procedure leads to a set of statistical mode power distributions which are equivalent to the continuous statistical configurations obtained in 2.3.[34]

Letting $\lambda_1, \lambda_2, \ldots \lambda_M$ be the eigenvalues of $\underline{h}$, and $\underline{U}$ the matrix of its eigenvectors, one can write

$$\underline{Q} = \underline{U}^{-1}\,\underline{C}\,(z_i)\,\underline{U}\,\underline{W} \tag{2.82}$$

where $\underline{W}$ and $\underline{Q}$ represent input and output power distributions respectively, and we have let

$$\underline{C}\,(z_i) = \text{diag}\,[\,\exp\,(\,-\lambda_1 z_i\,),\ \exp\,(\,-\lambda_2 z_i\,),\ldots,\ \exp\,(\,-\lambda_M z_i\,)\,]$$

z_i being the fiber length considered. The extension to the case of n cascaded fibers can be made combining (2.79) and (2.82). So, with the usual formalism, we are left with

$$\underline{Q}^{(n)} = \underline{U}^{-1}\,\underline{C}\,(z_n)\,\underline{U}\underline{K}^{(n-1)}\,\underline{U}^{-1}\,\underline{C}\,(z_{n-1})\,\underline{U}\ldots\underline{K}^{(1)}\,\underline{U}^{-1}\,\underline{C}\,(z_1)\,\underline{U}\,\underline{W}^{(1)} \tag{2.83}$$

This procedure can also be applied to the time dependent case, considering a matrix $\underline{h}$ whose elements of the main diagonal are $(-\gamma_m - i\omega\tau_m)$. Developing a second order perturbative solution, analogous to that presented in 2.3.2, an equation similar to (2.83) can be derived, in which matrices $\underline{C}$ contain real and imaginary exponentials.

2.4.5 Power Loss of a Connector between Two Unperturbed Fibers

Till now only formal relations have been derived for the power loss of a connector. Nevertheless, exact expressions can be obtained from simple geometrical considerations, depending on the particular imperfections (errors) of the connector. In this section some of these expressions will be determined, assuming unperturbed fibers. In the next section, this analysis will be extended also to perturbed fibers.

The geometrical position of the receiving fiber (fiber 2), with respect to the launching fiber (fiber 1), imposes a transformation of the coordinates which are present, as independent variables, in the radiance distribution. From (1.47), the total optical power which is guided at the input section of fiber 2 can be written as

$$P^{(2)} = \int_0^{2\pi} d\,\phi_2 \int_0^{2\pi} d\,\theta_{\phi 2} \int_0^a r_2\,dr_2 \int_0^{\theta'_M} B^{(2)}\,(r_2,\ \phi_2,\ \theta_2',\ \theta_{\phi 2})\sin\theta_2'\cos\theta_2'\,d\,\theta_2' \tag{2.84}$$

while the total optical power which is guided at the output section of fiber 1 is

$$P^{(1)} = \int_0^{2\pi} d\,\phi_1 \int_0^{2\pi} d\,\theta_{\phi 1} \int_0^a r_1\,dr_1 \int_0^{\theta'_M} B^{(1)}\,(r_1,\ \phi_1,\ \theta_1',\ \theta_{\phi 1})\sin\theta_1'\cos\theta_1'\,d\,\theta_1' \tag{2.85}$$

where θ_1' and θ_2' are considered in a medium (air or index matching oil) whose refractive index is n'. A transformation of coordinates, which can be formally expressed as

$$(r_2, \phi_2, \theta_2', \theta_{\phi 2}) = \mathbf{J}\,(r_1, \phi_1, \theta_1', \theta_{\phi 1})$$

is responsible for the variation in the radiance distribution, which results in

$$B^{(2)}[(r_2, \phi_2, \theta_2', \theta_{\phi 2})] = B^{(1)}[\mathbf{J}\,(r_1, \phi_1, \theta_1', \theta_{\phi 1})]$$

Any Fresnel reflection at the end faces is ignored, considering that the excess Fresnel loss, which usually turns out to be about 0.35 dB, can be greatly reduced using a suitable index matching oil. If the two fibers are not identical, we must set a_1, θ_{M1}', in (2.84), and a_2, θ_{M2}' in (2.85). The power loss of the connector $L_o = P^{(1)}/P^{(2)}$ can be evaluated from the knowledge of the radiance distribution $B^{(1)}$ in the first fiber, and of the transformation $\mathbf{J}$, which depends on the axial separation Δz, on the lateral displacement Δx, and on the angular misalignment $\Delta\theta$. These three errors will be separately considered. The same index profile $n_1(r)$ is assumed for the two fibers. Finally, for the sake of simplicity, we will consider a uniform excitation in fiber 1, characterized by the constant radiance distribution B_o.[35]

When an axial seperation affects the connector, we have $\phi_1 = \phi_2 = \phi$, $\theta_1' = \theta_2' = \theta'$, $\theta_{\phi 1} = \theta_{\phi 2} = \theta_\phi$, and it results in

$$1 + \frac{\Delta z^2}{r_1^2 + r_2^2 \cos 2\theta_\phi - 2 r_2 \cos\theta_\phi \sqrt{r_1^2 - r_2^2 \sin^2\theta_\phi}} = \frac{1}{\sin^2[\theta_M'(r_1, \theta_\phi)]} \tag{2.86}$$

where

$$\sin^2[\theta_M'(r, \theta_\phi)] = \frac{1}{n'^2} \min\left\{ \frac{n_1^2(r) - n_1^2(a)}{1 - \left(\frac{r}{a}\right)^2 \sin^2\theta_\phi}, 1 \right\} \tag{2.87}$$

Through (2.86) r_1 and r_2 are related. Since this equation does not depend on θ' and ϕ, (2.84) is reduced to

$$P^{(2)} = B_o \int_0^{2\pi} d\,\theta_\phi \int_0^a \sin^2[\theta_M'(r, \theta_\phi)]\, r_2\, dr_2$$

where r is defined as

$$r = \begin{cases} r_2 & \sin^2[\theta_M'(r_2, \theta_\phi)] \le \sin^2[\theta_M'(r_1, \theta_\phi)] \\ r_1 & \sin^2[\theta_M'(r_2, \theta_\phi)] > \sin^2[\theta_M'(r_1, \theta_\phi)] \end{cases} \tag{2.88}$$

When a lateral displacement is present, we have $\phi_1 = \phi_2 = \phi$, $\theta_1' = \theta_2' = \theta$, $\theta_{\phi 1} = \theta_{\phi 2} = \theta_\phi$, and it is

$$r_1 = \sqrt{r_2^2 + \Delta x^2 - 2 r_2 \Delta x \cos \phi} \tag{2.89}$$

Expression (2.84) becomes

$$P^{(2)} = \frac{B_o}{2} \int_0^{2\pi} d\phi \int_0^{2\pi} d\theta_\phi \int_0^a \sin^2 [\theta_M'(r, \theta_\phi)] \, r_2 \, dr_2$$

where r must be chosen between r_1 and r_2 according to (2.87) and (2.88), which again hold.

For the case of an angular misalignment, neglecting the axial separation due to the misalignment itself, if the two fiber end faces are perpendicular to their own axes, we have $\phi_1 = \phi_2 = \phi$, $\theta_{\phi 1} = \theta_{\phi 2} = \theta_\phi$, $r_1 \cong r_2 = r$, and, letting $\gamma_\phi = \phi + \theta_\phi$,

$$\cos \theta_2' \cos \Delta \theta - \sin \gamma_\phi \sin \theta_2' \sin \Delta \theta = \sqrt{1 - \sin^2[\theta_M'(r, \theta_\phi)]} \tag{2.90}$$

where $\sin^2 [\theta_M' (r, \theta_\phi)]$ is given again by (2.87). Two solutions hold for (2.90), θ_{2a}', θ_{2b}', being $\theta_{2a}' \leq \theta_{2b}'$. Therefore, letting

$$s_a (r, \theta_\phi, \gamma_\phi) = \max \{ 0, \sin [\theta_{2a}' (r, \theta_\phi, \gamma_\phi)]\}$$

$$s_b (r, \theta_\phi, \gamma_\phi) = \min \{ \sin [\theta_M'(r, \theta_\phi), \sin [\theta_{2b}' (r, \theta_\phi, \gamma_\phi)]\}$$

we obtain

$$P^{(2)} = B_o \int_0^{2\pi} d\phi \int_{-\pi/2}^{+\pi/2} d \gamma_\phi \int_0^a [s_a^2 (r, \theta_\phi, \gamma_\phi) - s_b^2(r, \theta_\phi, \gamma_\phi)] \, r \, dr$$

Lateral displacement loss does not depend on the numerical aperture of the fiber, whereas, for axial separation, L_o increases with increasing NA; on the contrary, for angular misalignment, L_o decreases with increasing the NA. Letting $NA_a = \sqrt{n_1^2(0) - n_1^2(a)}$ be the axial numerical aperture, the above properties suggest the definition of the following normalized errors: $(\Delta z/a)(NA_a/n')$ for axial separation, $\Delta x/a$ for lateral displacement, $\Delta\theta(n'/NA_a)$ for angular misalignment, where a represents the core radius.

It is possible to demonstrate that, with the same normalized errors, lateral displacement represents the worst cause of power loss. Fig. 2.14 shows a typical behavior of L_o *versus* the normalized errors, for axial separation and lateral displacement. For small values of a given normalized error e, L_o can be approximated by

$$L_o (e) \cong 1 + C_e e$$

where C_e is a constant depending on the particular error and on the index profile considered. Under this approximation, $L_o(e)$ can be

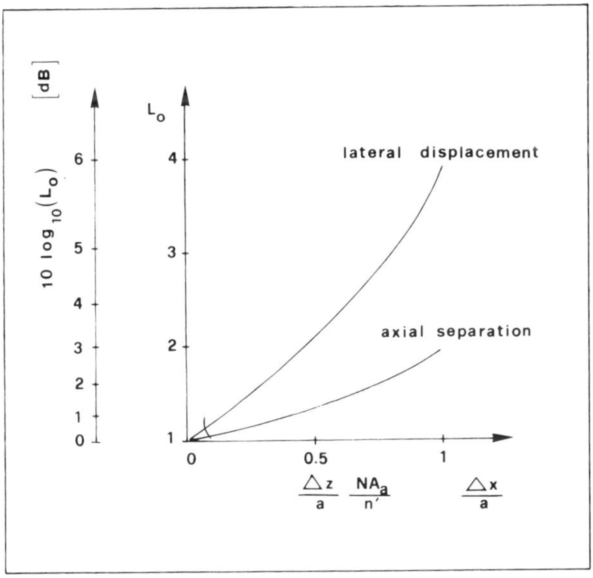

Fig. 2.14 - Power loss L_o *vs.* normalized errors for axial separation and lateral displacement

Table 2.3 - Approximate values of constant C_e [from (35)].

Cause of loss	*Step/Index fibers*	*Parabolic/index fiber*
axial separation	$\frac{4}{3\pi}$	$\frac{1}{2}$
lateral displacement	$\frac{2}{\pi}$	$\frac{8}{3\pi}$
angular misalignment	$\frac{2}{\pi}$	$\frac{8}{3\pi}$

obtained also in analytical form. Table 2.3 reports the expressions of the coefficients C_e, for step index and parabolic index fibers.[(35),(36),(37)] Though angular misalignment and lateral displacement have approximately equal values of C_e, the effects of the first one are usually much smaller, because, in practical cases, $\Delta\theta\,(n'/NA_a)$ is small with respect to $\Delta x/a$.

The superimposition of different errors in the same connector leads to different effects depending on the particular index profile considered. For instance, in (35) it has been shown that lateral displacement and angular misalignment, on the same plane, add linearly for step index fibers, and quadratically for parabolic index fibers.

2.4.6 Power Loss of a Connector between Two Perturbed Fibers

The reaching of the steady-state along a perturbed fiber, in the sense introduced in 2.2, is influenced by the presence of a connector between two fiber pieces. In order to simplify the following discussion, we will restrict ourselves to the case of a lateral displacement between two fibers characterized by equal parameters. Being $W_o^{(1)}(E, h)$ the optical power distribution in domain (E, h), at the output section of fiber 1, the correspondent total optical guided power, from (1.48), can be written as

$$P^{(1)} = \int_G W_o^{(1)}(E_1, h_1)\, dE_1\, dh_1$$
$$E_1 = n_1(r_1)\cos\theta \tag{2.91}$$
$$h_1 = n_1(r_1)\, r_1 \cos\theta \sin\theta_\phi$$

where 1 denotes the variables in fiber 1. The integral has been extended to the domain G of only guided rays, because we consider here a certain length of fiber, along which the contribution of leaky rays is practically negligible.

A lateral displacement, as shown in the previous section, leads to a variation in the variable r only, which is given by (2.89). So, the new pair of constants of the motion, which hold in fiber 2, result in

$$E_2 = n_1(r_2)\cos\theta$$
$$h_2 = n_1(r_2)\, r_2 \cos\theta \sin\theta_\phi$$

Comparing with the above definitions of E_1 and h_1, we have

$$E_2 = \frac{n_1(r_2)}{n_1(r_1)} E_1 \tag{2.92}$$
$$h_2 = \frac{n_1(r_2)\, r_2}{n_1(r_1)\, r_1} h_1$$

Letting $W_i^{(2)}(E_2, h_2)$ the optical power distribution at the input section of fiber 2,

$$P^{(2)} = \int_G W_i^{(2)}(E_2, h_2)\, dE_2\, dh_2 \tag{2.93}$$

In this case the power loss of the connector cannot be expressed simply through the ratio between (2.91) and (2.93), but it depends on the distance z at which the transmitted powers, with and without the connector, are compared. Letting z_1 and z_2 be the lengths of the two fibers, we can set $z = z_1 + z_2$. So we have

$$W_o^{(1)}(E_1, h_1) = \sum_n c_n^{(1)} Q_n(E_1, h_1) \exp(-\lambda_n z_1)$$

$$c_n^{(1)} = \int_G W_i^{(1)} (E_1, h_1) Q_n (E_1, h_1) dE_1 dh_1$$

where Q_n is the nth two variable statistical distribution, λ_n is its attenuation coefficient, and c_n are constants depending on the launching condition, expressed through the optical power distribution $W_i^{(1)}(E_1, h_1)$ at the input of fiber 1. The optical power distribution at the output section of fiber 2, in similar way, can be written as

$$W_o^{(2)} (E_2, h_2) = \sum_n c_n^{(2)} Q_n (E_2, h_2) \exp(-\lambda_n z_2)$$

$$c_n^{(2)} = \int_G W_i^{(2)} (E_2, h_2) Q_n (E_2, h_2) dE_2 dh_2$$

Therefore, in the presence of the connector, at the output section of fiber 2, we have

$$P^{(2)} (z_1, z_2) = \sum_n c_n^{(2)} \exp(-\lambda_n z_2)] \int_G Q_n (E_2, h_2) dE_2 dh_2$$

where the dependence on z_1 is in the coefficients $c_n^{(2)}$. On the contrary, without the connector, at the output section of the fiber 1, we have

$$P^{(1)} (z_1 + z_2) = \sum_n c_n^{(1)} \exp[-\lambda_n (z_1 + z_2)] \int_G Q_n (E_1, h_1) dE_1 dh_1$$

The power loss of the connector is then defined as

$$L_o (z_1, z_2) = \frac{P^{(1)} (z_1 + z_2)}{P^{(2)} (z_1, z_2)}$$

Nevertheless, since z_1 measures how $W_i^{(1)}(E_1, h_1)$ depends on $W_o^{(1)}(E_1, h_1)$, when z_1 is so high to entail the steady-state in fiber 1, L_o depends only on z_2. Generally, L_o is a function of the lateral displacement, of the optical power distribution at the output section of fiber 1, of the perturbation which affects fiber 2, and of the distance z_2. Comparing $L_o(z_1, z_2)$, as defined here, with the power loss L_o for unperturbed fibers discussed in the previous section, $L_o(z_1, z_2)$ may be either greater or smaller than L_o, depending on the optical power distribution for which L_o has been calculated.[38] In particular, considering the steady-state distribution, we have $L_o(z_1, z_2) < L_o$.

The results obtained in this section are based on a launching condition independent of ϕ in both fibers. This is not rigorously true for fiber 2, owing to the lateral displacement itself, which occurs along a given direction x of the transverse plane. Nevertheless, after a few meters, the propagation will certainly lose the memory of this direction, and the radiance distribution can be considered as effectively independent of ϕ. Assuming that it is also independent of θ_ϕ, it is possible to eliminate variable h, and to study the perturbed fibers through a single-variable model, like those discussed in 2.2 and 2.3.

2.4.7 Outline of Connector Time Dispersion Properties

Considering a connector characterized by a lateral displacement Δx between two equal unperturbed fibers, we will discuss here the modifications in the time dispersion properties which it introduces in the link. Assuming for simplicity that the radiance distribution in fiber 2, after a few meters, can be considered as independent of ϕ and θ_ϕ, one can express $B^{(2)}$ as a function of r_2 and θ only, or equivalently of the single constant of the motion E_2. The relation between E_1 and E_2 is given by the first of (2.92). Finally, considering α profile fibers, we can obtain also the group delay τ of each ray as a function of E only.

Performing the numerical calculation of the *rms* pulsewidth σ of the two cascaded fibers, it is possible to obtain a function of the type $\sigma(\Delta x/a, \alpha, z_1, z_2)$. Fixing a reference distance $z = z_1 + z_2$ equal to 1 km, we can compare $\sigma(\Delta x/a, \alpha, z_1/z)$ so obtained with the *rms* pulse width per km of a single fiber $\sigma_0(\alpha)$, given by (1.91). This comparison shows a reduction of time dispersion, more and more evident with increasing $\Delta x/a$, except for a narrow interval of α-values close to 2-4Δ.[39] This behavior can be easily explained, replacing variable E with variable ϵ through (1.75), as follows.

The lateral displacement entails a conversion of rays having ϵ_1 close to Δ, in part into rays having $\epsilon_2 < \epsilon_1$, and in part into rays having $\epsilon_2 > \Delta$ (leaky and radiating rays). This process can lead to an increase in the time dispersion, when $\tau(\epsilon)$ is slowly varying in the neighborhood of $\epsilon = \Delta$. This occurs for $\alpha \cong 2 - 4\Delta$, as shown by Fig. 1.17. In such a case, in fact, the optical power which is exchanged (and in part becomes lost for the propagation) does not belong to a tail of the single fiber impulse response, but its peak.

In conclusion, we can state that generally fiber joints produce a reduction in the time dispersion. In fact, even for a nominal value of α close to the optimum for single fibers, i.e., very near to $2 - 4\Delta$, the presence of many cascaded fibers, and the differences among their actual profiles, contribute to the relaxation of the critical situation discussed above. In this sense, connectors and splices, which operate a concentrated mode coupling, act as the fiber perturbation, leading to an increase of the attenuation and to a reduction of the time dispersion. But the main advantage in their presence is in the contribution to make the baseband response of the whole link more equalizable, and less critically dependent on index profile and launching condition.

REFERENCES

1. Di Vita, P., and Vannucci, R. "Geometrical Theory of Light Scattering in Dielectric Optical Waveguides," *Alta Freq.*, 43, pp. 789-794 (1974).
2. Urbach, F. "The Long Wavelength Edge of Photographic Sensitivity and of the Electronic Absorption of Solids," *Phys. Rev.*, 92, pp. 1324-1324 (1953).
3. Keck, D.B., Maurer, R.D., and Schultz, P.C. "On the Ultimate Lower Limit of Attenuation in Glass Optical Waveguides," *Appl. Phys. Lett.*, 22, pp. 307-309 (1973).
4. Maurer, R.D. "Glass Fibers for Optical Communications," *Proc. IEEE,* 61, pp. 452-462 (1973).
5. Newns, G.R., *et al*. "Absorption Losses in Glasses and Glass Fiber Waveguides," *Optoelectron.*, 5, pp. 289-296 (1973).
6. Irven, J. "Long Wavelength Performance of $SiO_2/GeO_2/P_2O_5$ Core Fibers with Different P_2O_5 Levels," *Electron. Lett.*, 17, pp. 2-3 (1981).
7. Takahashi, H., Sugimoto, I., and Sato, T. "Germanium Oxide Glass Optical Fiber Prepared by VAD Method," *Electron. Lett.*, 18, pp. 398-399 (1982).
8. Pinnow, D.A., Rich, T.C., Ostermayer, F.W., and Di Domenico M. "Fundamental Optical Attenuation Limits in the Liquid and Glassy State with Application to Fiber Optical Waveguide Materials," *Appl. Phys. Lett.*, 22, pp. 527-529 (1973).
9. Hannay, S.H. "Mode Coupling in an Elastically Deformed Optical Fiber," *Electron. Lett.*, 12, pp. 173-174 (1976).
10. Marcuse, D. *Theory of Dielectric Optical Waveguides,* New York: Academic Press (1974).
11. Cancellieri, G. "Mode Coupling in Graded Index Optical Fibers due to Perturbation of the Index Profile," *Appl. Phys.*, 23, pp. 99-105 (1980).
12. Rousseau, M., and Arnaud, J. "Ray Theory of Microbending," *Opt. Comm.*, 25, pp. 333-336 (1978).

13. Papoulis, A. *Probability, Random Variables, and Stochastic Processes,* New York: McGraw-Hill (1965).
14. Stratonovich, R.L. *Topics in the Theory of Random Noise,* Vol. 1, New York: Gordon & Beach (1963).
15. Olshansky, R., and Nolan, D.A. "Mode Dependent Attenuation of Optical Fibers: Excess Loss," *Appl. Opt.,* 15, pp. 1045-1047 (1975).
16. Cancellieri, G. "Mode Coupling in Graded Index Optical Fibers Due to Microbending," *Appl. Phys.,* 26, pp. 51-57 (1981).
17. Olshansky, R. "Mode Coupling Effects in Graded Index Optical Fibers," *Appl. Opt.,* 14, pp. 935-945 (1975).
18. Marcuse, D. "Losses and Impulse Response of a Parabolic Index Fiber with Random Bends," *Bell Syst. Tech. J.,* 52, pp. 1423-1437 (1973).
19. Rousseau, M., and Arnaud, J. "Ray Theory of the Impulse Response of Randomly Bent Multimode Fibers," *Opt. & Quant. Electron.,* 10, pp. 53-60 (1978).
20. Arnaud, J., and Laberre-Rousseau, M. "Theorie de la Propagation dans les Fibres Multimodales Deformees Aleatoirment. I. Theorie Generale," *Ann. Telec.,* 35, pp. 61-73 (1980).
21. de Fornel, F. "Role of Manadjacent Mode Coupling in the Theory of Microbending," *Opt. Comm.,* 28, pp. 55-58 (1979).
22. Eve, M., and Hannay, J.H. "Ray Theory and Random Mode Coupling in an Optical Fiber Waveguide: I," *Opt. & Quant. Electron.,* 8, pp. 503-508 (1976).
23. Morse, P.M., and Feshback, H. *Methods of Theoretical Physics,* New York: McGraw-Hill (1953).
24. Gloge, D. "Optical Power Flow in Multimode Fibers," *Bell Syst. Tech. J.,* 51, pp. 1767-1783 (1972).
25. Cancellieri, G., and Fantini, P. "Frequency Dependent Steady-State Distribution of Optical Power in Multimode Fibers," *Opt. & Quant. Electron.,* 13, pp. 229-239 (1981).
26. Cancellieri, G., and Fantini, P. "Mode Coupling Effects in Optical Fibers: Perturbative Solution of the Time Dependent Power Flow Equation," *Opt. & Quant. Electron.,* 15, pp. 119-136 (1983).
27. Cancellieri, G., and Fantini, P. "Optical Fiber Cabling Proc-

esses: Discussion of an Optimization Procedure," *Appl. Opt.*, 25, pp. 2651-2652 (1982).

28. Someda, C.G. "Simple, Low-Loss Joints between Single Mode Optical Fibers," *Bell Syst. Tech. J.*, 52, pp. 583-596 (1973).
29. Schroeder, C.M. "Accurate Silicon Spacer Chips for an Optical fiber Cable Connector," *Bell Syst. Tech. J.*, 57, pp. 91-97 (1978).
30. Le Noane, G., and de Vecchis, M. "Further Developments on Compartmented Fiber Optic Cable Structure and Associated Splicing Technique," IOOC 77, Tokyo, Japan, pp. 355-358 (1977).
31. Boerner, M., and Maslowski, S. "Single Mode Transmission Systems for Civil Telecommunication," "Nonlinear Distortions and Noise in Optical Communication Systems Due to Fiber Connectors," *IEEE J. Quant. Electron.*, QE-16, pp. 761-770 (year).
32. Petermann, K. "Nonlinear Distortions and Noise in Optical Communication Systems due to Fiber Connectors," *IEEE J. Quant. Electron.*, QE-16, pp. 761-770.
33. Personick, S.D. "Baseband Linearity and Equalization in Fiber Optic Digital Communication Systems," *Bell Syst. Tech. J.*, 52, pp. 1175-1194 (1973).
34. Unger, H.G. "Modes of Propagation in Optical Fibers," *Ann. Telec.*, 32, pp. 101-108 (1977).
35. Di Vita, P., and Rossi, U. "Theory of Power Coupling between Multimode Optical Fibers," *Opt. & Quant. Electron.*, pp. 107-117 (1978).
36. Gloge, D. "Offset and Tilt Loss in Optical Fiber Splices," *Bell Syst. Tech. J.*, 55, pp. 905-916 (1976).
37. Millar, C.M. "Transmission *vs.* Transverse Offset for Parabolic Profile Fiber Splices with Unequal Core Diameters," *Bell Syst. Tech. J.*, 55, pp. 917-927.
38. Di Vita, P., and Rossi, U. "Realistic Evaluation of Coupling Loss between Different Optical Fibers," *J. Opt. Comm.*, 1, pp. 26-32 (1980).
39. Cancellieri, G., and Fantini, P. "Time Dispersion Effects of Optical Fiber Joints," *Opt. Comm.*, 42, pp. 30-35 (1983).

CHAPTER 3
TRANSDUCERS

Optical communications are characterized by the requirement of transducers for the transformation of electric into optical signals and *vica versa*. So, we have electro-optical transducers and opto-electric transducers. An electro-optical transducer consists of an optical source whose light can be modulated (usually intensity modulated) by an electric signal. An opto-electric transducer is a photodetector whose electric output reproduces the intensity modulation of the optical power, which incides on its active surface.

The main requirement that any transducer must satisfy is the linearity of response. Errors in the linearity, like second or third order distortions, must be as small as possible to reduce the noise affecting transmission. A large baseband response is useful, especially when a high capacity signal has to be transmitted. Finally, a small intrinsic noise, particularly for photodetectors, is another desirable quality of transduction.

As for the spatial dimension of the emitting or receiving surfaces, we can state that, to simplify coupling with optical fiber ends, small area optical sources and large area photodetectors are preferable. To simplify coupling operations, the production of transducers with a short piece of fiber (pigtail) directly attached to the front of their emitting or receiving surfaces is very diffused. This way, only joints between fibers are made in a link.

3.1 OPTICAL SOURCES

The only optical sources of interest for transmission systems are:

i) light emitting diodes (LEDs)

ii) laser diodes

This is mainly due to the easy modulation of such optical sources by simply varying the injection current. For metrological purposes, sometimes a more collimated beam should be launched into the fiber under test. In this case, a greater amount of optical power can become guided power. This is usually done by certain types of lasers, which are not of the diode type. They will be called collimated beam lasers in the following paragraphs.

The main characteristic of an optical source is the radiance, defined as density of power per unit area and unit solid angle, as introduced in 1.2. The greater the radiance, the easier the coupling procedure between source and fiber. This is particularly true for monomode fibers. If an optical source is characterized by a beam with a very small divergence, its radiance is high and it does not decay rapidly with increasing distance. A large emitting area is necessary for attaining a very small beam divergence, owing to diffraction limits. This is impossible for laser diodes, but other types of lasers can be characterized by this property.

3.1.1 Collimated Beam Lasers

There are three main kinds of collimated beam lasers used in fiber characterization:

i) gas lasers

ii) crystalline ion lasers

iii) organic dye lasers

Any type of laser is based on a stimulated emission of light, caused by an external source of energy.[1] The process of energy transfer from this external source to the atoms or molecules of the active material of the laser is called pumping. Two main types of pumps exist:

i) optical pump, based on the light of a primary optical source

ii) electrical pump, consisting of exciting the active material by an electric current

The absorption bands in gas lasers are so thin that an optical pump is not usually convenient, therefore an electric pump is preferred. The atoms, ions, or molecules of the gas pass to an excited state, owing to a direct collision with an electron, or another atom, ion or molecule. We can have:

i) neutral atom gases, e.g., He-Ne, Kr, Ar, Xe

ii) ion gases, e.g., Ar^+, Kr^+

iii) molecule gases, e.g., CO_2

The most diffused gas lasers in fiber characterization are He-Ne lasers, argon ion lasers, and krypton ion lasers. He-Ne lasers can emit at three main wavelengths: 0.633 μm, 1.15 μm, and 3.39 μm.

Argon ion lasers are characterized by several lines in the green and blue regions of the visible spectrum, while krypton ion lasers emit in the red region.

Crystalline ion lasers are obtained by an optical pump employing a high power lamp in CW operation,.or in pulsed behavior. The active material consists of transition ions, e.g., Cr^{3+} (ruby laser) or Nd^{3+}- (neodymium laser), which are diffused as impurities in other materials. Ruby lasers are made of Al_2O_3 crystals, in which Cr^{3+} ions are diffused. Neodymium lasers can be divided into two types:

i) Nd: YAG lasers, based on $Y_3Al_5O_{12}$ crystals in which Nd^{3+} ions are diffused

ii) Nd: glass lasers, which consist of glassy materials in which Nd^{3+} ions are diffused.

The small thermal conductivity of the glass material entails that a Nd: glass laser can be used only in pulsed behavior. The main lines of ruby lasers are 0.694 μm, 0.55 μm, 0.42 μm. Neodymium lasers emit at λ=1.06 μm, but the wavelength λ=0.53 μm can be obtained by frequency doubling. An organic dye laser is based on organic molecules diffused in a liquid solution. An optical pump is employed, and the emission wavelength can be tuned within an interval of several nm. Using different organic dyes, it is possible to cover the whole visible spectrum.

With regard to the level of emitted power, we must distinguish between CW operation and pulsed behavior. In the CW condition, ruby lasers and dye lasers are characterized by maximum emitted powers on the order of 1 W, krypton ion and argon ion lasers can reach emitted powers of about 100 W, and Nd: YAG lasers ususlly exceed 100 W. In pulsed behavior, the peak optical power which can be obtained is greater by several orders of magnitude. He-Ne lasers are characterized by very small emitted powers in CW operation (a few mW); nevertheless, they are employed for fiber testing, due to their low cost and small size, to the point where even field measurements are performed with such laser sources. He-Ne lasers are not practical for use in pulsed behavior. If they were required, an external electro-optical modulator, e.g., a Pockel's cell (see Appendix 1), could be employed.

Collimated beam lasers have a radiance, whose value can be extremely high, which is slowly decreasing along the beam. For instance, a He-Ne laser, characterized by 1 mW of CW emitted power, with a beam diameter of about 0.8 mm and a typical beam divergence of 1 mrad, exhibits a radiance of the order of $5 \cdot 10^4$ W/sr·cm². A Nd: YAG laser, characterized by 100 W of CW emitted power, with a beam diameter of about 2 mm, and a typical beam divergence of 2 mrad, exhibits a radiance as large as $2 \cdot 10^7$ W/sr·cm².

The spectral widths of laser lines are usually very small (fractions of Å). It is more convenient to consider the corresponding optical frequency spectrum, whose width $\Delta\nu_\lambda$ is related to the width of the wavelength line $\Delta\lambda$ by

$$\Delta\nu_\lambda = \frac{c}{\lambda^2}\Delta\lambda$$

To reduce its spectral width, a laser can be forced to run on a single mode. This is done by properly adjusting the laser cavity.

The temporal coherence, Δt, of a radiation is related to the spectral width, $\Delta\nu_\lambda$ of the optical source which generates it by the approximate equation

$$\Delta t \cdot \Delta\nu_\lambda \cong 1$$

which expresses an indetermination principle. Therefore, the greater the temporal coherence, the smaller the spectral width. For example, a ruby laser in single mode operation exhibits a Δt on the order of 10^{-8} s, leading to a $\Delta\nu_\lambda$ of about 100 MHz. Gas lasers in single mode operation can have a Δt as long as 10^{-4} s, leading to a $\Delta\nu_\lambda$ of about 10 kHz. Spectral widths that thin are very interesting to use when performing fiber time dispersion measurements that are unaffected by material dispersion. The spatial coherence, on the other hand, is not related to the spectral properties of the light source. However, collimated beam lasers are characterized by a remarkable degree of spatial coherence. An ideal laser beam exhibits an intensity distribution (irradiance) of the type

$$I(r) = I_o \exp\left(-2\frac{r^2}{w^2}\right)$$

where r is the radial coordinate that measures the distance from the beam axis, I_o is a constant and w is a parameter usually called spot diameter. The main properties, which make this intensity distribution very attractive for metrological purpose, are

i) near and far field patterns having equal forms

ii) absence of phase shift over near and far field

iii) possibility of focusing the beam into a spot, whose diameter is approximately λ/θ, where θ is the angle of focalization

Yet, a too high degree of temporal and spatial coherence may lead to speckle patterns,[2] which give rise to modal noise, as discussed in 2.4.2. In metrological use, it is sometimes convenient to destroy the spatial coherence of a laser light by the insertion of a thin ground glass in the laser beam before it enters the fiber.

In pulsed behavior, the most interesting property of the optical signal is its pulsewidth. In fact, the smaller the pulsewidth, the better the resolution in transit time measurements. Two main techniques are employed to obtain very short pulses:[3]

i) Q switching, which is based on preventing laser action during a time interval that is short with respect to the relaxation time of excited states

ii) mode locking, where a train of pulses arises from a Fourier superimposition of a large number of independent modes, which are forced to run with an exact frequency and phase relationship.

Both these techniques can be realized in practice by inserting another active or passive medium in the laser cavity. An active medium, for example, is an electro-optical or an acousto-optical modulator. A passive medium generally consists of a saturable absorber.

Mode locking gives shorter pulsewidths. However, the following relationship holds between the pulse width ΔT and the bandwidth Δf of its spectrum

$$\Delta f \cdot \Delta T = K$$

where K is a constant depending on the pulse shape. Since Δf cannot exceed the linewidth $\Delta\nu_\lambda$ of the laser, it is possible to conclude that lasers having a large spectral width can exhibit very short pulses in mode locking operation. Organic dye lasers, characterized by the largest $\Delta\nu_\lambda$ (several GHz), have yielded the shortest pulses to date: a ΔT as small as 0.1 ps.

3.1.2 Top Emitting LEDs

When an external perturbation, e.g., an applied voltage, gives rise to electron hole pairs, their recombination may produce a radiation process called spontaneous emission. Letting E_g equal the band gap energy of the semiconductor, we can have two types of radiative transitions:[4]

i) from band to band: characterized by a spectrum of optical frequencies, ν_λ, sharply peaked around the value E_g/h, where h is Planck's constant;

i) from one band to the impurity states close to the other band, which are characterized by $\nu_\lambda < E_g/h$

The emission spectrum of a given device can be explained by considering the energy produced by a recombination is sometimes divided into two parts: the first results in a photon with less energy than E_g, the second produces a phonon, and therefore does not lead to any radiation emission.

Since the recombination process has a random nature, the light which is obtained by spontaneous emission is not coherent. Quantum efficiency is the fraction of radiative recombinations with respect to all the recombinations. It decreases with increasing temperature.

A LED is a semiconductor device based on a p-n junction, which is designed specifically for the production of radiation by spontaneous emission. This radiation can be intensity modulated by varying the voltage applied to the p-n junction. A very simple LED structure is shown in Fig. 3.1. It is characterized by point contacts on its upper surface. From all points on such a surface which are not covered by

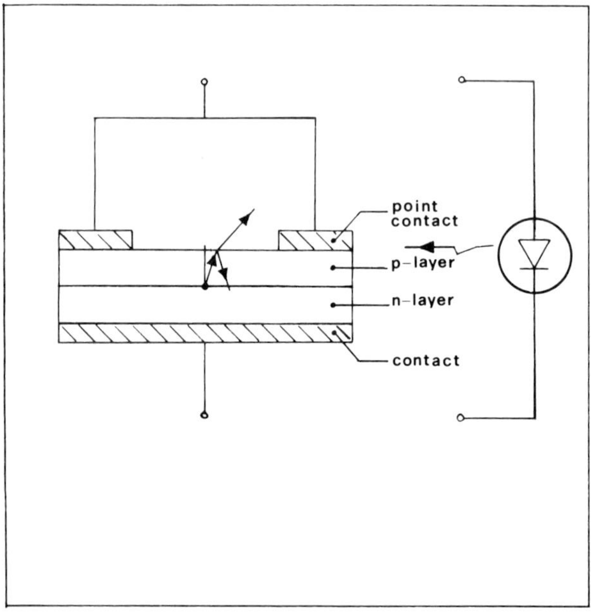

Fig. 3.1 - Conceptual scheme of a LED structure.

the contacts, radiation is emitted. A LED of this type is called top emitting.

The efficiency of external coupling, which is the fraction of optical power passing to the air at a given produced radiation, is limited by reflection at the interface and by absorption in the upper layer. To enhance the transmission at the interface, and in particular, to improve coupling efficiency with optical fibers, different geometries have been studied. The most suitable upper surface consists of a hemispherical surface. Another way to increase coupling efficiency is to employ a self-aligned sphere lens.[5]

Too large a spatial extension of the lower contact, which entails a limitation in the current density across the junction is the major limiting factor of radiance from this kind of device. In other words, the optical power is produced over a too extended region of the junction plane.

Burrus[6] proposed a structure with circular symmetry, whose scheme is shown in Fig. 3.2. It was characterized by a small lower contact and by an annular upper contact. Optimizing widths and dopant concentrations of the two layers, a value of radiance as large as 10 W/sr·cm^2 was obtained, which is very high when compared

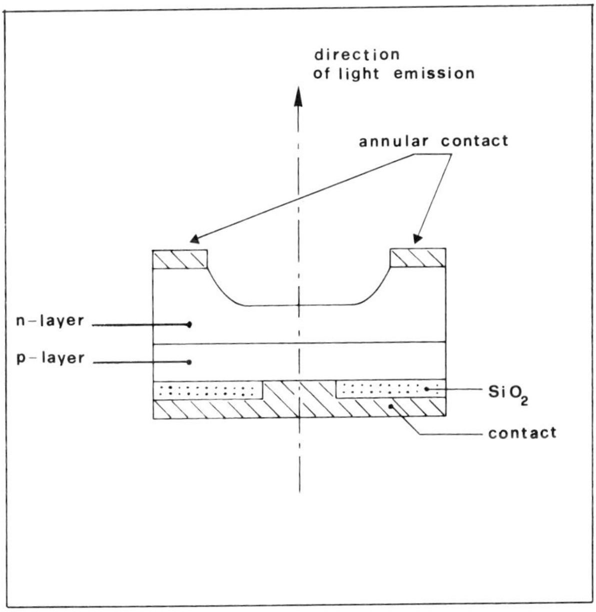

Fig. 3.2 - Burrus-type LED.

with the radiance of traditional top emitting LEDs. The LED structure proposed by Burrus was called a high radiance LED.

The far field pattern of this LED has a circular symmetry around the axis of the structure; mover, it is approximately uniform within a cone whose aperture is comparable with the *NA* of typical fibers. Therefore, a LED of this type can be used efficiently when a uniform excitation of all the guided modes is required, e.g., for attenuation measurements under a uniform launching condition or for index profiling with the near field technique.

The use of heterojunctions, i.e., pairs of semiconductor layers having different band gap energies, allows one to reach an even greater value of radiance.[7] A LED structure of this type is usually made of three layers: n-GaAs, p-GaAs, p-GaAlAs. The heterojunction between the two p-layers is characterized by a potential barrier which repels the injected electrons, preventing them from a nonradiative recombination at the p-contact.[8] Since an excellent lattice match can be obtained between the two materials, the interface turns out to be nearly free of defects which could act as nonradiative recombination centers, assuring a high efficiency in the light generation. A LED of this type, which is based on a single heterostructure, can exhibit a radiance as large as 70 $W/sr \cdot cm^2$.[7] Adding another heterojunction, one obtains a LED structure made of four or five layers, called a double heterostructure. In this case, a better optimization in the choice or the materials can be performed, to the point where a further increase of the radiance can be obtained. In (7) a radiance up to 100 $W/sr \cdot cm^2$ was measured on an optimized double heterostructure Burrus-type LED.

The main limit to achieving higher radiance values, for all the LEDs discussed here, is due to the current density across the junction, and consequently to the increase of temperature. This increase reduces the quantum efficiency and can also damage the device. A current density of about $1.5 \cdot 10^4$ A/cm^2 represents a typical value for devices with contact diameters of 50 μm. With a smaller contact, an increase in the current density and the radiance value, all other construction parameters remaining fixed, can be achieved. For instance, passing from a 50 μm contact diameter to a 13 μm contact diameter, the current density increases to about $7.5 \cdot 10^4$ A/cm^2, and the radiance turns out to be multiplied by approximately a factor of three.[7]

Other top emitting devices have recently been proposed.[9,10] Characterized by a broad lower contact, they have an annular upper contact, whose outer diameter is smaller than that of the lower contact.

In these structures, the current density is much smaller that in Burrus-type LEDs. They are designed to obtain a trade-off between a high radiance (however, always smaller than that of a Burrus-type LED) and a high speed response.

3.1.3 Edge Emitting LEDs

In order to obtain a greater radiance value, a quite different LED structure must be employed. It can be considered a modified laser diode structure, and consists of several layers, usually formating a pair of heterojunctions. A heterojunction, as introduced in the previous section, is realized by joining two semiconductors with different band gap energies. When the two layers have the same conductivity type, e.g., n-n or p-p, an isotype heterojunction is obtained. When they have different conductivity types, i.e., n-p, we are with an anisotype heterojunction.

The main properties of a heterojunction can be listed as follows: [(11)]

i) anisotype heterojunctions increase the carrier injection efficiency from the higher bandgap region

ii) isotype heterojunctions provide a potential barrier to confine minority carriers within the lower bandgap region

iii) both types of heterojunctions produce an index difference between the two layers, which leads to a radiation confinement process

This radiation confinement process can be efficiently exploited when the light is emitted from a surface perpendicular to the junction plane. So, this structure is called an edge emitting LED.

The choice of the materials for realizing a structure of this type depends on the region of wavelength spectrum considered. For instance, LEDs emitting in the first window are usually made of $Al_xGa_{1-x}As$ structures, where x represents the aluminum concentration. In Fig. 3.3 an example of edge emitting LED is schematically shown. The index difference at the two heterojunctions depends on the aluminum concentration difference $\Delta x = x_1 - x_2$. The stripe contact under the junction is designed to reduce the spatial width of the region crossed by the current, which becomes the primary emiting area.

In reference (12), a value of radiance of about 10^3 W/sr·cm^2 was obtained from a LED structure of this type, which was then called ultra-high-radiance LED. The far field pattern obtained has different properties depending on the direction. A guiding effect is present when the two heterojunctions realize an appreciable index difference and the thickness d of the active layer between these two heterojunctions is smaller than the wavelength λ. In this case, the divergence of the radiation pattern in the direction perpendicular to the junction plane can be rather small. The following approximate expression holds precisely [11] for the half intensity angle $\theta_\perp$ of the radiation pattern in this direction. For GaAlAs LEDs,

$$\theta_\perp \cong 20\, \Delta x \frac{d}{\lambda} \tag{3.1}$$

where Δx is the aluminum concentration difference at the heterojunctions. Typical values of Δx and d are respectively 0.2 and 0.1 μm, so that, for $\lambda = 1\,\mu$m, we have $\theta_\perp \cong 25^\circ$. In the direction parallel to the junction plane the divergence is greater, since no guiding effect is present.

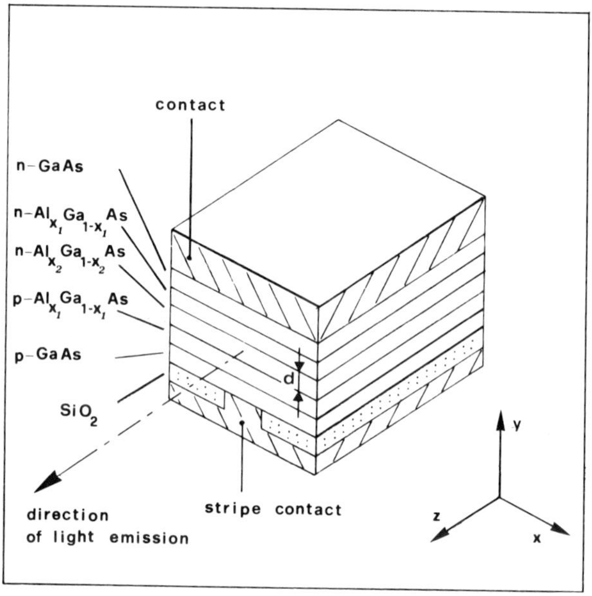

Fig. 3.3 - Scheme of an edge-emitting LED.

The current density which characterizes these devices is much smaller than that of a top emitting Burrus-type LED, being typically a few kA/cm^2. Although the total optical power obtained from an edge emitting LED is usually smaller than that obtained from a top emitting LED, the former is characterized by a smaller extension of the primary emitting area and by a smaller mean value of the divergence angle. This explains the much greater value of radiance which is obtained.

3.1.4 Laser Diodes

In a laser diode, the electric pump is obtained by the injection of carriers in the recombination region. So a laser of this type is also called injection laser. A lasing threshold exists, which is reached when the gain coefficient, produced by the injected carriers, equals the cavity loss. The structure of such a laser is very similar to that of an edge emitting LED. The laser cavity is made by cleaving the crystal along two parallel crystalline planes. The variation in the refractive index at the semiconductor-air interface provides a reflectivity of about 30%, which is enough for lasing behavior. The cavity is typically 300 μm long.

The injected carrier pair density is proportional to the current density that flows across the junction. Therefore, a threshold current density J_{th} can be defined, with decreases with increasing reflectivity of the two facets of the cavity, and with reducing thickness d of the active layer. For GaAs materials, it is typically $J_{th} = 10^3$ A/cm^2. The correspondent threshold current I_{th} can be reduced, improving the lateral confinement of the carriers, i.e., optimizing the sheet resistivities of the layers and the width of the stripe contact. An I_{th} as low as 50 mA has been obtained.

In this sense, an edge emitting LED can be considered as a laser diode with a very high threshold current, and which always operates below threshold. For both lasers and LEDs, the current-voltage relation below threshold can be written as

$$I = I_r \{ \exp [a [V - R_s I]] -1 \}$$

where I and V are, respectively, current and voltage extrenally applied to the device; a is a constant depending on the particular component; I_r is the reverse current; and R_s is the series resistance of the diode. For GaAs materials, typically, $a = 30\ V^{-1}$, and R_s ranges from 1 to 5 Ω. In 6.2 a model of the laser diode electric behavior will be discussed to explain some experimental results, and the origin of this series resistance will be investigated with more detail.

Above threshold, the voltage across the junction should remain constant at its threshold value. However, practical devices exhibit a current-voltage characteristics above threshold which can be approximated as [11]

$$V \cong \frac{E_g}{q} + R_s I$$

where E_g denotes the bandgap energy and q is the electron charge. For what concerns the relation between the emitted optical power P and the current I, we have a very fast increase of P just above the threshold. Next, the slope of this curve decreases, due to the reduction of the probability of radiative recombination at higher temperatures. An example of the transduction characteristics of a laser diode is given in Fig. 3.4, where the relation between voltage and current is also shown.

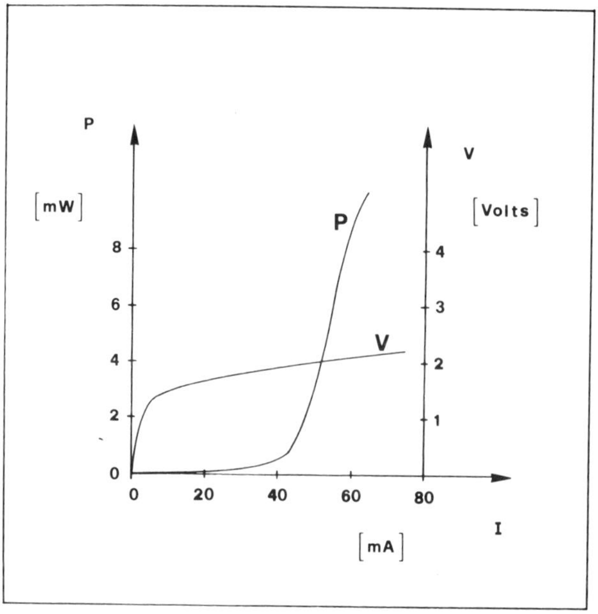

Fig. 3.4 - Emitted optical power *P* and voltage *V* as functions of the current *I* [from (11)].

The emission properties of a laser diode depend on the modes which are present in the cavity. We have three mode numbers: m,s,q, which measure the number of antinodes in the optical field along the three axes. Considering the rectangular coordinates introduced in Fig. 3.3, the direction y defines the transverse mode number m, the direction x the lateral mode number s, the direction z the longitudinal mode number q. Transverse modes depend on the index variations at the two heterojunctions and on the value of the distance d between them. For typical values of Δx and d, i.e., $\Delta x = 0.2$ and $d = 0.1$ μm, there is only the fundamental family of transverse modes m = 1. Lateral modes depend on the extension of the active region along x, that is, on the lateral confinement of the injection current. For a small width of the stripe contact and a suitable choice of the sheet resistivities of the layers, only the low order families of lateral modes are present; therefore s can be a very small integer. Finally, longitudinal modes depend on the length of the cavity and on the two reflecting facets which confine it along the z direction. The excitation of these modes is strongly influenced by the level of injected current I. The fundamental family of longitudinal modes ($q = 1$) is more and more strongly excited with respect to the other families as I increases.

Considering a laser diode, characterized by the fundamental family of transverse modes ($m = 1$), $\Delta\lambda_s$ and $\Delta\lambda_q$ are the wavelength distance, approximately uniform, which separate the emission lines of the modes, varying respectively s and q. For GaAs devices, $\Delta\lambda_s$ is a fraction of an Å, whereas $\Delta\lambda_q$ is on the order of 1 nm. In practice. two or three emission lines characterized by the same q and different values of s are partially superimposed, forming an envelope. When $s = 1$, the line width for a given q is about 0.1 Å . In this case, if the longitudinal mode selection allows one to neglect the contributions of the modes having $q > 1$, the laser diode can be considered to be in a single mode operation. In 3.1.7 we will discuss some techniques to reduce further the spectral width in single mode operation.

The far field pattern in the direction perpendicular to the junction plane exhibits one lobe if the modes having $m = 1$ are present only. In this case, if d is small enough, expression (3.1) holds again for $\theta_\perp$. The far field pattern in the direction parallel to the junction plane exhibits one lobe if the modes having $s = 1$ are present only. This fundamental lobe has a half intensity angle of radiation pattern $\theta_{//}$ as small as few degrees. Also the lobes of the other lateral modes are characterized by a very small extension, so that, even when more than one family of lateral modes is present, coupling with multimode fibers is not particularly difficult in this plane. Longitudianl modes do not affect appreciably the far field pattern.

A laser diode can operate in Q switching or in mode locking, as well as other types of lasers. Here, however, as a consequence of the very small size of the laser cavity, the device must be coupled with an external resonater.[13] This operation is attractive especially for metrological purpose. Pulse widths as small as 10 ps can be obtained in mode locking operation.[14]

3.1.5 Some Imperfections of the Laser Emission

In this section, we will consider some imperfections in the transduction of electric current into optical power performed by a laser diode. They are related essentially with nonlinearities in its dynamic operation. Three main effects of this type can be listed as follows:

i) non linearity in the transduction characteristic

ii) oscillating behavior after a step current pulse

iii) self-oscillations

In this section we will consider again stripe contact laser structures, though the effects which will be briefly described here are not restricted to this kind of laser diode only. In the next section, we will consider other types of laser structures.

The excitation of a mode of the laser emission gives rise to a rapid variation in the slope of the curve $P(I)$, where P is the emitted optical power and I is the drive current. These rapid variations are called kinks, and may produce a remarkable nonlinear distortion of the signal when a modulating current $i(t)$ is superimposed on a fixed current I_o. It occurs especially when $i(t)$ is a signal with more than two amplitude levels and if it is an analog signal. Fundamental lateral operation can reduce this effect.[11] So, to have a sufficiently large dynamic range for $i(t)$, the two thresholds of the fundamental lateral mode ($s = 1$) and of the first higher order lateral mode ($s = 2$) should be as far apart as possible. An increase in the difference between two such thresholds is usually achieved by restricting the width of the stripe contact up to values of a few μm.

The kinks are associated with a filamentary structure of the near field pattern, which characterizes the laser emission along the direction parallel to the junction plane.[15] Particularly in pulsed operation, the random nature of this filament emission can produce a remarkable noise, which must be carefully considered, due to the optical source. This noise is relevant not only for transmission systems employing laser diodes as optical sources, but also some measurement techniques based on the use of pulsed laser diodes.

The remaining two cases of distortion in the transduced signal can be described by studying the following pair of differential equations, called rate equations,[15]

$$\frac{dn}{dt} = \frac{J}{qd} - GS - \frac{n}{\tau_e} \tag{3.2}$$

$$\frac{dS}{dt} = GS + \beta\frac{n}{\tau_e} - \frac{S}{\tau_p}$$

where n and S are electron and photon densities, J is the injected current density, G is the gain of a given mode, τ_e and τ_P are electron and photon lifetimes, β is a spontaneous emission factor, and d is the thickness of the active layer. As a first approximation, the gain G can be considered proportional to the electron density n itself. These equations describe the evolution of the variables n and S as functions of the time t for a single mode. In order to take into account the presence of different modes, several pairs of rate equations must be superimposed.

The steady state solution of equations (3.2) explains the laser operation without modulation, and gives substantially the value of the threshold current density J_{th}. On the contrary, a transient solution, e.g., the response to a step current pulse, shows an oscillating behavior, though rather smoothed, before reaching the steady state values of n and S. This behavior produces a nonlinear distortion. When a sinusoidal current signal $i(t)$ is applied, harmonic distortion takes place. In,(16) the calculation of the second and third order harmonic distortions derived from this model has been reported. A strong dependence on the level of the continuous current I_o characterizes this behavior. The comparison with experimental results has shown good agreement with the theoretical predictions, provided the laser is not affected by kinks. This cause of nonlinearity may produce a remarkable harmonic distortion.

The baseband response of a laser diode, as a consequence of the dynamic behavior explained by (3.2), exhibits a resonance modulation frequency. It depends mainly on the value of I_o, but is usually in the neighborhood of 1 GHz. Sometimes the intrinsic noise of the device, in the presence of a given I_o without any modulating signal, may lead to self-oscillations corresponding to this resonance modulation frequency. This process limits severely the modulation performances of laser diodes. Some variations in the emission spectrum, depending on self-oscillations, have been measured, and even a relationship with the filament emission has been proved,[17] showing the strict interdependence of different imperfections in the laser emission.

When a transmission system or a measurement setup is designed, great care must be devoted to the evaluation of possible imperfections in the emission of a laser diode. To the three main causes of imperfections previously described, one can add the presence of modal noise and the effect of the temperature. The latter effect is not a direct cause of usefull signal deterioration, like noise or distortions, but it may lead to remarkable variation in the threshold current. Junction heating has a great influence on all the imperfections in the laser emission discussed above. To avoid the temperature effect, which usually entail drifts in the optical power launched into the fiber, a compensation in the laser driver is required.

Such a compensation can be realized by exploiting the temperature dependence of a silicon diode in the driver circuit. More frequently, an external opto-electric feedback is used, detecting a fraction of the emitted optical power.[18] Since the bias current is quasistatic, it is a sufficient low speed feedback. The primary emission represents the effective optical source, and the secondary emission, much smaller but proportional to the previous one, can be used as a signal monitor.

There are other imperfections in the laser emission, which have little importance for traditional transmission systems, but do have some importance for coherent transmission systems (which will be introduced in 4.2), and for metrological inspections. They are:

i) the influence of the injection current on the spectral behavior of the laser diode

ii) the influence of an external reflection on the linearity of the response

iii) the presence of a fundamental intrinsic noise, due to quantum fluctuations inside the laser cavity

The first imperfection leads to dynamic properties of the spectral distribution of the emitted optical power, depending on the level of the transmitted signal, during the same transmission. This may influence traditional transmissions on monomode fibers at a very high bit rate, where the wavelength of zero material dispersion must be well approximated by the laser emission. The most important consequence of this imperfection is the possibility of directly modulating the optical frequency emitted by the laser diode, which may be very useful for practical coherent transmissions.

The second imperfection is due to an undesired optical feedback, which influences the laser behavior. This reflection is typically produced by the input fiber end face. The main result of this effect is an increase in the nonlinearity of the transduction characteristics, leading to a large harmonic distortion in the intensity modulation. This may deteriorate the quality of traditional analog transmissions. The possibility of an external reflection forcing the device to run under an exact frequency and phase relationship can be efficiency exploited for mode locking operation reducing the spectral width of emission. This will be briefly discussed in 3.1.7. This feature could be useful for coherent transmissions, particularly direct phase modulation of the optical frequency.

The intrinsic noise, caused by the quantum processes inside the laser cavity, fixes an upper limit on the achievable signal to noise ratio SNR. The SNR also depends on the detection conditions, such as the receiver noise bandwidth B_N.[19] For B_N = 10 MHz the maximum SNR turns out to be about 70 dB, but it decreases very rapidly with increasing B_N. Some experimental results confirm the theoretical predictions derived from including a short noise term in (3.2).

3.1.6 Laser Structures

Until now, we have considered only stripe-contact lasers, but other structures have been proposed for the lateral confinement of the carriers. We will briefly describe how a stripe-contact laser can be realized in practice, followed by an outline on other laser structures, which can be based again on a stripe geometry or can be completely different.

Considering stripe-contact lasers, whose structure has been already described in Fig. 3.3, we can observe that the p-GaAs layer is usually lowly doped to reduce current spreading. GaAs is needed to make a good contact, which would be much more difficult using GaAlAs. Once the SiO_2 has been applied, a stripe window is opened up by a photolithographic process. The width w of the stripe is chosen to have single transverse mode operation. To choose the best value of w, one must take into account that for a large w (typically, greater than 20 μm), the carrier density is not uniform along the x axis, exhibiting the smaller value in the central region. This effect is called spatial hole burning.[20] Instabilities, as self-oscillations, become more evident with reducing w.

There are some methods for obtaining a stripe geometry which are not based on a stripe contact. In Fig. 3.5 we have reported some of these modified structures.[21] In (a), an insulation performed by SiO_2 laterally confines a p-GaAs region near the contact. Structure (b), called a shallow mesa stripe laser, is characterized by the whole p-GaAs layer which is confined laterally. In this case, even a highly doped p-GaAs material can be employed, since reduction of the current spreading is not required. In structure (c), a p-n junction between the n-GaAs layer and a Zn-p$^+$ zone, obtained by planner Zn-diffusion, gives rise to the carrier confinement. In (d), a similar confinement is obtained using a junction between n-GaAs and p-GaAs. In (e), another confinement, based on a p-n junction, is realized in a n-GaAs substrate.

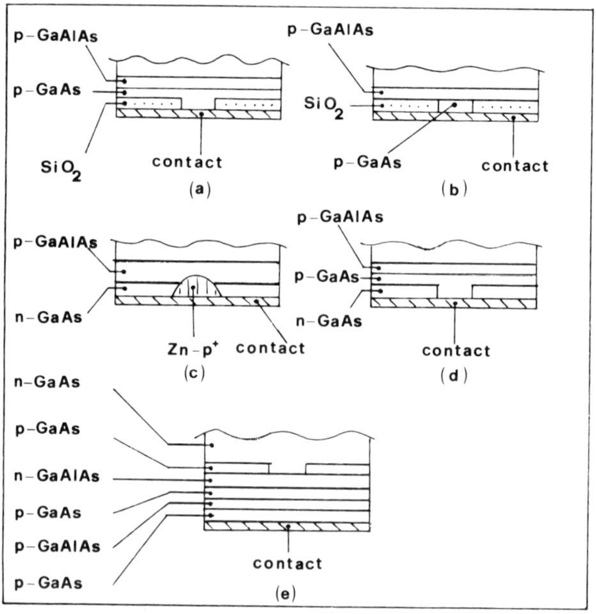

Fig. 3.5 - Modified stripe geometries.

Another technique for the lateral carrier confinement consists of realizing a high resistivity region around the stripe by proton bombardment or ion implantation. The proton bombardment may cause nonradiative recombination centers. Ion implanation seems to be more convenient, since it does not cause appreciable increase in the number of nonradiative recombination centers. Buried heterostructures are called particular stripe lasers, in which the same heterojunction configuration used for y confinement is employed for x confinement. In this case, there is a region of lower bandgap GaAl-As entirely embedded in a higher bandgap GaAlAs region. The p-n junction passes across this embedded region.

Some stripe geometry laser structures schematically described here can be combined to obtain new structures characterized by particular features. Zn diffusion near the contact, frequently used in several kinds of stripe lasers, increases the conducibility in the y direction, leading to a better carrier confinement.

Among the laser structures without a stripe are lasers based on a deep Zn diffusion, forming a guide for the carriers along the y axis, including the active region.[22] Another laser structure, completely different from those described, is the diffused feedback laser. It does not have a pair of reflecting facets to bind the cavity, but is based on many longitudinal resonances. These resonances are obtained by a periodic variation of the geometrical properties of the cavity along the z axis. The feedback occurs because the energy of a wave propagation in one direction is continuously fed back in the opposite direction by Bragg reflection.[21]

The electric properties of all the lasers described in the present section are not strongly different. On the contrary, radiation and carrier confinements can depend very critically on the particular laser structure considered. Threshold currents typically range between 50 and 120 mA. The overall output power can reach the maximum values of a few mW in CW operation, and about 1 W for pulsed operation with very low duty cycles. Finally, stimulated emission is capable of giving a radiance two orders of magnitude greater than spontaneous emission. For example, assuming an average beam divergence of about 0.1 rad, the emitting surface has a typical area on the order of 10^{-7} cm^2, therefore, in CW operation we can have a radiance as large as 3.10^5 W/sr·cm^2.

3.1.7 Spectral Properties of LEDs and Laser Diodes

LEDs and laser diodes designed for optical communications are usually made of GaAs and GaAlAs if the device must emit in the first window, i.e., for $0.8 \leq \lambda \leq 0.9$ μm. InGaAs and InGaAsP are used for obtaining longer emission wavelengths, precisely between 1.0 and 1.6 μm, covering the second and the third windows. Recently other components have been proposed for the longest wavelengths, e.g., GaAlAsSb or GaInAsSb, which operate between 1.5 and 2.0 μm.[23]

For the spectral width $\Delta\lambda$ of the emitted optical power, a Burrus-type LED typically has $30 \leq \Delta\lambda \leq 50$ nm, whereas an edge emitting LED is characterized by a spectral width which is remarkably smaller.

The emitted power of a laser diode has a spectral distribution strongly dependent on its modal configuration. As discussed in 3.1.4, in single-mode operation the laser spectral width can be as small as 0.1 Å. In this case, a noticeable temporal coherence characterizes the radiation. This is of some interest for time dispersion measurements in optical fibers, where a quasimonochromatic optical source must be employed. Yet, for use in transmission systems, single mode operation may suffer a remarkable level of modal noise.

When a reflection of suitable amplitude is given to the output end of a single mode laser from a proper distance, the spectral width of the light produced can be reduced by several orders of magnitude. In this case, a phase modulation of the optical frequency can also be obtained. A laser diode, inserted in a suitable long cavity external resonator, can exhibit a spectral width $\Delta\nu_\lambda$ as small as 100 kHz.[24] Conventional spectroscopy techniques, based on monochromators or scanning interferometry (see Appendix 1), cannot offer a spectral resolution fine enough to resolve the extension of such an ultranarrow spectrum. In 6.2 we will describe a technique characterized by the resolution required for such measurements.

3.1.8 Modulation Properties of LEDs and Laser Diodes

For both LEDs and laser diodes, we essentially have two processes which limit the baseband response of the transduction:[11]

i) at low current levels, the space charge capacitance of the junction C_s

ii) at high current levels, the lifetime of the minority carriers injected in the recombination region

The latter effect is sometimes taken into account through the diffusion capacitance of the junction C_J, which is a parameter directly measurable by means of an electric experimental setup.

In both cases, the baseband response of the device has a low pass form. This baseband response is defined as follows: applying a sinusoidally modulated signal, whose amplitude is constant and whose modulation frequency f is varied, one can measure a phasor $\bar{P}(f)$ which represents the total optical emitted power. If a lower cutoff frequency is negligible, the ratio $\bar{P}(f)/P(0)$ can be assumed to be the baseband response $\bar{H}(f)$ of the electro-optical transducer. Such a measurement can be performed only by means of an electro-optical experimental setup, in which also an opto-electric transducer is present.

Since the operation with high current levels is the most diffused one for both transmission and metrological purpose, the effect of the lifetime τ of the minority carriers is predominant. The amplitude of the baseband response can be written as

$$|\overline{H}(f)| = \frac{1}{\sqrt{1 + (2\pi\tau f)^2}} \tag{3.3}$$

C_J and τ are related by

$$C_J = \tau \left(\frac{dI}{dV}\right) \Bigg|_{I = I_o}$$

where the constant current I_o defines the point of the current-voltage characteristic in which the device operates. The baseband response can be experimentally obtained from a completely electric measurement, provided that it gives C_J and the diffusion resistance of the junction, R_J:

$$R_J = \frac{1}{\left(\frac{dI}{dV}\right)\Bigg|_{I = I_o}}$$

There are two main ways to increase the modulation bandwidth, and reduce τ.[25] One consists of increasing the dopant concentration of the recombination region. Yet, very high dopant levels entail an increase in the number of nonradiative recombination centers, decreasing quantum efficiency. The other way consists of increasing the current density, e.g., by a better carrier confinement. Burrus-type LEDs typically have a 3 dB bandwidth smaller than 100 MHz. Modified top emitting LEDs can reach 500 MHz. Edge emitting LEDs usually do not exceed 400 MHz.

The explanation of the baseband response for real laser diodes is much more complex than for LEDs. In principle, the carrier lifetime should impose the baseband response for lasers and LEDs. Considering that a stimulated emission at low threshold current occurs only when good carrier confinement is realized, one could expect a 3 dB bandwidth from 0.5 to 1 GHz. Yet, an exact explanation of the baseband response of a laser diode can be obtained only taking into account the rate equations (3.2). A resonance modulation frequency exists as shown in 3.1.5, significantly extending the laser bandwidth, leading to a nonuniform baseband response. In most cases, we first have a depression of a few dB, then a peak in correspondence with the resonance modulation frequency. Other causes which contribute to the modification of the actual laser baseband response, with respect to the expected one, are the parasitics, which come to be not negligible at some GHz.

The amplitude of the total baseband response of a laser diode can be written as [26]

$$|\overline{H}(f)| = F(f)\,|\overline{H}_e(f)|$$

where $F(f)$ is the intrinsic baseband response of the device, as derived from the rate equations, and $\overline{H}_e(f)$ takes into account the electric coupling between the source of the modulating signal and the device itself. The source is assumed to be characterized by a given internal impendance R_g (typically 50 Ω).

To obtain $\overline{H}_e(f)$ between 0.1 and about 5 GHz, the equivalent circuit of a laser diode can be represented as shown in Fig. 3.6, where the current $i(t)$ represents a sinusoidal modulating signal, obtained from a sinusoidal applied voltage $v(t)$. C_P and L_s account for wire and contracts, and should play a negligible role for the lower modulation frequencies. $\overline{H}_e(f)$ can be written as

$$\overline{H}_e(f) = \frac{R_g + R_J}{R_g + \overline{Z}_I(f)}$$

where

$$\frac{1}{\overline{Z}_I(f)} = i2\pi f C_p + \frac{1}{i2\pi f L_s + \overline{Z}_J(f)}, \quad \frac{1}{\overline{Z}_J(f)} = i2\pi f C_J + \frac{1}{R_J}$$

Note that R_J and C_J depend on the level of the continuous current I_o. With typical values of I_o, we have R_J from 1 to 10 Ω, and C_J from 10 to 50 pF.

3.2 PHOTODETECTORS

There are many types of photodetectors. They can be classified on the basis of characterizing parameters such as the responsivity, the speed of response, the noise equivalent power (NEP), and the detectivity. The responsivity $\mathcal{R}$ represents the constant of transduction of the photodetector. It is measured in A per W of incident optical power when the output is a current, and in V per W of incident optical power when the output is a voltage. In this definition, it has been assumed that a proportionality relationship holds between incident optical power and output current or voltage. $\mathcal{R}$ depends on the wavelength of the optical power. In this sense, another very important characteristic of the photodetector is the spectrum region in which it can operate.

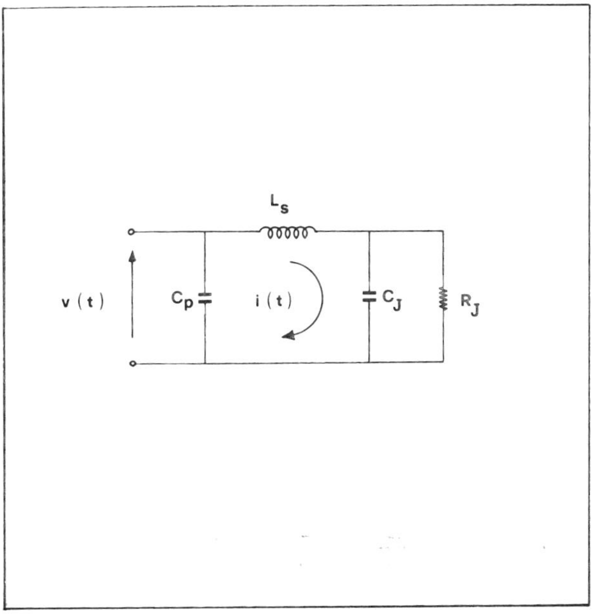

Fig. 3.6 - Equivalent circuit of a laser diode.

The speed of response is usually measured by the time constant τ of the detected signal, when a step radiation pulse incides on the photodetector. Alternatively, it can be expressed through the 3 dB bandwidth B of its baseband response. Both these figures depend upon the load on which the photodetector is closed.

The noise superimposed to the output signal is measured in A^2 or V^2. Assuming that the noise spectral power distribution is uniform at least over a given interval of modulation frequencies, the noise spectral density p_n can characterize the photodetector independently of its bandwidth. In this case, p_n is measured in A^2/Hz or V^2/Hz. Through the responsivity, which transforms the output current or voltage into the correspondent incident optical power, p_n can be transformed into a noise equivalent power. The NEP is then measured in $W/\sqrt{Hz}$.

The inverse of the NEP, is called detectivity. The normalized detecivity is the parameter $D^* = D\sqrt{A}$, where A is the active area of the photodetector. D^* measures the sensitivity of a photodetector placed in front of an optical source characterized by a given radiance at a given distance.

A high responsivity value allows one to detect even very low levels of optical power without the use of sophisticated electric instruments like picoampmeters or nanovoltmeters. However, the minimum level of detectable incident power is imposed by the NEP. The response of any photodetector to intensity modulated optical power can always be determined. If the output of the photodetector is dc measured, the response is always continuous and positive. The property is independent of the photodetector time constant, τ , which measures the capability of the photodetector to follow the modulation of the optical power.

3.2.1 Review of the Existing Photodetectors

The existing photodetectors can be divided into two classes:

i) thermal detectors, which exploit the temperature increase due to the absorption of photons

ii) electric detectors, in which the absorption of photons originates electric charges

Thermal detectors are characterized by an active area which is as small as possible, to enhance the temperature variation. The increase of the temperature is always transduced into an electric signal. For instance, thermistor bolometers are based on variations in the values of a temperature-dependent resistance, thermopiles employ a thermocouple, pryoelectric detectors exploit the spontaneous polarization which characterizes some crystals by increasing their temperature. Thermal detectors have a NEP of the order of 10^{-10} $W/\sqrt{Hz}$, a time constant τ whose minimum values are about 1 μs, and their responsivity $\mathscr{R}$ can vary widely depending on the applications. The spectral response is, however, very wide (from the ultraviolet to the far infrared), with a practically flat behavior. This property makes such photodetectors particularly suitable for use as calibrated radiometers.

Electric detectors can be divided into:

i) photoelectric detectors, which exploit the extraction of electrons from a cathode by photoelectric effect

ii) photoconductive detectors, in which a homogeneous semiconductor exhibits an increase in its conductivity, owing to the presence of photoelectrons in its conduction band

iii) photodiodes, in which a hole electron pair is generated by the absorption of a photon near a p-n junction

The responsivity of a photoelectric cell is about 1 mA/W. In order to increase this value, which appears too small, secondary emissions of electrons are exploited by a set of dynodes, realizing a complex photodetector called a photomultiplier. The stream of electrons increases exponentially with the number of dynodes. So amplifications as large as 10^6 can be easily obtained. Yet, this process requires a noticeable overall potential difference (about 100 V or more between two subsequent dynodes), to the point where photomultipliers are efficiently used only for laboratory applications. The time constant of a photomultiplier can be very short (about 0.1 ns),[27] while its spectral response depends on the cathode material. In Table 3.1, some cathode materials are listed separately, on the basis of the spectral region to which they are devoted. Each material exhibits a spectral response whose width ranges typically between 100 and 500 nm.

Table 3.1 - Cathode materials for photomultipliers.

Spectral Region	Cathode Material
ultraviolet	Cd, U, Th-Ni, Ce, Ta, Pt, Ti-Ni, W-Ni, Zr, Ni, Mo, Zn
visible	Cs-Mg, Sr, Ba, Se-Fe, Cu_2O, K, Rb, Li, Na-Pt, K-Pt, Pt, Li-W
infrared	Cs-Pt, BaO-Pt, Ba-Ag, Cs-Cs_2O-Ag

Photoconductive detectors must be externally polarized, so that variations in the applied voltage can measure variations in the conductivity induced by the incident light. To increase the responsivity of a photoconductive detector and to reduce its NEP, it is usually cooled by a cryogenic dewar filled with liquid N (77° K), or liquid He (10° K). In this condition, we typically have $\mathscr{R} \cong 10^4$ V/W and NEP $\cong 10^{-11}$ W/$\sqrt{\text{Hz}}$. The spectral response is sometimes very extended, though not flat, depending on the semiconductor employed. The main photoconductive semiconductors are CdS and CdSe for the visible, and PbS and InSb for the near and far infrared. The time constant for this class of photodetectors can be rather short, typically $\tau \cong 10$ ns if they are not cooled, but in this case, the responsivity and the NEP are not optimized.

Photodiodes appear as the most suitable photodetectors for optical communications. They can be divided into several classes, depending on the mechanisms of photogeneration of the hole electron pairs. The most employed classes are:

i) simple p-n junctions (solar cells)

ii) PIN photodiodes

iii) avalanche photodiodes (APD)

There are also metal semiconductor, point contact, and heterojunction devices,[4] but their use is not so diffused. In the following, we will describe with more detail the above three main classes of photodiodes.

3.2.2 Solar Cells and PIN Photodiodes

The separation of electrons and holes, after their generation, is more likely to occur when an electric field is present. Since the electric field is at a maximum in the depletion region of a junction, to have the maximum efficiency in the transduction, photons must be absorbed in this region. The depth to which a photon penetrates into the semiconductor depends on the absorption coefficient α of the material, which is a function of the wavelength λ. For a given material, $\alpha(\lambda)$ has a decreasing behavior, so that the photons of shorter wavelength are absorbed near the surface, while those of longer wavelength may penetrate deeply into the crystal. In this sense, the depletion region must be properly positioned under the receiving surface, and its extension must be as large as possible to absorb the maximum number of photons.

For a given structure, the extension of the depletion region can be increased by applying an external reverse bias voltage. However, the depletion region exists even if no reverse bias voltage is applied (photovoltaic operation), because of the diffusion of minority carriers across the junction. The simple scheme of a solar cell is shown in Fig. 3.7. The p-layer is thin, to enhance the short wavelength response. Yet a too thin p-layer leads to a very low resistivity of the whole device, and then the reverse bias voltage to be applied for extending the depletion region becomes high.

A better optimization is possible in PIN structures. In this case, the high resistivity of the material between the p- and the n-layer makes it possible to obtain an extended depletion region with a very low reverse bias voltage. The circuital representation of both solar cells and PIN photodiodes is reported in Fig. 3.8, where I_s represents the photocurrent, R_J the junction resistance, C_d the depletion capacitance, and finally R_s is a series resistance. R_s is due in part to the ohmic contacts, and in part to an internal space charge effect. This equivalent circuit is correct up to modulation frequencies of about

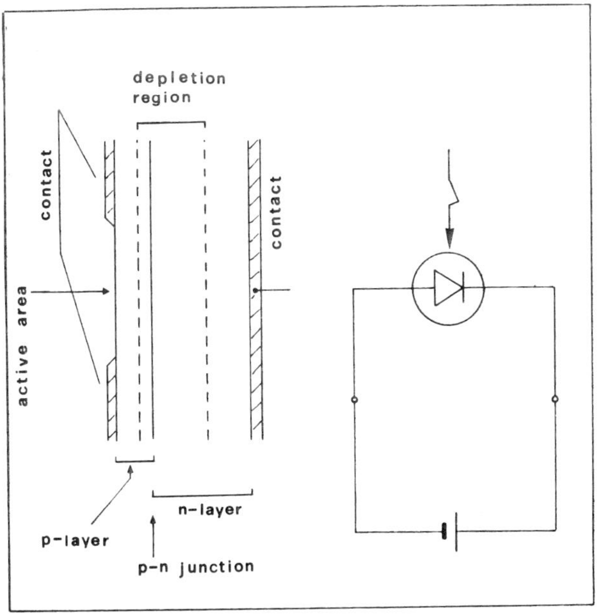

Fig. 3.7 - Simple scheme of a solar cell and bias circuit.

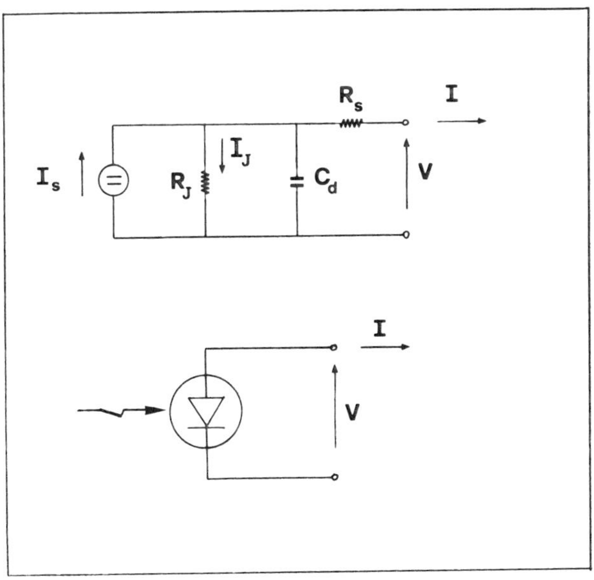

Fig. 3.8 - Circuital representation of a solar cell, or a PIN photodiode.

1 GHz, at which point the reactive effects of the electric coupling with the receiver circuit must also be taken into account. R_s is typically on the order of a few Ω, and although it leads to a reduction in the electric power which can be extracted from the device, for communication purposes it can usually be neglected. On the other hand, R_J and C_d have great importance for studying the linearity of transduction and of the baseband response, respectively.

The responsivity of a solar cell or a PIN photodiode depends on the semiconductor employed. It can be written precisely as

$$\mathcal{R} = \eta \frac{q}{h\nu_\lambda}$$

where q is the electron charge, h is Planck's constant, ν_λ is the optical frequency, and finally η is the quantum efficiency. $\mathcal{R}$ increases with increasing λ owing to the ratio $q/h\nu_\lambda$. But η, which is imposed by the absorption properties of the material, decreases with increasing λ. So $\mathcal{R}$ exhibits a maximum. For Se, Si and Ge, which are the most diffused single semiconductors employed, the maximum of $\mathcal{R}$ is ordinarily at about 0.6, 0.9 and 1.4 μm correspondingly; $\mathcal{R}$ is approximately 0.3, 0.5 and 1.0 A/W.

Considering a continuous optical power P incident on the active area of the device, the photocurrent $I_s = \mathcal{R} P$ can flow through R_j or through a load resistance R_L. R_j depends on the value of the voltage V. In fact, letting I_j be the current that flows through R_j, we can write

$$R_J = \frac{dV}{dI_J} \quad \frac{V_k}{I_r \exp(V/V_k)}$$

where I_r is the reverse current of the diode, and V_k is a constant voltage. For the photovoltaic operation, R_J depends on the level of P itself, while for operation with a reverse bias voltage, this dependence can be avoided.

Typical current-voltage characteristics are reported in Fig. 3.9. The first quadrant represents photovoltaic operation. For a small R_L, it is possible to write

$$V \cong R_L I_s = R_L \mathcal{R} P$$

which defines a linear transduction (points O,A,B in Fig. 3.9). For a large R_L, we have

$$V \cong V_k \ln\left(\frac{I_s}{I_r}\right) = V_k \ln\left(\frac{\mathcal{R} P}{I_r}\right)$$

that is a logarithmic transduction (points O,A',B' in Fig. 3.9). Since for communication purposes a linear transduction is required, the value of R_L in photovoltaic operation must be a trade-off: a high value of R_L entails a large voltage response, but R_L must be small enough to satisfy the condition $R_L << R_J$.

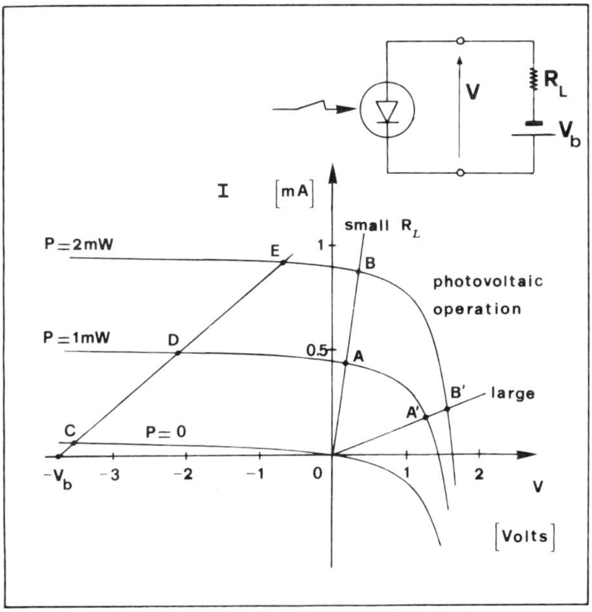

Fig. 3.9 - Typical *V* (*I*) curves with various *P* for a solar cell or a PIN photodiode at low modulation frequencies.

Note that for very low values of P, provided that $R_L << V_k/I_r$, we always have $R_L << R_J$. Yet R_J decreases with increasing P. Therefore, for a fixed R_L, we can define the linear dynamic range of a photovoltaic detector as the range between the minimum detectable optical power and the maximum level of P at which the output V is proportional to P. It is extremely important to observe that an ambiguous definition of linear dynamic range could be given, since in linear operation a proportionality relationship occurs between a current or voltage signal and the corresponding optical power signal. In fact, two optical powers, characterized by a ratio of X dB, lead to a pair of currents or voltages, characterized by a ratio of 2X dB. In this sense, when we define a linear dynamic range we will always consider ratios between optical powers.

The value of V_k/I_r, for a given semiconductor, depends on the junction extension of the active area. Photodiodes with smaller active areas exhibit higher values of V_k/I_r, and hence, more extended linear dynamic ranges employing the same R_L.

Consider, now, the operation with a reverse bias voltage V_b: if V_b is large enough, for any incident power P we have $V < 0$, and it is always $R_J > V_k/I_r$ (points C,D,E in Fig. 3.9). Here the only condition $R_L << V_k/I_r$ must be satisfied at any level of the incident optical power, extending by far the linear dynamic range of the photodetector. In this case, the voltage output can be written as

$$V = -V_C + R_L \mathcal{R} P$$

where $V_C \cong V_b$ is imposed by the presence of the reverse current I_r. This current gives rise to an undesired signal, also present when no optical power is incident on the photodetector, called dark current. The only limitation on the maximum achievable linear dynamic range, in operation with a reverse bias voltage, can be imposed by the breakdown voltage. This, for a given semiconductor, increases as the active area decreases.

For example, for Si small area photodiodes, V_k/I_r is typically on the order of $5.10^7\ \Omega$; therefore, choosing $R_L = 1\ \text{M}\Omega$, and $V_b = 10\ \text{V}$, we have an overall responsivity $\mathcal{R}_v = V/P$ of about $5.10^5\ V/W$, and the maximum detectable optical power turns out to be on the order of $20\ \mu\text{W}$. In section 3.2.4 we will discuss the NEP of solar cells and PIN photodiodes, with and without reverse bias voltage. As for the baseband response of the photodetector, we limit ourselves here to the following observations. For a given semiconductor, C_d increases with the extension of the active area. For a given device, a reverse bias voltage leads to a reduction of C_d. However, the greater the R_L, i.e., the overall responsivity, the smaller the bandwidth, which in linear transduction turns out to be

$$B = \frac{1}{2\pi R_L C_d}$$

3.2.3 Avalanche Photodiodes (APD)

Avalanche photodiodes are characterized by a region of high electric field, in which an avalanche multiplication of the photogenerated carriers takes place. A high reverse bias voltage must be applied in order to have avalanche multiplication, which is also a function of the temperature. In Fig. 3.10, a typical behavior of the photocurrent multiplication M (avalanche gain) is plotted *vs* the reverse bias voltage V_b, with the temperature T as a parameter.[28] A very stable

V_b and even a system for the temperature control are frequently employed to avoid possible drifts in the responsivity. This can be written as

$$\mathscr{R} = M\mathscr{R}_o$$

where the intrinsic responsivity $\mathscr{R}_o$ results approximately that of a PIN photodiodes, characterized by the same material. As for the choice of the load resistance R_L, considerations very similar to those performed for PIN photodetectors can be repeated. So, APD photodetectors can exhibit an overall responsivity $\mathscr{R}_v$, M times higher than that of the correspondent PIN photodetectors. In 3.2.4 a comparison between the noise properties of these two classes of photodiodes will be carried out.

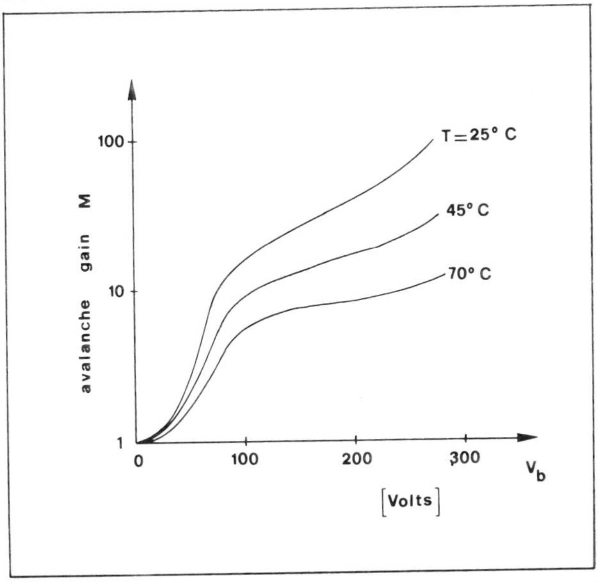

Fig. 3.10 - Typical behavior of the photocurrent multiplication [from (28)].

As regards the baseband response of APD photodetectors, we must observe that the avalanche gain M exhibits an intrinsic limitation in the speed of response, imposed by the transit times of the carriers τ_e through the avalanche region. So, M is a function of the modulation frequency f of the type [(29)]

$$M(f) = \frac{M(0)}{\sqrt{1 + [2\pi f h M(0) \tau_e]^2}}$$

where h is a constant varying from 1/3 to 2. The higher the avalanche gain M(O) the smaller the bandwidth. Thus, a multiplication bandwidth product characterizes the avalanche gain process. In practice, silicon devices are limited more by the $R_L\,C_d$ time constant than by this process. But both the effects can contribute to limit the baseband response of the photodetector for germanium devices. In 3.2.6 the modulation properties of PIN and avalanche photodiodes will be investigated with more detail.

The use of an APD is also characterized by possible errors of linearity. They are caused by saturation effects, due to the presence of the series resistance R_s, which depends in part on the level of the incident optical power, and on the heating of the junction.[28] The first cause leads to a reduction of the reverse bias voltage effectively applied across the junction. The second cause entails a reduction of M owing to an increase of the temperature. These linearity errors limit the maximum achievable avalanche gain to values lower than 300. However, even for M as low as 100, the linear dynamic range of an APD photodetector is always less extended than that of a PIN photodetector. Typically, an APD has 60 dB of linear dynamic range, whereas PIN photodetectors can exceed 100 dB.

3.2.4 Some Considerations about the Noise of Photodiodes

The photodiode is the first block of any optical receiver, therefore, in order to enhance the overall signal to noise ratio SNR, it should exhibit the maximum responsivity and the minimum noise. A solar cell or a PIN photodiode gives a useful signal power which increases as R_L. The fundamental lower limit of noise is the thermal noise of the load resistance, whose power is independent of R_L. Therefore the SNR increases as R_L.

The spectral density of thermal noise current results:

$$\frac{d}{df} < i_L^2 > = \frac{4KT}{R_L}$$

where k is Boltzmann's constant and T is the absolute temperature. The current I_s of the useful signal, in linear operation, is $I_s = \mathcal{R}\,P$. To the thermal noise, one must add a shot noise arising from the random nature of the photon arrival times. The spectral density of this shot noise current [30] is

$$\frac{d}{df} < i_{sh}^2 > = 2\,q\,I_s$$

where q is the electron charge. At very low modulation frequencies, there is also a contribution of Flicker noise, which is practically negligible, however, when f>100 Hz.

Thus, the SNR over a noise bandwidth B_N at the output of a solar cell or a PIN photodiode can be written as

$$SNR_{PIN} = \frac{I_s^2}{< i_{sh}^2 > + < i_L^2 >} = \frac{\mathscr{R}^2 P^2}{\left(2 q \mathscr{R} P + \frac{4 K T}{R_L}\right) B_N} \qquad (3.4)$$

In photovoltaic operation, since there is no dark current, the NEP can be calculated as

$$NEP_{pv} = \frac{1}{\mathscr{R}} \sqrt{\frac{4 K T}{R_L}}$$

In the operation with a reverse bias voltage, the fundamental limit of noise is imposed not only by thermal noise, but also by the shot noise which corresponds to the dark current I_r. Therefore it is

$$NEP_{rb} = \frac{1}{\mathscr{R}} \sqrt{2 q I_r + \frac{4 K T}{R_L}}$$

In Table 3.2 the values of NEP_{rb} and NEP_{pv}, for some typical R_L and I_r are reported. It is possible to observe that when $I_r \leq 10^{-1}$ nA, we always have $NEP_{rb} \cong NEP_{pv}$. However, when $I_r \geq 10$ nA, we have NEP_{rb} appreciably greater than NEP_{pv}, and the difference increases with increasing R_L.

Table 3.2 – NEP_{rb} and NEP_{pv} expressed in W, for some typical values of R_L and I_r

	NEP_{rb}				NEP_{pv}
	10^{-2} nA	10^{-1} nA	1 nA	10 nA	
0-1 MΩ	$8 \cdot 10^{-13}$	$8 \cdot 10^{-13}$	$8 \cdot 10^{-13}$	$8.8 \cdot 10^{-13}$	$8 \cdot 10^{-13}$
1 MΩ	$2.5 \cdot 10^{-13}$	$2.5 \cdot 10^{-13}$	$2.5 \cdot 10^{-13}$	$2.7 \cdot 10^{-13}$	$2.5 \cdot 10^{-13}$
10 MΩ	$8 \cdot 10^{-14}$	$8 \cdot 10^{-14}$	$2.7 \cdot 10^{-14}$	$1.4 \cdot 10^{-13}$	$8 \cdot 10^{-14}$
100 MΩ	$2.5 \cdot 10^{-14}$	$2.7 \cdot 10^{-14}$	$4.2 \cdot 10^{-14}$	$1.1 \cdot 10^{-13}$	$2.5 \cdot 10^{-14}$

To summarize, at very low levels of the useful signal, it can be more convenient to operate in the photovoltaic mode. Nevertheless, when the photogenerated current is much greater than I_r, equation (3.4), which holds for both photovoltaic and reverse bias modes, shows that these two modes of operation are perfectly equivalent. However, we recall that by using a reverse bias voltage a greater R_L can be employed, as discussed in 3.2.2. In conclusion, the operation with a reverse bias voltage is preferable, except for an incident optical power so small as to be comparable with the noise, especially when I_r is not too small, like for large area photodetectors.

For an avalanche photodiode, we must add another cause of noise, which is due to the statistical nature of the avalanche process itself. In other words, the avalanche gain M is a random variable, characterized by a mean value $<M>$ and a mean square value $<M^2>$. So the spectral density of shot noise current has to be modified as follows [31]

$$\frac{d}{df} < i_{sh}^2 > = 2q < M^2 > \mathscr{R}_o P$$

This expression is usually replaced by

$$\frac{d}{df} < i_{sh}^2 > = 2q < M >^2 F(< M >) \mathscr{R}_o P \tag{3.5}$$

where $F(<M>) = <M^2> / <M>^2$ represents an excess noise factor due to the avalanche process. For $F(<M>)$ the following expression holds

$$F(< M >) = < M > \left[1 - (1 - k) \left(\frac{< M > - 1}{< M >} \right)^2 \right]$$

where k is a constant which expresses the ratio between the smallest and the largest ionization coefficients. $F(<M>)$ increases with increasing k. For silicon, k typically ranges between 0.02 and 0.1, whereas for germanium it is typically about 0.5. When $<M>$ is greater than 10, the above expression can be well approximated by

$$F(< M >) \cong 2(1 - k) + k < M > \tag{3.6}$$

The SNR over a noise bandwidth B_N, at the output of an APD, can be written as

$$SNR_{APD} = \frac{I_s^2}{< i_{sh}^2 > + < i_L^2 >} = \frac{\mathscr{R}_o^2 < M >^2 P^2}{\left[2q < M >^2 F(< M >) \mathscr{R}_o P + \frac{4KT}{R_L} \right] B_N} \tag{3.7}$$

Finally the NEP results

$$NEP_{APD} = \frac{1}{\mathscr{R}_o < M >} \sqrt{2q < M >^2 F(< M >) I_r + \frac{4KT}{R_L}}$$

We can conclude from this expression that, for APD photodetectors, shot noise can be predominant even at very low values of the useful signal. For instance, letting $I_r = 1$ nA and $<M> = 10^2$, with Si-APD we typically have $F(<M>) \sim 3$, and the thermal noise turns out to be negligible, provided that $R_L > 1$ MΩ. With Ge-APD we typically have $F(<M>) = 50$, and the thermal noise is negligible provided that $R_L > 0.1$ MΩ.

Before concluding the present section, we will briefly compare the performances of PIN and avalanche photodetectors.[32] When a small value of SNR is sufficient, as for digital communications or for roughly approximated measurements, the overall noise is dominated by the circuit noise; i.e., that produced by R_L and by the preamplifier which follows the photodiode. In this condition, the avalanche gain can reduce the noise contribution of the preamplifier. Therefore, there is a convenience in the use of an APD. Nevertheless, the photodiode noise increases with increasing $<M>$, and hence an optimum value of $<M>$ exists, after which the overall noise is dominated by shot noise. When a large value of SNR is required, as for analog communications or accurate measurements, the shot noise of the photodiode normally dominates. Therefore, there is no convenience in the use of an APD, and PIN photodiodes are preferable, considering also their greater linearity.

3.2.5 Spectral Properties of Photodiodes

Solar cells are usually made of Se, Si, or Ge. Typical normalized spectral response of these three materials are reported in Fig. 3.11. These curves depend on the intrinsic absorption properties of the semiconductors, particularly their bandgap energies, and are not symmetric about their maximum valves, because of the depletion region extension. Similar spectral responses characterize PIN and avalanche photodiodes made of the same materials.

Recently, some compound semiconductors have been employed for photodiode fabrication to exploit the features of the second and third windows of optical fibers. (For example: $Ga_xIn_{1-x}As$, $Al_xGa_{1-x}Sb$, and $Ga_xIn_{1-x}As_yP_{1-y}$.) In Table 3.3 some optical and electric characteristics of photodiodes made of these materials are compared with those of Si and Ge photodiodes. The wavelength range of spectral response for compound semiconductors can be controled through the concentrations x and y. However, it is always between 1 and 2 μm. We have reported the minimum values practically achievable for dark currents. The responsivities ($\mathscr{R}$ for PIN and $\mathscr{R}_0$ for APD) are given at two wavelengths, and k is the constant which is present in expression (3.6) for avalanche photodiodes. The comparison of this table refers to an equal active area of $4 \cdot 10^{-5}$ cm^2.[33]

We can observe that a dark current is always greater for compound semiconductors than for Si, and sometimes it is even greater than for Ge. So one can expect a high NEP from these devices, both as PIN and as APD. With an APD in particular, we must consider the large value of factor *k*, which is present in (3.6), for both Ge and compound semiconductors. One can conclude that the excess noise due to the avalanche process has greater effects at longer wavelengths. In (33), a comparative study of the application of different kinds of photodiodes has been carried out.

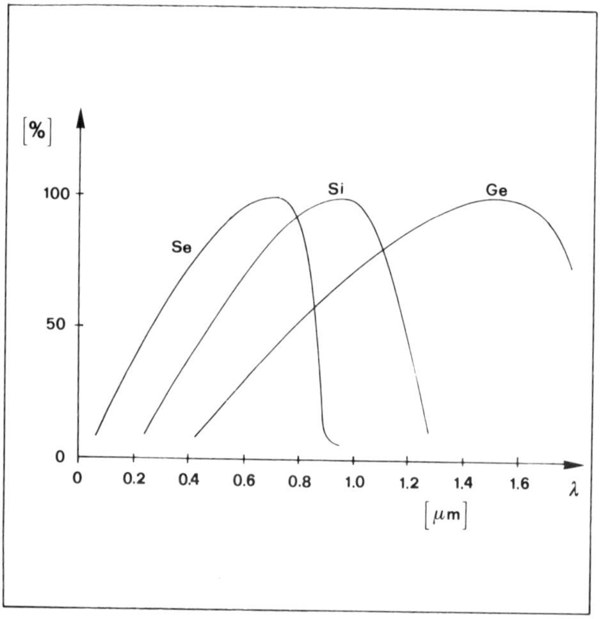

Fig. 3.11 - Normalized spectral responses of Se, Si and Ge.

Table 3.3 - Optical and electric characteristics of some materials for photodiode fabrication [from (33)].

Material	Wavelength Range	Dark Current	Responsivity $\lambda = 0.8\mu m$	Responsivity $\lambda = 1.3\mu m$	k
Si	0.4 - 1.1μm	10^{-13}A	0.5 A/W	-	0.02
Ge	0.7 - 1.8μm	10^{-8} A	0.4 A/W	1 A/W	0.25
$Ga_xIn_{1-x}As$	1 - 2μm	10^{-10}A	-	1 A/W	1.3
$Al_xGz_{1-x}Sb$	1 - 2μm	10^{-7} A	-	1 A/W	0.25
$Ga_xIn_{1-x}As_yP_{1-y}$	1 - 2μm	10^{-11}A	-	1 A/W	2.0

3.2.6 Modulation Properties of Photodiodes

In 3.2.2 and 3.2.3 we have shown that, except for Ge APD photodetectors which exhibit an intrinsic cutoff modulation frequency due to the avalanche gain cutoff effect, the main limitation in the photodetector bandwidth B is imposed by the parallel R_L, C_d. Since R_L determines the overall responsivity $\mathscr{R}_v$ of the photodetector, it leads to a limitation in the product $B\mathscr{R}_v$. This type of limitation can be reduced by the use of a transimpedance preamplifier. The electric scheme of a photodetector based on this type of preamplifier is shown in Fig. 3.12, where the capacitor C_L is necessary for avoiding instabilities when the gain G of the amplifier is very high. In this way the output voltage V results

$$V = \mathscr{R}\, R_L\, P$$

but the input impendance, seen by the photodiodes, turns out to be

$$\frac{1}{\overline{Z}_i(f)} = \mathrm{G}(f)\left(\frac{1}{R_L} + i\,2\,\pi f\, C_L\right)$$

where f is the modulation frequency.

Therefore, the baseband response of the photodetector can be written as

$$|\overline{H}(f)| = \frac{1}{\sqrt{1 + \left\{ \dfrac{R_L}{G(f)} \left[C_d + G(f)\, C_L \right] 2\,\pi f \right\}^2}}$$

At higher frequencies, it is dominated by the decay of $G(f)$, but at lower frequencies, it is certainly $C_d << G(f) C_L$. Therefore, the 3 dB bandwidth is

$$B \cong \frac{1}{2\,\pi\, R_L\, C_L}$$

As an example, a photodiode characterized by $R_J \geq V_k/I_r = 50\ \mathrm{M\Omega}$ and $C_d = 10$ pF can be used with a transimpendance preamplifier, when G is greater than 10^2 up to 10 MHz, $R_L = 100\ \mathrm{k\Omega}$, and $C_L = 1$ pF. In this way, we have an overall responsivity

$$\mathscr{R}_v = \mathscr{R}\, R_L = 5 \cdot 10^4\ \frac{V}{W}$$

and a bandwidth $B \cong 1.5$ MHz instead of 150 kHz. To enhance the responsivity, one can choose $R_L = 100\ \mathrm{M\Omega}$ and $C_L = 10$ pF, leading to $\mathscr{R}_v = 5.10^7$ V/W and $B \cong 150$ Hz. The linearity is assured because the input resistance seen by the photodiode is smaller than 1 MΩ. Yet, in the later case, the linear dynamic range is limited by the saturation voltage of the amplifier. The maximum output voltage is typically 15 V, which corresponds to an optical power level as low as 30 nW.

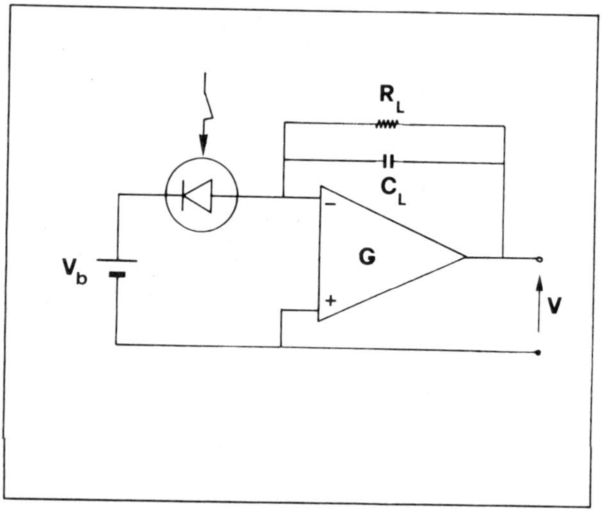

Fig. 3.12 - Electric scheme of a transimpedance preamplifier.

The NEP of the photodetector can be enhanced by the use of a transimpedance preamplifier when thermal noise is the dominant cause of signal deterioration.[34] However, we must also consider the noise contribution of the preamplifier. Fig. 3.13 shows an equivalent circuit in which all the sources of noise are reported. i_a and e_a represent the noise current and voltage of the differential amplifier, respectively. Assuming that all these noises are uncorrelated, we can write

$$NEP = \frac{1}{\mathcal{R}_v}\sqrt{\frac{d}{df}<v_L^2>+\frac{d}{df}<e_a^2>+\left(\frac{d}{df}<i_{sh}^2>+\frac{d}{df}<i_a^2>\right)R_L^2}$$

where

$$\frac{d}{df} < v_L^2 > = 4\,K\,T\,R_L$$

For commercially available integrated circuit amplifiers, it is typically

$$\sqrt{\frac{d}{df} < i_a^2 >} = 1.10^{-14}\ \frac{A}{\sqrt{Hz}}$$

$$\sqrt{\frac{d}{df} < e_a^2 >} = 2.10^{-8}\ \frac{V}{\sqrt{Hz}}$$

So, in photovoltaic operation, for very high values of R_L, the dominant term in the NEP may be $<i_a^2>$. On thc other hand, inreverse bias operation, especially for APD photodetectors, $<i_{sh}^2>$ is usually greater than $<i_a^2>$; i.e., the preamplifier introduces an excess noise which is practically negligible.

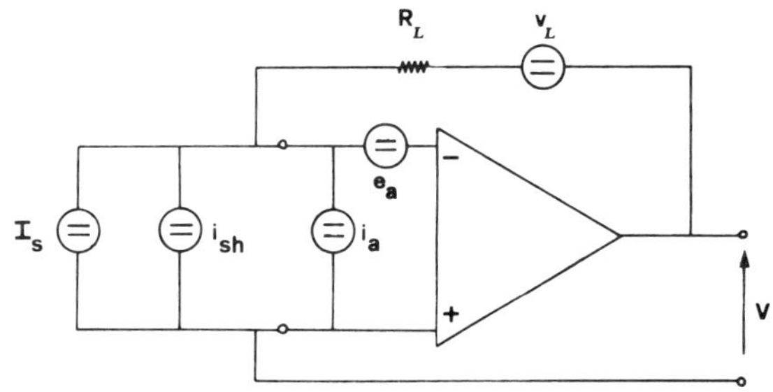

Fig. 3.13 - Sources of noise in a transimpedance photodetector.

When the capacitance C_d has a very low value (smaller than 1 pF) —also for PIN photodiodes — an ultimate limit to the speed of response arises from the transit time of the carriers across the depletion region. Note that to have good quantum efficiency, the extension of the depletion region must be large, yet the speed of response decreases with increasing this extension. Carrier velocity approaches the typical value 10^7 cm/s. Therefore, depletion regions as thin as 1 μm can lead to rise times of about 10 ps.[32]

REFERENCES

1. Young M. *Optics and Lasers,* Berlin: Springer Verlag (1977).
2. Daino, B., De Marchis, G., Piazzolla, and S., "Analysis and Measurement of Modal Noise in an Optical Fibre, "*Electron. Lett.*, 15, pp.755-756 (1979).

3. Faubert, D., and Chin, S.L. "Short Laser Pulse Generation: Part One," *Opt. & Laser Technol.*, 14, pp.197-207 (1982).
4. Sze, S.M. *Physics of Semiconductor Devices,* New York: John Wiley & Sons (1969).
5. Horiuchi, S., Ikeds, K., Tanaka, T., and Susaki, W. "A New LED Structure with a Self-Aligned Sphere Lens for Efficient Coupling to Optical Fibers," *IEEE Trans. Electron. Dev.*, ED-24, pp.986-990 (1977).
6. Burrus, C.A., and Dawson, R.W. "Small Area High-Current-Density GaAs Electroluminescent Diodes and a Method of Operation for Improved Degradation Characteristics," *Appl. Phys. Lett.*, 17, pp.97-99 (1970).
7. Burrus, C.A. "Radiance of Small Area High-Current-Density Electroluminescent Diodes," *Proc. IEEE*, 60, pp. 231-232 (1972).
8. Burrus, C.A., and Ulmer, E.A. "Efficient Small Area GaAs-Ga_{1-x} Al_x As Heterostructure Electroluminescent Diodes Coupled to Optical Fibers," *Proc. IEEE*, 59, pp. 1263-1264 (1971).
9. Harth, W., Heinen, J., and Huber, W. "Influence of Active Layer Width on the Performance of Homojunction and Single Heterjunction GaAs Light Emitting Diodes," *Electron. Lett.*, 11, pp.23-24 (1975).
10. Heinen, J., and Harth, W. "High Speed n-AlGaAs-p-GaAs Electoluminescent Diodes," *Electron. Lett.*, 11, pp. 512-513 (1975).
11. Kressel, H., Ettemberg, M., Wittke, J.P., and Ladany, I. "Laser Diodes and LEDs for Fiber Optical Communication," from *Semiconductor Devices for Optical Communication*, H. Kressel, editor, Berlin: Springer Verlag (1980).
12. Ettemberg, M., Kressel, M., and Wittke, J.P. "Very High Radiance Edge Emitting LED," *IEEE J. Quant. Electron.*, QE-12, pp. 360-365 (1976).
13. Ho, P.T., Glasser, L.A., Ippen, E.P., and Haus, H.A. "Picosecond Pulse Generation with a CW GaAlAs Laser Diode," *Appl. Phys. Lett.*, 33, pp. 241-242 (1978).
14. Inaba, H. "Ultrashort Optical Pulse Generation from Semiconductor Diode Laser," from *Optical Devices and Fibers — 1982*, Y. Suematsu, editor, Tokyo: OHM (1982).

15. Arnold, G., and Russer, P. "Modulation Behavior of Semiconductor Injenction Lasers," *Appl. Phys.*, 14, pp. 255-268 (1977).

16. Stubkjr, K.E., "Nonlinearity of DH GaAlAs lasers," *Electron. Lett.,* 15, pp.61-63 (1979).

17. Arnold, G., K. Petermann, "Self-pulsing phenomena in (GaAl)As double-heterostructure injection lasers," *Opt. & Quant. Electron.*, 10, pp.311-322 (1978).

18. Shumate, P.W., Chen, F.S., and Dorman, P.W. "GaAlAs Laser Transmitter for Lightwave Transmission Systems," *Bell Syst. Tech. J.*, 57, pp. 1823-1836 (1978).

19. Arnold, G., and Petermann, K. "Intrinsic Noise of Semiconductor Lasers in Optical Communication Systems," *Opt. & Quant. Electron.*, 12, pp. 207-219 (1980).

20. Dixon, R.W. "Current Directions in GaAs Laser Device Development," *Bell Syst. Tech. J.*, 59, pp. 699-722 (1980).

21. Thompson, G.H.B. *Physics of Semiconductor Laser Devices,* New York: John Wiley & Sons (1980).

22. Ueno, M., and Yonezu, H. "Stable Transverse Mode Oscillation in Planar Stripe Laser with Deep Zn Diffusion," *IEEE J. Quant. Electron.*, QE-15, pp. 1189-1196 (1979).

23. Kano, H., and Sugiyama, K. "2.0 μm C.W. Operation of GaInAsSb/GaSb D.H. Lasers at 80 K," *Electron. Lett.*, 16, pp. 146-147 (1980).

24. Okoshi, T., and Kikuchi, K. "Heterodyne-Type Optical Fiber Communications," *J. Opt. Comm.*, 2, pp. 82-88 (1981).

25. Bergh, A.A., and Copeland, J.A. "Optical Sources for Fiber Transmission Systems," *Proc. IEEE*, 68, pp. 1240-1247 (1980).

26. Figueroa, L., Slayman, C.W., and Yen, H.W. "High Frequency Characteristics of GaAlAs Injenction Lasers," *IEEE Trans. Mic. Th. & Tech.*, MTT-30, pp.1706-1715 (1982).

27. Anderson, L.K., and Murtry, B.J. "High Speed Photodetectors," *Proc. IEEE*, 54, pp. 1335-1349 (1966).

28. Schinke, D.P., Smith, R.G., and Hartman, A.R., "Photodetectors," from *Semiconductor Devices for Optical Communica-*

tion, H. Kressel, editor, Berlin: Springer Verlag (1980).

29. Emmons, R. B. "Avalanche Photodiode Frequency Response," *J. Appl. Phys.*, 38, pp. 3705-3714 (1967).

30. Smith, R.G., and Personick, S.D., "Receiver Design for Optical Fiber Communication," from *Semiconductor Devices for Optical Communications*, H. Kressel, editor, Berlin: Springer Verlag (1980).

31. Mc Intyre, R.J. "Multiplication Noise in Uniform Avalanche Diodes," *IEEE Trans. Electron. Dev.*, ED-13, pp. 164-168 (1966).

32. Smith, R.G. "Photodetectors for Fiber Transmission Systems," *Proc. IEEE*, 68, pp.1247-1253 (1980).

33. Pearsall, T.P. "Photodetectors for Optical Communication," *J. Opt. Comm.*, 2, pp. 42-48 (1981).

34. Hamstra, R.H., and Wendland, P. "Noise and Frequency Response of Silicon Photodiode Operational Amplifier Combination," *Appl. Opt.*, 11, pp. 1539-1547 (1972).

CHAPTER 4

TRANSMISSION SYSTEMS

We can distinguish between two classes of modulation techniques for the transmission of information via optical fibers:

i) traditional techniques, based on an intensity modulation of the optical source

ii) coherent techniques, based on the field modulation of a laser light and on homodyne or heterodyne detection

In this chapter we mainly describe traditional techniques, because they appear to be the most suitable for practical transmission systems, and have been tested for several cases during the last ten years.

We will briefly treat transmitter and receiver designs for both analog and digital transmission systems, with particular attention to their application in measurement setups. We will then discuss the convenience of performing equalization. Finally, we will give an outline of coherent communications, which appear interesting for monomode fibers using single-mode laser operation, in order to attain very high transmission capacities and very long repeater spacings.

4.1 TRANSMISSION SYSTEMS BASED ON INTENSITY MODULATION

An optical transmission system consists of the following blocks:

i) a transmitter, which couples the modulating signal to the optical source

ii) a transmission medium, which is a sequence of cascaded optical fibers

iii) a receiver, which returns the modulating signal detected by the photodetector

A tramsmitter consists of a LED or a laser diode and its driver circuit. The signal propagation along the transmission medium has been discussed in Chapters 1 and 2. In a receiver we have a photodiode, a preamplifier, whose input circuit is usually called front end, and sometimes an equalizer circuit. LEDs, laser diodes, and photodiodes have been treated in Chapter 3, therefore we will only study driver circuits on the transmitter side, and preamplifiers and equalizer circuits on the receiver side.

The repeater spacing Δz of a system depends on its transmission capacity C. For given transducers and transmission media, Δz decreases with increasing C. At a lower C we have an attenuation limitation due to the signal to noise ratio SNR. At a higher C we have a bandwidth limitation, due to distortion effects on the transmitted signal. Equalization is necessary only when C is so high that the bandwidth limitation is predominant. Discussion of equalization is different for analog and digital transmission systems.

To attain greater transmission capacities, several optical channels can be derived in the same fiber by using more than one transmission wavelength. This technique, called wavelength division multiplexing (WDM), requires a launching apparatus for injecting the beams of different laser diodes into the fiber, and a selecting apparatus for separating the channels at the fiber output. The latter apparatus is sometimes directly integrated in a multiple photodiode, having as many output currents as the optical channels.

4.1.1 LED and Laser Drivers

A typical LED or laser driver, preferably used for analog transmission is shown in Fig. 4.1. R_s represents the resistance of the source of the modulating signal $v(t)$. Capacitance C_1 imposes the lower cutoff modulation frequency. Resistances R_1, R_2, R_3 and the applied bias voltage v_b are used for adjusting the continuous current I_o across the transducer. When this is a laser diode, I_o must be high enough to always maintain the modulated optical signal above threshold. In this way, such a circuit converts an input base voltage into a collector current.

With a driver of this type, depending on the transistor selection and on the transducer considered, 3 dB bandwidths as large as 100 MHz, with 80% modulation indices can be obtained. Harmonic distortions as small as 45 dB under the useful signal typically occur.[1] The latter characteristic might not be sufficient for a particular analog transmission, which commonly requires a very high SNR. To reduce harmonic distortion, one must choose a smaller modulation index.

For digital systems using LEDs as optical sources, it is sometimes more convenient to employ the simplified driver scheme presented

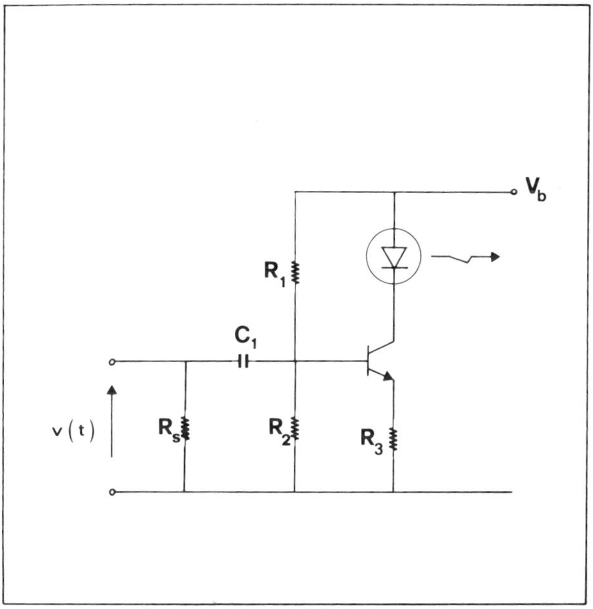

Fig. 4.1 - Typical LED or laser driver, suitable for both analog and digital transmissions.

in Fig. 4.2. This is a common emitter saturating switch, where resistance R_2 limits the current through the LED, and capacitance C_1 is used for pre-emphasizing (overdriving) the base current during the turn-on period. Finally, resistance R_3 together with the applied bias voltage V_b, imposes a suitable value for the continuous current I_o. Using this driver, with suitable LED and transistor, bit rates as large as 300 Mbit/s can be obtained.

When a laser diode is employed as the optical source in a digital transmission system, we have to supply this device with a continuous current I_o, which allows us to operate just below threshold in the off state. Maintaining this state very close to the threshold is necessary to reduce turn-on delays and possible overshoots in the optical output. Moreover, I_o just below threshold can be easily controlled to compensate for a changing ambient temperature, and even for device aging.[1] A driver circuit which can be used specifically for laser diodes, at bit rate up to 1 Gbit/s, is that reported in Fig. 4.3. It is based on a MESFET, typically made of GaAs. Resistance R_1 is set to allow the transistor operation well inside the pinch-off region. For both LEDs and laser diodes employed in digital transmissions, the

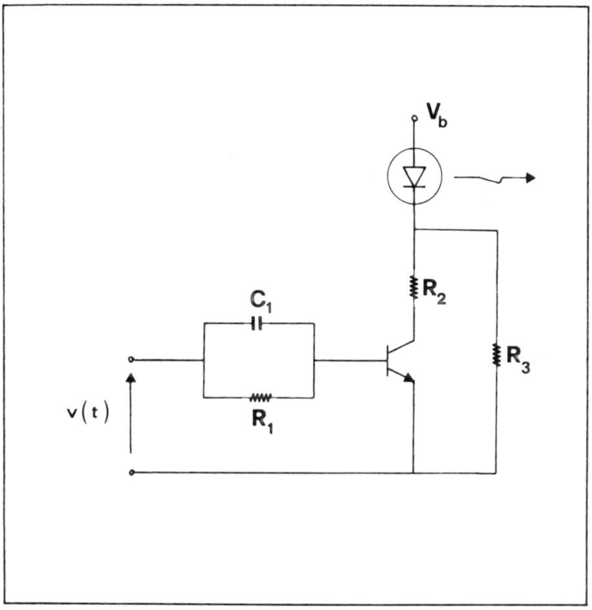

Fig. 4.2 - Simplified scheme of a LED driver for digital transmissions.

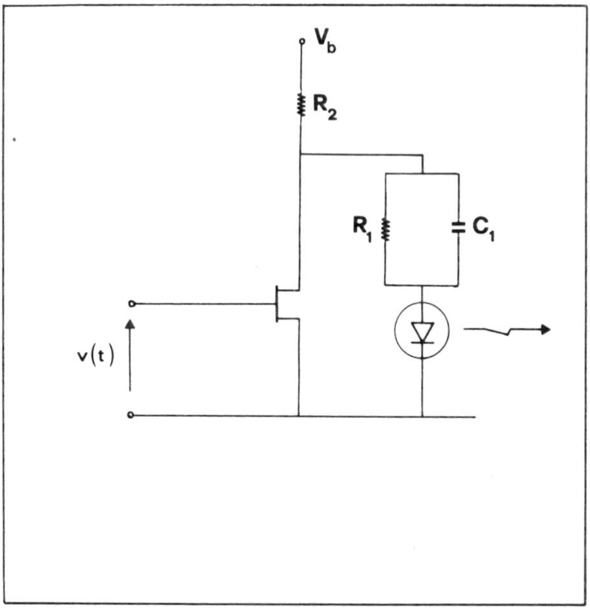

Fig. 4.3 - Example of laser driver for digital transmissions.

extinction ratio r (the ratio between the output powers in the off and on states) influences the SNR. Shot noise, especially when APD photodetectors are used, produces a larger value of r, yielding a lower SNR. This is another reason to maintain the current I_o just below threshold. Nevertheless, to obtain a very high modulation speed, it is sometimes possible to bias the laser diode so that it operates slightly above threshold in the off state, but then the SNR worsens. The dependence of SNR on r can be determined as suggested in reference (2).

4.1.2 Types of Front Ends

All sources of noise present in a receiver circuit are schematically represented in Fig. 4.4, where I_s is the photogenerated circuit, i_{sh} the photodiode shot noise current, C_d the photodiode depletion capacitance, R_L and C_L are the bias parameters, including parasitics, and i_L is the related noise current. Finally $\overline{Y_a}(f)$ denotes the input admittance of the preamplifier, i_a and e_a are its noise current and voltage. The junction resistance R_J of the photodiode has been neglected because it is shunted by R_L to have a linear operation. Letting $\overline{A}(f)$ equal the voltage transfer function of the preamplifier, one can

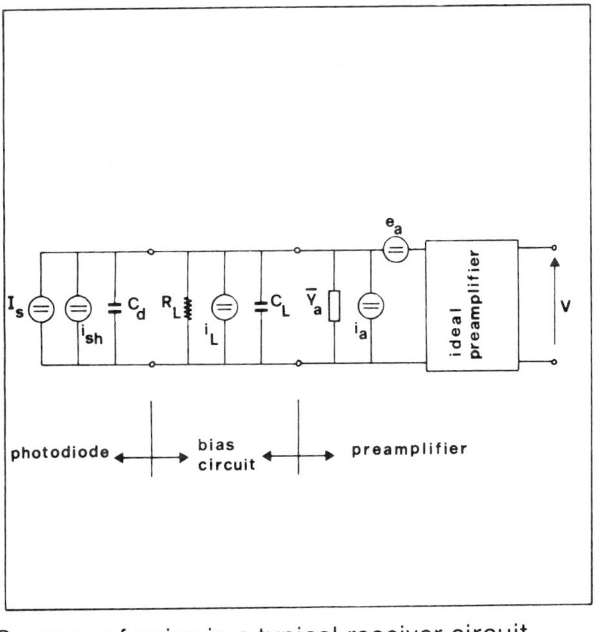

Fig. 4.4 - Sources of noise in a typical receiver circuit.

assume that the overall transfer function between I_s and the output voltage V has the dimension of an electric impedance, and results in

$$\overline{Z}_T(f) = \frac{\overline{A}(f)}{\overline{Y}(f)}$$

where

$$\overline{Y}(f) = \frac{1}{R_L} + i\,2\pi f(C_d + C_L) + \overline{Y}_a(f)$$

Letting $\overline{Y}_a(f) = 1/R_a + 2\pi fC_a$, it is useful to introduce the following new parameters

$$C_i = C_d + C_L + C_a$$

$$\frac{1}{R_i} = \frac{1}{R_L} + \frac{1}{R_a}$$

So, the total output noise voltage can be written as

$$< v_{nT}^2 > = \frac{d}{df} < i_{sh}^2 > \int_0^\infty |\overline{Z}_T(f)|^2 df + \int_0^\infty \frac{d}{df} < i_c^2 > |\overline{Z}_T(f)|^2 df$$

where the circuit noise current density $d/df < i_c^2 >$ results

$$\frac{d}{df} < i_c^2 > = \frac{d}{df} < i_L^2 > + \frac{d}{df} < i_a^2 > + \frac{1}{R_i^2}\frac{d}{df} < e_a^2 > + (2\pi f C_i)^2 \frac{d}{df} < e_a^2 > \tag{4.1}$$

Finally $d/df < i_{sh}^2 >$ has been given in 3.2.4 both for PIN and avalanche photodiodes, as well as $d/df < i_L^2 >$.

When $d/df < i_a^2 >$ and $d/df < e_a^2 >$ are independent of the modulation frequency f, the circuit noise current can be written as

$$< i_c^2 > = aB_N + bB_N^3 \tag{4.2}$$

where B_N represents the noise equivalent bandwidth of the receiver, that is the bandwidth of $|\overline{Z}_T(f)|$ where a and b are constants. So, one can conclude that two terms must be considered in the circuit noise current density:

i) the first, uniform in f, to which the shot noise of the photodiode $< i_{sh}^2 >$ must be added

ii) the second, which is proportional to f^2, due to the presence of C_i.

Therefore we can distinguish two types of front ends: high impedance and low impedance front ends, depending on the value of R_i. Since

$$\frac{d}{df} \quad < i_L^2 > = \frac{4KT}{R_L}$$

for high impedance front ends, the circuit noise density is usually dominated by the quadratic term, whereas for low impedance front ends it is usually dominated by the uniform term. High impedance front ends are used when a high sensitivity is required over a limited bandwidth. Low impedance front ends provide a wider bandwidth, sacrificing sensitivity.

From the view point of practical realizations, we can distinguish two main solutions:

i) a FET front-end

ii) a bipolar front-end

Their circuital schemes are shown in Fig. 4.5

It is possible to demonstrate [3] that, for a FET front end, we have a circuit noise current effectively on the type (4.2), in which constant a

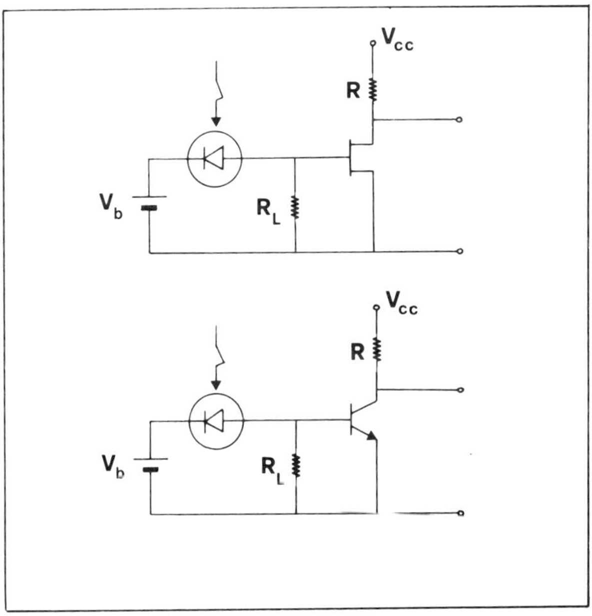

Fig. 4.5 - Circuital scheme of FET and bipolar frontends.

coincides with $4KT/R_L$, since the gate current is usually small. On the other hand, for a bipolar front end, we can approximate

$$< i_c^2 > \cong a' B_N + b' B_N^2 + c' B_N^3 \tag{4.3}$$

where a', b' and c' are new constants. In particular, we also have a quadratic term. Moreover, constant a' depends not only on R_L, but also on the transistor base resistance $r_{bb'}$. At low frequencies, the first two terms are dominant. However, when $R_L \to \infty$, either for FET or for bipolar front ends, the first term vanishes. A transimpedance preamplifier, like that introduced in 3.2.6, can be realized either with a FET or with a bipolar front end, giving the advantages of a high impedance front end accompanied with a more extended bandwidth.

4.1.3 Analog Receivers

Let us consider a sinusoidally intensity modulated signal

$$P(t) = P_a[1 + m\sin(2\pi f t)]$$

where P_a is the average received optical power, and m is the modulation index. We have the photogenerated current

$$I_s(t) = I_a[1 + m\sin(2\pi f t)]$$

where, for PIN photodiodes, it is $I_a = \mathscr{R} P_a$, and for avalanche photodiodes it is $I_a = \mathscr{R}_o <M> P_a$, as discussed in 3.2. Then, the input mean square current can be written as $\frac{1}{2} m^2 I_a^2$, and hence the SNR turns out to be

$$SNR = \frac{1}{2} \frac{m^2 I_a^2}{< i_{sh}^2 > + < i_c^2 >}$$

From (4.2), and recalling the expressions of the shot noise current density $d/df <i_{sh}^2>$ given in 3.2.4 for PIN and APD, we have

$$SNR_{PIN} = \frac{1}{2} \frac{m^2 \mathscr{R}^2 P_a^2}{2q\mathscr{R} P_a B_N + < i_c^2 >} \tag{4.4}$$

$$SNR_{APD} = \frac{1}{2} \frac{m^2 \mathscr{R}_o^2 < M >^2 P_a^2}{2q < M >^2 F(< M >) \mathscr{R}_o P_a B_N + < i_c^2 >} \tag{4.5}$$

The above expressions are very similar to (3.4) and (3.7), except for the substitution of $(4KT/R_L) B_N$ with $<i_c^2>$ and for the multiplication by the factor $\frac{1}{2} m^2$, which takes into account the sinusoidal modulation.

In 3.2.4 we have observed that PIN photodiodes are more suitable for analog systems. We restrict ourselves here to the discussion of (4.4) only. When we are with a very low level of incident optical power, circuit noise dominates, and hence

$$SNR_{PIN} \cong \frac{1}{2} \quad \frac{m^2 \mathscr{R}^2 P_a^2}{< i_c^2 >}$$

which increases and P_a^2. This expression holds typically when SNR_{PIN} is small. On the other hand, when P_a is larger, shot noise dominates, and thus we have

$$SNR_{PIN} \cong \frac{m^2 \mathscr{R} P_a}{4 q B_N}$$

which increases as P_a. This expression typically holds when SNR_{PIN} is large: that is, for usual operations.

4.1.4 Digital Receivers

Digital receivers can use a high impedance or a low impedance front end, and can be realized either with FET or with bipolar transistors. Expressions very similar to (4.2) and (4.3) hold in this case, with the only difference of replacing B_N with the bit rate C. Finally, the constants a, b, a', b', c' also depend on the pulse shape. Considering an optimal choice of design parameters, and assuming rectangular pulses, it is possible to show [(3)] that the minimum circuit noises, respectively for FET and bipolar front ends, can be obtained from the following practical formulae

$$< i_c^2 >_{FET} \cong \left[\frac{8 \cdot 10^{-21}}{R_L} \right] C + 8 \cdot 10^{-41} C^3 \qquad [A^2]$$

(4.6)

$$< i_c^2 >_{BJT} \cong \left[\frac{8 \cdot 10^{-19}}{R_L^2} \right] C + 8 \cdot 10^{-33} C^2 + 3 \cdot 10^{-41} C^3 \; [A^2]$$

where R_L is expressed in ohms, and C in bit/s. For high impedance front ends the first term can be neglected.

It is possible to demonstrate that for $C < 100$ Mbit/s it is more convenient to use a FET front end, while for $C > 100$ Mbit/s a bipolar front end is preferable. Finally, a transimpedance preamplifier, such as for analog systems, allows one to obtain the advantages of a high impedance front end without a large penalty of bandwidth. However, to attain very large speeds of transmission, e.g., 800 Mbit/s or more, it is necessary to employ a low impedance front end.

The probability of error, or bit error rate, gives the quality of transmission for digital systems and can be calculated as follows, considering a binary transmission for simplicity. If we assume that all the sources of noise are uncorrelated, and these noises can be expressed as gaussian random processes, they lead to an overall noise variance σ^2_{nk}, k = 1,2, which in general is different for the two transmitted levels, owing to shot noise. Nevertheless, when the SNR is small, $<i_c^2>$ dominates, and therefore

$$\sigma_{n1} \cong \sigma_{n2} \cong \sigma_n$$

So, considering two equiprobable transmission levels, the bit error rate can be written as

$$P_E = \frac{1}{2} \text{ erfc } \frac{|D - s_k|}{\sigma_n}$$

where erfc represents the complementary error function,[4] D is the voltage decision level, s_k is the output voltage corresponding to the level k. In this expression, σ_n represents the output noise voltage corresponding to $<i_c^2>$, given by (4.6).

For a pair of equiprobable transmission levels, it is convenient to set the decision level in the middle between them. Therefore, considering the input currents instead of the output voltages, we are left with

$$P_E = \frac{1}{2} \text{ erfc } \frac{I_s(1) - I_s(0)}{2\sqrt{<i_c^2>}}$$

where $I_s(1)$ and $I_s(0)$ are the photocurrents associated respectively to the two received power levels. The quantity

$$Q = \frac{[I_s(1) - I_s(0)]^2}{4 <i_c^2>}$$

has the meaning of SNR. It is possible to show that, in order to obtain $P_E \leq 10^{-9}$, which is a typical value for telephone PCM transmission systems, it is necessary to have $Q \geq 36$.

The attenuation limitation Δz_a in the repeater spacing of a transmission system characterized by a bit rate C, using a PIN photodiode, can then be calculated from

$$\mathscr{R}^2 [P(1) - P(0)]^2 \exp(-2\gamma \Delta z_a) \cong 144 <i_c^2>$$

where γ is the attenuation coefficient of the fiber, considered for simplicity without perturbations, $P(1)$ and $P(0)$ are the two launched levels of optical power. Finally, the depencence on C is in the expression of $<i_c^2>$, given by (4.6) for FET and bipolar front ends. Assuming $\mathcal{R}$ = 0.5 A/W, γ = 0.4 km^{-1}, which corresponds to an attenuation per km of about 2 dB/km, and $P(1) - P(0)$ = 2mW, and taking into account a high impedance FET front end, we are left with[5]

$$\Delta z_a \cong 1.25 \ln \left(\frac{10^{32}}{1.2 \cdot C^3} \right) \quad [\, km \,] \tag{4.7}$$

4.1.5 Equalization

When a bandwidth limitation is present, intersymbol interference adds to the other causes of noise, leading to a further reduction of the repeater spacing. Two main expedients can be exploited to reduce this bandwidth limitation:

i) a transmission with more than two levels

ii) equalization

The use of more than two transmitted signal levels entails a reduction of difference $|D - s_k|$, therefore there is an increase in the minimum value of Q, required for a given bit error rate. This leads to a reduction of the repeater spacing again. Moreover, the simplicity of a binary system, which is its most attractive feature, is lost.

As for equalization, the main problem in performing it efficiently is in the uncertain prediction of the fiber baseband response. It is possible to use adaptive equalizer circuits,[6] like those employed in radio relay links. Nevertheless, since the baseband response of an optical channel is constant in time, this solution is too sophisticated and expensive; furthermore, it is difficult to optimize. When a long repeater spacing is considered, the presence of different cascaded fibers separated by imperfect joints and the effects of the fiber perturbation, as discussed in Chapter 2, contribute to an easy prediction of the baseband response shape for the optical link.

A single unknown remains:.the overall bandwidth, which can be difficult to predict. Quasi-standard passive equalizer circuits can be employed, characterized by a few parameters to be adjusted at the time of the installation, depending on the actual fiber bandwidth.[5] Yet in this case we have an increase of the noise superimposed on the useful signal, and therefore a further reduction in the repeater spacing.

Many authors have discussed whether equalization is or is not convenient. The conclusions depend on several parameters in the system design, i.e., the use of a PIN or avalanche photodiode, the choice of a high impedance or a low impedance front end, its realization by FET or bipolar transistors, and the extinction ratio and pulse shape. From (4.2)it's possible to state that equalization gives rise to a modest increase in the overall noise when a low impedance front end is employed, therefore there is an effective convenience in performing it.[7] But when a high impedance front end is used, equalization entails a remarkable increase in the overall noise, to the point where it might be greater than that caused by the intersymbol interference without any equalization.[8]

To compare equalizing and nonequalizing receivers, we will consider a PIN photodiode and a FET high impedance front end. All the design parameters which have been assumed in the previous section for the prediction of the attenuation limitation (4.7) are taken here too. For both equalizing and nonequalizing receivers we introduce the optical power penalty, which is the increase of the transmitted optical power necessary to maintain a fixed bit error rate. This leads to a new limitation in the repeater spacing Δz_b, which depends also on the decay of the fiber bandwidth $B(z)$. Such a decay, from what was discussed in Chapter 2, can be expressed by

$$B(z) = \frac{B_o}{z^q} \tag{4.8}$$

where q is a constant which ranges between 0.5 (steady-state condition) and 1 (unperturbed fiber).

Letting L_p be the optical power penalty, expressed in dB, it can be reported as a function of the ratio C/B. So, from (4.8), one can obtain L_p as a function of C and z, and the new limitation in the extension of the repeater spacing is

$$\Delta z_b = \Delta z_a - \frac{1}{A} L_p(C, \Delta z_b) \tag{4.9}$$

where A is the attenuation per km of the fiber, and Δz_a is given by (4.7). The actual limitation, for $L_p(C, \Delta z_b)$ not negligible, is obtained as $\Delta z_b(C)$, from (4.9).

We first consider the optical power penality of nonequalizing receivers. Assuming that circuit noise and intersymbol interference are uncorrelated, we can write

$$\frac{N_o + N_i}{S} = \frac{N_o}{S_o} \tag{4.10}$$

where N_i is the noise power due to intersymbol interference, N_o is the remaining noise power, S and S_o are the required useful signal powers respectively with and without intersymbol interference. The ratio N_i/S can be obtained depending on the line code used, and on the actual fiber baseband response. However it is function of the ratio C/B. Equation (4.10) can be rewritten as

$$\frac{S}{S_o} = \frac{1}{1 - \frac{S_o}{N_o} \frac{N_i}{S}}$$

and the optical power penalty without equalizer becomes

$$L_p^{(ne)} = 10 \log_{10} \left(\sqrt{\frac{S}{S_o}} \right) \tag{4.11}$$

Considering now the case of an equalizing receiver, we assume as a reference the noise power caused by a noise power density of the type bf^2, which results in

$$N_o = b \int_0^{(1+\beta)\frac{C}{2}} H_{rc}^2(f) f^2 \, df$$

where $H_{rc}(f)$ is a raised cosine transfer function, characterized by the roll-off factor β and Nyquist frequency $C/2$. Adding an equalizer circuit, whose transfer funcltion is $1/H(f)$, where $H(f)$ is the fiber baseband response (for simplicity, only in amplitude), we have the new noise power

$$N_e = b \int_0^{(1+\beta)\frac{C}{2}} \frac{H_{rc}^2(f)}{H^2(f)} f^2 \, df$$

N_e depends on the roll-off factor β and on the actual fiber baseband response. However, it is a function of the ratio C/B. The optical power penalty with the equalizer then results

$$L_p^{(e)} = 10 \log_{10} \left(\sqrt{\frac{N_e}{N_o}} \right) \tag{4.12}$$

Substituting (4.11) or (4.12) into (4.9), one obtains Δz_b as a function of C in the two cases. For typical values of the design parameters, e.g., a roll-off factor $\beta = 0.2$, and a 3B4B line code, finally assuming for the fiber baseband response a gaussian shape and a coefficient of bandwidth decay $q = 1$, it is possible to obtain the behavior reported in Fig. 4.6, where two values of B_o in (4.8) have been assumed. Dash lines refer to nonequalizing receivers. For small values of C, the attenuation limitation (4.7) dominates. For intermediate values of

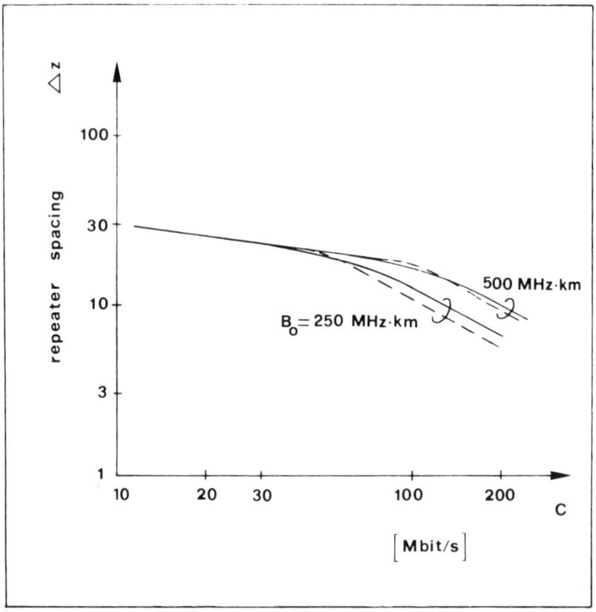

Fig. 4.6 - Maximum repeater spacing for equalizing and non-equalizing receivers [from (5)].

C, it is more convenient to use nonequalizing receivers. For high values of C, on the contrary, there is a noticeable convenience in the equalization.[5]

The repeater spacing for analog systems is limited by attenuation and bandwidth, and the use of an equalizer circuit to relax the bandwidth limitation worsens for the case of attenuation limitation, as well as for digital systems. The high SNR usually required is limited mainly by the shot noise of the photodiode, especially when an APD is employed. So the deterioration introduced by the presence of the equalizer circuit is normally less remarkable than for digital systems.

4.1.6 Ultimate Limits of Transmission Capacity and Repeater Spacing for Intensity Modulation Systems

Using monomode fibers and laser diodes operating at the wavelength of zero material dispersion, with very narrow spectral width, a transmission capacity C as large as 2.24 Gbit/s can be achieved.[9]

Moreover, a very long repeater spacing Δz can be obtained for these ultra-high-capacity systems. The maximum value till now has been about 50 km.[10] Only two levels of transmitted signal are usually employed, owing to the very large bandwidth offered by the transmission medium. In fact, the effect of the residual material dispersion, connected with the laser spectral fluctuations by undesired optical feedback, can be estimated to be about 5 ps/km, and birefringence exhibits even smaller effects. Very fast Ge-APD photodectors are usually employed.

Equalizatin is required to compensate for the bandwidth limitations of both transducers. Keeping in mind the deterioration of the attenuation limitation when more than two transmission levels are employed, the above value of the product $C \cdot \Delta z$ can be considered the maximum for practical high-capacity traditional systems, based on the intensity modulation of the optical source.

Coherent transmission systems, whose features will be briefly described in the next paragraph, owing to a remarkable improvement in the attenuation limitation, appear very attractive for the possibility of reaching higher values of the product $C \cdot \Delta z$. Although at present they are only at the first *practical* demonstrations of feasibility, it is possible to foresee that their importance will increase very rapidly in the next ten years.

4.2 COHERENT TRANSMISSION SYSTEMS

Traditional optical transmission systems are based on an intensity modulation of the optical source and on a direct detection of the received signal. So, in the following, they will be denoted briefly as IM-DD. From the theory of radio relay links, it is well known that this technique, though very simple, is not optimized. Nevertheless, since in the first years of optical communications laser diodes with a high spectral purity and a good frequency stabilization were not available, the IM-DD technique has been practically the only one employed.

Yet great advantages, in terms of the minimum level of received optical power for a given bit error rate, can be obtained by the use of transmission systems in which amplitude, frequency or phase of the optical carrier are directly modulated. Thus we will have ASK, FSK, or PSK transmission systems at optical frequencies, as well as nowadays they are commonly used at radio frequencies. These systems can be generally called coherent transmission ones, referring to their modulation technique. Next, considering the detection process, they must be divided also into coherent or noncoherent detection systems. Among coherent detection systems we can have only a heterodyne detection for FSK , whereas for ASK and PSK either a heterodyne or a homodyne detection can be applied.

Heterodyne detection is based on combining the received wave front with a locally generated wave front. A conceptual scheme of this procedure is shown in Fig. 4.7. The combination of the two wave fronts can be made by a simple photodiode, usually an APD, which operates as a mixer. One can select the difference between the two optical frequencies, received and locally generated, [11] from among the interference products with a band-pass filter at radio frequencies, set at the photodiode output. The lower cutoff modulation frequency of the RF filter must be greater than the spectral widths of the two optical sources, whereas its upper cutoff modulation frequency depends on the transmitted signal. Homodyne detection can be seen as a particular case of heterodyne detection, characterized by a locally generated wave front having the same optical frequency as the received wave front. In this case, the frequency conversion gives a baseband signal whose spectrum is translated with respect to the transmitted signal, and not a modulated signal, directly. This leads to a reduction by a factor of two in the noise equivalent bandwidth, with respect to a simple heterodyne detection. However, homodyne detection appears to be difficult to achieve in practice.

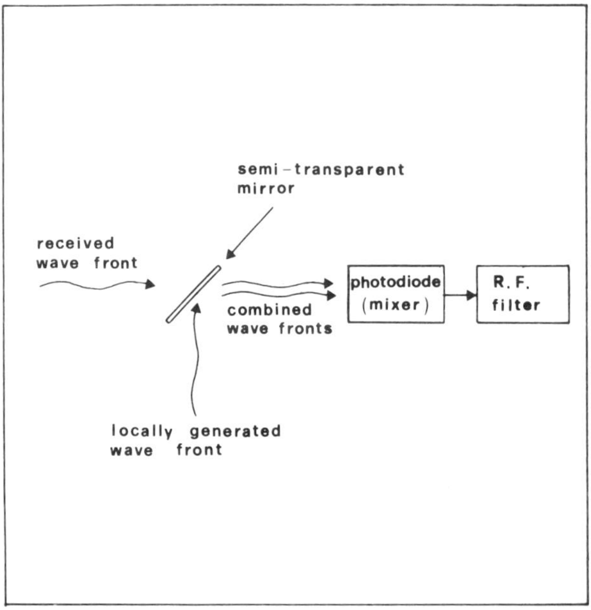

Fig. 4.7 - Conceptual scheme of heterodyne detection.

After a heterodyne frequency conversion, we can perform the demodulation with a noncoherent technique. Noncoherent demodulation techniques for ASK and FSK are usually called envelope detection, while for PSK we have the differential coherent detection. In this paragraph, after a brief outline of the theoretical aspects of different types of coherent systems, we will discuss the characteristics required by the optical sources to be suitable for such applications.

4.2.1 Advantages of Coherent Systems over Traditional Systems

To compare the features of traditional IM-DD systems and coherent systems, we will consider a single parameter called sensitivity of the receiver, which expresses the minimum level of optical power to be received to obtain a given bit error rate P_E. This parameter depends on the value of P_E, so we restrict ourselves to the typical value $P_E = 10^{-9}$. A 3 dB improvement of the sensitivity, by reducing the noise bandwidth, characterizes homodyne detection with respect to coherent heterodyne detection when it is possible. The comparison of different types of heterodyne detection systems with IM-DD systems will be carried out in the following, considering an avalanche photodiode and a FET high impedance front end. This comparison is nearly independent, when there is no bandwidth limitation, of the transmission capacity.

Heterodyne detection is characterized by a high level of shot noise, mainly due to the locally generated signal, which is usually much greater than the received signal. This noise can be considered predominant with respect to the remaining noises, present also in IM-DD systems. For typical values of the transmitter and receiver parameters, the improvement in the receiver sensitivity can be attained by an ASK coherent detection over a traditional IM-DD system range between 5 and 20 dB.[12] The lower extreme holds for a Si-APD at $\lambda = 0.8\mu m$, the upper extreme for an APD made of Ge or of compound semiconductors at longer wavelengths.

FSK and PSK coherent detections allow a receiver sensitivity improvement of 3 dB and 6 dB respectively over ASK coherent detection. Finally, ASK and FSK envelope detections exhibit a decline in the receiver sensitivity with respect to the corresponding coherent detections of about 1 dB; as do PSK differential coherent detection with respect to PSK coherent detection. As a consequence of the above receiver sensitivity improvements, which appear particularly attractive at the longer wavelengths where the fibers have better performances, transmission with more than two levels is convenient. For FSK and PSK, the receiver sensitivity may be improved by 3 dB and 5 dB for a four level and an eight level transmission respectively, provided that the frequency and phase fluctuations of the optical source can be neglected.[12]

In Table 4.1 the advantages of receiver sensitivity improvement, which differentiate coherent transmission from traditional IM-DD transmission, are summarized. The maximum improvements, reached at the longer wavelengths, have been considered. Another advantage of a heterodyne detection over a direct detection is the better selectivity for separating different channels in WDM transmissions. This is due to the possibility of performing this separation at radio frequencies, instead of optical frequencies. Thus, a very small carrier separation can be employed, leading to a more efficient use of the wavelength region of zero material dispersion. *

4.2.2 Main Problems of Practical Coherent Transmissions

The coherent combination of two wave fronts, as shown in Fig. 4.7, is possible only when there is a matching between the polarizations of the two fields. A monomode fiber is necessary for transmission in coherent systems. Fluctuations in the polarization state of the propagating field due to birefringence is possible, and may seriously affect the mixer efficiency. A technique which has been proposed to overcome this problem is based on a separate detection of the two polarization states of the received field and on a later summation of the two detected signals.[13]

Table 4.1 Receiver Sensitivity Improvements of Different Types of Coherent Systems Based on Heterodyne Detection (Maximum Theoretical Predictions).

Type of System	Noncoherent Detection	Coherent Detection 2 levels	Coherent Detection 4 levels	Coherent Detection 8 levels
IM-DD	0 dB	—	—	—
ASK	19 dB	20 dB	—	—
FSK	22 dB	23 dB	26 dB	28 dB
PSK	25 dB	26 dB	29 dB	31 dB

*See notes pg. 497

Other problems of practical coherent transmissions are in the spectral purity of the optical source, i.e., on the thickness of its spectral width, and on its frequency stabilization. In 3.1.7 we have seen that the smallest spectral width of a laser diode emission, when it is inserted in a suitable long cavity external reasonator, can be on the order of 100 kHz. Possible shifts in the emission spectrum may, however, be present during operation. Their extension, when the ambient temperature and the injected current are rigorously controlled, can be as small as a few MHz.[13] These data encourage experiments on coherent transmission. *

Direct frequency modulation of the laser diode has been realized by varying the injection current.[14] A current variation as small as 1 mA induces a frequency deviation of about 200 MHz. This current modulation does not disturb the single mode operation of the laser diode. Optical phase modulation has also been proven, using a pair of laser diodes suitably coupled.[15] The possibility of coherent FSK transmission, with heterodyne detection, has been already proven experimentally.[16] However, PSK transmissions appear to be more difficult to realize in practice at the moment, especially for modulation signals with high capacity.

REFERENCES

1. Shumate, P. W., and Di Domenico, M. "Lightwave Transmitters," from *Semiconductor Devices for Optical Communication*, H. Kressel, editor, Berlin: Springer Verlag (1980).
2. Personick, S. D. "Receiver Design for Digital Fiber Optic Communication Systems, II," *Bell Syst. Tech. J.*, 52, pp. 875-886 (1973).
3. Smith, R. G., and Personick, S. D. "Receiver Design for Optical Fiber Communication Systems," *Semiconductor Devices for Optical Communication*, H. Kressel, editor, Berlin: Springer Verlag (1980).
4. Abramowitz, M., and Stegun, I. A. *Handbook of Mathematical Functions,* New York: Dover Publishers (1968).
5. Cancellieri, G., and Frosini, F. "Discussion on Equalization in Optical Fiber Transmission Systems," *J. Opt. Comm.,* 3, pp. 146-151 (1982).
6. Kasper, B. L. "Equalization of Multimode Optical Fiber Systems," *Bell Syst. Tech. J.,* 61, pp. 1367-1388 (1982).

*See notes pg. 497

7. Paladin, G., and Pietroiusti, R. "Considerations on the Receivers of Optical Fiber Digital Transmission Systems," *Alta Freq.,* 49, pp. 8017 (1980).
8. Personick, S. D., "Comparison of Equalizing and Nonequalizing Repeaters for Optical Fiber Systems," *Bell Syst. Tech. J.,* 55, pp. 957-971 (1976).
9. Albrecht, W., *et al.,* "Experiences with an Optical Long-haul 2.24 Gbit/s Transmission System at a Wavelength of 1.3 μm," *Electron. Lett.,* 18, pp. 746-748 (1982).
10. Yamada, I., and Kimura, T. "Characteristics of Gbit/s Optical Receiver Sensitivity and Long-span Single Mode Fiber Transmission of 1.3 μm," *IEEE J. Quant. Electron.,* QE-18, pp. 718-727 (1982).
11. Gagliardi, R. M., and Karp, S. *Optical Communication,* New York: John Wiley & Sons (1976).
12. Okoshi, T., Emura, K., Kikuchi, K., and Kersten, R.T. "Computation of Bit-error Rate of Various Heterodyne and Coherent-type Optical Communication Schemes," *J. Opt. Comm.,* 2, pp. 89-96 (1981).
13. Okoshi, T., Kikuchi, K. "Heterodyne-type Optical Fiber Communication," *J. Opt. Comm.,* 2, pp. 82-88 (1981).
14. Kobayashi, S., Yamamoto, Y., and Kimura, T. "Modulation Frequency Characteristics of Directly Optical Frequency Modulated AlGaAs Semiconductor Laser," *Electron. Lett.,* 17, pp. 350-351 (1981).
15. Kobayashi, S. and Kimura, T. "Optical Phase Modulation in an Injection Locked AlGaAs Semiconductor Laser," *IEEE Trans. Mic. Th. & Tech.*, MTT-30, pp. 1650-1657 (1982).
16. Saito, S., Yamamoto, Y., and Kimura, T. "Optical FSK Heterodyne Detection Experiments Using Semiconductor Laser Transmitter and Local Oscillator," *IEEE J. Quant. Electron.,* QE-17, pp. 681-689 (1981).

CHAPTER 5
SCANNING MEASUREMENTS

Scanning measurements are devoted to investigation of the spatial and angular properties of emitting and receiving surfaces. They are employed both for fibers and for transducers. In fiber measurements, the main purpose of scanning is to test the behavior of single families of modes or rays. In this case, the selection can be done either at the fiber input end, through a selective excitation, or at the fiber output end, through a selective detection. The most interesting kinds of fiber measurements are:

i) differential mode attenuation (DMA)

ii) differential mode delay (DMD)

Even measurements of numerical aperture, core diameter, and index profile are often performed by a scanning technique.

The measurements on transducers can be divided into:

i) measurements on optical sources, in which the emitted optical power is detected in a selective way

ii) measurements on photodetectors, in which a selective excitation of the active area is performed.

In this case, both amplitude and delay measurements are possible. In the following, we will refer to far field measurements when angular selection is required, and to near field measurements when the scanning is made spatially over the surface under test. The meanings of far and near field are obvious.

5.1 SCANNING MEASUREMENTS ON FIBERS

The objects of a scanning measurement on a multimode optical fiber can be several. If mode coupling is negligible, it is possible to obtain an experimental evaluation of the mode dependent part of the intrinsic attenuation coefficient from a differential mode attenuation measurement. When material dispersion can also be considered negligible, a differential mode delay measurement may help to estimate the fiber baseband response, and even to determine the value of parameter α which characterizes the fiber index profile. Finally, a scanning measurement, when it is present, can give an insight into the mechanisms which lead to mode coupling.

After consideration of the diffraction limit of a scanning technique, we will briefly describe DMA and DMD measurements, and give an introduction to indirect tests which can be performed on mode coupling effects via scanning measurements. These subjects will be investigated with more detail in Chapters 9, 10 and 11, respectively, especially for the electric part of the arrangements.

5.1.1 Diffraction Limit

A diffraction limit, which relates angular and spatial resolution of scanning procedures, must be considered in fiber and transducer measurements. In 1.1.3 we have discussed the existence of an electromagnetic mode on an aperture of cross section area A within a solid angle Ω, A and Ω being related by

$$A\Omega = \lambda^2 \tag{5.1}$$

where λ represents the wavelength. We consider for the measurement a spatial resolution δx and an angular resolution $\delta\theta_a$. Assuming an elementary circular aperture and an elementary solid angle, we can write

$$A = \frac{\pi}{4}\,\delta x^2$$

$$\Omega = \pi\,\delta\theta_a^2$$

Substitution into (5.1) leads to the following difraction limit

$$\delta x \cdot \delta\,\theta_a \cong 0.62\,\lambda \tag{5.2}$$

in which $\delta\theta_a$ is measured in radians, and δx and λ are usually expressed in μm.

From the experimental point of view, it is possible to select even with higher resolution than that imposed by (5.2), but in this case the measurement cannot be univocally related to the point selected. The diffraction limit is useful in order to fix the maximum number of independent measurement points which can be selected. Performing only a far field scanning, we have $\delta\,x \to \infty$, whereas, performing only a near field scanning, we have $\delta\theta_a \to \infty$. In these two cases, a very dense selection in θ_α and in x can be made, but any information on the conjugate variable is completely lost.

5.1.2 Optical Arrangements

A very simple optical apparatus,[(1),(2)] which allows a near and far field scanning selection, is shown in Fig. 5.1. It is based on a microscope objective, with magnification M and numerical aperture NA_m, which must be greater than the fiber numerical aperture NA to excite or detect all the fiber guided modes. For typical values of M and NA_m of commercially available microscope objectives, see Appendix 1. When the end face of the fiber is placed in the focus of the microscope objective, its image, which is formed on the observation plane behind the objective itself, exhibits any linear dimension as magnified by M. At the same time, the angle θ'_a results M times smaller than the angle θ_a. So, the magnification M must be chosen considering both the spatial and the angular resolutions which have been fixed.

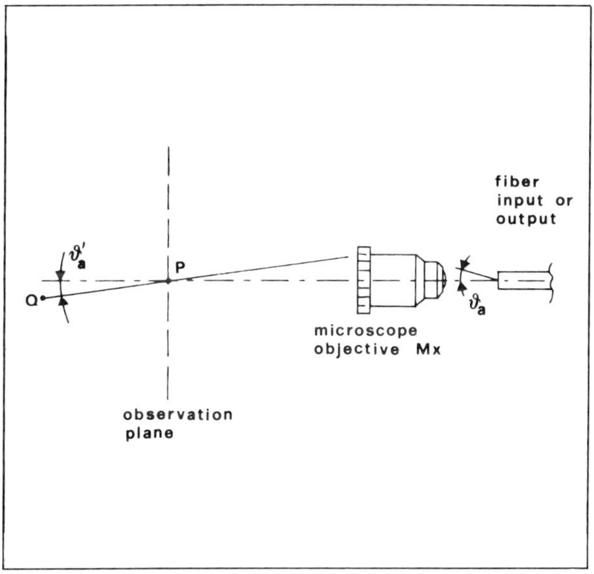

Fig. 5.1 - Optical apparatus for scanning measurements.

A point P on the observation plane is usually selected by a pinhole driven by micropositioners. The optical source or the photodetector is placed in the point Q for the input or output selection, respectively. When the dimensions of such devices are too large, so that the uncertainty of θ'_a is too great, another pinhole is placed in Q, either in front of the emitting surface or the active area. Since θ'_a is usually a very small angle, Q can vary on a plane instead of on a spherical surface.

Typical values of the parameters involved in this optical apparatus, for standard multimode fibers, with core diameter $2a$ = 50 μm and NA = 0.2, are: magnification M = 20 and pinhole diameter $d \doteq 100$ μm, so that the spatial resolution of a near field measurement is 5 μm. The angular resolution depends on the distance D between Q and P. Denoting d as the second pinhole diameter, we have the angular resolution of a far field measurement

$$\Delta \theta_a \cong \frac{Md}{D} \quad [rad]$$

For example, with D = 10 cm, $\Delta\theta_a \cong 0.04$ rad, i.e., about 1.5 degrees.

Although the optical apparatus of Fig. 5.1 can be used either for exciting or for detecting single families of modes, it is most frequently employed in the latter case, while other modified schemes can be practically employed for a selective excitation. A small number of modes can be launched into a graded index fiber following a technique described in (3), based on the optical apparatus shown in Fig. 5.2. A laser beam is focused on the fiber input end face to a small spot of light. A trade-off must be considered between two extreme tuations:

i) a plane wave with an infinite spot size, which can be obtained in principle by using a zero magnification objective

ii) a null spot size with an infinite divergence, which can be obtained in principle by using an infinite magnification objective

The optimum spot size to launch the minimum number of modes into a near-parabolic fiber, for the usual laser beam sizes, is reached with a magnification $M \cong 10$, and the resulting spot has a diameter smaller than 10 μm.[3]

In 1.3.7 we have shown that for $\theta \cong 0$, i.e., for an excitation perpendicular to the fiber cross section, there is a relationship between the radial coordinate r at which a mode is excited or detected and its variable ϵ. This relationship is expressed, for near-parabolic fibers, by equation (1.97). Using the apparatus of Fig. 5.2, it is possible to select modes with different values of ϵ, maintaining the laser beam parallel to the fiber axis. The main advantage of this technique, over that of Fig. 5.1, for launching selections is in the possibility of exploiting the ultra-high radiance of a laser beam. In reference (4), the selective excitation of groups of modes in near-parabolic fibers in directions not parallel to the fiber axis has been theoretically investigated.

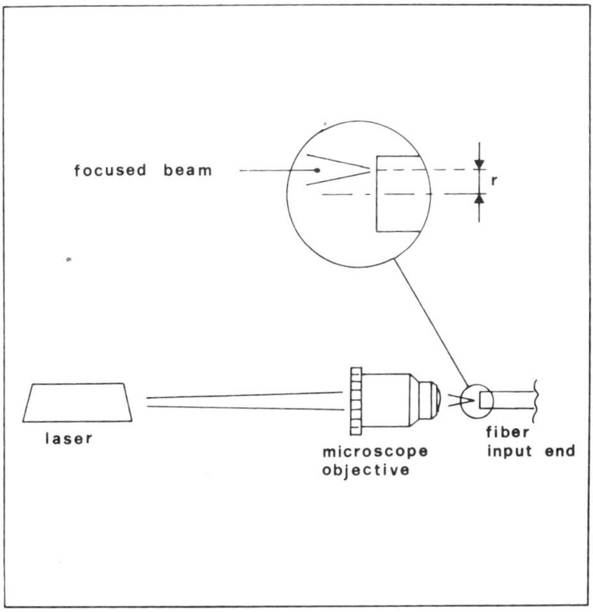

Fig. 5.2 - Optical apparatus for the excitation of a small number of modes.

The optical arrangement based on the use of a pair of lenses, shown in Fig. 5.3, allows one to select modes having a different ϵ.[5] It is preferrable to perform an output selection. The traces of rays corresponding to modes with $\theta \cong 0$ and with $\theta \neq 0$ have been reported in continuous lines and in dashed lines respectively. Three remarkable planes are shown in Fig. 5.3:

i) a far field plane behind the first lens

ii) a near field plane before the second lens

iii) a far field plane behind the second lens

Maintaining the photodetector in the position Q on the second far field plane allows one to obtain all the modes having $\theta \cong 0$. Next, the selection of the radial coordinate, the variable ϵ, can be performed by placing a series of concentric annular rings on the near field plane. In this way it is possible to collect the whole family of modes characterized by a given value of ϵ in Q, leading to a noticeable increase in the level of the detected optical power with respect to the previously described techniques.

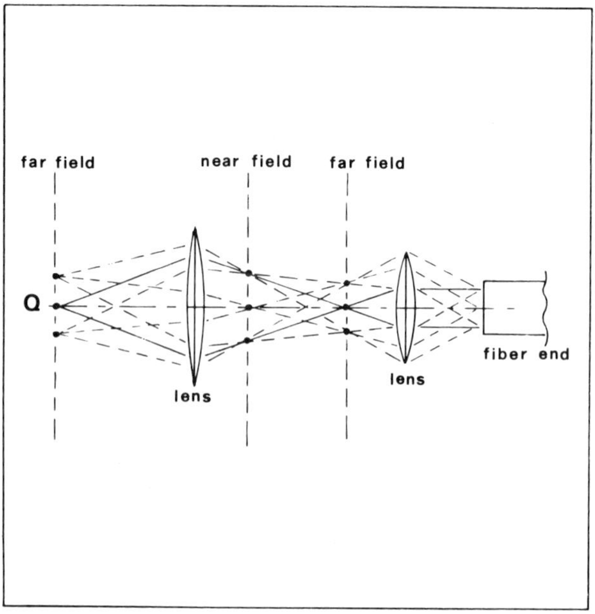

Fig. 5.3 - Scanning apparatus for the selection of modes having different ε [from (5)].

To have near and far field selection on the same plane, a microscope objective can be used, followed by a pair of lenses: the first one is a conventional round symmetry lens, and the other one is a cylindrical lens. Two slits are also employed in this setup: the first one is vertically mounted on the far field plane behind the microscope objective, the other one is horizontally mounted on its observation plane in front of the first lens. On the vertical axis of a given plane, behind the cylindrical lens, the output angle θ_a can be measured. On the horizontal axis of the same plane, it is possible to measure the eccentricity ρ/a. Here the two slits are fixed, whereas for the previously described techniques the movements of pinholes or annular rings were necessary for the selection. This arrangement has been proposed in (6) for an output selection, i.e., for plotting the optical power distribution in a two dimension domain (ρ/a, θ_a) directly from an experimental test.

In any scanning measurement on optical fibers, no matter what technique is employed, it is extremely important to prepare the fiber properly (see Appendix 3). In fact, even very small microfractures or dust may alter the measurement results completely. Before concluding the present section, we will list the kinds of selection usually employed for geometrical and index profile measurements. A measurement of the fiber *NA* is typically based on a far field scanning. Some techniques for the measurement of the index profile, and for the determination of the core diameter, are based on a near field selection. Scanning may be performed either at the output[7] or at the input[8] of the fiber. Measurements of index profile will be described in Chapter 7. Measurements of core diameter and *NA* will be treated in Chapter 8. We will next discuss differential mode measurements on fibers.

5.1.3 Differential Mode Attenuation Measurements

A DMA measurement is usually based on the assumption that mode coupling has negligible effects. Such a measurement is particularly significant only for uncabled fibers. The intrinsic attenuation coefficient, a function of the two constants of motion, E and h, characterizes the fiber. It can be written as

$$\gamma\,[E,h] = \gamma_o + \Delta\,\gamma\,[E,h]$$

where γ_o is common to all modes and $\Delta\gamma(E, h)$ is the mode dependent part. In DMA measurements performed by mode selection at the fiber input, rays having the same pair of (E, h) are injected into the fiber each time, and the total output power is detected. For DMA measuremnts performed by an output mode selection, the whole domain of guided rays is filled at the input fiber end, through a uniform excitation, and each time a ray with a particular pair (E, h) is selected.

In both cases, it is necessary to verify that the launching condition does not influence the measurement. For this test, the fiber must be cut close to its input section. For an input selection, the differential excitation is repeated, after cutting, and the whole guided power is detected. For an output selection, the differential detection is repeated, after cutting, maintaining the same excitation of the whole guided power. A good repeatability in fiber end preparation and a noticeable stability of the optical source are required. Also the stability and repeatability of selection assured by the optical apparatus must be considered in the evaluation of the accuracy of such measurements.

For step index fibers, a selection of variable E can be easily obtained by varying the propagation angle θ through a far field scanning. For this type of fiber a weak h dependence of $\Delta\gamma$ is expected, except for leaky rays. However, h can be selected by varying the eccentricity ρ/a. But for graded index fibers, a selection in E requires both near and far field scannings. Nevertheless, assuming that $\Delta\gamma$ is practically independent of h, one can let $\theta \cong 0$, and thus the measurement can be performed varying only the radial coordinate r, as shown in Fig. 5.2. Examples of input and output selections in step index fibers are given in (9) and in (2), respectively. For graded index fibers, similar examples are given in (3), (10) and in (1), (11). In 9.4 the results of these techniques will be discussed in more detail.

5.1.4 Differential Mode Delay Measurements

The objectives of a DMD measurement, by which the relative delay distribution is obtained, can be:

i) the evaluation of the fiber baseband response, when material dispersion is negligible[(2)]

ii) the determination of the index profiled at a given wavelength[(12)]

iii) the evaluation of the effects of profile dispersion[(5)]

In all these cases, it is necessary to assume that mode coupling has negligible effects. Either input or output mode selection can be performed, as well as DMA measurements.

The selection of the propagation angle θ alone is usually sufficient for step index fibers, whereas near and far field scannings are required for graded index fibers, except when $\theta \cong 0$. In references (9) and (2) respectively, examples of input and output selections are given for step index fibers, in references (12), (13) and in (5), (14) for graded index fibers. The measurement of the relative delay distribution can be made using:

i) a pulse modulated signal and a sampling oscilloscope

ii) a sinusoidally modulated signal and a vector voltmeter

In the first case, the relative delay distribution of the pulse peaks is directly evaluated. In the other case, the differential phase ϕ_i is measured

$$\phi_i = 2\pi f_o \tau_i$$

where f_o is the modulation frequency, and τ_i represents the relative delay of the ith mode group, with respect to a given mode group, assumed as characterized by $\tau = 0$. The first technique is discussed in (5), (12), (13), the other one in (2), (14), (15). As will be shown in 10.4, a better sensitivity is usually obtained through a sinusoidally modulated signal.

Recalling the results discussed in 1.3.8 about the influence of the spatial relative delays of emission of the optical source on the baseband response of a multimode fiber, one can conclude that DMD measurements on step index fibers are not affected by the launching condition. In fact, the relative group delays are considerably larger than the spatial relative delays which characterize the optical sources. For graded index fibers, on the contrary, especially when an output selection is employed, this phenomenon may greatly limit the accuracy of the measurement. In particular, as will be shown in 6.2, spatial relative delays as large as some hundreds of ps may occur, and near parabolic fibers, on distances of the order of 1 km, exhibit relative group delays of the same order of magnitude. So, another measurement is required for comparison, cutting a short length of fiber, and maintaining the same excitation condition.

5.1.5 Indirect Tests for Investigating Mode Coupling Effects

An experimental evaluation of mode coupling effects is possible through input or output scanning measurements, considering the mode power distribution only (16), (17), (18), or, more frequently, also the mode delay distribution (2), (19), (20), (21). In (16) a selective excitation and a selective detection have been performed on a step index fiber. From the comparison among different mode power distributions measured at various distances from the input section under the same selective excitation, it is possible to investigate the diffusion of the optical power in the domain of guided modes. This allows the experimental verification of the assumption made in 2.2.5. Once the model is verified, it is sufficient to compare the amplitude of input and output optical power distributions to determine the behavior of the coupling coefficient $h_\infty(n)$. Although this is probably the most accurate method for a direct evaluation of mode coupling effects, it is rather complex, requiring two scanning equipments. Its implementation is practically limited to step index fibers, which can be excited under different propagation angles θ by a collimated beam laser.

In reference (18) a technique has been proposed which requires only one mode selection for step index fibers. The amplitude of the mode power distribution, detected after reaching steady-state is used to evaluate the behavior of $h_\infty(n)$ indirectly. It is interesting to note that this evaluation is possible only if any mode dependent intrinsic attenuation is assumed negligible. In other words, this technique is completely the opposite of a DMA measurement, where the mode dependent part of γ should be evaluated assuming that mode coupling has quite negligible effects.

The techniques proposed in (19) (20) aim to evaluate mode coupling coefficients and differential mode attenuation by the measurement of the response of several groups of modes to a pulse modulated signal. An output mode selection is performed, and a comparison is made between input and output pulses, the former evaluated again cutting a short length of fiber. This method has been implemented only for step index fibers, and is based on an optical power flow equation of the type (2.45), where the independent variable to be considered is the propagation angle θ, and γ itself depends on θ. However, in principle, it can be extended to graded index fibers.

In reference (2) a comparison has been proposed between the measured mode delay distribution on a step index fiber and the results of a theoretical prediction, in which mode coupling has been ignored. From such a comparison, the effects of mode coupling can be qualitatively evaluated. This test has been extended to graded index fibers in (14). Finally, quantitative results, both for step and for graded index fibers, have been presented in reference (21). An accurate measurement of the index profile is also necessary when graded index fibers are considered, for the theoretical prediction of the mode delay distribution without mode coupling. The method suggested in reference (21) is not based on an optical power flow equation, but on simple summations of the optical powers carried by different groups of modes, and multiplied by a set of unknown coupling coefficients. The purpose of the measurement is to determine these coupling coefficients and a possible mode dependent attenuation. The optical source is sinusoidally modulated, and an output mode selection is performed.

Recently,[22] the following test has been proposed to verify the validity of the assumption that the mode group number n of equation (2.39) is the only independent variable of mode coupling processes. The beam of a He-Ne laser is launched into the test fiber, exciting a group of modes characterized by $\nu \cong 0$ and $\mu \cong \mu_0$, where μ_0 is a given permitted value of μ. To avoid speckle patterns due to the coherence of the laser light, a mechanical vibration has been transmitted to the fiber. Performing a series of near and far field scanning measurements, to select ν and μ at different distances from the input section, it has been shown that a noticeable distance is necessary to make negligible any optical power transfer among quasi-degenerate modes denoted by the same mode group number n. This distance depends on the kind and the amplitude of the fiber perturbation; for practical cases, it is on the order of several hundred meters.

Chapter 11 is devoted to describing and discussing with greater detail some of the techniques schematically presented here. In the next section we will stress the convenience of using the domain of guided modes (r_1, θ_1), introduced in 1.2.10, to perform complete mode selections.

5.1.6 Advantages of Domain $(\mathbf{r}_1, \theta_1)$

In 5.1.2 we have shown the main optical arrangements for mode selection in graded index fibers. We can have:

i) a simplified mode selection, based on the single mode variable ϵ

ii) a complete mode selection, based on a pair of mode variables

The first approach is justified only for near-parabolic fibers when the index profile distribution is not known. In this case, equation (1.97) can be used to relate variable ϵ and the radial coordinate r. Alternatively, if the profile parameter α is known, it can be applied to any graded index fiber, using the general equation (1.96). Recently[(23),(24)], measuring the optical power distribution, as a function of one independent mode variable, with a simple near field scanning procedure has been experimentally demonstrated. It is impossible to derive a unique two variable optical power distribution from only a near field scanning, even if the index profile is known.[(25)] Nevertheless, for optical power distributions depending on a single mode variable, it is effectively possible.

The second approach, on the other hand, is completely general, but, except for $\alpha \cong 2$, it also requires the knowledge of the index profile. In fact, the resolution of mode selection, and thus the accuracy of the method, depends on the number of modes for each selected mode family. Since mode distribution in the domain G of guided modes is influenced by the index profile, its knowledge is useful. For example, the mode group number n increases with increasing the radial coordinate r, and the spatial resolution must be set considering not only the particular value of r as in reference (4), but also the value of α which imposes this dependence.[(26)] Furthermore, when the single contributions of the various groups of modes must be added up to obtain the resultant behavior of the fiber, each contribution must be multiplied by a suitable weight, which accounts for geometrical and guidance properties. A detailed analysis of the geometrical properties is reported in reference (27). As for the guidance properties, they depend again on the index profile.

From the above considerations, we can conclude that the knowledge of the index profile distribution is of fundamental importance for a correct interpretation of scanning measurements. This is true, independent of the mode selection (simplified or complete) performed. It is possible to ignore the index profile only if the selection is not based on guidance properties, but solely on geometrical properties. In this case, more than two coordinates must be scanned. An example of this procedure is reported in reference (21). We will discuss it in 11.2.

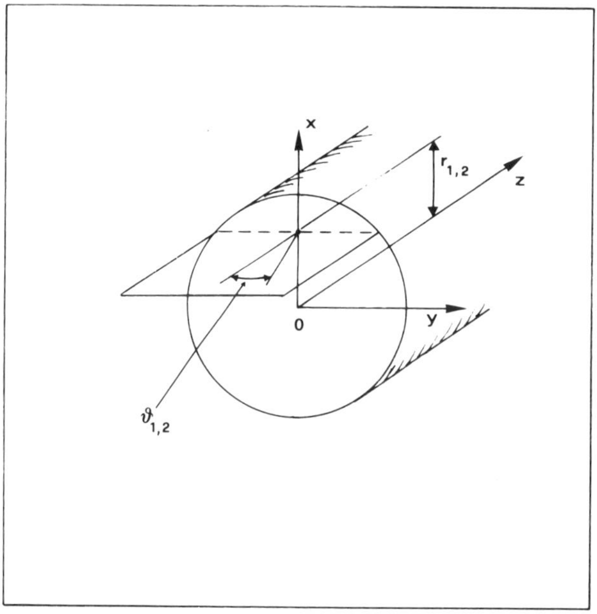

Fig. 5.4 - Mode selection in domain (r_1 , θ_1).

For what concerns the choice of the pair of mode variables for a complete selection, considering also the fiber guidance properties, it is not advisable to employ the domain (E, h) or (ϵ, h), because relating them to the pair of coordinates r and θ, obtained directly by near and far field scanning, would be cumbersome. A better choice is the pair (r_1, θ_1), which can be measured directly by means of the optical arrangement of Fig. 5.1, whose observation plane has been denoted xy in Fig. 5.4. The selection, made through a near field scanning along the radial direction x, and a far field scanning on a plane perpendicular to x, can give a pair (r_1, θ_1) or a pair (r_2, θ_2). Recall that θ_2 is the propagation angle which occurs when the ray trajectory reaches its maximum distance $r = r_2$ from the fiber axis.

The curve e of the helical rays, which separates domain (r_1, θ_1) and domain (r_2, θ_2), as discussed in 1.2.10, can be obtained from the knowledge of the index profile $n(r)$. Then, for a given radial coordinate r, to perform the scanning only in domain (r_1, θ_1), the angle θ must be greater than the helical ray propagation angle corresponding to r. So, for each near field point (selection of $r = r_1$), several far field points (selection of $\theta = \theta_1$) are considered, provided that the angle selected is greater than a minimum angle, given by

$$\sin^2\theta_1 = -\frac{r_1}{2}\left.\frac{dn^2}{dr}\right|_{r=r_1} \tag{5.3}$$

which usually increases with increasing r_1.

For near-parabolic fibers, the domains (r_1, θ_1) and (r_2, θ_2) appear as shown in Fig. 5.5, where the helical ray curve e (practically a straight line) is dashed. The former domain lies under the curve e, the latter above. From the conservation of the constants h and E, it is

$$r_1\,\theta_1 \cong r_2\,\theta_2 \tag{5.4}$$

$$\theta_1^2 - \theta_2^2 \cong 2\,\Delta\left[\left(\frac{r_2}{a}\right)^2 - \left(\frac{r_1}{a}\right)^2\right]$$

and from the expression of curve e, given by (5.3), we have

$$\theta_e \cong \sqrt{2\Delta}\,\frac{r_e}{a} \tag{5.5}$$

Now, let us consider that a given pair (r_1, θ_1) corresponds to a point A in Fig. 5.5. The two points on curve e, determined from (5.5), letting $r_{e1} = r_1$ and $\theta_{e2} = \theta_1$, have corresponding coordinates given by

$$\theta_{e1} \cong \sqrt{2\Delta}\,\frac{r_1}{a}$$

$$r_{e2} \cong \frac{a}{\sqrt{2\Delta}}\,\theta_1$$

θ_{e1} and r_{e2} satisfy equations (5.4) respectively as θ_2 and r_2, therefore the point in domain (r_2, θ_2) which corresponds to the given point A in domain (r_1, θ_1) is that denoted by A' in Fig. 5.5.

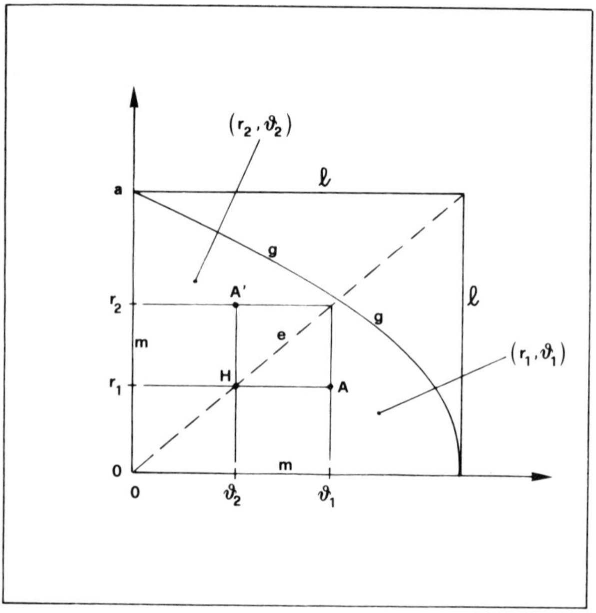

Fig. 5.5 - Domains (r_1 , θ_1) and (r_2 , θ_2) for near-parabolic fibers.

Thus, for this class of fibers a double scanning procedure can be made, collecting two measurements for each group of modes. This allows an easy approximate evaluation of the accuracy of the procedure itself. In fact, the behavior of the measurement over the line HA (obtained for a fixed $r = r_1$, varying θ_1), should be equal to that over the line HA' (obtained for a fixed $\theta = \theta_2$ varying r_2) because the same rays are considered in the two cases.

To summarize, we can conclude that the use of either domain (r_1, θ_1) or domain (r_2, θ_2) is very useful for performing scanning measurements. Practical applications of this technique have been described for an output selection in reference (28) where good spatial and angular resolution can be achieved easily. However, in principle, an input selection appears possible.

5.2 SCANNING MEASUREMENTS ON TRANSDUCERS

As already introduced, scanning measurements on optical sources (LEDs and laser diodes) detect the emitted optical power in a selective way, while scanning measurements on photodetectors are performed through a selective excitation of the photodiode active area. In both cases, either an amplitude or a relative delay can be measured. However, for the optical sources, especially top emitting LEDs, there are noticeable spatial relative delays of emission, whereas in photodiodes this effect is practically negligible.

Scanning measurements are sometimes performed with a spectral investigation. A very good sensitivity is necessary, in this case, owing to the low level of emitted or received optical power measured after the wavelength selection. Optical sources and photodiodes realized expressly for use in transmission systems are frequently produced with a short fiber pigtail directly attached, this simplifies coupling to the fiber. In this case, scanning measurements on transducers turn out to be very similar to those on fibers. Nevertheless, to improve the technological processes of device fabrication, it is necessary to test the geometrical properties of the light generation and the light detection. This is done by scanning measurements made directly on the emitting surface of LEDs and laser diodes, or on the active area of photodiodes.

5.2.1 Scanning Measurements on LEDs and Laser Diodes

The optical apparatus reported in Fig. 5.1, substituting the emitting surface for the fiber output end, is commonly employed for a selective detection of the emitted optical power from a LED or a laser diode. Varying point P, we obtain the near field pattern, and varying point Q, the average far field pattern. If we select, for instance, by means of two pinholes, both in P and in Q, we obtain the far field pattern of a precise point on the near field distribution. This is the most complete investigation of the radiation behavior that can be done on an emitting surface. However, near and far field properties are often separable, especially for spontaneous emission. The use of a pinhole in Q, followed by a photodiode, has been suggested in reference (29). Alternatively, the input section of an optical fiber, provided that its NA is greater than the divergence of the beam whose power must be measured, can be employed as well.[30]

The diffraction limit discussed in 5.1.1 for optical fiber measurements must be considered again, for device measurements, when the far field patterns of some near field points are investigated. The measurement of top emitting LEDs is very similar to an output scanning measurement on optical fibers owing to the cylindrical symmetry of their spatial and angular emission. On the other hand, edge-emitting LEDs and laser diodes have emitting surfaces which are approximately rectangular, with one dimension much greater than the other, and characterized by a far field radiation pattern with different apertures in the two directions. First of all, the position of the emitting surface, with respect to the scanning apparatus, must be determined by a near field scanning procedure. Next, it is necessary to place the emitting surface as it is shown in Fig. 5.6, so the radiation pattern of the device can be completely characterized by scanning the near field along x and y and the far field along θ_x and θ_y. θ_x and θ_y correspond, respectively, to $\theta_{//}$ and $\theta_{\perp}$ introduced in 3.1 for edge emitting devices.

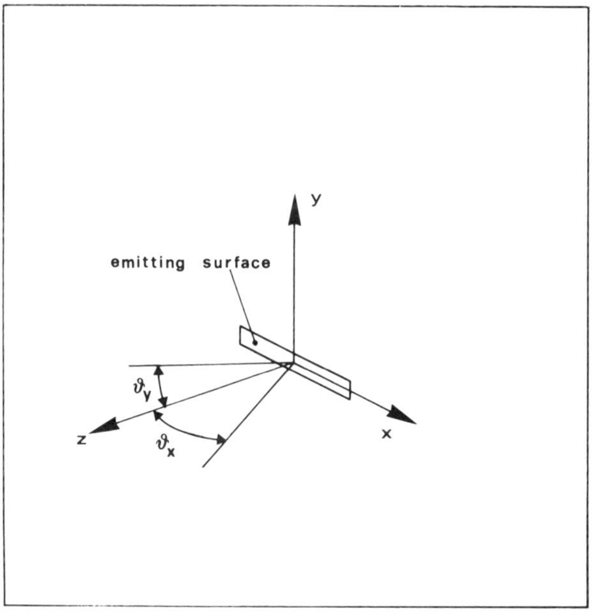

Fig. 5.6 - Scanning measurements on edge-emitting LEDs and laser diodes.

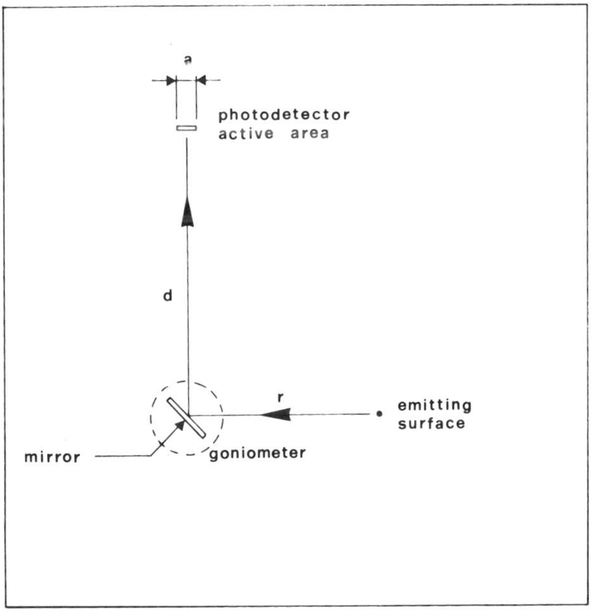

Fig. 5.7 - Simplified technique for far field scanning [from (32)].

In particular, for a laser emission, the number of filaments of near field pattern along x is related to the number of families of lateral modes ($s = 1, 2, \ldots$). The number of lobes in the far field pattern along θ_y is related to the number of families of transverse modes ($m = 1, 2, \ldots$), as discussed in 3.1.4. An interesting experimental test has been reported in reference (31), where a combined spectral and angular investigation has been performed on laser diodes.

A technique which is simpler from the experimental point of view for far field scanning not accompanied by a near field selection, is that based on a mirror mounted on a goniometer, without any microscope objective.[32] In this case, as shown in Fig. 5.7, both the device under test and the detector are fixed, and along the distance d even a monochromator may be inserted to add a spectral investigation. To maintain the vertical branch practically fixed, we must let $r \ll d$. The angular resolution $\Delta\theta$ becomes

$$\Delta\theta \cong \frac{a}{d + r} \cong \frac{a}{d} \tag{5.6}$$

where r and d are the lengths of the two branches, before and after the mirror, and a is the lateral extension of the photodetector active area. If this were too large, a pinhole could be inserted in front of it. This, however, reduces the amount of collected power.

For laser diodes, a value of $\Delta\theta$ fine enough for investigating the far field pattern in the junction plane (direction x), characterized by the smallest divergence, is 0.01 rad, i.e., about 0.5 degrees, Letting $a = 2$ mm, from (5.6) we have $d = 20$ cm. Near field scanning procedures on edge emitting LEDs and laser diodes along the y direction are the most critical, because the overall width of the active region in this direction can be as small as 0.1 μm. Therefore, even using an objective magnification $M = 50$, in order to attain a spatial resolution along y comparable with this very small distance, the pinhole diameter to select near field points must not be greater than a few microns.

5.2.2 Scanning Measurements on Photodiodes

Scanning measurements on photodiodes test the dependence of the responsivity on the spatial and angular coordinates of the received light. We must distinguish between two classes of photodiodes:

i) small-area photodiodes, which are those employed in transmission systems and in many experimental setups for fiber characterization, where the speed of response must be as fast as possible

ii) large-area photodiodes (e.g., solar cells), used essentially in fiber characterization, without particular requirements on the speed of response

Small-area photodiodes are usually covered by a transparent protective coating. They have shown a substantial uniformity of response, except when the angle of incidence exceeds 30°. Only for large-area PIN photodiodes or solar cells, like those employed for scattering loss measurements on uncabled fibers (which will be shown in 9.1), responsivity fluctuations as large as several percent may be expected. For this kind of device, the spatial position on the active area is more critical than the angle of incidence.

In reference (33), the effects of detection inhomogeneity on fiber attenuation measurements, performed by a cutback technique, have been stressed. For all measurements on fiber transmission properties based on comparison, the uniformity of response over the active area is essential to good repeatability, and hence acceptable accuracy. Some tests used to verify the uniformity of response of a solar cell have been presented in reference (34). For one of these tests, the active area is scanned by a small-spot beam and the output signal is recorded. This test is usually performed to detect the possible presence of surface defects. Another test consists of varying the spot size of a beam using different focusing procedures. The output signal should remain unchanged, provided that the whole spot is collected by the active area. This test assures that the same output signal is given by a photodetector when the optical power at the output end of a fiber incides on its active area, independent of the distance between the fiber end and the photodiode surface.

To determine the effect of variations in the angle of incidence, the light of a collimated beam laser can be sent to the photodiode under a variable angle. In this way, it has been shown in reference (34) that the two polarizations (parallel and perpendicular to the surface plane) exhibit different behaviors. However, in both cases, even for angles of incidence as large as 60°, the reduction in the responsivity is only a few per cent, showing that the reflectivity of the surface, which belongs to one of the two semiconductor layers, is very small.

REFERENCES

1. Olshansky, R., Oaks, S. M.,and Keck, D. B., "Measurement of Differential Mode Attenuation in Graded Index Fiber Optical Waveguides," *Top. Meet. Opt. Fiber Transm.*, Williamsburg, 1977, TuE5.
2. Cancellieri, G. "Time-dispersion Measurement in Optical Fibre by Near and Far Field Scanning Technique," *Electron. Lett.*, 14, pp. 465-467 (1978).
3. Jeunhomme, L. and Pocholle, J. P. "Selective Mode Excitation of Graded Index Optical Fibers," *Appl. Opt.*, 17, pp. 463-468 (1978).
4. Saijonmaa, J., Sharma, A. B., and Halme, S. J. "Selective Excitation of Parabolic-Index Fibers by Gaussian Beams," *Appl. Opt.*, 19, pp. 2442-2452 (1980).
5. Costa, B., Esposto, F., and Sordo, B. "Wavelength Dependence of Differential Group Delay in Graded Index Optical Fibers: Application to Fiber-Links Characterization," *Opt. Fiber Comm.*, Washington, pp. 122-125, (1979).
6. Stewart, W. J., "Method of Measuring Power Distributions in Graded and Step Index Fibers," *Opt. Fiber Comm.*, Washington, pp. 116-119, (1979).
7. Sladen, F.M.E., Payne, D.N., and Adams, M.J. "Determination of Optical Fiber Refractive Index Profiles by Near Field Scanning Technique," *Appl. Phys. Lett.*, 28, pp. 255-258 (1976).

8. Arnaud, J.A., and Derosier, R.M. "Novel Technique for Measuring the Index Profile of Optical Fibers," *Bell Syst. Tech. J.*, 55, pp. 1489-1508 (1976).

9. Keck, D.B. "Spatial and Temporal Power Transfer Measurements on a Low-Loss Optical Waveguide," *Appl. Opt.*, 13, pp. 1881-1888 (1974).

10. Olshansky, R., Oaks, S.M. "Differential Mode Attenuation Measurements in Graded Index Fibers," *Appl. Opt.*, 17, pp. 1830-1835 (1978).

11. Eve, M. "Novel Technique for Measurement of Differential Mode Attenuation in Graded Index Fibres," *Electron. Lett.*, 13, pp. 744-745 (1977).

12. Olshansky, R., and Oaks, S.M. "Differential Mode Delay Measurement," *4*th E.C.O.C., Geneva 1978, pp. 128-132.

13. Jeunhomme, L., Pocholle, J.P., and Raffy, J. "Wavelength Dependence of Modal Dispersion in Graded Optical Fibres," *Electron. Lett.*, 14, pp. 364-366 (1978).

14. Cancellieri, G., Mezzetti, M., Someda, C.G., and Zoboli, M. "Simplified Procedure for Indirect Tests of Optical Fibre Transfer Functions," *Electron. Lett.*, 15, pp. 234-236 (1979).

15. Jeunhomme, L., and Lamouler, P. "Intermodal Dispersion Measurements and Interpretation in Graded Index Optical Fibres," *Opt. & Quant. Electron.*,12, pp. 57-64 (1980).

16. Gambling, W.A., Payne, D.N., and Matsumura, H. "Mode Conversion Coefficients in Optical Fibers," *Appl. Opt.,* 14, pp. 1538-1542 (1975).

17. Jeunhomme, L., and Pocholle, J.P. "Experimental Determination of the Radiation Pattern of Optical Fibres," *Opt. Comm.,* 12, pp. 89-92 (1974).

18. Jeunhomme, L., and Pocholle, J.P. "Angular Dependence of the Mode-Coupling Coefficient in a Multimode Optical Fiber," *Electron. Lett.*, 11, pp. 425-426 (1975).

19. Kawakami, S. "New Principle to Measure Loss and Mode Conversion Parameters of Multimode Fibers," *Electron Lett.*, 13, pp. 706-707 (1977).

20. Kitayama, K., and Ikeda, M. "Mode Coupling Coefficient Measurements in Optical Fibers," *Appl. Opt.*, 17, pp. 3979-3983 (1978).

21. Cancellieri, G., and Mezzetti, M. "Scanning Technique for Investigating Optical Fibre Dispersion," *Opt. & Quant. Electron.*, 11, pp. 305-317, (1979).

22. Kawakami, S., and Tanji, H. "Evolution of Power Distribution in Graded Index Fibres," *Electron. Lett.*, pp. 100-102 (1983).

23. Daido, Y., Miyanchi, E., Iwama, T., and Otsuka, T. "Determination of Modal Power Distribution in Graded Index Optical Waveguides from Near Field Patterns and Its Application to Differential Mode Attenuation Measurements," *Appl. Opt.*, 18, pp. 2207-2213 (1979).

24. Calzavara, M., Di Vita, P., and Rossi, U. "Reliability of a New Method for Measurements of Modal Power Distribution in Optical Fibres with Application to Mode Scrambler Testing," *Electron. Lett.*, 17, pp. 543-545 (1981).

25. Leminger, O. G., and Grau, G. K. "Near Field Intensity and Modal Power Distribution in Multimode Graded Index Fibres," *Electron. Lett.*, 16, pp. 678-679 (1980).

26. Saijonmaa, J., Sharma, A.B., and Halme, S.J. "Optimal Excitation of Multimode Graded Index Fibres in D.M.D. and D.M.A. Measurements," *Electron. Lett.*, 16, pp. 690-692 (1980).

27. Lambert, A.M., Goure, J.P., and Massot, J.N. "Commentaires sur les misures de flux dans le fibres optiques,"*Opt. & Quant. Electron.* , 13, pp. 509-513 (1981).

28. Cancellieri, G., Fantini, P., and Mezzetti, M. "Misura degli effetti dell'accoppiamento tra i modi in fibre ottiche a profilo d'indice graduale," *3.a Riun. Naz. Elettrom. Appl.*, Bari, pp. 277-279, (1980).

29. Ettemberg, M., Hudson, K.C., and Lockwood, H.F. "High-Radiance Light Emitting Diodes," *IEEE J. Quant. Electron.*, QE-9, pp. 987-991 (1973).

30. Dyment, J.C., Kapron, F.P., and Springthorpe, A.J. "Top and Side Emission from Double Heterostructure LEDs and Lasers," *IEEE Trans. Electron. Dev.*, ED-24, pp. 995-1000 (1977).

31. Sommers, H.S. "Experimental Properties of Injection Lasers, IV: Modes of Large Cavity with Sawed Sides," *J. Appl. Phys.*, 44, pp. 3601-3608, (1973).

32. Sommers, H.S. "Experimental Properties of Injection Lasers: Modal Distribution of Laser Power ," *J. Appl. Phys.*, 44, pp. 1263-1275 (1973).

33. Sharma, A.B., Hubach, E.J.R., Lahteenoja, M., and Halme, S.J. "Technique for Minimizing Effect of Detector Inhomogeneity on Fibre Attenuation Measurements," *Electron. Lett.*, 18, pp. 56-57 (1982).

34. Tynes, A.R. "Integrating Cube Scattering Detector," *Appl. Opt.*, 9, pp. 2706-2710 (1970).

CHAPTER 6
MEASUREMENTS ON TRANSDUCERS

LEDs and laser diodes give an emitted optical power which depends on the spatial and angular coordinates of the emission pattern, and on the wavelength. Photodiodes, in a similar, though less critical way, give a current response which depends on the spatial and angular properties of the collected optical power, and on its wavelength. This suggests the integrated and differential measurements should be discussed separately. Integrated measurements consider the total optical power emitted or collected by a transducer, whereas differential measurements evaluate exclusively the transducer properties, varying spatial, angular, and spectral coordinates.

Among integrated measurements, there is also an electric characterization of the transducer: a measurement of the parameters of its electric equivalent circuit. Another very important integrated measurement is the baseband response evaluation of the transduction process. This is characterized in practice by lowpass behavior. From these two integrated measurements, it is possible to design the circuit in which a given transducer must be inserted for transmission systems realization or fiber testing setups.

Particular measurements then test the reliability of the devices, especially the laser diodes, which suffer facet damages and gradual internal degradations. These phenomena are due respectivety to the high densities of optical power and to the electric current which characterize the laser emission above threshold. For LEDs, only the latter kind of degradation is sometimes present.

Differential measurements can be performed in the wavelength domain or in the geometrical domain. The latter are mostly scanning procedures, very similar to those performed on fibers. Precisely, scanning tests on LEDs and laser diodes are like fiber output selections, scanning tests on photodiodes are like fiber input selections. Combined spectral and geometrical selections are useful only for the optical sources.

6.1 INTEGRATED MEASUREMENTS

The most important integrated measurements which can be performed on a transducer are related to its electric characterization, transduction behavior, and baseband response. The electric characterization of a LED or a laser diode consists of the determination of its current-voltage characteristic, and of all the parameters which are present in its equivalent circuit, including parasitics. Therefore, it is performed without any optical measurement. For measurement of the equivalent circuit parameters, the continuous current I_o, onto which the modulating signal $i(t)$ is superimposed, must be high enough to allow a good reproduction of $i(t)$ by the optical emitted signal. Especially, for laser diodes, I_o is necessary for operation above threshold. On the other hand, the electric characterization of a photodiode depends on the incident optical power. So, its current-voltage characteristic and all the parameters present in its equivalent circuit must be determined at a given level of incident optical power (sometimes in the dark). Moreover, these properties are strongly influenced by the bias voltage V_b which is used. Therefore, when the tests are performed, V_b must also be varied from zero (photovoltaic operation) to a given upper limit.

The baseband response of a LED or a laser diode is measured by applying a sinusoidal modulating current signal, and detecting the corresponding optical power response. The bias condition and the circuit parameters between the signal source and the device must be specified, because they may influence the overall baseband response measured. The baseband response of a photodiode is evaluated by collecting a sinusoidal intensity modulated optical power, and revealing the corresponding electric signal (current or voltage). Similar to the previous case, the bias condition and the circuit parameters, between the device and the instrument, must be specified. In both cases, the modulation frequency is varied from zero (at which one can test the transduction behavior without any reactive affect), to a given upper limit. The operation at zero, or at very low modulation frequency, usually tests the linear dynamic range of the transducer in a given configuration. As for the very high speed of modulation, to measure directly some typical properties of digital transmission, other types of modulating signals are occasionally employed. In this case, one can obtain rise times or turn-on delays by simply applying a sequence of rectangular pulses to the transducer under suitable conditions.

6.1.1 Electric Characterization of LEDs and Laser Diodes

The current-voltage characteristic of a LED or a laser diode can be obtained, as for any diode, by applying a variable bias voltage V_b through a given resistance R_b, and measuring the current which flows in the circuit with a milliampermeter, as shown in Fig. 6.1. Usually, optical monitoring is added to avoid operations with optical power levels which are too high; e.g., far above the laser threshold, which may damage the device.

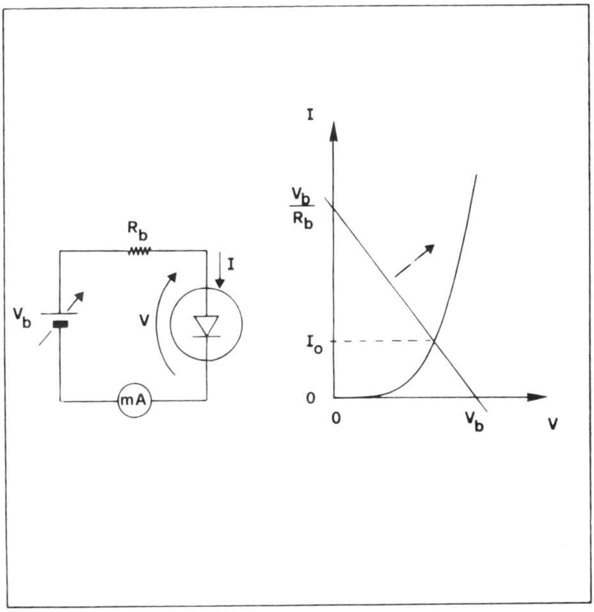

Fig. 6.1 - Electric setup for evaluating the current-voltage characteristic of a LED or a laser diode.

As discussed in 3.1.4, the equation which describes this characteristic can be written as

$$I = I_r \left\{ \exp \left[a\,(V - R_s I) - 1 \right] \right\} \tag{6.1}$$

where I and V are, respectively, the current and voltage, a is a constant depending on the particular device, I_r is the reverse current, and R_s the series resistance of the diode. Since R_s is rarely too small, the knowledge of its value is of some importance for impedance matching. R_s can be easily determined from the position of the knee of the characteristic, and from knowledge of I_r and a, which depend mainly on the device material and size. Precisely, for laser diodes above threshold, the characteristic approaches a straight line having equation

$$V \cong \frac{E_g}{e} + R_s I$$

where E_g is the bandgap energy and e the electron charge. This equation allows an easy evaluation of R_s.

The electric characterization at high modulation frequencies is rather complex. The complete equivalent circuit of a LED or a laser diode at these frequencies is shown in Fig. 6.2, where C_p and L_s account for wire and contacts, R_J is the nonlinear junction resistance defined by equation (6.1), and C_J and C_S are, respectively, diffusion and space charge capacitances. Since C_S decreases with increasing bias voltage, its effect is nearly negligible when a small sinusoidal signal is superimposed on a high V_b. However, it may have some effect when an on-off modulation is applied.

The equivalent circuit reported in Fig. 6.2 can be characterized in two ways:

i) through the measurement of the reflection coefficient, as a S-parameter, by the use of a network analyzer

ii) studying the emitted optical power under a suitable electric input

An example of the former approach is reported in reference (1), where a small sinusoidal signal with variable modulation frequency has been superimposed on a bias. From the experimental data collected, no appreciable variation in the modulus of the reflection coefficient, $|S_{11}|$, have been observed between 0.1 and 10 GHz. On the other hand, its phase changes strongly. Different laser diodes have shown an equivalent impedance which is capacitive up to about 7 GHz, and then becomes inductive, owing to the presence of L_s. The simplified equivalent circuit reported in Fig. 3.6 is sufficient for an

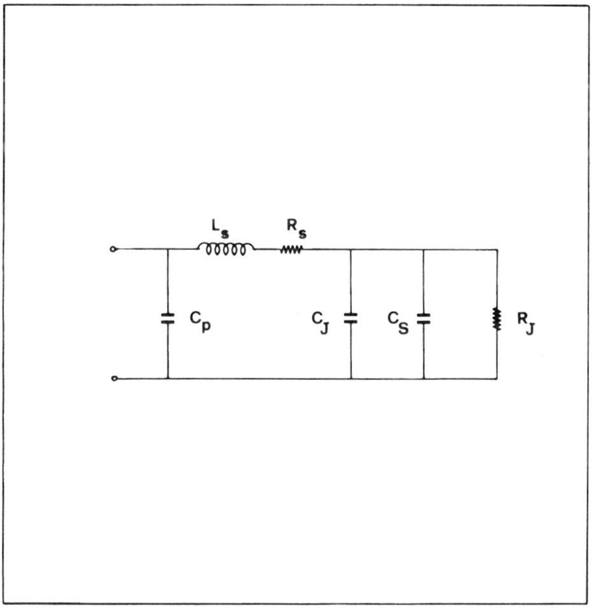

Fig. 6.2 - Complete equivalent circuit of a LED or a laser diode at high modulation frequencies.

explanation of results collected. In Table 6.1 the values of the circuit parameters, obtained from the experimental data of S_{11} on stripe contact (SC) and buried heterostructure (BH) lasers, are reported. Note that in this simplified equivalent circuit, R_J represents not only the junction resistance, but the combined effect of the junction and the series resistances.

The latter approach is based on optical measurements, and will be discussed in the next section. For LEDs or laser diodes below threshold, the baseband repsonse can be considered as practically imposed by the circuit parameters of Fig. 6.2. Therefore, an evaluation of the baseband response can lead to an indirect measurement of some circuit parameters. For laser diodes above threshold, other processes influence, sometimes very strongly, the modulaion behavior of the device. So this indirect evaluation appears to be of doubtful reliability. In this sense, when an electric characterization is required for an optimized design of the laser driver, the determination of the reflection coefficient S_{11} is preferable.

Table 6.1 - Equivalent circuit parameters obtained from experimental data on S_{11} [from (1)].

Laser Type	C_p	L_s	C_J	R_J
SC	0.3 pF	1.3 nH	40 pF	2.7 Ω
BH	0.3 pF	1.5 nH	10 pF	10 Ω

6.1.2 Measurement of the Electro-Optical Transduction and Its Baseband Response

The characteristic *P(I)*, where *P* is the emitted optical power and *I* the continuous drive current, can be evaluated adding a photodetector to the simple experimental setup shown in Fig. 6.1. For LEDs, this test allows one to obtain the nonlinearity due to junction heating at high current levels. For laser diodes, it gives the value of the threshold current I_{th}, moreover it denounces possible kinks and again the effect of junction heating.

To appreciate the nonlinearities of the electro-optical transduction, it is convenient to use a sinusoidal modulating signal superimposed on a continuous current I_o. In this case, one can obtain the relative second and third order harmonic distortions as functions of I_o. The experimental setup, based on the use of a spectrum analyzer, is schematically shown in Fig. 6.3. In reference (2), such a measurement technique, which is usual for the evaluation of kinks, has also been applied for appreciating nonlinearities due to the dynamic processes described by the rate equations. Due to these effects, for small modulation currents, there is an increase in the second order harmonic distortion which is almost proportional to the value of the modulation current for many commercially available laser diodes.

The measurement of electro-optical transducer baseband responses, whose definition has been discussed in 3.1.8, can be performed by a spectrum analyzer, adding a RF sweep generator to the experimental setup reported in Fig. 6.3. [3] It is very important that the driver of

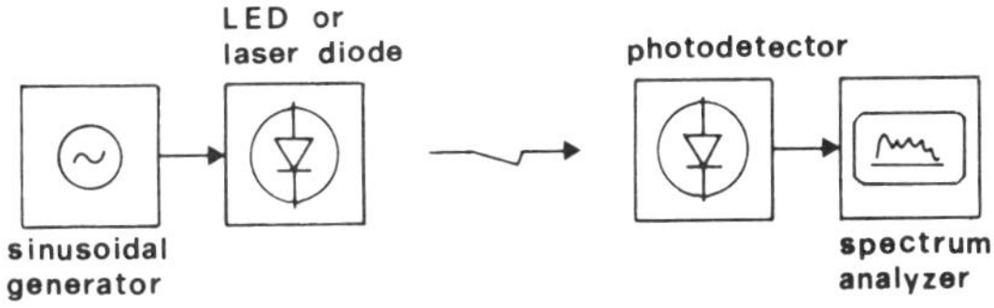

Fig. 6.3 - Experimental setup for evaluating second- and third-order harmonic distortions of the electro-optical transduction.

the LED or laser diode and the photodetector are not intrinsically band-limited. Replacing the spectrum analyzer with a vector voltmeter, it is possible, in principle, to measure both amplitude and argument of the baseband response $\overline{H}(f)$. However, only the amplitude $|\overline{H}(f)|$ is usually of interest.

For LEDs, $|\overline{H}(f)|$ exhibits a simple expression of the type (3.3). Therefore, from the measurement of $|\overline{H}(f)|$ it is possible to derive the value of the lifetime τ of the minority carriers injected in the recombination region. τ can also be determined by a completely electric measurement, as described in 3.1.8; but the evaluation through $|\overline{H}(f)|$ is usually more accurate. The values of τ depend not only on the level of the continuous current I_o, but also on the amplitude of the modulating signal $i(t)$. In this sense, we have a nonlinear phase distortion, as well as nonlinear amplitude distortion.[4] This effect leads to an appreciable deterioration of the transducted signal only when it is an analog signal with a high quality requirement, e.g., a video signal.

For Laser diodes, $|\overline{H}(f)|$ consists of two contributions:

i) the effect of the dynamic processes described by the rate equations (3.2)

ii) the effect of the electric parameters of the laser diode equivalent circuit, including parasitics

It is very difficult to separate the two contributions. A measurement performed with the experimental setup, based on a spectrum analyzer already discussed for LED characterization, can give the overall effect. This measurement must be prepared with great care, because of the very high frequencies involved (up to several GHz). In particular, the photodetector must exhibit an extremely wide baseband response.

To evaluate the resonance modulation frequency, a consequence of the laser dynamic processes, it is necessary to exclude the presence of a resonance due to the circuit parameters of the laser diode equivalent circuit reported in Fig. 6.2. In the previous section, we have shown how, using a network analyzer, it is possible to measure these circuit parameters. From the comparison between an electric measurement of this type and an electro-optical measurement to determine the overall $|\overline{H}(f)|$, it is possible to obtain a first approximation of the intrinsic dynamic effect of the laser operation.

In reference (1), a very interesting method for determining some of the constants in the rate equations has been proposed. Through an electric measurement with a network analyzer, the values of C_p, L_s, C_J and R_J have been obtained (see Table 6.1). Considering a model based on the rate equations (3.2) and on the numerical values of such circuit parameters, a prediction of the baseband response, with I_o/I_{th} as a parameter, has been carried out. This prediction has been compared with the overall baseband response measurement, obtained electro-optically with the same values of I_o/I_{th}, to estimate the role of some unknown quantities. For example, for experimental data and theoretical predictions to agree, the spontaneous emission factor β, present in the second part of (3.2), can be as large as 10^{-2}, while from theory it was expected to be about 10^{-4}. A possible explanation for this discrepancy is that a lateral carrier out-diffusion is neglected in the rate equations. From the overall baseband response measurement, it is possible to derive that the resonance modulation frequency increases with I_o, approximately as $\sqrt{I_o/I_{th} - 1}$.

The accuracy of the measurement of the overall baseband response can be enhanced by taking extra precautions. For instance, it is sometimes useful to insert a short piece of optical fiber between the LED or laser diode being tested and the photodetector, to avoid electric interference. [3] A fiber length of about 10 m is long enough, and yet it is sufficiently short to make any time dispersion effect due to the fiber itself negligible. To reduce possible saturation or nonlinear distortion of the photodetector, one can insert an optical attenuator, such as a variable neutral density filter, in front of the photodetector itself. [1] The whole emitted optical power can be measured without any selection on the far field pattern of the device, which could be influenced by the direction selected, especially for laser diodes. A lens can sometimes be mounted in front of the emitting device to obtain a beam focused either on the fiber input end, or directly on the photodiode active area.

Before concluding the present section, we will describe briefly some measurements in the time domain. They are performed by applying a step current pulse to the LED or laser diode,[5] and using an experimental setup of the type shown in Fig. 6.4. LEDs are characterized by a rise time, imposed mainly by C_J, whereas laser diodes exhibit a turn on delay, during which the emitted optical power is nearly negligible. Often, some oscillations due to the laser dynamic processes occur. The use of a sampling oscilloscope is required when rise times and turn on delays as short as a few ns must be appreciated. These measurements test given devices for suitability in communication systems with a fixed speed of transmission.

In particular, a high speed photodetector with an excellent linearity of response must be employed. In this sense, a PIN photodiode is preferred. Connecting the output of the photodiode directly to the 50 Ω input resistance of the oscilloscope, it is possible to obtain a photodetector rise time as short as 200 ps. For measurements of this type, one can test the effect of pulse shaping, whose purpose is to reduce rise times and turn on delays. In reference (5), it has been shown that giving a sharp spike at the beginning of each driving current pulse entails an appreciable reduction of such undesired effects, especially for laser turn on delays. This technique, called overdriving, is realized in practice by the type of driver schemes shown in Fig. 4.2. Yet, it is necessary to observe that too large a spike may damage the device.

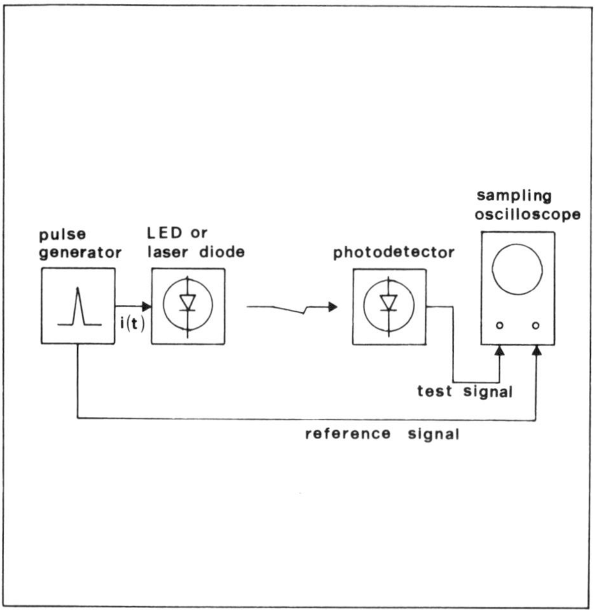

Fig. 6.4 - Experimental setup for measuring rise times and turn-on delays in LEDs and laser diodes.

6.1.3 Electric Characterization of Photodiodes

The current-voltage characteristic of a photodiode is usually obtained without any incident optical power. The simple electric setup reported in Fig. 6.1 can be employed, taking $V_b > 0$ for testing photovoltaic operation, and $V_b < 0$ for appreciating the reverse (dark) current I_r and the position of the breakdown voltage, which are useful parameters for reverse bias operation.

As discussed in 3.2.2, since the series resistance R_s is nearly negligible, the current which flows across the junction can be written as

$$I_J = I_r \left[\exp\left(\frac{V}{V_k}\right) - 1 \right]$$

where V_k is a constant voltage. From the above described measurement, both I_r and V_k can be evaluated. In particular, their ratio, V_k/I_r, allows one to discuss the linearity of transduction, depending on the load resistance R_L, in photovoltaic operation. I_r is useful for the analysis of the minimum shot noise, which characterizes solar cells and PIN photodiodes, as discussed in 3.2.4, when a reverse bias voltage is applied.

The electric characterization at high modulation frequencies can be performed usually two ways:

i) through the measurement of the reflection coefficient, as a S-parameter, by the use of a network analyzer

ii) studying the output current under an optical power which is intensity-modulated with a variable modulation frequency

The former technique allows one to determine the role of the depletion capacitance C_d, and even of parasitics. An interesting example of such measuremets has been reported in reference (6), where commercially available devices have been tested in the following four configurations: anode connected to case, cathode connected to case, and grounded case. The results obtained are shown in Fig. 6.5. The last technique is the one most frequently employed, although it gives information on C_d only when it is the dominant limiting agent of photodiode baseband response. It belongs to the class of electro-optical measurements on photodiodes, and therefore will be discussed in the next section.

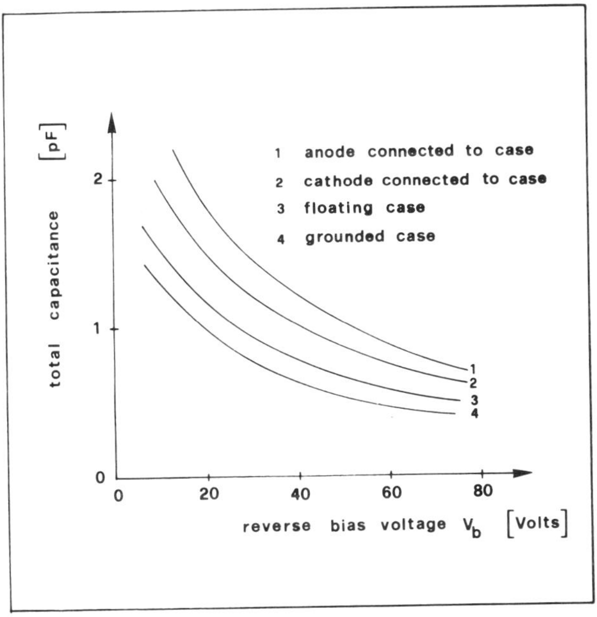

Fig. 6.5 - Total capacitance of a photodiode in different electric configurations [from (6)].

6.1.4 Measurements of the Responsivity and Its Baseband Response

A measurement of the responsivity of a photodiode can be performed by inserting a milliampermeter between the photodiode and the load resistor, and collecting an optical power of known intensity on the photodiode active area. For avalanche photodiodes, varying also the reverse bias voltage V_b, it is possible to measure the relative behavior of the mean avalanche gain $<M>$ as a function of V_b. It is important to fix the temperature during a test of this type, because M depends on the temperature T. Repeating different tests at various values of T, one obtains a family of curves like those shown in Fig. 3.10.

This sort of procedure on an avalanche photodiode is usually accompanied by a noise measurement evaluating the constant k in the expression of the excess noise factor $F(<M>)$ given by (3.6). This can be done by a spectrum analyzer, keeping the photodiode in the dark.[7] An alternative technique to measure the shot noise of an avalanche photodiode employs a true *rms* responding meter.[8]

Linearity tests are frequently performed on APD photodetectors. In this case, the optical power of a high stability source is varied, over some decades, by an optical attenuator. A PIN photodiode having a very wide linear dynamic range is used as a reference. The scheme of this experimental setup is shown in Fig. 6.6. In reference (9), a pyroelectric detector has been used as a reference, owing to its good linearity and nearly flat response over an extended wavelength spectrum, as described in 3.2.1. Pyroelectric detectors are also used for the absolute evaluation of the responsivity $\mathscr{R}$ of a test photodiode.[10] However, any photodetector whose overall responsivity $\mathscr{R}_v$ is known as a function of the wavelength can be assumed to be a calibrated photodetector. This is very useful for absolute measurements of radiance, as will be shown in 6.2, or for the calibration of other photodetectors.

The baseband response of a photodiode can be measured directly if the modulation bandwidth of the optical source is large enough. In this case, the experimental setup appears as shown in Fig. 6.3. Alternatively, the spectral distribution of the shot noise can be evaluated by means of a spectrum analyzer.[11] The former technique has a limitation in the extension of the optical source modulation bandwidth. Moreover, the baseband response of the source, even if very wide, is often not flat, leading to poor measurement accuracy. The latter technique suffers some uncertainties, due to the presence

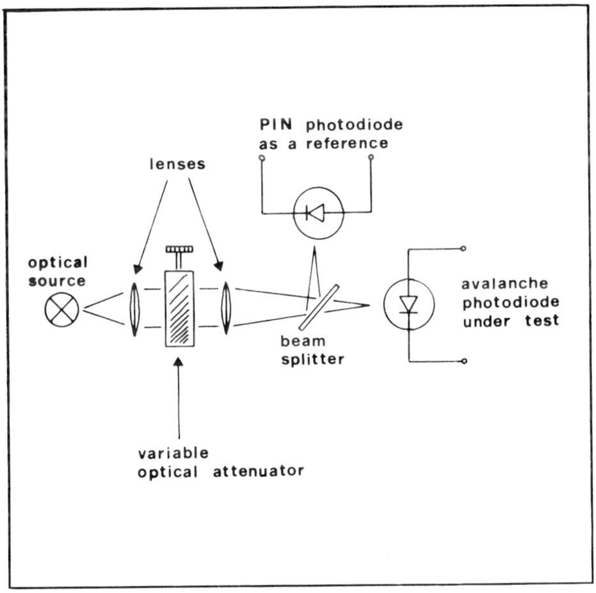

Fig. 6.6 - Experimental setup for testing the linearity of an APD photodetector.

of an amplifier. An amplifier must be employed because of the very low level of the useful signal, and its baseband response is superimposed on that of the photodiode.

In reference (12), an interesting method for obtaining an optical source with a very flat baseband response over several GHz has been proposed. It is based on a pair of single mode laser diodes, whose beams are made coincident in both polarization and wave front using the optical apparatus shown in Fig. 6.7. The temperature of one laser diode is fixed, while the other is varied, to obtain emission optical frequency variations. The beat frequency obtained can be varied from 0 to about 5 GHz without appreciable amplitude variations. This technique is also efficiently employed for fiber time dispersion measurements, as will be discussed in 10.3.

Before concluding the present section, it is necessary to stress that when the baseband response of the photodiode is not limited by the time constant $R_L C_d$, the electric measurement of C_d is not enough for a complete characterization of the photodiode behavior. Therefore an electro-optical measurement, like those described in the present section, is required. This occurs typically for Ge APDs, and also PIN photodiodes when the load resistance R_L is very small.

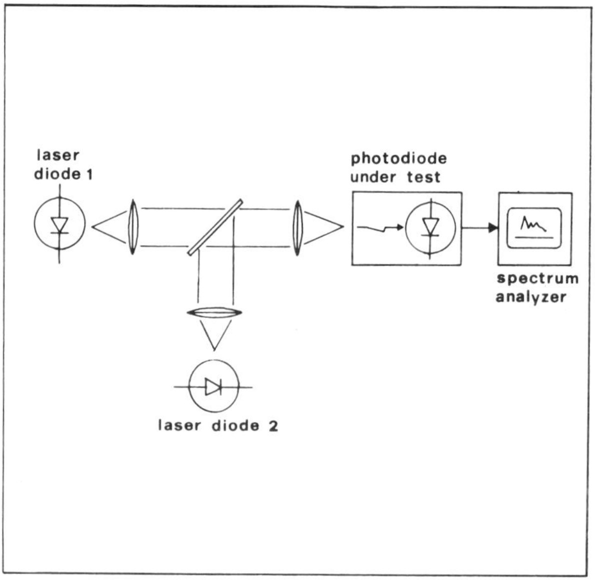

Fig. 6.7 - Optical apparatus for obtaining an optical source with baseband response practically flat up to several GHz [from (12)].

6.1.5 Reliability Tests on Laser Diodes

Two main kinds of degradation occur in laser diodes: [13]

i) damages in the mirror facets

ii) gradual internal degradation

The former effect is due to the high level of optical power density which crosses the output mirror facets; the latter effect is due to the high current density across the junction, and also occurs sometimes for LEDs.

These phenomena are often related to pulsed operation, especially when overdriving is used to shorten turn on delays. Letting Δt equal the pulsewidth, from experimental data, [14] the minimum optical power density which damages the output mirror of a given device decreases as $\Delta t^{-1/2}$. For stripe-contact laser diodes, facet damages are usually initiated in the active layer, over the stripe-contact, where the optical flux is highest. The gradual internal degradation is probably due to defects in the recombination region acting as non-radiative recombination centers. They contribute to the increase in the threshold current density J_{th}. Finally, a laser diode exhibits a threshold so high, that, with the usual current levels, it gives only a spontaneous emission.

From a near field scanning technique performed on the laser emitting surface, it is possible to test for the presence of facet damages. They appear as dark regions. When a gradual internal degradation is very remarkable, this test shows dark regions which are concentrated nonradiative centers. However, normally, gradual internal degradation can be discovered by repeated measurements of the threshold current, performed during the laser life.

Accelerated aging tests allow one to predict life times of laser diodes. Since the degradation processes are temperature dependent, it is possible to perform tests in which the device operates at high temperatures, and the measured life times are projected back to room temperature. The thermal acceleration factor, however, must be experimentaly determined. The linearity, the lateral mode stability, and the spectral stability are also investigated through aging tests. In particular, single mode operation is more and more difficult towards the end of whose life, and the spectral width may increase widely.

In reference (15), an aging measurement has been reported which shows that the main cause of laser degradation, and even of the increase in J_{th}, is due to facet damages. This test has demonstrated that the self-oscillation phenomena, described in 3.1.5, are strictly related to imperfections of the mirror surface. The use of a half-wave thick dielectric facet coating, made of Al_2O_3, for example, reduces the probability of facet imperfections a great deal. The amplitude of self-oscillations has been reported as a function of the aging time for laser diodes with and without mirrors coatings, placed in 70°C nitrogen. The laser diodes without mirror coatings have exhibited a remarkable increase in the amplitude of self-oscillations *versus* aging time, whereas those with mirror coatings were characterized by an almost negligible increase.

6.2 DIFFERENTIAL MEASUREMENTS

There are complex relations between the spectral and geometrical properties of the emissions of LEDs and, especially, of laser diodes. We will discuss some experimental arrangements which can be used for separately testing these properties, and for deriving these relations. This is very useful in understanding the mechanisms of light emission, and in improving the technology of device fabrication. Some theoretical models for the explanation of the experimental data will be presented.

For photodiodes, the spectral and geometrical dependence of the responsivity is not as critical. However, care must be taken, in experimental tests on optical fibers, especially when good accuracy is required, to avoid systematic errors and to obtain repeatable results. An exhaustive characterization of the spectral and geometrical properties of the responsivity is sometimes necessary.

6.2.1 Measurement of the Spectral Distribution of the Emitted Power in LEDs and Laser Diodes

The experimental set up for the measurement of the spectral distribution which characterizes the power emission in a LED or a laser diode is schematically shown in Fig. 6.8. The output $v(\lambda, t)$ of the photodetector represents the response of the system source-monochromator-photodetector to the signal $i(t)$ when the monochromator selects the wavelength λ. Usually $i(t)$ is a sinusodial modulating current, superimposed on a continuous current I_o, and $v(\lambda, t)$ is sinusoidal with the same modulation frequency. With this intensity modulation the measurement can be performed at the ambient light, and the accuracy enhanced, especially when a lock-in amplifier is used (see Appendix 2).

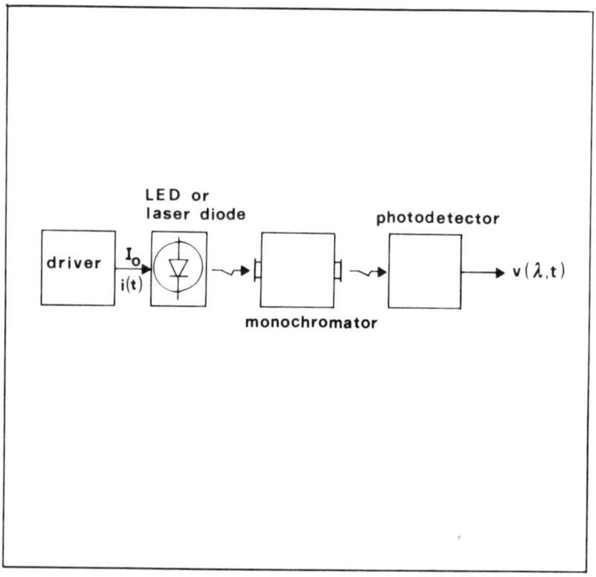

Fig. 6.8 - Experimental setup for the measurement of the spectral response of LED's and laser diodes.

The spectral response obtained depends also on I_o. This is particularly true for laser diodes, whose spectrum tends to become narrower with increasing I_o, as discussed in 3.1.4, because the fundamental family of longitudinal modes becomes more excited with respect to the other modes. So, one can plot a family of curves of spectral responses with I_o as a parameter.

It has been shown that the spectral distribution of a laser diode depends also on the modulating signal $i(t)$. In particular, for digital transmissions at high bit rates, using laser diodes characterized by very narrow line widths, the spectral behavior of the laser output depends on the point of the waveform $i(t)$ considered. In reference (16), time resolved dynamic spectra have been measured by means of a sampling oscilloscope in a suitable arrangement. For certain laser structures, noticeable spectral changes occur when the signal consists of a series of pulses. In this case the central wavelength can have a shift of more than 1 nm, towards the higher values of λ. Also, the pulse duration has a strong influence on the laser spectral distribution. The photodetector and the monochromator should have a flat spectral response in the wavelength region interested by the source emission. In this case, the relative spectral response of the LED or laser diode is proportional to the amplitude of $v(\lambda,t)$. If there is not a flat spectral response, the measurement must be performed by comparison with a reference source characterized by a uniform spectrum. Letting $v'(\lambda,t)$ equal the output of the photodetector when this reference source is set in front of the monochromator, the relative spectral response of the LED or laser diode under test can be written as

$$P(\lambda) = \frac{V(\lambda)}{V'(\lambda)}$$

where $V'(\lambda)$ and $V(\lambda)$ represent, respectively, the amplitudes of the signals $v'(\lambda,t)$ and $v(\lambda,t)$.

To resolve the fine structure of a laser diode spectrum, and to investigate modal configurations, it is necessary to use a Fabry-Perot scanning interferometer (see Appendix 1). Yet, to appreciate the line width of laser diodes in single mode operation, whose spectrum is sharpened by suitable external reflections, like for coherent transmissions, it is necessary to employ a technique whose spectral sensitivity is noticeably enhanced. A new scanning method has been recently proposed, called the self-heterodyne technique.[17] It is based on the experimental setup reported in Fig. 6.9. Branch 1 of this interferometer produces a frequency shift f_s in the spectrum of the emitted optical power. f_s must be much larger than the spectral width $\Delta\nu_\lambda$ to be measured. Branch 2 introduces only a time delay τ_d of the order of several μs.

The electric field through branch 1, considering a given optical frequency ν_λ, can be written as

$$E_1(t) = A_1 \exp\left\{i\left[\phi_n(t) + 2\pi'(\nu_\lambda + f_s)\, t\right]\right\}$$

where A_1 is a constant and $\Phi_n(t)$ represents a random phase fluctuation. The electric field through branch 2 similarly results in

$$E_2(t) = A_2 \exp\left\{i\left[\phi_n(t-\tau_d) + 2\pi\,\nu_\lambda(t-\tau_d)\right]\right\}$$

where A_2 is another constant. The output current of the avalanche photodiode, which acts as a mixer, letting $\mathscr{R}_o\langle M\rangle$ be its overall responsivity, is

$$I_s(t) = \mathscr{R}_o\langle M\rangle \left\{A_1^2 + A_2^2 + 2A_1A_2\cos\left[2\pi f_s t + 2\pi\nu_\lambda\tau_d + \phi_n(t) - \phi_n(t-\tau_d)\right]\right\} \tag{6.2}$$

Assuming that the spectral distribution of the laser emission has a Lorentization shape with central optical frequency $\nu_{\lambda o}$, this spectral distribution can be written as

$$S(\nu_\lambda) = \frac{\delta\nu_\lambda}{2\pi\left[(\nu_\lambda - \nu_{\lambda o})^2 - \left(\frac{\delta\nu_\lambda}{2}\right)^2\right]} \tag{6.3}$$

where $\delta\nu_\lambda$ is proportional to the spectral width $\Delta\nu_\lambda$. From (6.2) and (6.3) it is possible to demonstrate that, when $\tau_d \gg 1/\delta\nu_\lambda$, the spectrum $S_{IF}(f)$ of the IF signal, which can be measured by the spectrum analyzer, results in

$$S_{IF}(f) \cong \frac{\delta\nu_\lambda}{\pi(f^2 + \delta\nu_\lambda^{\,2})}$$

which is approximately twice as large as the original laser output spectrum. The spectral resolution of this technique depends on τ_d.

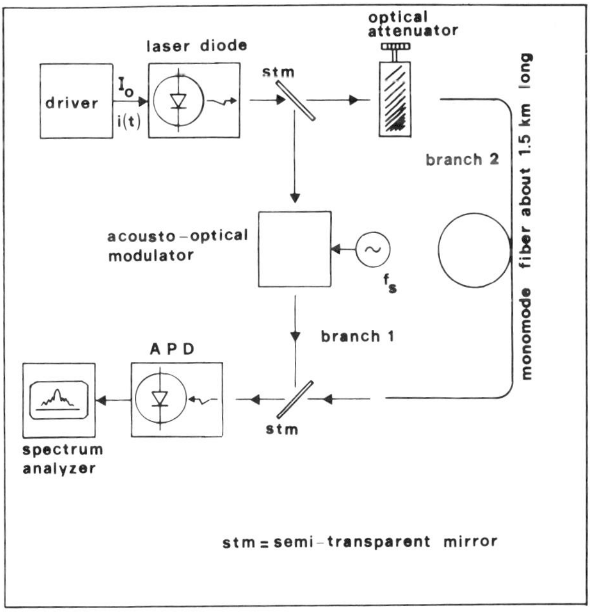

Fig. 6.9 - Self-heterodyne technique for ultra-high resolution measurements of spectral distribution [from (17)].

The longer the τ_d the better the spectral resolution. However, other causes of uncertainty become predominant when $\delta\nu_\lambda$ is approximately 100 kHz, which represents the ultimate limit of spectral resolution of this interferometer. [17] In this case, the monomode fiber in branch 2 must be as long as 1.5 km.

6.2.2 Measurements of Near and Far Field Patterns in LEDs and Laser Diodes

In 5.2.1 we have introduced scanning measurements of optical sources employed for transmission systems and fiber testing equipment, especially those concerning the optical arrangements. Here we will devote mote attention to the practical aspects of such measurements. First of all, we can distinguish between two main classes of measurement techniques on the basis of their purposes:

i) technological measurements, which are of fundamental importance for improving the processes of device fabrication

ii) transmission measurements, which interest mainly the design of transmission systems, to predict the behavior of a device in a given system configuration

In principle, both classes of measurements are designed essentially for testing radiation confinement, the absolute value of the radiance obtained, and the possible presence of spatial relative delays of emission.

Radiation confinement is considered either in the near or in the far field pattern, and the objective of this test is to predict coupling efficiency with respect to a given type of fiber. The devices realized expressly for use in optical transmission systems are frequently produced with a short fiber pigtail directly attached in front of their emitting surface. So, transmission measurements of radiation confinement are reduced to test the dimensional compability between pigtails and optical fibers. Technological measurements of radiation confinement are performed directly on the emitting surface. One of the objectives of such measurements is to help the fiber pigtail design to be mounted on the devices, matching the spatial and angular properties of the optical sources and those of the fiber.

The absolute value of the radiance, emitted by a given source, must be known for the evaluation of the optical power level which can be launched into a fiber. For commercially available optical sources, especially when provided with pigtails, the maximum radiance is frequently replaced with the maximum optical power coupled with a standard multimode fiber, having core diameter $2a = 50\ \mu m$ and $NA = 0.2$. Therefore, radiance measurements are currently of interest for technological purpose only. Finally, the possible presence of spatial relative delays of emission in the near field radiation pattern is of some importance, because they can modify the baseband response of the optical fiber, owing to a nonuniform phase excitation of the guided modes, as discussed in 1.3.8. This effect is particularly evident when very large bandwidth fibers are employed over short distances. The presence of spatial relative delays of emission may become very cumbersome when DMD measurements are performed, selecting at the fiber output end. In this case, as stressed in 5.1.4, the relative delays of the guided modes to be measured may be of the same order of magnitude as the spatial relative delays of emission which affect the optical source, especially when this is a LED. The near field scanning measurement on the LED emitting surface, devoted to the experimental evaluation of this phenomenon, requires a vector voltmeter, and will be presented in the next section. Here, we will restrict ourselves to describing scanning measurements on LEDs and laser diodes, performed only in amplitude, for the determination of radiation confinement and the absolute value of the radiance.

These measurements are carried out with an optical arrangement of the type shown in Fig. 5.1. Near field scanning is performed using a single pinhole, placed on the observation plane of the microscope objective. The overall far field scanning is made by a single pinhole, placed behind this observation plane. Tests on far field patterns of given points of the near field distribution can be achieved too, if the sensitivity of the optical detection is good enough, by means of a pair of pinholes. The evaluation of the radiance B is made through the following expression

$$B = \frac{P}{\Omega' A'} \tag{6.4}$$

where P is the collected optical power, Ω' the solid angle selected by the far field scanning procedures, and A' the area of the pinhole in the observation plane.

Considering circular pinholes, and letting a′ be the diameter of the circle selected in the near field, and $\Delta\theta'/2$ be the aperture of the solid angle selected in the far field, it is possible to write

$$A' = \frac{\pi}{4}\, a'^2 \qquad\qquad \Omega' \cong \pi \left(\frac{\Delta\theta'}{2}\right)^2$$

where the latter equation holds only for very small values of $\Delta\theta'/2$, like those effectively employed. Recalling that any linear dimension is multiplied by the magnification M and any angle is divided by the same quantity, passing from the object to its image, we can conclude that the area A and the solid angle Ω before the microscope objective, which correspond respectively to A' and Ω' behind it, can be written as

$$A = \frac{A'}{M^2} \qquad\qquad \Omega = M^2\, \Omega'$$

Therefore expression (6.4) effectively gives the radiance of the optical source, independently of M.

The relative radiance distribution can be obtained with a maximum error on the order of one percent, without particular difficulties. This distribution is a function of the radial coordiante r and of the angle θ from the device axis for top emitting LEDs, whereas it is a function of two orthogonal coordinates x, y, and two angles θ_x, θ_y (see Fig. 5.6) for edge emitting LEDs and laser diodes. When a large number of points must be selected, the emission pattern is very finely separated, and some problems in the sensitivity of the photodetection may occur. In this case, the use of a lock-in amplifier (see Appendix 2) is necessary.

On the other hand, the absolute evaluation of the radiance is more difficult than a simple measurement of its relative behavior. A calibrated photodetector is required, moreover some corrections must be taken into account for the reflection and absorption losses of the microscope objective. In conclusion, this absolute measurement cannot be extremely accurate, and an overall maximum uncertainty $\Delta B/B$ on the order of 10% always remains.

When the average radiance over the whole far field pattern has to be measured, only a near field scanning is necessary. In this case, expression (6.4) holds again provided that Ω' is considered the average solid angle of emission. For top emitting LEDs, whose radiation pattern is approximately lambertian, i.e., proportional to cos θ, and employing a microscope objective whose numerical aperture NA_m is not too large, Ω' can be considered imposed by the whole NA_m. For this kind of device, the radiance in the direction of the maximum can be obtained by simply applying a corrective factor to the average value obtained by (6.4). Such a corrective factor accounts for the directivity.[18]

The absolute value of the radiance, as well as the level of total emitted optical power, depends on the current which flows through the device. In laser diodes, there is also a change in the emission pattern, related to the modal configuration. On the other hand, either top or edge emitting LEDs exhibit quite negligible changes of the emission pattern varying the drive current. So repeated scanning measurements, performed varying this current, are of optical interest for laser diodes only.

6.2.3 Measurements of Spatial Relative Delays of Emission in LEDs

The optical arrangement used for the measurement of spatial relative delays of emission is a simple near field scanning apparatus. The electric arrangement is schematically shown in Fig. 6.10. The vector voltmeter gives the relative phase distribution along the emitting surface $\phi(x,y)$ from a phase comparison between the detected signal and a reference signal. Letting f_o equal the modulation frequency employed, the corresponding distribution of spatial relative delays of emission becomes

$$\tau(x,y) = \frac{\phi(x,y)}{2\pi f_o}$$

Assuming a phase resolution of about 0.1 degrees, for f_o = 27.7 MHz, we have a time resolution of the order of 10 ps.

This measurement was first proposed in (19) for top emitting LEDs with some point contacts on their upper layer. It has been successively extended to Burrus-type LEDs in (20). The interest in optical sources with a nearly uniform near field radiation pattern, at least within the core of typical fibers, is due to the verification requirement that the uniform amplitude of excitation in fiber testing procedures is accompanied by a uniform phase. The relative delay distribution depends on the modulation frequency f_o. For example, experimental data of the relative delay difference $\Delta\tau(f_o)$ between the central point of emission and a point characterized by half-emitted power, with respect to the maximum, is reported in Fig. 6.11. In the next section we will discuss a theoretical model for the interpretation of both radiation confinement, which is almost independent of f_o, and spatial relative delays of emission.

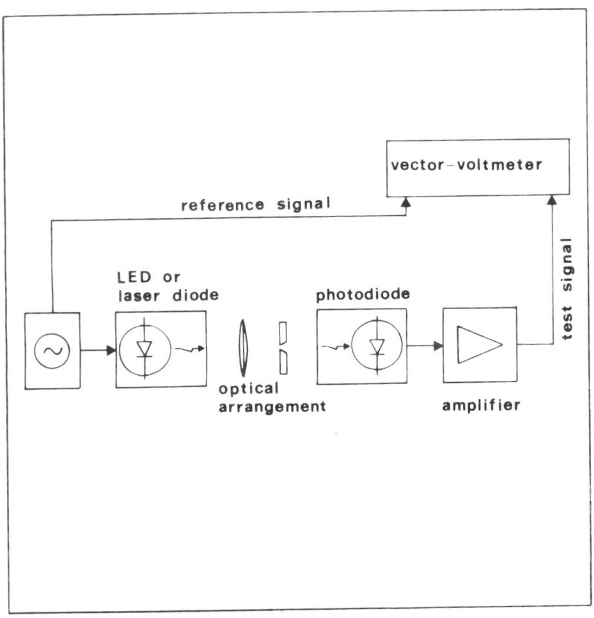

Fig. 6.10 - Electric scheme for the measurement of spatial relative delays of emission.

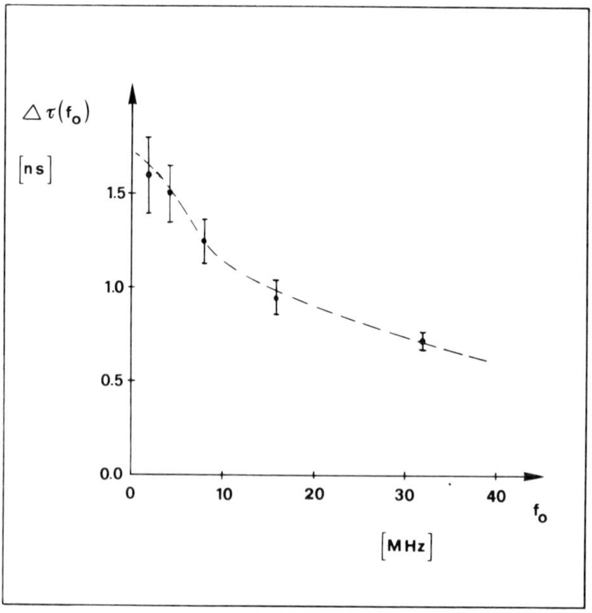

Fig. 6.11 - Example of relative delay difference *vs.* the modulation equency [from (20)].

6.2.4 Theoretical Model of Radiation Confinement and Spatial Relative Delays of Emission.

Carrier confinement can be studied by considering the set of layers of a LED or a laser diode as a transmission line. So, top emitting and, especially, Burrus-type LEDs are characterized by round symmetry,[20] whereas edge emitting LEDs and laser diodes are characterized by a longitudinal geometry parallel to the junction plane.[21] In the first case a radial transmission line must be considered, in the other case an ordinary transmission line. The technique for achieving carrier confinement, especially in laser diodes such as, stripe contact or burried heterostructure, leads to suitable boundary conditions which must be added to this model.

Anyway, considering a m layer structure, the set of m differential equations which describes this process can be written as

$$\hat{D}\,\underline{V} = \underline{\underline{Z}}\,\underline{\underline{Y}}\,\underline{V} \qquad (6.5)$$

where $\hat{D}$ is a suitable second order differential operator, V is the voltage vector, having one voltage for each layer of the structure, and the matrices $\underline{Z}$ and $\underline{Y}$ can be written from the following relationships

$$Z = \text{diag}\left[\rho_i\right] \qquad Y_{ii} = \sum_{j=i}^{m} y_{ij} \qquad Y_{ij} = y_{ij}$$

in which ρ_i are the sheet resistivities of the layers, and y_{ij} the layer to layer surface admittance. As for $\hat{D}$, it can be written

$$\hat{D} = \frac{1}{r}\frac{d}{dr}\left(r\frac{d}{dr}\right) \qquad \hat{D} = \frac{d^2}{dx^2}$$

respectively, for top emitting and edge emitting devices.

The resolutions of (6.5) are given in terms of Bessel functions in the first case, and of exponential functions in the second case. Considering a zero modulation frequency, y_{ij} are simple, real, quantities and coincide with the conductances g_{ij}, which in general are not linear, between different pairs of layers. Since radiation cofinement is nearly independent of the modulation frequency, because it is imposed by a greater or smaller resistance along a given carrier trajectory, the simplified model based on the conductances g_{ij} is sufficient for studying this property. However, to investigate the spatial dependence of the relative delays of emission, the imaginary part of y_{ij} also has to be taken into account. In this case, Bessel functions or exponential functions having a complex independent variable are solutions of (6.5).

In (20), the model of a two layer Burrus-type LED has been presented. The voltage applied across the junction, which can be considered proportional to the emitted optical power, is

$$V_{12}(r) = \begin{cases} A\, I_o\left[(\rho_n y)^{1/2} r\right] - V_{cont} & (0 \le r \le a) \\ B\, K_o\left\{\left[(\rho_n + \rho_p)\, y\right]^{1/2} r\right\} & (r > a) \end{cases}$$

where V_{cont} is the uniform voltage of the round contact under the junction, with diameter $2a$, ρ_n and ρ_p are the sheet resistivitics of the two layers (respectively, n and p type), y is the surface admittance between them, A and B are constants, and I_o and K_o are, respectively, zero order modified Bessel functions of the first and second kind. Over the contact, i.e., for $0 \le r \le a$, we have a nearly uniform distribution of $V_{12}(r)$. Outside the contact, we have decreasing behavior. The stronger its slope, the better the radiation confinement.

In Fig. 6.12, the theoretical prediction of radiation confinement with ρ_p as a parameter is reported for typical values of the remaining construction parameters. The role of the sheet resistivity of the layer close to the round contact (here ρ_p) is fundamental, provided that the other sheet resistivity is much smaller. In this case, a high value of ρ_p provides a good carrier confinement over the round contact.

Considering the reactive part of y, we can write

$$\tau(r) = \frac{1}{2\pi f_o} \left\{ \arg \left[V_{12}(r)\right] - \arg \left[V_{12}(a)\right] \right\}, r > a$$

where f_o is the modualtion frequency employed. It is possible to demonstrate, from the expression of $y(f_o)$ and from the properties of a function K_o of complex variable, that the relative delay $\tau(r)$ is always negative. In other words, the outer region emits before. As for the absolute value of $\tau(r)$, it becomes larger with increasing ρ_p and hole diffusion time τ_p, which is present in y. Furthermore, $|\tau(r)|$ decreases with increasing f_o, as shown by the dashed line of Fig. 6.11, to which experimental data has been fitted through the theoretical model discussed here.

Since the dependence on f_o is not strong for high modulation frequencies, it has been neglected in (1.99), considering the phase of the excitation condition obtained by a given optical source instead. The upper sign in (1.99), which holds for top emitting devices, is justified by the negative sign of $\tau(r)$.

This model has been verified experimentally, both in amplitude and in phase, by measurement of near field scanning measurements. The agreement is very good,[20] to the point where some construction parameters, which are not known with high precision, like ρ_p and τ_p, can be experimentally determined through a suitable fitting of the measurement data.

Edge emitting devices (LEDs or laser diodes) generally have more than two layers. Somehow, this makes their study less easy. Moreover, carrier confinement can be obtained in different ways (stripe contact, buried heterostructure, ...). In (21), a model based on a two wire transmission line has been presented for stripe-contact devices, in which the sheet resistivity of a set of layers has been assumed as the average value of the single sheet resistivities of the different

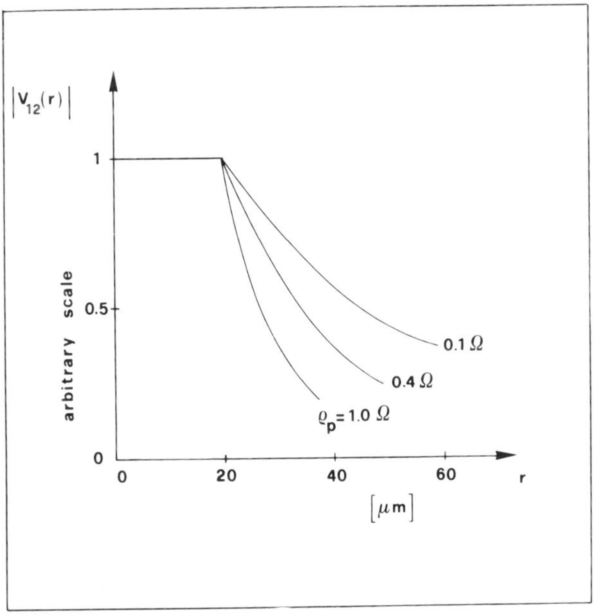

Fig. 6.12 - Theoretical prediction of radiation confinement in two-layer Burrus-type LEDs [from (20)].

layers. In (22), from a three-dimensional diagram of the potential distribution around the stripe contact, an immediate representation of this model has been proposed for three layer devices. Buried heterostructure lasers have been investigated in (22).

In (23), radiation confinement has been studied in stripe-contact laser diodes having four layers, assuming a four wire transmission line. Under some simplifying assumptions, to obtain a closed from analytical solution, it is possible to give an easy interpretation of the presence of a more than negligible series resistance R_s in the equivalent circuit of the device, and its effect on the transmission characteristic. Assuming that the *p-n* junction is between layers 3 and 4, we have a decomposition of the total applied voltage per unit junction area, V_a, as schematically shown in Fig. 6.13, where g_{34} represents the nonlinear surface conductance of the junction, and it is

$$r_s \cong \frac{g_{12} + g_{23}}{g_{12} g_{23}}$$

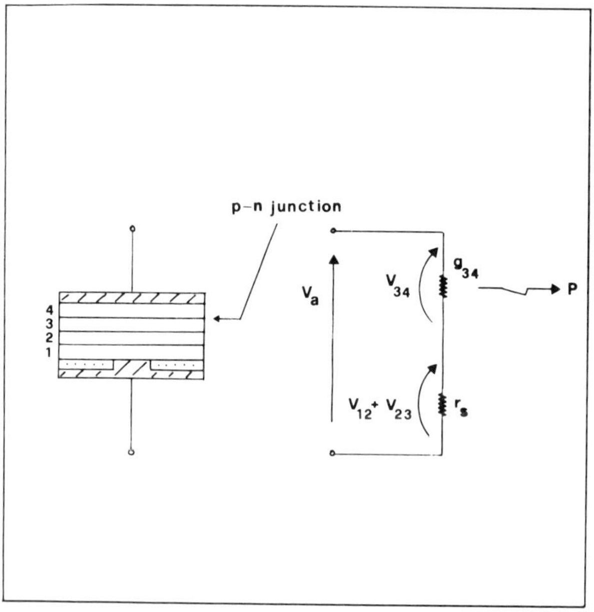

Fig. 6.13 - Simplified equivalent circuit of a four-layer stripe-contact laser diode.

in which g_{12} and g_{23} are the surface admittances, respectively, between layers 1 and 2, and 2 and 3. R_s represents the value of r_s after integration over the region crossed by the carriers, i.e., the approximately area of the stripe contact.

As for the r_s effects, since g_{12} and g_{23} can be considered nearly constant for varying V_{12} and V_{23}, whereas g_{34} increases appreciably when V_{34} increases, the voltage V_{34} effectively applied to the junction represents a fraction of V_a, which decreases when V_a itself increases. Assuming that the emitted optical power is proportional to the voltage V_{34}, this effect contributes, together with a reduction of the internal quantum efficiency due to junction heating, to the saturating behavior which occurs in the transduction characteristic of the device.

For radiation confinement, assuming the sheet resistivity ρ_4 is much smaller than ρ_1, ρ_2, ρ_3, we have a reduction of the carrier spreading with increasing ρ_1 and ρ_2, with respect to ρ_3. In Fig. 6.14, the voltage distribution $V_{34}(x)$ is reported, with ρ_3 as a parameter, for typical values of the other constants involved in the model. The width of the

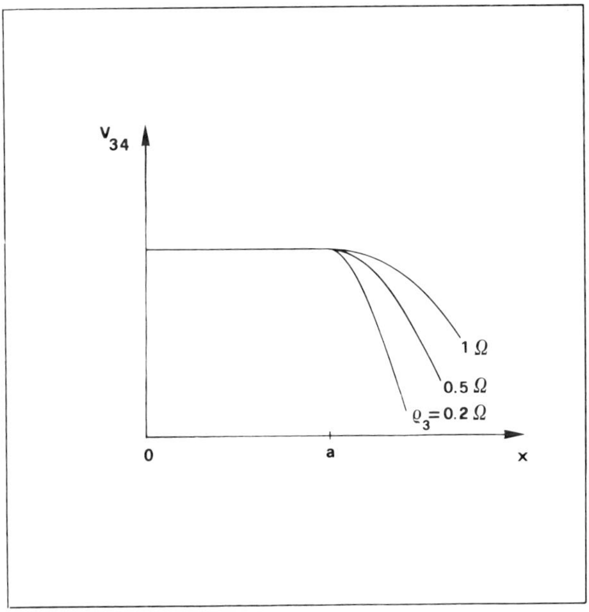

Fig. 6.14 - Voltage distribution across the junction for typical values of the design parameters, in a four-layer stripe-contact laser diode [from (23)].

stripe contact is $2a$. A high value of g_{34} contributes to improved radiation confinement. A trade-off must be considered for the static point of operation on the transduction characteristic. The continuous voltage V_a must be sufficiently high to have g_{34} small enough for a good radiation confinement, but not too high, to limit nonlinear distortions due to saturation.

Finally, the spatial relative delay distribution $\tau(x)$, which can be evaluated considering the reactive part of the surface admittance y_{34}, is always positive with respect to the central delay τ (0). This is the reason for the lower sign in (1.99), which holds for edge emitting devices. In LEDs this theoretical prediction is well verified by experiment. In laser diodes, other processes, e.g., the modal nature of the radiation emission, add their contributions to the effect discussed here. However, spatial relative delays of emission are rather small for laser devices (typically 100 ps or less), so that the remaining imperfections of the laser emission, described in 3.1.5, are usually much more evident. Just for differential mode delay investigations on large bandwidth graded index fibers, such an effect leads to unquestionable problems. They can be overcome performing a double measurement: comparing the differential mode delay distributions measured at the fiber output and at a fiber section a few meters after the optical source.

Before concluding the present section, we will briefly discuss equation (1.98), which gives the amplitude distribution of a typical fiber launching condition. It is well justified for top emitting LEDs, which exhibit round symmetry, but somehow appears to be unjustified for edge emitting devices characterized by different properties along the two directions x, y. Nevertheless, especially for devices with fiber pigtails, but even for those without any pigtails, by considering the light propagation along the first few meters of fiber, one can assume that the round symmetry takes place owing to the properties of the fiber itself.

In a similar way for fiber-to-fiber joints affected by lateral displacements, a single mode variable can be used in the study of the power loss and the time dispersion properties of a connector. In this sense, expression (1.98) can be considered justified both for top emitting and for edge emitting devices. The only value of parameter A will be different in the two cases. Precisely, edge emitting devices, and in particular laser diodes, will be characterized by smaller values of A.

6.2.5 Combined Spectral and Geometrical Differential Measurements in LEDs and Laser Diodes

The emitted optical power is a function of the wavelength λ, and of some geometrical coordinates. The latter ones can be divided into spatial coordinates (r for top emitting devices, x, y for edge emitting devices), and angular coordinates (θ or θ_x, θ_y). For the choice of the directions x, y, see Fig. 5.6. In (24), it has been shown that remarkable differences occur in the far field radiation pattern selected at different wavelengths for a Burrus-type LED. This measurement has been performed by means of the scanning apparatus of Fig. 5.7, inserting a monochromator along the branch denoted d. In (25), a similar test has been made on laser diodes, showing a strict interdependence between spectral and angular properties of the radiation emission. This measurement had been carried out to investigate the laser modal configuration at different current levels.

In Fig. 6.15, the angular dependence of the central wavelength obtained by this technique for some longitudinal mode families is reported. The lateral angle θ_x has been selected, the number q denotes different longitudinal mode families, and each point represents the position of one lobe on the lateral far field radiation pattern. This experimental test shows that the emission lobes of a given longitudinal mode family exhibit a reduction in their angular separation with increasing wavelength. The pair of straight lines which interpolate such points intersect on the axis of the device ($\theta_x = 0$) at a fixed $\lambda = \lambda_q$. For $\lambda > \lambda_q$ the modes of this family cannot exist.

From an approximate theory, based on a rectangular cavity with abrupt, perfectly reflecting walls, the central wavelength $\lambda(q,s,m)$ of the spectrum of a given mode (q,s,m) can be written as

$$\lambda\,(q,s,m) = \lambda_o - \frac{\lambda_o}{2\gamma}\left[\frac{q}{nL}\,\lambda_o + \frac{(s^2-1)\,\lambda_o^2}{(2nW)^2} + \frac{(m^2-1)\,\lambda_o^2}{(2nD)^2}\right] \tag{6.6}$$

where λ_o is the wavelength in the air, n is the refractive index inside the cavity, L, W, D are its longitudinal (along z), lateral (along x), and transverse (along y) dimensions, respectively. Finally,

$$\gamma = 1 - \frac{\lambda}{n}\frac{dn}{d\lambda}$$

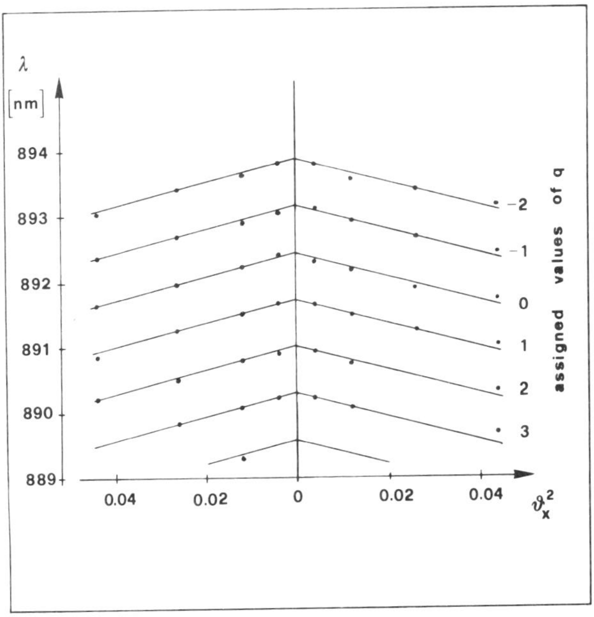

Fig. 6.15 - Angular dependence of the central wavelength of emission for different logitudinal mode families[from (25)].

For rather small angles θ_x, the following relationship holds between the mode number s and the angle $\theta_x(s)$, which characterizes the far field pattern of modes of a given family,

$$\theta_x(s) \cong \frac{s\lambda_o}{2W} \tag{6.7}$$

This is exact, as long as the penetration of the internal wave into the reflecting side wall is negligible compared to W itself.

Assuming s high enough, so that $s^2 \gg 1$, substitution of (6.7) into (6.6) yields

$$\lambda(q, s, m) = \lambda_o - \frac{\lambda_o}{2\gamma}\left[\frac{q}{nL}\lambda_o + \frac{\theta_x^2(s)}{n^2} + \frac{(m^2-1)\lambda_o^2}{(2nD)^2}\right]$$

Therefore, for a fixed mode number m we can write

$$\frac{\Delta\lambda}{\Delta q} = -\frac{\lambda_o^2}{2n\gamma L}, \quad \theta_x = constant$$

$$\frac{\Delta\lambda}{\Delta\theta_x^2} = -\frac{\lambda_o}{2n^2\gamma}, \quad q = constant$$

$$q = -\frac{L\,\theta_x^2}{n\lambda_o}, \quad \lambda = constant$$

where $\Delta\lambda$ represents the wavelength difference, varying q or θ_x. Considering these relationships, it is possible to determine the refractive index n from the experimental results summarized in Fig. 6.15, the dispersion of n through parameter γ, and the effective cavity width W (25).

The angular and spectral resolutions of the technique employed in (24) and (25) depend on the sensitivity of the photodetector and on the level of the total emitted optical power which characterizes the device being tested. A laser diode certainly exhibits a higher level of emitted optical power; nevertheless, it has a much more complex interdependence between spectral and angular power distribution so that a finer resolution is necessary. Typical values of the angular resolution are 0.2 degrees for laser diodes, and 0.5 degrees for LEDs, whereas the spectral resolution is typically 0.1 Å and 1 nm in these two cases.

No appreciable interdependence has been revealed between spectral and spatial optical power distributions. In the preceding sections we have discussed the spatial dependence of relative delays of emission in LEDs. In these devices there is also a spectral dependence of relative delays of emission. This effect, called chromatic delay distribution, has been experimentally measured either by a pulse modulation [26] or by a sinusoidal modulation. [27] In the former case a sampling oscilloscope was used, and in the latter case a vector voltmeter. In (28) an interpretation of the results reported in (26), based on the effect of carrier transit times and on the wavelength-dependence of the absorption coefficient, has been given.

A quite different model has been suggested in (27), where a fraction of the emitted optical power has been assumed to be reemitted after an absorption. This absorption occurs in the part of the semiconductor that the radiation crosses before reaching the emitting surface. Therefore, the phenomenon is influenced by the type and the width of the semiconduct layer between the region of radiation generation and the emitting surface. This model might explain the following experimental results: top-emitting devices exhibit a very small chromatic delay when the light is generated and emitted in the same layer. This delay is practically negligible for edge-emitting devices.

An example of a chromatic delay distribution obtained with a sinusoidal modulating signal, having modulation frequency $f_o = 27$ MHz, is reported in Fig. 6.16, where the continuous current I_o has been assumed as a parameter. The Burrus-type LED under testing is characterized by light generation in the p layer and emission from the n layer. The shift of the minima with increasing I_o is accompanied by a shift of the wavelength of peak emission. Delay differences as large as 600 ps may be present between spectral components of typical devices. This value establishes a lower limit for material dispersion measurements on optical fibers performed by such optical sources.

6.2.6 Tests on the Spectral and Geometrical Dependence of the Responsivity

The spectral dependence of the responsivity in a photodiode is imposed mainly by the semiconductor employed, but its dopant concentration and the depth of the junction under the receiving surface also have important roles, especially out of the maximum. In this sense, it is of some importance to experimentally test the spectral behavior of the responsivity. This test is performed by the experimental setup shown in Fig. 6.17. The light source (usually a halogen lamp) and the monochromator should generate a flat spectral power

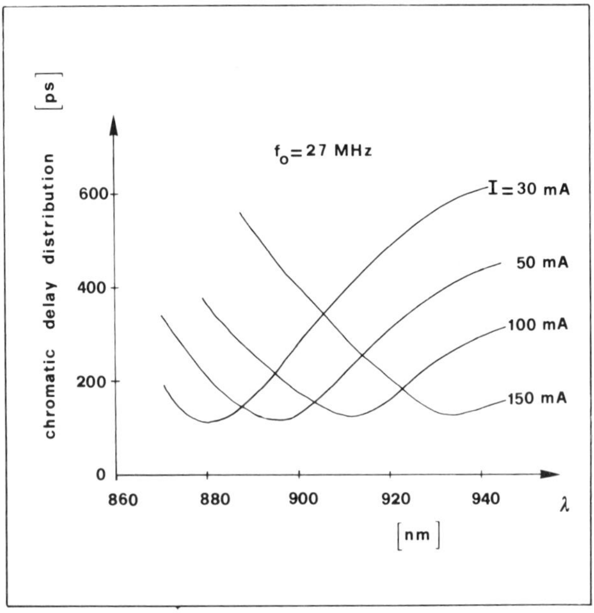

Fig. 6.16 - Chromatic delay distribution measured on a Burrus-type LED with the continuous current I_o as a parameter [from (27)].

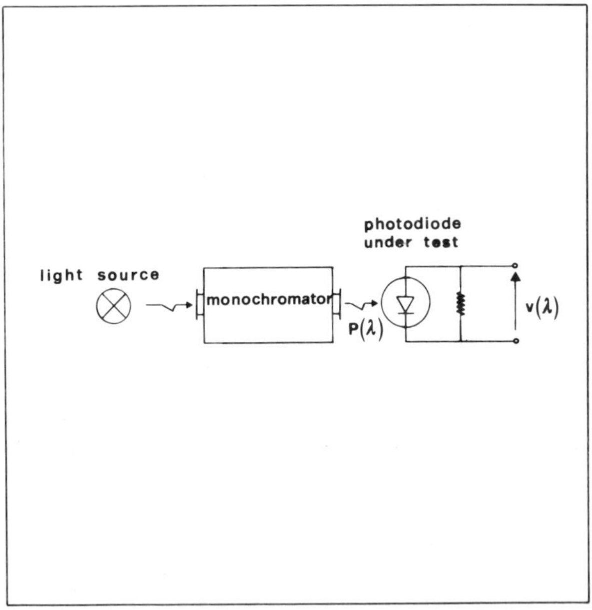

Fig. 6.17 - Experimental arrangement for the measurement of the spectral response of photodiodes.

distribution $P(\lambda)$ in the wavelength region interested by the response of the photodiodes. In this case, the relative behavior of the photodiode responsivity *vs.* λ is proportional to the output voltage $v(\lambda)$.

The photodetector usually consists of the photodiode being tested, simply loaded by a suitable resistance R_L, so that the output voltage results in

$$v(\lambda) = R_L \mathscr{R}(\lambda) P(\lambda) \tag{6.8}$$

where $\mathscr{R}(\lambda)$ is the responsivity, the object of the measurement. If $P(\lambda)$ is not a flat spectral power distribution, the measurement must be performed by comparison with a reference detector, such as a pyroelectric one.[10] Letting $v'(\lambda)$ equal the output signal of this reference detector, it is possible to write

$$v'(\lambda) = \mathscr{R}_v P(\lambda) \tag{6.9}$$

where $\mathscr{R}_v$ is the detector's uniform responsivity, expressed in V/W. Dividing (6.8) by (6.9), we obtain

$$\mathscr{R}(\lambda) = \frac{\mathscr{R}_v}{R_L} \frac{v(\lambda)}{v'(\lambda)}$$

This relationship, when R_L and $\mathscr{R}_v$ are known, also allows one to determine the absolute value of $\mathscr{R}(\lambda)$. The maximum uncertainty in the evaluation of the relative behavior of $\mathscr{R}(\lambda)$ is typically a few percent. On the other hand, its absolute measurement requires a very accurate knowledge of $\mathscr{R}_v$, and the overall maximum uncertainty can be practically written as

$$\frac{\Delta\mathscr{R}}{\mathscr{R}} = 0.01 + \frac{\Delta\mathscr{R}_v}{\mathscr{R}_v}$$

in which 0.01 accounts for the uncertainties on $v(\lambda)$, $v'(\lambda)$, and R_L.

For avalanche photodiodes, as discussed in 3.2.3,

$$\mathscr{R} = M\mathscr{R}_o$$

where $\mathscr{R}_o = \eta q / h\nu_\lambda$ is the intrinsic responsivity and M the avalanche gain. It is of some interest, especially for technological purposes, to separate the spectral behaviors of these two contributions. The quantum efficiency η is practically independent of the bias voltage V_b, whereas M is strongly influenced by V_b. We can assume that M and V_b[29] are related by

$$M = \frac{1}{1 - \left(\dfrac{V_b}{V_{br}}\right)^n} \tag{6.10}$$

where V_{br} is the breakdown voltage and n a suitable coefficient.

In (30) some spectral measurements on Ge avalanche photodiodes have been reported, between 1.1 and 1.7 μm. From simple ratios of photocurrents, knowing V_b and V_{br} and using (6.10), the spectral dependence of n has been derived. Once $M(\lambda)$ is obtained, through a measurement of the total responsivity $\mathscr{R}(\lambda)$ performed by a pyroelectric detector, the intrinsic responsivity and the quantum efficiency $\eta(\lambda)$ have been determined.

Scanning measurements on photodiodes, as discussed in 5.2.2, are of interest only when large-area devices are employed in fiber testing equipments. In fact, for transmission systems, only small-area photodiodes with fiber pigtails are used in practice. Some of the tests devoted to the investigation of the geometrical (spatial and angular) dependence of the photodiode responsivity have been described in 5.2.2. Combined tests on spectral and geometrical properties have not shown an appreciable interdependence.

REFERENCES

1. Figueroa, L., Slayman, C.W., Yen, H.W. "High-Frequency Characterization of GaA1As Injection Lasers," *IEEE Trans. Mic. Th. & Tech.,* MTT-30, pp. 1706-1715 (1982).
2. Stubkjr, K.E. "Nonlinearity of D.H. GaA1As Lasers," *Electron. Lett.,* 15, pp. 61-63 (1979).
3. Lockwood, H.F., Wittke, J.P., Ettemberg, M. "LEDs for High Data Rate Optical Communications," *Opt. Comm.,* 16, pp. 193-196 (1976).
4. Asatani, K., Kimura, T. "Nonlinear Phase Distortion and Its Compensation in L.E.D. Direct Modulation," *Electron. Lett.,* 13, pp. 162-163 (1977).
5. Lee, T.P. "Effects of Junction Capacitance on the Rise Time of LEDs and on the Turn-On Delay of Injection Lasers," *Bell Syst. Tech. J.,* 54, pp. 53-68 (1975).
6. Carni, P.L. "Photodetectors," from *Optical Fiber Communication by the Technical Staff of CSELT,* New York: McGraw-Hill (1981).
7. Melchior, H., Linch, W.T. "Signal and Noise Response of High Speed Germanium Avalanche Photodiodes," *IEEE Trans. Electron. Dev.,* ED-13, pp. 829-838 (1966).

8. Brain, M.C. "Absolute Noise Characterization of Avalanche Photodiodes," *Electron. Lett.*, 14, pp. 485-487 (1978).
9. Iwasaki, T., Yokoshima, I., Nakamura, K. "Pyroelectric Substitution Method for Calibrating Optical Fibre Components Attenuation," *Electron. Lett.*, 18, pp. 385-386 (1982).
10. Doyle, W.M., McIntosh, B.C., Geist, J. "Implementation of a system of Optical Calibration Based on Pyroelectric Radiometry," *Opt. Eng.*, 15, pp. 541-548 (1976).
11. Andersson, T., Johnston, A.R., Eklund, A. "Temporal and Frequency Response of Avalanche Photodiodes from Noise Measurement," *Appl. Opt.*, 19, pp. 3496-3499 (1980).
12. Piccari, L., Spano, P. "New Method for Measuring Ultrawide Frequency Response of Optical Detectors," *Electron. Lett.*, 18, pp. 116-118 (1982).
13. Kressel, H., Ettemberg, M., Wittke, J.P., Ladany, I. "Laser Diodes and LEDs for Fiber Optical Communication," from *Semiconductor Devices for Optical communication*, H. Kressel, editor, Berlin: Springer Verlag (1980).
14. Kressel, H., Butler, J.K. *Semiconductor Lasers and Heterojunction LEDs,* New York: Academic Press (1977).
15. Nash, F.R., Hartman, R.L., Paoli, T.L., Dixon, R.W. "Stabilization of Aging Induced Self-Pulsations and Elimination of an Initial Mode of Degradation in (Al,Ga)As Lasers by Means of Facet Coatings," *Appl. Phys. Lett.*, 35, pp. 905-907 (1979).
16. Wenke, G., Enning, B. "Spectral Behaviour of InGaAsP/InP 1.3 μm Lasers and Implications on the Transmission Performance of Broadband Gbit/s Signals," *J. Opt. Comm.*, 3, pp. 122-128 (1982).
17. Kikuchi, K., Okoshi, T. "High Resolution Measurement of the Spectrum of Semiconductor Lasers," from *Optical Devices and Fibers — 1982,* U. Suematsu, editor, Tokyo: OHM (1982).
18. Ettemberg, M., Hudson, H.C., Lockwood, H.F. "High-Radiance Light-Emitting Diodes," *IEEE J. Quant. Electron.*, QE-9, pp. 987-991 (1973).
19. Daino, B., Piazzolla, S., Spano, P., Bertolotti, M. "Spatial Dependence of Time Delay in Light-Emitting Diodes," *J. Appl. Phys.*, 47, pp. 2773-2775 (1976).

20. Cancellieri, G., Mezzetti, M. "Design Parameters for Reducing Radiation Confinement Errors and the Spatial Dependence of the Emission Delays in Light-Emitting Diodes," *J. Appl. Phys.,* 51, pp. 1813-1817 (1980).
21. Dumke, W.P. "Current Thresholds in Stripe-Contact Injection Lasers," *Solid State Electron.,* 16, pp. 1279-1281 (1973).
22. Thompson, G.H.B. "Physics of Semiconductor Laser Devices," New York: John Wiley & Sons (1980).
23. Cancellieri, G., Ganzetti, A., Fantini, P., Ravaioli, U. "Confinamento Della Radiazione in Laser con Contatto a Striscia per Comunicazioni Ottiche," 4.a Riun. Naz. Elettrom. Appl., Firenze, pp. 247-249 (1982).
24. Bassi, P., *et al.* "Caratterizzazione di Fibre e Componenti per Sistemi di Transmissione in Fibra Ottica," 2.a Riun. Naz. Elettrom. Appl., Pavia, pp. 247-253 (1978).
25. Sommers, H.S. "Experimental Properties of Injection Lasers-IV: Modes of Large Cavity with Sawed Sides," *J. Appl. Phys.,* 44, pp. 3601-3608 (1973).
26. Holden, S.W., Hubbard, W.M., Personick, S.D. "Chromatic Delay in Light Emitting Diodes," *Appl. Opt.,* 13, pp. 1050-1052 (1974).
27. Spano, P. "Chromatic Delay in Light-Emitting Diodes, Measurements and Theory," 3rd E.C.O.C., Munich, pp. 151-153 (1977).
28. Misek, J., Kucera, L., Srobar, F., Kortan, J. "Origin of Chromatic Delay in Electroluminescent Diodes," *Solid State Electron.,* 20, pp. 333-334 (1977).
29. Miller, S.L. "Ionisation Rates for Holes and Electrons in Silicon," *Phys. Rev.,* 105, pp. 1246-1249 (1957).
30. Brain, M.C. "Responsivity and Noise Characterization of Ge Avalanche Photodiodes Throughout Wavelength Range 1.1-1.7 μm," *Electron. Lett.,* 15, pp. 821-823 (1979).

CHAPTER 7
INDEX PROFILE MEASUREMENTS

The bandwidth performance of an optical fiber, as shown in 1.3, is mainly determined by its index profile distribution. An accurate measurement of this profile distribution can lead to the exact computation of the bandwidth, provided that material dispersion and mode coupling have negligible effects. Yet, the bandwidth evaluation obtained in such a way is very critical, especially for near-optimum profile fibers. Recalling the class of α profiles introduced in 1.3.2, an error in α of about 3% with respect to the optimum gives rise to a bandwidth reduction by a factor of 10, and even heavier reductions may be caused by the presence of an axial dip or a sinusoidal perturbation superimposed on the optimum profile. This means that reliable forecasts about the fiber bandwidth need high precision in the measurement of refractive index, and that a fine spatial resolution is required. Nevertheless, the bandwidth evaluation from an index profile measurement may be largely incorrect,[1] and should only be taken as just a qualitative test.

Fiber profiling is extremely useful to manufacturers, who need to improve the processes of fiber fabrication, and also to users, who need a quick way to monitor the fiber performances and geometrical dimensions. The index profile of the preform is easily available to manufacturers, before the fiber is pulled. Comparing preform and fiber profiling results together with accurate geometrical measurements seems to be a suitable way to set the best drawing conditions.

7.1 FIBER PROFILING

Among the proposed methods of index profile measurements, the ones based on the fiber guidance properties seem to be the most appropriate to an industrial environment because of their low cost and minimum time commitment. Moreover, they give a measurement which is averaged over the fiber length considered, while other techniques refer only to the fiber section under test. The main methods based on the fiber guidance properties are:

i) the near field scanning technique

ii) the refracted near field tecnhique

The former method correlates the refractive index distribution to the intensity of guided light at a point of the fiber output end. The latter method makes it possible to determine the index profile through a scanning of the light which is lost sideways from the fiber core. We could say that such a method is based on nonguidance properties, and appears to be the complement of the previous one.

Some other methods are not based on the optical properties of the fiber core, but do make use of an external light source. Interferometry methods investigate the interference fringes in a thin slice of fiber. The reflection method is performed observing the light reflected by the input fiber end. Other methods which involve a transverse illumination of the bare fiber are the transverse interferometry and the transverse focusing methods.

Many nonoptical profiling techniques have been developed. They are mainly based on the measurement of the amount of dopants which are added to $Si0_2$, modifying its refractive index to achieve the desired profile shape. Among them are Xray analysis,[2] Abbe refractometry,[3] and scanning electron microscopy.[4] Yet, such methods are not used much in practice, because they are difficult to implement. They will not be detailed further in this discussion.

7.1.1 Near Field Scanning Technique

The total guided power P_o in given fiber section can be written as

$$P_o = \int_0^a I(r)\, r\, dr$$

where a is the core radius, and $I(r)$ represents the intensity distribution of this optical power on the core transversion section. Note that $I(r)$ can be obtained as an irradiance distribution integrated over the whole 2π azimuthal angle ϕ. Comparing with (1.51), we obtain

$$I(r) = \frac{2\pi^2}{P_o} B_o [n^2(r) - n^2(a)] \tag{7.1}$$

which takes into account guided rays only. Also considering leaky rays, from (1.52), we have

$$I(r) = \frac{2\pi^2}{P_o} B_o \frac{n^2(r) - n^2(a)}{\sqrt{1 - \left(\frac{r}{a}\right)^2}} \tag{7.2}$$

These expressions hold when the optical source is characterized by a uniform radiance in air B_o. Under this condition, scanning the near field output pattern of the optical fiber, the index distribution *n(r)* can be obtained.[5] The uniform excitation of all the rays, required by this technique, is not easily obtained. In fact, the excitation of guided rays only is practically impossible; and the excitation of all the guided and leaky rays with the same optical power distribution appears to be difficult even for very short distances, owing to the greater attenuation which characterizes leaky rays. In (6), some corrective formulae are proposed to take this problem into account, but the accuracy of the measurement may be poor. Since the knowledge of the exact optical power distribution among leaky rays is difficult, it would be better to strip the leaky rays away from the fiber by the aid of a mode stripper, using (7.1) for the computation. The relative distribution of the intensity *I(r)* can be written

$$\frac{I(r)}{I(0)} = \frac{n^2(r) - n^2(a)}{n^2(0) - n^2(a)} = \frac{n^2(r) - n^2(a)}{NA^2(0)} \tag{7.3}$$

where *NA(0)* is the numerical aperture on the fiber axis. Therefore, by knowing *NA(0)* and *n(a)*, even the absolute value of *n(r)*, and not only its relative distribution, can be determined.

The experimental setup can be arranged as shown in Fig. 7.1. The output fiber end is magnified and the scanning procedure is performed on the observation plane of the microscope objective, as discussed for general near field investigations in 5.1.2. The spatial resolution is set by the diameter of the pinhole in front of the active area of the photodetector. Typical values, for multimode fibers having a standard core diameter $2a = 50$ μm, are of the order of 1 μm, so that about 50 independent measurement points can be obtained on a diameter. Usually, two perpendicular directions on the core transverse section are tested, to denote possible errors in the round symmetry. Also the step-motor driven micropositioner, which moves the photodetector, transversally must have good precision, at least on the order of the pinhole diameter. The index profile resolution, i.e., the minimum index difference which can be appreciated, is imposed

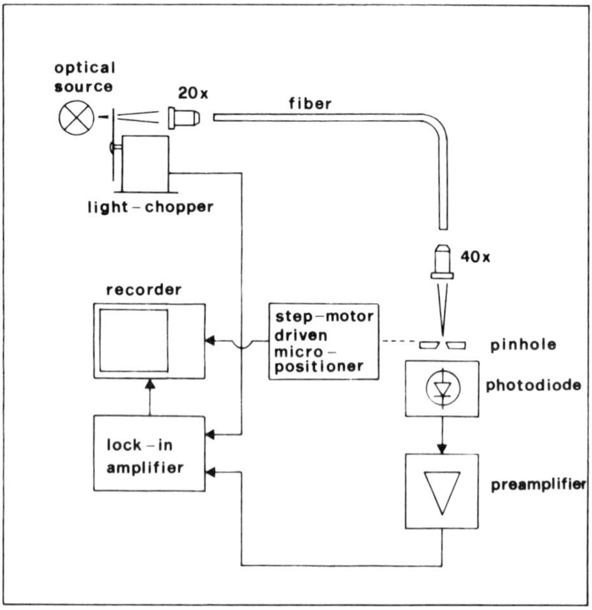

Fig. 7.1 - Experimental setup for index profiling via near field scanning at the fiber output end.

by the optical source employed through the level of transmitted power, and by the sensitivity of the photodetector. The use of a lock-in amplifier (see Appendix 2) allows the achievement of a better sensitivity, and it is useful for performing the measurement at the ambient light.

Considering a general index profile having the form

$$n^2(r) = n_o^2\,[1 - f(r)]$$

where n_o is the on axis index and $f(r)$ is a suitable function of the radial coordinate, (7.3) can be rewritten as

$$I(r) = I(0)\,\frac{n_o^2}{NA^2(0)}\,[f(a) - f(r)]$$

in which the quantity between square brackets is to be measured. Therefore, the maximum error on $I(r)$ is coincident with the maximum error on the profile function $f(r)$. Assuming that the optical power level is high enough to have only a reading uncertainty as maximum error $\Delta I(r)/I(r)$, we can consider a maximum uncertainty of about 10^{-3} affecting $f(r)$. This leads to an overall maximum error on $n(r)$ of the order of a few parts per 10^5.

As for the optical source, a uniform excitation of all the guided rays can be well approximated using a LED, especially if top emitting. When the measurement must be repeated at different wavelengths, e.g., to test profile dispersion effects, a lamp, followed by a monochromator, can be employed. In this case, a suitable optical apparatus has to be arranged to simulate the uniform launching condition required. However, the optical source must be characterized by a large wavelength spectrum; in fact, the finite number of the guided modes may cause a ripple in the measured distribution $I(r)$, if the source spectral width is too small.[7,8] The period of this spatial ripple decreases with increasing the quantity

$$V = \frac{2\pi a}{\lambda}\sqrt{n^2(0) - n^2(a)}$$

and the amplitude of the ripple is inversely proportional to V. Yet, when many spectral contributions are superimposed, the different ripples are averaged, and their effect turns out to be negligible. A source spectral width greater than 50 nm is then necessary.

Instead of an output scanning, an equivalent input scanning can be performed.[9] In this case, a small spot of light, characterized by a numerical aperture greater than that of the fiber, is focused on the fiber input end face. This spot is moved transversely along a core diameter, and the total output power is recorded. To have a very small spot, a high power well-collimated beam must be employed before the optical arrangement. This is easy only when using a collimated beam laser, whose light is characterized by a high degree of coherence and a very narrow spectrum. In this sense, the spatial ripples due to the modal nature of the propagation may affect more strongly this technique than that based on an output scanning.

To summarize, a fiber length of the order of some meters appears suitable for performing the measurement, essentially to avoid the effects of mode coupling or a mode dependent attenuation, which could vary the launching condition. Both fiber end faces must be accurately cleaved and polished (see Appendix 3), but the one used for the scanning procedure especially so. As for the index profile resolution, it is porportional to the local numerical aperture, assuming perfect imaging optics. Since the local numerical aperture decreases from the inner core region to the outer one, nonnegligible errors may be made for large values of the radial coordinate, where profile imperfections have a greater influence on the fiber time dispersion.[10] Finally, the spatial resolution can be fine enough to resolve the thickness of the layers which characterize the CVD technique for fiber fabrication, provided an incoherent wide-spectrum optical source is employed.

7.1.2 Refracted Near Field Technique

Let us consider a ray which is launched into the fiber at an angle in air θ_a, which exceeds the local value of the fiber *NA*. It is a radiating ray, which crosses the core-cladding interface with an angle ψ, with respect to the fiber axis. By simple geometrical considerations, θ_a and ψ are related by

$$\sin^2 \theta_a = n^2(r) - n^2(a) + \sin^2 \psi$$

where, as shown in Fig. 7.2, r is the value of the radial coordinate at which the rays enter the fiber.

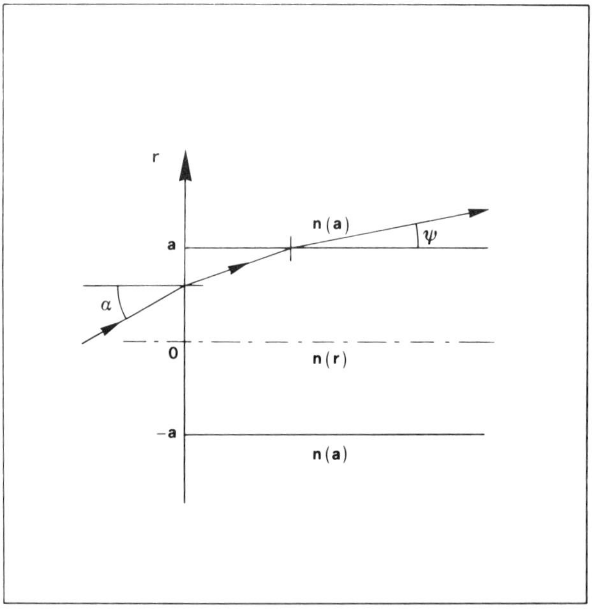

Fig. 7.2 - Example of a radiated ray.

Considering ψ a fixed angle imposed by the receiving optics, since the difference $n^2(r) - n^2(a)$ decreases with increasing r, and the angle θ_a decreases with increasing r. This suggests measuring the difference $n^2(r) - n^2(a)$. This measures the total optical power which passes into the cladding, varying the input radial coordinate r, and for a near field technique with input scanning. The fiber must be imbedded in an index-matching oil with refractive index $n(a)$, so that rays which reach the outer cladding surface can escape without deflection. The problem involved with the presence of leaky rays, which also leave the cladding outer surface, is easily overcome, since they emerge at low angles. Placing a circular screen on the far field pattern, the inner part of the solid cone of light can be removed, suppressing leaky rays. A large area photodetector, behind the screen, collects the power of the radiating rays.

In Fig. 7.3 the experimental setup is schematically shown. The angle ψ_m which is imposed by the screen does not influence the measurement, provided it remains constant. It is not important that some of the radiated power is stopped by the screen together with

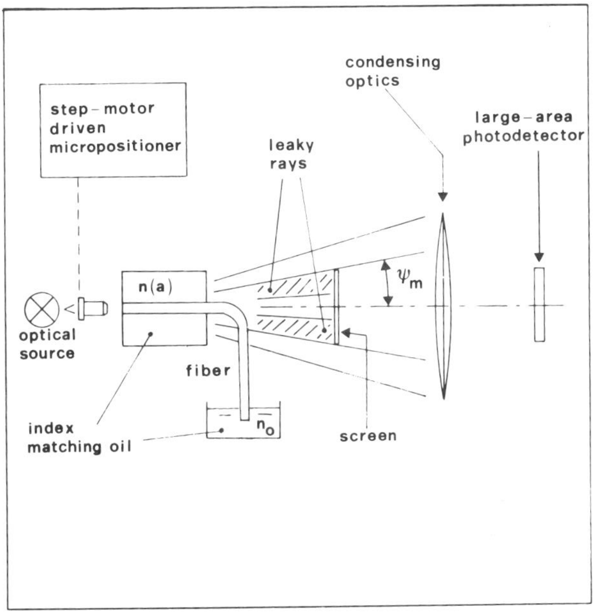

Fig. 7.3 - Experimental setup for refracted near-field profiling.

that carried by leaky rays, but it is necessary that for $\psi > \psi_m$ only radiated power is present. When the surface required for collecting radiated power is too large, it is convenient to use a condensing lens in front of the photodetector active area. Since the radiating modes are a continuum, which is different from guided modes which are finite in number, this method is not affected by the modal nature of the propagation, so that no ripples occur even when a collimated beam laser is employed as an optical source. A quarter wave plate should be inserted before the input microscope objective to obtain a circular polarization, avoiding the influence of a linear polarization on the reflectivity at the glass-air interface. A very large numerical aperture is required for the input microscope objective to exceed that of the fiber, in any condition. A high precision micropositioner, whose spatial resolution must be of the order of 1 μm, is to be employed for the input scanning. Special care must be devoted to the alignment between the fiber axis and that of the incident cone of light. The tilt angle should not exceed 0.2 degrees, to avoid asymmetry of the received power. Once this requirement is fulfilled, the results obtained from such a technique can also give reliable information about core-cladding eccentricity, core diameter, and ellipticity.

The refracted near field profiling was first proposed in reference (11). Its experimental implementation has been detailed in (12). The estimated accuracy, in principle, appears better than that of the near field technique, either on the determination of the profile function $f(r)$, or on the absolute value of $n(r)$. Nevertheless, some uncertainties may affect the results, owing to the micropositioner and possible errors in the alignment. In conclusion, this method is probably one of the most accurate, provided that it is implemented with all the care requested. In particular, it allows one to resolve even a very narrow axial dip produced in CVD multimode fibers,[(13)] and can also be applied to monomode fibers.

7.1.3 Slice Interferometry

A thin slice of fiber, about 50 μm thick, inserted in one branch of an interference microscope, gives rise to a series of fringes on the image obtained while a homogeneous slice is placed in the other branch as a reference. The fringes are straight in the regions of the test slice having a constant value of the refractive index, whereas they are bent where the index exhibits a variation. The direct observation of the fringes through a microscope gives only a qualitative evaluation of the fiber index profile. For a quantitative calculation, it is necessary to consider the following relation between the fringe displacement $S(r)$ at a given radial position on the core transverse section and the refractive index distribution:

$$S(r) = \frac{hD}{\lambda}\left[n(r) - n(a) \right] \tag{7.4}$$

where λ is the wavelength of the light source, h the slice thickness, and D the cladding fringe spacing. An example of fringe pattern is shown in Fig. 7.4.

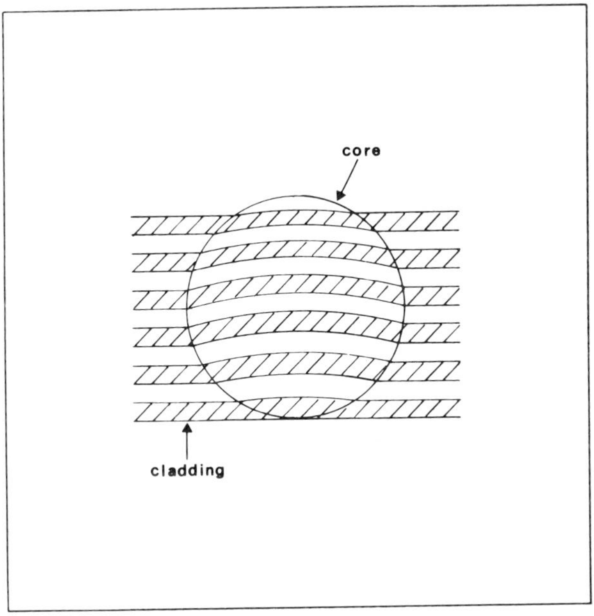

Fig. 7.4 - Example of fringe pattern in slice interferometry.

This technique can be implemented either with a reflected light interferometer,[14] or with a transmitted light interferometer.[15] A scheme of the latter method is shown in Fig. 7.5. A reflected light interferometer is essentially a Michelson interferometer, in which the fiber sample is set close to a mirror, so that the light transverses it twice, and its effective thickness is doubled. The reference sample is replaced by another plane mirror. Scanning the fringe pattern, the minima of the intensity allow one to determine the index distribution applying (7.4). Automatic analysis of the interferograms obtained is convenient, reducing the time spent collecting the data. It is possible to digitize a microphotograph, to record the encoded data on a magnetic tape, and to process them by a numerical computer.[16]

Slice interferometry was the earliest method proposed for index profiling, yet it is probably still the most accurate. But the difficulties in the fiber sample preparation make it unsuitable for an industrial environment. In particular, there are some practical problems in cutting and polishing two fiber end faces, only 50 μm apart. However, for laboratory measurements, especially with technological purposes, this method is very useful. Its ultimate accuracy on the determination of $n(r)$ can be estimated to be one part per 10^5, or better.

7.1.4 Reflection Method

The amount of optical power reflected by a glass surface depends on its refractive index. For perpendicular incidence on a point of the core transverse section, we have

$$\frac{P_r(r)}{P_i(r)} = \left[\frac{n(r) - 1}{n(r) + 1}\right]^2$$

where $P_r(r)$ and $P_i(r)$ represent, respectively, the reflected and the incident optical powers. This property can be exploited for the measurement of $n(r)$.[17,18] The experimental setup can be arranged in two ways: measuring the total reflected power under a selective illumination, or equivalently, scanning a magnified image reflected by the core region under a uniform illumination. The latter technique appears better for avoiding micropositioner errors. In both cases, a suitable optical apparatus must be employed for separating the incident and the reflected powers.

A collimated beam laser can be used as optical source. A quarter wave plate should be inserted in the beam to obtain a circular polarization, avoiding the influence of a linear polarization on the reflection. The preparation of the fiber end must be accurate. The end face should not be polished, since polishing introduces variations of the index profile immediately under the surface, influencing reflection. A good cleaving tool is recommended, like that described in Appendix 3, and then the measurement should be performed as soon as possible, to avoid the absorption of impurities, particularly H_2O vapor.

The accuracy of this method is comparable with that of the near field scanning technique. We have here a very small level of reflected signal, in fact P_r is about $3.10^{-2} P_i$ and therefore some care must be taken to avoid spurious reflections, e.g., from the beam splitters employed in the optical apparatus. In conclusion, the reflection method is attractive for its simplicity, and can be useful for a quick check of the fiber behavior. In fact the access to only one fiber end is required. It does not suffer the modal nature of the light propagation along the fiber, yet it reproduces only the properties of the fiber section under test, while the other techniques perform an average of these properties over the fiber length considered. In this sense, the reliability of this technique is not excellent, and its results cannot be employed to predict the fiber baseband response with high precision.

7.1.5 Transverse Interferometry

An interferometric technique which does not require an accurate sample preparation like that discussed in 7.1.3 is based on a transverse illumination of the sample. This is immersed in an index matching oil having the same refractive index of the cladding, and is placed in one branch of an interference microscope. In the other branch the same index matching oil is placed as a reference. In this way an experimental setup very similar to that reported in Fig. 7.5, except for the direction of the sample,[19] is used. Alternatively, a double pass interferometer can be employed,[20] as shown in Fig. 7.6. The interferogram is collected by a vidicon camera through the eyepiece, and is automatically digitized.

An example of fringe pattern is reported in Fig. 7.7. The refractive index distribution $n(r)$ can be obtained[21] by solving numerically the following integral equation

$$n(r) = n(a) + \frac{\lambda}{\pi D} \int_0^{\infty} \left| \frac{dS(\rho)}{d\rho} \right| \frac{d\rho}{\sqrt{\rho^2 - r^2}} \qquad (7.5)$$

where D is the uniform fringe spacing in the cladding, $S(r)$ is the fringe shift in the core, and ρ is the distance from the fiber axis of the trajectory element considered. It can be seen that a numerical differentiation is required before the integration. A numerical calculation based on an iterative technique with a set of free parameters[22] is employed, starting from an a priori profile function. The accuracy of the method is strongly influenced by the particular algorithm implemented. However, it is noticeably less than that of the slice interferometry. Moreover, a perfectly round symmetry has been assumed in (7.5), making it impossible to evaluate errors of symmetry. This limit can be overcome by observing the sample in several directions, but this takes a remarkable amount of time.

The main advantages of this technique are that it is nondestructive and is performed on fiber length of about 100 μm, allowing an interesting check on axial uniformity. Results given in reference (20) show that the double pass interferometer, though more practical, is not as accurate as the single pass one, owing to the lower magnification available for interference objectives. In conclusion, this method appears suitable for experimental tests during fiber fabrication, essentially because the fiber does not have to be cut and carefully prepared.

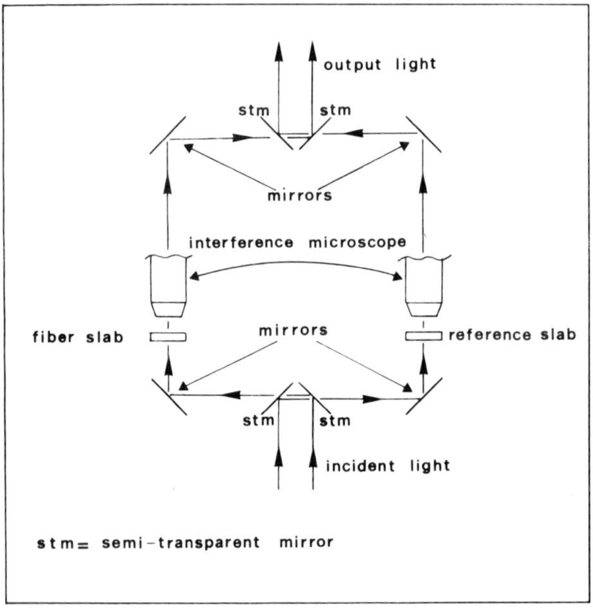

Fig. 7.5 - Slice interferometry by a transmitted light interferometer [from (13)].

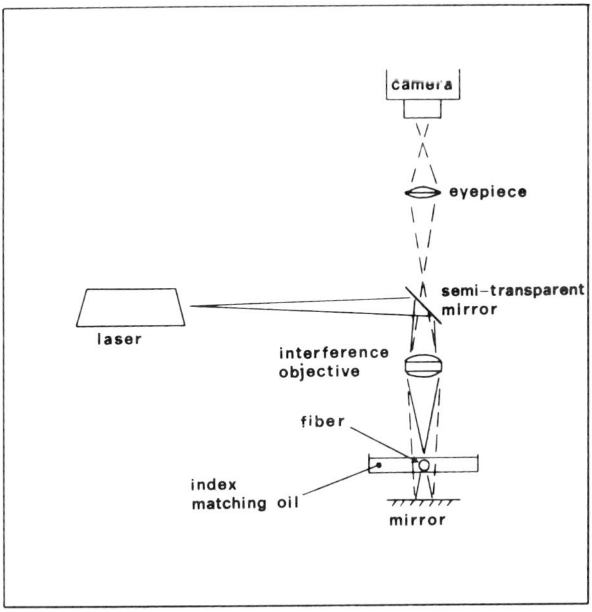

Fig. 7.6 - Transverse interferometry by a double-pass interferometer [from (20)].

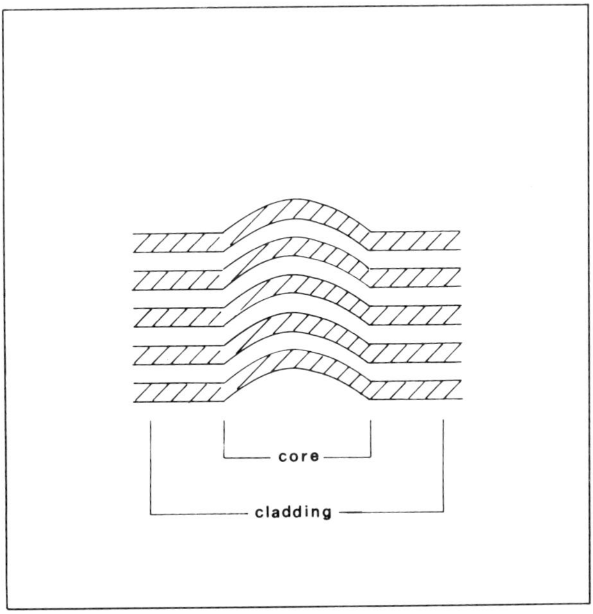

Fig. 7.7 - Example of fringe pattern in transverse interferometry.

7.1.6 Transverse Focusing Technique

Considering an incoherent light that is incident on the fiber outer surface normal to its axis, the fiber core acts as a cylindrical lens focusing the light. It is necessary to immerse the sample in an index matching oil to avoid deflections due to the cladding outer surface. Observing the focused optical power distribution on a plane parallel to the fiber axis, sufficiently close to the core region so that the rays do not cross each other, allows one to determine the index distribution $n(r)$.[23] A microscope objective is used to magnify this image. This technique has been extended to perform profiling. In this case, a vidicon camera can replace the microscope objective.[24] As well as transverse interferometry, this method is nondestructive and makes use of a transverse illumination, testing a finite fiber length, but is does not require any interferometric equipment. Considering Fig. 7.8, $n(r)$ can be obtained as[25]

$$n(r) = n(a) + \frac{1}{\pi L} \int_r^\infty \frac{t - y(t)}{\sqrt{t^2 - r^2}}\, dt$$

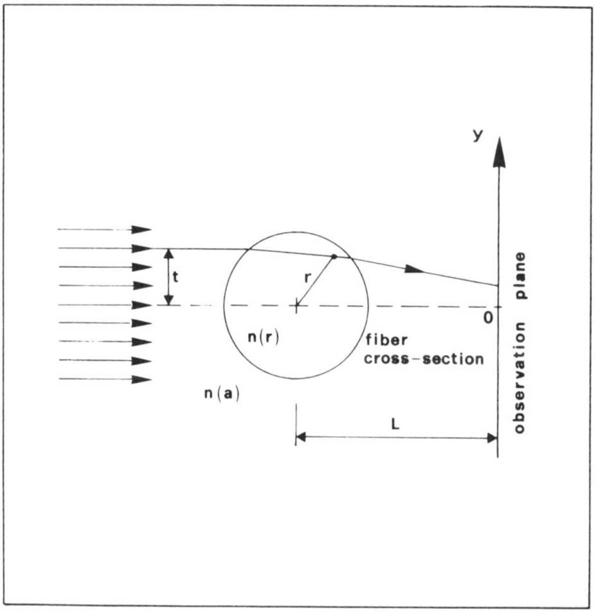

Fig. 7.8 - Fiber illumination in transverse focusing method [from (23)].

where t is the eccentricity of a ray before crossing the fiber core, and $y(t)$ the resulting eccentricity of the same ray after the deflection, as detected on the observation plane. Finally, L is the distance between the fiber axis and the observation plane. A perfectly round symmetry has been assumed. Some uncertainties can arise about the exact knowledge of the distance L, especially when errors of concentricity between core and cladding are present.

Very fine details near the core axis are difficult to reveal because a too stong focusing occurs for very sharp index change, as it is in the neighborhood of an axial deep. But when this technique is applied to preform profiling, because of the greater size, even very fine details can be measured. Particular care must be devoted to cleaning the index matching oil, the optics, and the fiber or preform under test. In fact, large errors may occur owing to diffusion, dust, or contamination. When all these factors are taken into account the resulting accuracy can reach a considerable value, of the same order as that characterizing slice interferometry.[25]

7.1.7 Brief Review of Other Techniques

There are several other proposed techniques for measuring index profile distributions. In this section we will give a brief review of some of them. They can be divided into the following classes:

i) far field scanning techniques

ii) scattering methods

iii) diffraction methods

Far field techniques are based on a longitudinal illumination of the fiber, whereas the other methods exploit a transverse illumination.

In (26), it has been shown that it is possible to determine an equivalent index profile distribution, with some profile parameters, which is monotonic; and leads, under a uniform excitation, to the the same far field pattern obtained experimentally. The main features of this technique are in its simplicity and repeatability. However, some doubts remain on possible differences in the time dispersion properties between the equivalent index profile obtained from the measurement and the actual one.

The transverse illumination of the fiber by a collimated beam and the measurement of the forward scattering distribution has been proposed in (27). The fiber must be immersed in an index matching oil to avoid the effect of the cladding. Since some simplifying assumptions have been made regarding the smallness of the product, $a[n(0) - n(a)]$, such a scattering method is practical just for monomode fibers. A modified version of the above technique has been suggested in (28), in which the index matching oil is not necessary, and the measurement is not limited to monomode fibers only.

A very similar technique[29, 30] is based on the analysis of the diffraction pattern of the fiber under a uniform transverse illumination. The determination of the refractive index distribution $n(r)$ is a fairly complex inversion problem. It can be solved by optical data processing, using the Fourier transform properties of a lens and a spatial filtering. The fiber must be immersed in a suitable index matching oil. This method can also be applied to monomode fibers.

The accuracy of the methods briefly described here can be considered on the same order of magnitude as that of the near field scanning technique. Some uncertainties are essentially due to the complex calculation involved in the data processing. The round symmetry cannot usually be tested by the techniques based on a transverse illumination. On the other hand, they can give information on the axial uniformity.

7.1.8 Comparative Discussion

In (13), an interesting comparison of some of the techniques described in the preceding sections has been reported. The same fiber samples have been measured with the following four methods: refracted near field, slice interferometry, transverse interferometry, and transverse focusing. The test has been developed considering both the accuracy in the determination of $n(r)$ and the spatial resolution. To test these two properties, samples having different characteristics, e.g., amplitude and extension of the axial dip, and other index perturbations, have been employed.

Transverse interferometry and transverse focusing usually give a more detailed analysis of the profile perturbation, though this is possible owing to the assumption of round symmetry. The errors in this assumption, on the other hand, can be revealed only by refractred near field and slice interferometry. In this sense, using a pair of methods, the one based on a transverse illumination and the other on a longitudinal illumination, seems the only way for a complete characterization of the fiber under test. However, the experimental results agree substantially, even in the determination of the core radius a, to the point where no scaling of the radial coordinate is necessary for the comparison of the relative behaviors.

In Table 7.1 the main features of the methods discussed in the previous sections are summarized. The estimated accuracy in the determination of $n(r)$ is always between one part per 10^4 and one part per 10^5, showing that all the techniques are very good in principle. Yet, different degrees of difficulties may arise in the sample preparation. Moreover, the possibilities of achieving information on the axial uniformity and on the round symmetry are important features to be considered. In this sense, a transverse illumination allows the former possibility and not the latter one, and *vice versa* for a longitudinal illumination. In (13) it has been shown that, assuming an elliptical core deformation, by transverse interferometry it is possible to evaluate the ellipticity of the fiber under test. The techinques based on the measurement of the optical guided power, i.e., near field and far field techniques, cannot be employed for monomode fibers.

The experimental evaluation of profile dispersion, i.e., of the dependence on the wavelength λ of the profile distribution $n(r)$, can be performed by different techniques. In any case, a wide spectrum lamp followed by a monochromator must be used. Since the level of optical power selected must not be too small, a tungsten halogen lamp is usually not suitable. A high radiance Xe arc lamp is generally preferred. In 10.1 we will discuss this point in greater detail, showing that not only wavelength-dependent profile measurements can be performed for evaluating profile dispersion. Among the measurement techniques of that class, transverse interferometry appears to be one of the most appropriate.[(32),(33)]

7.2 PREFORM PROFILING

It is very important to test the quality of a fiber profile before the fiber is even pulled. In fact, many properties of the fiber preform are also present in the fiber obtained from it. Both slice and transverse interferometries can be extended to preforms, too. The transverse focusing and the scattering methods, developed for fiber measurements, can be employed efficiently for preform measurements. Finally, there is a technique, called the ray tracing method, which has been expressly proposed for preform profiling.

Table 7.1-Main features of the methods for fiber profiling.

Method	Estimated Accuracy of $n(r)$	Problems in Sample Preparation	Fiber Cut	Information on Axial Uniformity	Information on Errors in the Round Symmetry	Suitable for Monomode Fibers
near field scanning	$5 \cdot 10^{-5}$	moderate	yes	average, over some meters	yes	no
refracted near field	$3 \cdot 10^{-5}$	moderate	no	average, over few centimeters	yes	yes
slice interferometry	$1 \cdot 10^{-5}$	great	yes	average, over $\sim 50\ \mu m$	yes	yes•
reflection	$1 \cdot 10^{-4}$	moderate	no	no	yes	yes
trans. interferometry	$3 \cdot 10^{-5}$	no	no	yes, over $\sim 100\ \mu m$	no	yes
transverse focusing	$1 \cdot 10^{-5}$	moderate	no	yes, over $\sim 100\ \mu m$	no	yes
far field scanning	$5 \cdot 10^{-5}$	no	yes	average, over some meters	yes	no
scattering	$5 \cdot 10^{-5}$	no	no	yes, over some millimeters	no	yes
diffraction	$5 \cdot 10^{-5}$	no	no	yes, over some millimeters	no	yes

7.2.1 Techniques Also Employed for Fiber Profiling

The preparation of a preform sample for slice interferometry or for fiber samples is rather difficult. The cutting procedure is rarely successful if applied to the body of the preform. Better results are achievable when the sample is taken from one tip of the preform.[34] An example of the interferogram obtained is shown in Fig. 7.9. The preform under test exhibits an external region of pure fused silica as cladding, a borosilicate layer whose refractive index is smaller than that of the fused silica, and a series of layers of Ge-doped silica, whose width decreases towards the axis. There is a dip, which is usual for preforms fabricated by the CVD technique.

Since in a preform we have a very slow index variation with respect to the variation of the corresponding fiber, extremely fine details can be revealed by preform inspection, which could not be recognized in the fringe pattern of the fiber sample. The preform sample is typically 200 - 300 μm thick, and the fringe shift is easily followed, whereas for fiber samples a thickness not greater than 50 μm must be assumed. The transverse dimension of a preform sample, even if taken from a tip, is on the order of several millimeters. Therefore, different magnifications must be used to test first the overall behavior, and then some details.

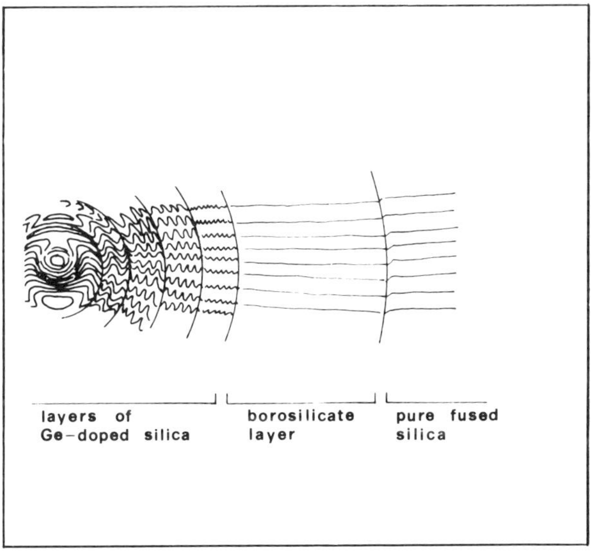

Fig. 7.9 - Example of interferogram obtained from slice interferometry for preform profiling.

Transverse interferometry can be applied to preform samples only with a single pass technique, because the fringe shift with a double pass technique may become too large to be observed. Equation (7.5) holds again. The accuracy is not excellent near the sample axis, owing to the numerical integration of (7.5), but it becomes better and better with increasing *r*. On the whole, an overall accuracy on the order of one part per 10^4 is achieved by this technique, about one order of magnitude less than for slice interferometry. The method suggested in reference (31) for taking into account a possible ellipticity can be applied to preform profiling, too.

Transverse focusing has been successfully applied to preform profiling[(24)] as well as to fiber profiling. The *y* plane of Fig. 7.8 is defined in this case as the object plane on which the camera is focused. An optimum distance can be determined for this plane from the sample axis. An interesting computer simulation for estimating the accuracy of this method has been reported in reference (35). It appears to be more suitable for graded index profiles than for step index profiles, whose measurement is more strongly influenced by ray crossover and the para-axial ray approximation. However, an overall accuracy comparable with that of slice interferometry is achievable, as the agreement of experimental results obtained by the two techniques on the same samples confirm. The computer simulation has also examined the effect of a deviation of the incident light from the ideal collimated beam condition. It has been shown that a divergence as large as 0.5 degrees does not introduce appreciable errors.

A modified scattering method can be applied for the determination of the index difference and the core diameter in step index preforms.[(36)] The measurement is based on a transverse illumination of the sample and on the analysis of the backscattered power distribution. Index matching is not required. The locations of the external maxima are functions of the cladding index, whereas the intensity distribution near the center depends on the index difference and the core diameter.

7.2.2 Ray Tracing Method

The transverse dimension of a preform sample is large enough to allow the following technique for the measurement of $n(r)$. Considering a transverse illumination of the preform by a collimated beam, normal to its axis, at a distance y from the axis itself, it is possible to record the exit angle $\psi(y)$, which is related to $n(r)$ through an Abel transformation. Two main practical implementations of this method have been proposed. The one uses a He-Ne laser beam,[37] which transversally scans the preform by means of a lens mounted on a step motor driven micropositioner. A photodetector is mounted on a goniometer, and for any incident beam position, it performs a far field scanning of the deflected pattern, determining the position of the maximum assumed to be $\psi(y)$. The other implementation[38] is based on a He-Ne laser beam which is reflected by a rotatable mirror. The rotation of this mirror moves a 20 μm spot size beam across the preform. The deflected pattern is focused by a cylindrical lens onto a linear position sensor, whose output is directly proportional to $\psi(y)$.

In both cases, the experimental data are processed by a minicomputer, which also controls the movements of the laser beam and the photodetector. Immersing the preform in an index matching oil allows the removal of the refraction at the preform boundaries, and the avoidance of possible errors due to the roughness of the preform surface. The accuracy of this technique is, as expected, the same order of magnitude as that of the transverse focusing technique, since the principles on which the two methods are based and the numerical computations employed are very similar.

7.2.3 Comparison between the Fiber and Its Parent Preform

It is very interesting to study the relation between the profile of a given fiber and the one which characterizes its parent preform. In,[13] this comparison has been carried out. First of all, two techniques for preform profiling have been compared: transverse focusing and ray tracing. Except for small differences close to the fiber axis, where a remarkable dip was present, the agreement was good. For fiber profiling, the refracted near field technique has been employed. Scaling of the radial coordinate, but not of the index value, was necessary to make a correct comparison.

In particular, the ray tracing technique, characterized by the finer spatial resolution, has shown more detail, but is in substantial agreement with the refracted near field fiber profiling. Comparing a large number of fibers and parent preforms, it is possible to conclude that an excellent agreement occurs for relatively smooth profiles. For instance, a difference in the estimated profile parameter α as low as 0.035 can be obtained.[13] However, the difference between preform and fiber profiles obtained from the measurement increases when the profiles exhibit rapid variations. Especially for near-optimum profiles, great errors may occur when the fiber baseband response is predicted from a measurement of the preform profile, even for very smooth profiles. This is due to the extremely critical role played by the profile parameter α on the fiber baseband response. In this sense, preform profiling must be considered not as a technique for bandwidth prediction, but only as a test procedure for checking the quality of the fiber during its fabrication. In other words, the main purpose of such a measurement is to improve the technology of preform fabrication, reducing some of its imperfections, e.g., extended dips or sinusoidal deformations.

References

1. Ramskov Hansen, J. J. "Propagation in Graded Index Fibers: Comparison between Experiments and Three Theories," *Appl. Opt.*, 17, pp. 2831-2835 (1978).
2. Burrus, C. A., and Standley, R. D. "Viewing Refractive Index Profiles and Small Scale Inhomogeneities in Glass Optical Fibers: Some Techniques," *Appl. Optl.*, 13 pp. 2365-2369 (1974).
3. Wemple, S. H., *et al.* "Binary $Si0_2$-B_2 Glass System: Refractive Index Behavior and Energy Gap Considerations," *J. Appl. Phys.*, 44, pp. 5432-5437 (1973).
4. Krumpholtz, O. "Measuring Parameters of Fibres and Fibre Cables," *3rd* E.C.O.C., Munich, pp. 469-471 (1977).
5. Sladen, F. M. E., Payne, D. N., and Adams, M. J. "Determination of Optical Fiber Refractive Index Profiles by a Near Field Scanning Technique," *Appl. Phys. Lett.*, 22, pp. 255-258 (1976).
6. Costa, B., and Sordon, B. "Fibre Characterization," *Optical Fibre Communication*, technical staff of CSELT, New York: McGraw Hill (1980).
7. Bianciardi, E., Rizzoli, V., and Someda, C. G. "Spatial Correlation of Field Intensity in Incoherently Illuminated Multimode Fibers," *Electron. Lett.*, 13, pp. 25-27 (1976).
8. Adams, M. J., Payne, D. N., Sladen, F. M. E., and Hartog, A. "Resolution Limit of the Near Field Scanning Technique," *3rd* E.C.O.C., Munich, pp. 25-27 (1977).
9. Arnaud, J. A., and Derosier, R. M. "Novel Technique for Measuring the Index Profile of Optical Fibers," *Bell Syst. Tech. J.*, 55, pp. 1489-1508 (1976).

10. Hazan, J. P. "Intensity Profile Distortion Due to Resolution Limitation in Fibre Index Profile Determination by Near Field," *Electron. Lett.*, 14, pp. 158-160 (1978).

11. Stewart, W. J. "A New Technique for Measuring the Refractive Index Profiles of Graded Index Optical Fibres," I.O.O.C. 77, Tokyo, pp. 395-398 (1977).

12. White, K. I. "The Measurements of the Refractive Index Profiles by the Refracted Near field Technique," *4th* E.C.O.C., Geneva, pp. 146-155 (1978).

13. Presby, H. M. "Profile Characterization of Optical Fibers — A Comparative Study," *Bell Syst. Tech J.*, 60, pp. 1335-1362 (1981).

14. Martin, W. E. "Refractive Index Profile Measurements of Diffused Optical Waveguides," *Appl. Opt.*, 13, pp. 2112-2116 (1974).

15. Presby, H. M. Mammel, W., and Derosier, R. M., "Refractive Index Profiling of Graded Index Optical Fibers," *Rev. Sci. Instr.*, 47, pp. 348-352 (1976).

16. Wonsiewicz, B. C., French, W. G., Lazay, P. D., and Simpson, J. R. "Automatic Analysis of Interferograms: Optical Waveguide Refractive Index Profiles," *Appl. Opt.*, 15, pp. 1048-1052 (1976).

17. Eickhoff, W., and Weidel, E. "Measuring Method for the Refractive Index Profile of Optical Glass Fibres," *Opt. & Quant. Electron.*, 7, pp. 109 -113 (1975).

18. Ikeda, M., Tateda, M., and Yoshikiyo, H. "Refractive Index Profile of Graded Index Fiber: Measurement by a Reflection Method," *Appl. Opt.*, 14, pp. 814-815 (1975).

19. Marhic, M. E., Sotto, P., and Epstein, M. "Nondestructive Refractive Index Profile Measurement of Clad Optical Fibers," *Appl. Phys. Lett.*, 26, pp. 574-575 (1975).

20. Boggs, M., Presby, H. M., and Marcuse, D. "Rapid Automatic Index Profiling of Whole Fiber Samples: Part I," *Bell Syst. Tech. J.*, 58, pp. 867-882 (1979).

21. Presby, H. M., Marcuse, D., Astle, H. W., and Boggs, L. M. "Rapid Automatic Index Profiling of Whole Fiber Samples: Part II," *Bell Syst. Tech. J.*, 58, pp. 883-902 (1979).

22. Saunders, M. J., and Gardner, W. B. "Nondestructive Interferometric Measurement of the Delta and α of Clad Optical Fibers," *Appl. Opt.*, 16, pp. 2368-2371 (1977).

23. Marcuse, D. "Refractive Index Determination by the Focusing Method," *Appl. Opt.*, 18, pp. 9-13 (1979).

24. Presby, H. M., and Marcuse, D. "Preform Index Profiling (PIP)," *Appl. Opt.*, 18, pp. 671-677 (1979).

25. Marcuse, D., and Presby, H. M. "Focusing Method for Nondestructive Measurement of Optical Fiber Index Profiles," *Appl. Opt.*, 18, pp. 14-22 (1979).

26. Frende, W. "Far Field Profiling of Multimode Optical Fibres," *Electron. Lett.*, 17, pp. 385-386 (1981).

27. Okoshi, T., and Hotate, K. "Refractive Index Profile of an Optical Fiber: Its Measurement by the Scattering Pattern Method," *Appl. Opt.*, 15, pp. 2756-2654 (1976).

28. Saekeang, C., and Chu, P. L. "Nondestructive Determination of Refractive-Index Profile of an Optical Fibre: Forward Light Scattering Methods," *Electron. Lett.*, 14, pp. 802-804 (1978).

29. Brinkmeyer, E. "Refractive-Index Profile Determination of Optical Fibers from the Diffraction Pattern," *Appl. Opt.*, 16, pp. 2802-2803 (1977).

30. Brinkmeyer, E. Refractive-Index Profile Determination of Optical Fibers by Spatial Filtering," *Appl. Opt.*, 17, pp. 14-15 (1978).

31. Chu, P. L. "Nondestructive Refractive-index Profile Measurement of Elliptical Optical Fibre or Preform," *Electron. Lett.* 11, pp. 357-358 (1979).

32. Gloge, D., Kaminow, I. P., and Presby, H. M. "Profile Dispersion in Multimode Fibres: Measurement and Analysis," *Electron. Lett.*, 11, pp. 469-471 (1975).

33. Presby, H. M., and Kaminow, I. P. "Binary Silica Optical Fibers: Refractive Index and Profile Dispersion Measurements," *Appl. Opt.*, 15, pp. 3029-3036 (1976).

34. Presby, H. M., and Marcuse, D. "The Index Profile Characterization of Fiber Preforms and Drawn Fibers," *Proc. IEEE*, 68, pp. 1198-1203 (1980).

35. Daido, Y., and Otsuka, T. "Estimation of the Accuracy of the Focusing Method for Measuring the Refractive Index Profiles of Fiber Preforms," *Fujitsu Sci. Tech. J.*, 18, pp. 24-48 (1982).

36. Presby, H. M., and Marcuse, D. "Refractive Index and Diameter Determination of Step Index Optical Fibers and Preforms," *Appl. Opt.*, 13, pp. 2882-2885 (1974).

37. Chu, P. L., Whitbread, T. "Measurement of Refractive Index Profile of Optical Fibre Preform," *Electron. Lett.*, 15, pp. 295-297 (1979).

38. Watkins, L. S. "Laser Beam Refraction Transversely through a Graded-Index Preform to Determine Refractive Index Ratio and Gradient Profile," *Appl. Opt.*, 18, pp. 2214-2222 (1979).

CHAPTER 8
GEOMETRICAL MEASUREMENTS

It is possible to separate all the measurements which can be performed on optical fibers into two classes:

i) transmission measurements

ii) geometrical measurements

Transmission measurements are related to the fiber behavior by serving as carriers for information transfer. In this sense, they are of interest for the design of a transmission system and the construction of the repeaters. The measurements of the latter class are useful for the layout of the transmission system.

As geometrical measurements, we have:

i) measurements on the fiber cross section, whose purpose is to test the geometrical compatibility of the two fiber ends to be joined by a splice or a connector, and the coupling properties with sources and detectors

ii) fiber length measurements, which allow one to test if a given fiber piece can cover the distance between two joint locations

The measurements of the fiber cross section interest spatial properties, e.g., core and cladding diameter, their concentricity and eccentricity, and also their angular properties, i.e., the numerical aperture *NA* of the fiber.

In past years some international standardization bodies, e.g., the IEC and CCITT, started work leading to a series of preliminary recommendations. In the following, we will report some of these recommendations, which involve not only the values of several fiber properties and dimensions, but also the experimental methods used to measure them.

8.1 MEASUREMENTS ON THE FIBER CROSS SECTION

Standardization of the fiber geometrical parameters which can be measured on the fiber cross section is still a matter of discussion. No standardization has been recommended for step index fibers, since their current use in telecommunications appears to be limited to local area networks, especially for computer communications. The geometrical parameters defined in the fiber cross section which are of interest in graded index fibers are:[1]

i) the core diameter, which is the diameter of the smallest circle within which the whole core region is contained

ii) the reference surface diameter, which is the diameter of the smallest circle within which the outer surface, taken as a reference for joints (usually the cladding outer surface), is contained

iii) the core eccentricity, defined as the distance between core and reference surface centers, divided by the core diameter

iv) the core noncircularity, which is the difference between the longest and the shortest chord of the core-cladding interface, crossing the core center, divided by the core diameter

v) the reference surface noncircularity, defined as above, but considering the outer fiber surface, instead of the core

iv) the maximum *NA*, which is equal to the square root of the difference between the square of the maximum core refractive index and of the cladding refractive index.

CCITT [2] has proposed the following recommendation for dimensions and tolerance of graded index fibers

i) core diameter 50 μm $\pm$ 6%

ii) reference surface diameter 125 μm $\pm$ 2.4%

iii) core eccentricity $<$ 6%

iv) core noncircularity $<$ 6%

v) outer surface noncircularity $<$ 2%

vi) maximum *NA* 0.18 - 0.23 $\pm$ 0.02

The CCITT document states that the shape of the index profile should be nearly parabolic. A tight control of fiber geometry is of great importance to obtain high, repeatable launching efficiencies and low joint losses. For low loss joints, it is crucial to keep the tolerances about the outer surface diameter, and the core eccentricity as small as possible.

Recently, an alternative method proposed to CCITT,[3],[4] of particular interest for manufacturers, suggests the acceptance of fibers whose core and reference surface fall within a template of four circles, as shown in Fig. 8.1. This would allow relative freedom for manufacturing processes, since it would not require the strict verification of all the tolerances listed above. The purposed maximum deviations for the core diameters $2\Delta a$ and for the reference surface diameter ΔD are, respectively, 4 μm and 5 μm, [5] which enable most of the tolerances given in i) to v) to be fulfilled.

8.1.1 Direct Fiber End Face Inspection

It is possible to directly control the geometrical parameters of the fiber cross section by using a microscope, provided that the end face is accurately cleaved (see Appendix 3). Visual techniques are very quick, but may cause remarkable subjective errors, so that they are convenient only for a qualitative check and not for quantitative measurements. It is then preferable to take a photograph of the fiber end face, or, when many fibers have to be routinely inspected, it is advantageous to use an automatic system which provides a scanning of the whole image [6] through a digitized video camera. The fiber must be illuminated by launching a suitable light into its far end face. The core region is bright because of guidance properties. The cladding region, which has a lower refractive index, appears as

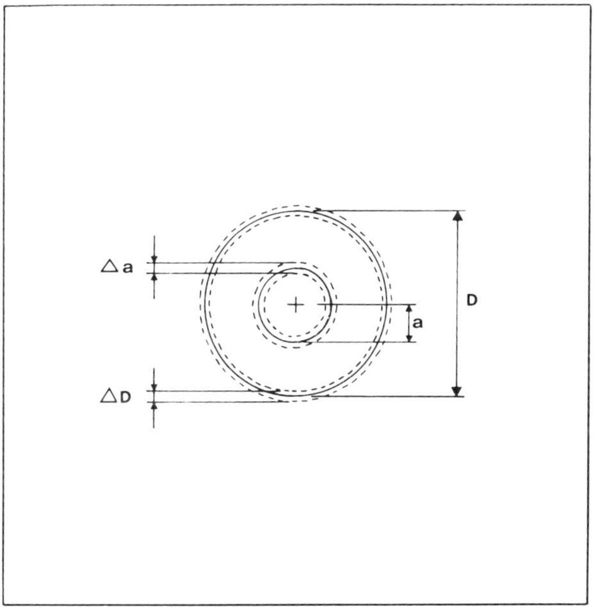

Fig. 8.1 - Four circles template.

a dark annular surface. If the background is illuminated too, an acceptable contrast occurs at the cladding outer surface.

Automatic inspection is particularly suited to industrial environments. High precision and good repeatability can be obtained because subjective errors are totally avoided, and it is possible to process a large number of image points. Thus even slight ellipticities can be easily resolved. The experimental setup is arranged as shown in Fig. 8.2. A visual measurement would give a comparable amount of information, provided that the technique of the four circle template is employed: inserting in the microscope ocular a thin glass layer, where the four circles are accurately drawn.

8.1.2 Near Field Techniques

The near field techniques used for index profiling, already described in Chapter 7, provide accurate measurements of some geometrical fiber properties. A modified near field scanning technique [7] is able to give core and cladding diameters, core ellipticity and eccentricity. To perform this measurement, any light which can propagate inside the cladding must not be stripped away. The near field pattern is measured in both the core and in the cladding regions. The defects

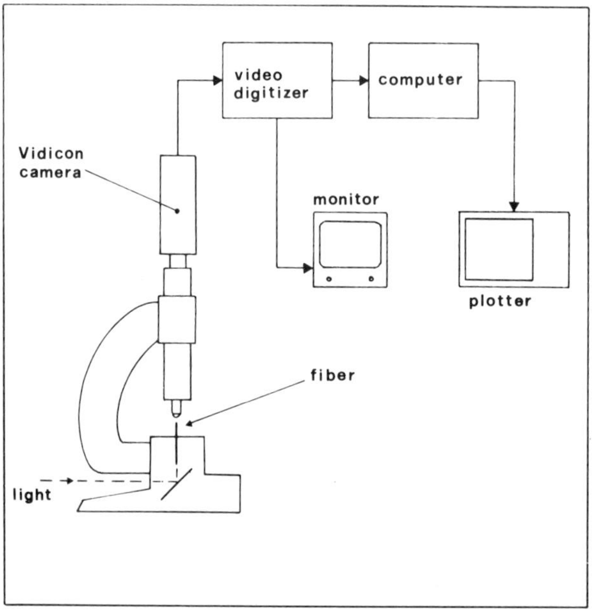

Fig. 8-2 - Setup of an automatic system for microscope geometrical measurements [from (6)].

on the cladding cross section surface, which are tolerated in conventional near field measurements, must be removed, and the whole output end face of the fiber polished. Source and detection optics must have a numerical aperture greater than that of the total fiber structure, which is typically about 0.45. In this way the total fiber structure is considered as a high *NA* step index fiber whose profile is perturbed by the graded index core.

The knowledge of the refractive index of the cladding, or of the primary coating if this exists, enables us to assign a scale to the measured profile. Thus, an accurate measurement of the fiber numerical aperture *NA* is also possible. Results obtained on the *NA* following this modified near field scanning technique and the traditional far field technique usually agree within 5%. An example of near field intensity distribution is shown in Fig. 8.3. The peaks at the extremities are due to leaky modes. Yet, to obtain information about core ellipticity and eccentricity, many scanning procedures must be repeated over different diameters.

The refracted near field technique, on the other hand, is able to give this information very quickly. Moreover, such a technique can be applied whether a single mode or a few modes propagate. Examples

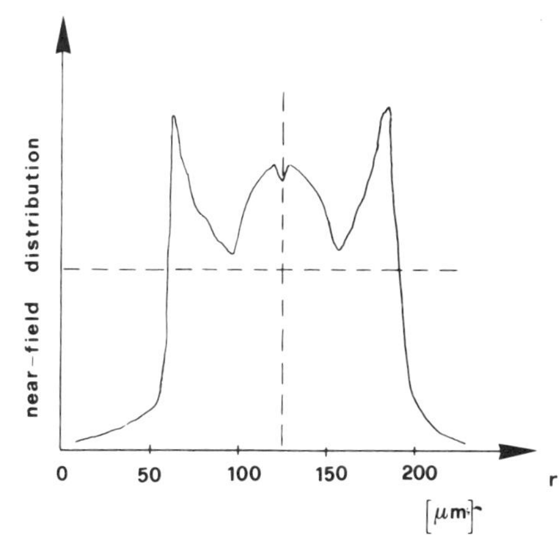

Fig. 8.3 - Example of near field intensity distribution [from (7)].

of fiber profiles obtained by this technique on a multimode graded index fiber and on a monomode fiber are reported in Fig. 8.4. The refracted near field technique is recommended by Standards Committees in Europe and the U.S.A.

Index profile measurements seem to be the most suited methods for determining the fiber core diameter. In fact, the meaning of fiber core, although obvious for step index fibers, might be ambiguous for graded index fibers, which exhibit a continuous refractive index transition between cladding and core. CCITT[2] has defined the core as the smallest cross sectional area of a fiber contained within the locus of points where, except for any central dip, the refractive index n_x is

$$n_x = n(a) + K[n_o - n(a)] \tag{8.1}$$

where n_o is the maximum refractive index of the core, $n(a)$ is the refractive index of the cladding, and K is a constant. The choice of the value of K is a constant. The choice of the value of K is still under study, and clearly it is strongly related to the choice of a standard manufacturing process and measurement. The value most likely to be standardized is 0.05.

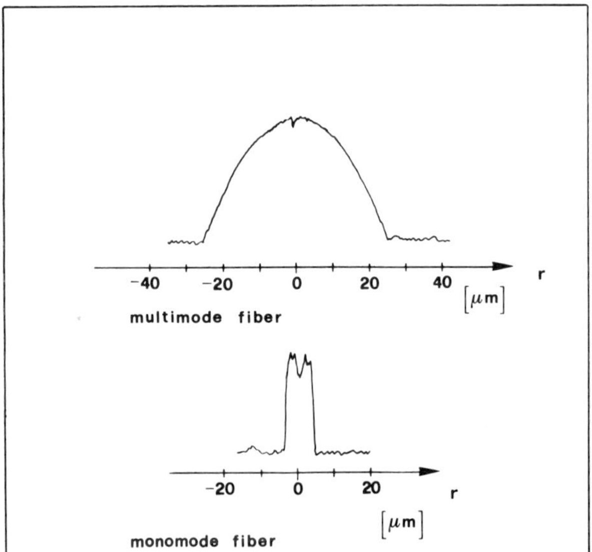

Fig. 8.4 - Example of refracted near field scannings on multimode and monomode fibers.

An alternative definition states that the fiber core is the area where a certain fraction, such as 95%, of guided optical power is contained. This definition, however, is strongly dependent on the launching condition and on a differential mode attenuation. In this sense, the launching condition and the fiber length, as well as to obtain $n^{(r)}$ from near field techniques, would be fixed.

In reference (8), an interlaboratory comparison of core diameter measurements has been reported. Definition (8.1) has been assumed, and a refracted near field technique has been employed to obtain $n^{(r)}$. The test samples were 2.0 ± 0.2 m long, and a uniform launching condition has been simulated with a numerical aperture larger than 0.3, employing an incoherent light characterized by a wavelength $\lambda = 0.85$ μm. Fibers chosen for the comparison were preselected for good core circularity. The results obtained are reported in Table 8.1, showing a good, substantial agreement.

8.1.3 Outer Diameter Measurements

The fiber outer diameter is a very important parameter, since the fiber outer surface is taken as a reference for jointing procedures. In most cases the outer surface coincides with the cladding outer surface, so that optical measurements which are based on the evaluation of scattered patterns can be exploited. An accurate control of the fiber outer diameter during the fiber pulling process is often necessary. In this case, an implementation which is quick and easy is desirable. Moreover, such on-line measurements must be noncontracting to avoid contaminations, which would greatly affect the fiber transmission properties. This excludes a priori any mechanical method. In (9), a scanning beam technique has been proposed

Table 8.1 Results of and interlaboratory comparison of core diameter measurements [from (8)], expressed in microns.

Participant	Fiber 1	Fiber 2	Fiber 3	Fiber 4	Fiber 5	Fiber 6
A	47.5	44.7	50.7	49.6	50.3	47.5
B	46.9	44.1	49.4	48.8	49.4	46.7
C	47.6	44.9	50.2	49.4	49.7	47.5
D	47.3	43.7	49.5	48.4	50.7	47.0
average	47.3	44.4	50.0	49.1	50.0	47.2
std. deviation	0.3	0.6	0.6	0.6	0.6	0.4

which is based on a narrow beam from a He-Ne laser, deflected by a rotating mirror. The measurement of the outer diameter is obtained by recording the time interval during which a large area photodetector behind the fiber is eclipsed by the fiber itself.

The behavior of the fiber is that of a cylindrical lens with a very short focal length. Therefore, it produces in practice almost the same shadow of an opaque wire. The experimental setup is schematically shown in Fig. 8.5, where a cylindrical lens has an axis parallel to the rotating mirror. The photodetector output is monitored on an oscilloscope trance, synchronized with the rotating mirror. Examples of the oscilloscope traces, obtained with fibers having different outer diameters, are reported in Fig. 8.6. The lower parts of the traces represent the uniform illumination of the photodetector, when the fiber is not intercepted by the light beam. The pulse occurs during the interception. A small fraction of the light reaches the photodetector passing through the fiber inner region. This causes the small dip in the center of the dark pulse, which is not influential for the measurement.

The finite slope of the pulse is imposed by the laser spot diameter (about 10 μm). The use of a large area photodetector enables one to perform valid measurements even when the fiber is slightly displaced from its nominal position. This measurement technique is not based on the transparency properties of the fiber, therefore it can also be applied for determining the diameter of an opaque primary coating.

A method based on a very similar approach, but whose accuracy can be even better, uses a commercially available instrument called the Milmaster Electronic Micrometer. The photodetector has a small active area limited by a slit. The width of the slit can be adjusted, so that a wide range of fiber diameters can be measured. An incandescent lamp is used. To avoid possible fluctuations of the lamp light, the photodetector is exposed alternately to the measurement signal and to the reference signal by means of a particular light chopper. [10]

Other techniques are based on the scattering properties of the fiber under a transverse illumination. The fiber scattering pattern can be measured either in the forward or in the backward directions, but the knowledge of the fiber refractive index is necessary for the computation of the fiber's outer diameter. [10] In this sense, these techniques can be applied only to uncoated fibers. The instrumentation required by such measurement techniques is sophisticated, to

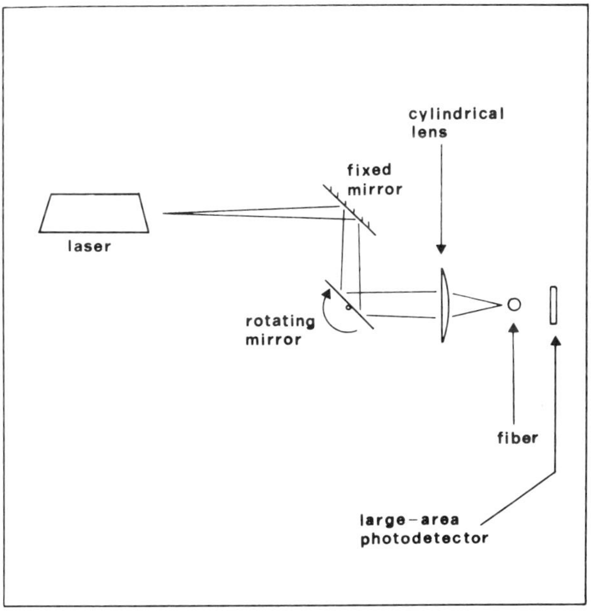

Fig. 8.5 - Experimental set up for scanning beam measurements of the fiber outer diameter.

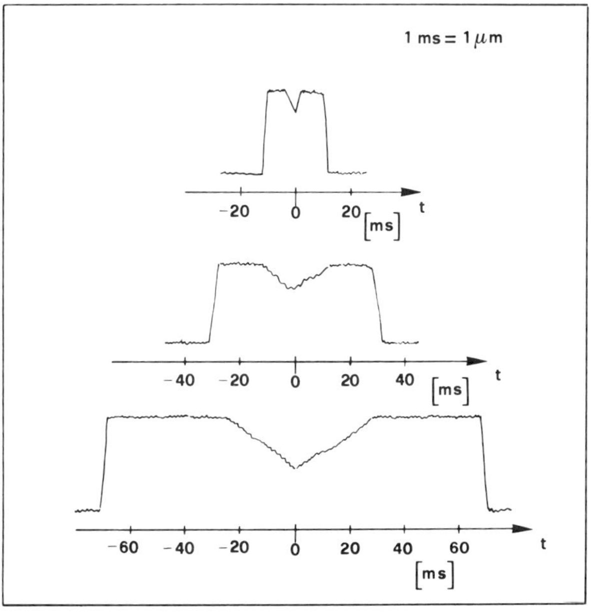

Fig. 8.6 - Oscilloscope traces of scanning beam technique for different fiber diameters.

the point where they do not appear to be suitable for routine measurements in industrial environments.

8.1.4 Measurements of the Ellipticity of the Fiber Outer Surface

A backward scattered pattern, obtained from a laser beam which is transversely incident on the fiber, can also be used for testing fiber ellipticity. Multiple measurements are needed to obtain the width of the scattering pattern as a function of the fiber orientation.[11] This method is accurate and gives repeatable experimental results, but it is slow and thus not suitable for measurements during fabrication. A simpler and rather fast method for measuring fiber ellipticity requires two simultaneous detections of backward scattered patterns using two laser beams, directed perpendicularly to the major and minor axes of the fiber ellipse, as shown in Fig. 8.7.[12]

The angle of minimum deviation, called the rainbow angle, ϕ_o characterizes the backscattered pattern and is related to the refractive index n according to

$$\phi_o(n) = 4 \sin^{-1} \sqrt{\frac{4 - n^2}{3n^2}} - 2 \sin^{-1} \sqrt{\frac{4 - n^2}{3}}$$

The ellipticity e is given by

$$e = \sqrt{1 - (b/a)^2}$$

a and b being the major and minor radii of the ellipse, respectively. Finally, it is possible to write

$$\phi_1 = \phi_o (n) + C_1 e^2$$

$$\phi_2 = \phi_o (n) + C_2 e^2$$

where ϕ_1 and ϕ_2 are the rainbow angles for the two laser beams, obtained from the measurement, and C_1, C_2 are known constants.

The ellipticity can be then calculated as

$$e = \sqrt{\frac{\phi_1 - \phi_2}{C_1 - C_2}}$$

Moreover, $\phi_o(n)$, which is given by

$$\phi_o(n) = \frac{C_1\phi_2 - C_2\phi_1}{C_1 - C_2}$$

allows the determination of the refractive index n.

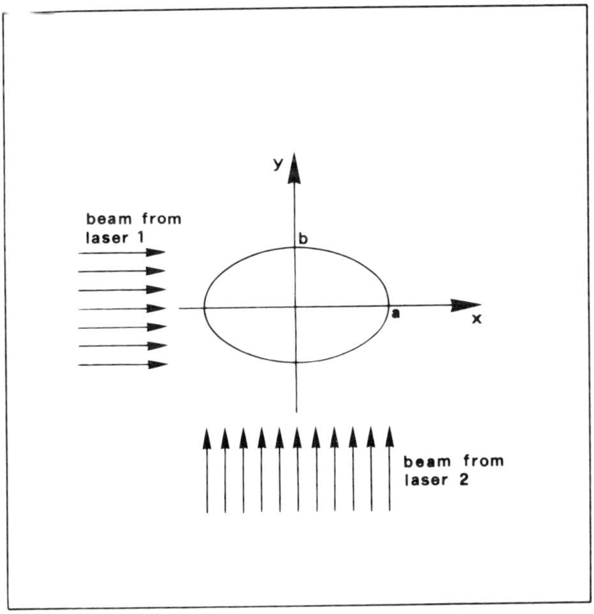

Fig. 8.7 - Principle of the ellipticity evaluation through the measurement of the rainbow angles [from (12)].

Both these techniques for the measurement of the ellipticity and the methods described in the previous section for the measurement of the outer diameter, which are based on transverse scattering, could be applied rigorously only to unclad fibers with a uniform value of the refractive index. Nevertheless, since the index difference between core and cladding is always rather small, they can be extended to real fibers, either step index or graded index.

8.1.5 Measurements of NA

The measurement of the numerical aperture *NA* of a step index fiber can be performed simply through a far field scanning at the fiber output, under a uniform launching condition. In this case, the numerical aperture of the launching beam must be greater than *NA*. This technique, however, gives just the effective *NA* of the fiber, i.e. that imposed not only by the index difference between core and cladding, but also by mode coupling and a mode dependent attenuation. To reduce these effects, a short piece of fiber is recommended. Nevertheless, the fiber piece must be long enough to eliminate radiating modes. A length of several meters appears to be a good trade-off.

In reference (13), a double far field scanning procedure, which scans both at the fiber input and at the fiber output, has been suggested to obtain the value of *NA* imposed strictly by the index difference. A He-Ne laser beam, weakly focused, illuminates the input fiber end face, selecting a given launching angle. The output far field pattern exhibits a maximum which corresponds to the launching angle, when this is within the fiber *NA*, whereas exhibits two maxima when the launching angle exceeds the fiber *NA*. One of the maxima coincides with the launching angle, and is due to leaky and radiating modes, the other one is approximately equal to θ_{aM}, where we have let $NA = \sin \theta_{aM}$. Repeated measurements, varying the launching angle by small quantities, provide good repeatability and precision.

For graded index fibers, whose *NA* is a function of the radial coordinate r, a method based on far field scanning gives only an average value of *NA*. Even selecting a precise value of r, by means of a suitable launching apparatus, the measurement turns out to be not very accurate. Moreover, a technique of this type would require a noticeable waste of time. Thus, it is necessary to employ one of the techniques used for fiber profiling, e.g., slice interferometry or the reflection method provided they are able to give both the indices of the core and of the cladding. Also the modified near field scanning procedure, shown in 8.1.2 for core diameter measurements, appears suitable for this purpose. Alternatively, a method based on a proper index matching and an interferometric inspection of the fiber cross section [14] can be employed. Assuming an α profile, such a method allows a very accurate determination of parameter Δ, which is quite independent of the value of α.

8.2 FIBER LENGTH MEASUREMENTS

It is sometimes useful to obtain a precise evaluation of the length of a fiber, when this is wound on a drum, without unwinding it or counting the coil number. Some instruments for the fiber attenuation measurement by backscattering also give a measurement of the fiber length, which is based on the reflection of the light at the far fiber end. Alternatively, it can be obtained using a lock-in amplifier. Both these techniques are founded on the measurement of the time delay of the light along the fiber length. The backscattering technique exploits a narrow light pulse, whereas the lock-in amplifier technique is based on a sinusoidal modulation. The advantage of the former method is in the requirement of the access to only one fiber end. Nevertheless, the latter method appears more accurate, though somehow more complex.

8.2.1 Fiber Length Measurement by Backscattering

Optical time domain reflectometry (OTDR), as will be shown in Chapter 9, is one of the most efficient methods for evaluating fiber attenuation. It is based on the detection of a backscattered signal, when a light pulse travels along the fiber, so that a correspondence exists between the attenuation along the fiber length and the time on the backscattering trace.

This correspondence can be expressed through the following relationship

$$z = \frac{1}{2}\frac{c}{n_e}\, t \tag{8.2}$$

where c is the free space velocity of light, n_e is an equivalent refractive index, t is the time delay from the launching instant, and z is the distance from the fiber input section. A factor of ½ takes into account the double distance covered, forward and backward. The equivalent refractive index n_e is necessary, since not all the rays travel axially. It depends on the index profile, but it can be considered coincident with the on-axis index n_o, except for an error smaller than 10^{-5}.

Equation (8.2) is affected by an intrinsic approximation which depends on the pulse duration. The shorter this duration the better the spatial resolution. The lower limit of pulse duration is imposed by the requirement on the minimum level of detected signal, and is about 10 ns. It corresponds to a spatial resolution on the order of 1 m.

On the backscattering trace, it is possible to reveal the signal reflection at the far fiber end. This appears as a sharp peak of signal, on the right-hand side of the trace. The peak width, owing to the low-pass filter behavior of the receiver and of the oscilloscope, is usually a little larger than the pulse duration. In this sense, this peak width imposes the accuracy of the present method for measuring fiber lengths. In conclusion, it is possible to perform measurements with maximum uncertainties of the order of few meters.

8.2.2 Fiber Length Measurement by a Lock-in Amplifier

Using a light source externally modulated by a sinusoidal signal, e.g., a LED or laser diode, and a photodetector, it is possible to compare directly the phases of the two signals obtained and to insert the test fiber between the optical source and photodetector. Letting $\Delta\phi$ be the phase difference, which can be measured by the use of a lock-in amplifier, the fiber length L can be written as

$$L = \frac{c}{n_e}\frac{1}{2\pi f}\,\Delta\phi$$

where f is the modulation frequency. $\Delta\phi$ must be smaller than 2π to perform an unambiguous measurement. Yet, too low a frequency allows only a roughly approximated evaluation of L. Therefore, more than one measurement is required to attain good accuracy, and is acomplished by increasing the modulation frequency f.

For example, considering L = 1.028 km, and letting n_e = 1.5, one can begin the measurement procedure assuming f = 10 kHz, so that

$$\Delta\phi = 0.323 + 2n\pi, \qquad n = 0,1 \ldots$$

It is clear that n = 0 must be chosen. The result of the measurement, considering a maximum error due to the uncertainty of 1 degree in the phase evaluation (see Appendix 2), is

$$L = 1030 \pm 55\ m$$

Next, one can assume f = 100 kHz, obtaining a new measurement result which can be written as

$$L = 1028 \pm 5.5\ m$$

The general expression of the uncertainty Δ L, as a function of the modulation frequency f, is

$$\Delta L = \frac{c}{n_e}\frac{1}{360}\frac{1}{f} \cong 5.5\cdot 10^2\ \frac{1}{f} \qquad [km]$$

where f is expressed in Hz. In this expression no errors in the frequency measurement have been considered, only the 1 degree uncertainty in the phase evaluation has been taken into account.

The upper limit of modulation frequency which can be used with commercially available lock-in amplifiers is on the order of several hundred kHz.

Therefore an uncertainty of about 1 m is achievable. To obtain a smaller uncertainty, a vector voltmeter, whose range of modualtion frequencies is extended typically from 1 MHz to several hundreds of MHz, can be used. In this way, the uncertainty can be reduced to few centimeters, which is also the lower limit imposed by the definition of the equivalent index n_e. However, the measurement procedure must be initiated with, at least, one test performed by a lock-in amplifier, at a low modulation frequency, to avoid ambiguities.

REFERENCES

1. Costa, B., Rossi, F., and Tosco, F. "Standardization and Measurement of Optical Fibers for Telecommunication Systems," 1st I.E.R.E, London 1982, pp. 173-183 (1982).
2. CCITT. "Characteristics of 50/125 μm Graded Index Optical Fiber Cables," Draft Recommendation G.651 (1980).
3. CCITT. "Proposals for Graded Index Optical Fibre Cables," German Document to CCITT, No. 210/XV (1978).
4. CCITT. "Geometrical and Optical Tolerances of Optical Fibre," Italian Document to CCITT, No. 35/XV (1981).
5. CCITT. "Correlation between the Definitions of the Geometrical Characteristics of the Fibre and the Template of Four Concentric Circles as Tolerance Fields," German Document to CCITT, No. 11/XV (1981).
6. Marcure, D., and Presby, H. M. "Automatic Geometric Measurements of Single Mode and Multimode Optical Fibers," *Appl. Opt.*, 18, pp. 402-408 (1979).
7. Sabine, P. V. H., Donaghy, F., Irving, D. "Fiber Refractive-index Profiling by Modified Near Field Scanning," *Electron. Lett.*, 16, pp. 882-883 (1980).
8. Kim, E. M., and Franzen, D. L. "An Interlaboratory Measurement Comparison of Core Diameter on Graded Index Optical Fibers," Symposium on Optical Fiber Measurements, Boulder, pp. 143-146 (1982).
9. Cohen, L. G., and Glynn, P. "Dynamic Measurement of Optical Fiber Diameter," *Rev. Sci. Instrum.*, 44, pp. 1749-1752 (1973).
10. Marcuse, D. "Principles of Optical Fibre Measurement," New York: Academic Press (1981).
11. Marcuse, D. "Light Scattering from Elliptical Fibers," *Appl. Opt.*, 13, pp. 1903-1905 (1974).
12. Chu, P. L., Seakeang, C., and Chow, Y. M. "Determination of Refractive Idex and Ellipticity of an Unclad Elliptical Optical Fibre," *Electron. Lett.*, 12, pp. 41-42 (1977).

13. Jeunhomme, L., and Pocholle, J. P. "Measurement of the Numerical Aperture of a Step Index Optical Fiber," *Electron. Lett.*, 12, pp. 63-64 (1976).

14. Stone, F. T. "Rapid Optical Fibre Delta Measurement by Refractive Index Tuning," *Appl. Opt.*, 16, pp. 2735-2742 (1977).

CHAPTER 9
ATTENUATION MEASUREMENTS

Attenuation measurements are probably the most important measurements which can be performed in optical fibers, in fact the main limitation of the extension of the repeater spacing in most transmission systems is due to the fiber attenuation. Only those systems which have a very high transmission capacity also suffer a bandwidth limitation. However, in Chapter 4, we have shown how a bandwidth limitation can be compensated for only through a worsening attenuation limitation. From all the previous considerations, it is possible to conclude that a great deal of attention must be devoted to performing reliable attenuation measurements with a high degree of accuracy.

In the following, we will distinguish between two classes of measurements on the basis of their purposes:

i) technological measurements, which are an essential support for improving the process of fiber fabrication

ii) transmission measurements, which are of interest mainly for the design of transmission systems

In the former class of measurements, it is important to evaluate only the intrinsic loss, and to separate absorption and scattering losses. In the latter class, on the contrary, great care must be devoted to the evaluation of the overall loss, due in part to intrinsic attenuation and in part to fiber perturbation. Separation between these two contributions to the overall loss allows one to extrapolate the results over any distance.

For technological measurements it is essential to use an optical source whose central wavelength of emission can be varied without appreciable change of the launching condition into the fiber. In this way, it is possible to appreciate spectral variations of loss not influenced by errors due to mode dependent attenuation and perturbation. The optical sources specifically devoted to these measurements, which satisfy the above requirement, are:

i) organic dye lasers, when a high level of launched power is requested

ii) a wide spectrum lamp followed by a monochromator, when a smaller amount of launched power is sufficient

One has to distinguish specific purposes in the case of transmission measurements. At first, the evaluation of the steady-state attenuation coefficient, which is important for the design of transmission systems characterized by very long repeater spacing. In this case, once the wavelength to be used in the transmission has been chosen, a measurement of attenuation at such a wavelength may be enough, provided that great care is devoted to the control of the reaching of the steady-state. The distance required for this reaching can be shortened by using an external strain concentrated over a short length of fiber, immediately after the launching section. This is achieved by means of an apparatus called a mode-scrambler.

Another purpose of transmission measurements can be to evaluate the amplitude of the perturbation affecting a particular type of cabled fiber. This may be a useful support in optimizing the cable manufacturing process, or in estimating the attenuation behavior of the link over a repeater spacing on the order of the distance required for the reaching of the steady state. In this case, different approaches to the steady state can be simulated varying the launching condition.

The present chapter is organized as follows. In an introductory paragraph, a general review of the main existing techniques for attenuation measurements is presented. The next two paragraphs are devoted to discussions about the practical implementation of the two most diffused techniques. The last paragraph deals with differential mode measurements.

9.1 REVIEW OF THE EXISTING TECHNIQUES

Attenuation measurements of the transmission type are performed sometimes on bare fibers (consisting of core and cladding without any coating), but more often on fibers already jacketed by a primary coating, and even on cabled fibers. The primary coating protects the outer cladding surface from external contaminations, which could influence the light propagation. Moreover, the primary coating gives a mechanical strength to the fiber, allowing bending radii as small as a few millimeters. On the other hand, technological attenuation measurements are mainly performed on samples of the bulk glass from which the fiber will be drawn, or on unclad fibers.

Once the fiber has been jacketed or cabled, only its two end faces are available for measurement. The most immediate transmission measurement on this type of fiber is an insertion loss measurement, which consists of launching a given amount of optical power $P(0)$ into the fiber and in measuring the optical power $P(L)$ at its output end, L being the fiber length. Unfortunately, it is very difficult to estimate the exact value of the guided optical power $P(0)$, because a considerable part of the optical power emitted by the source may exceed, in space or in angle, the fiber light acceptance. To know $P(0)$ with sufficient precision, the fiber can be cut a short length after its input section. To minimize the amount of fiber sacrificed, one would like to perform the cut very near the input end, but it leads to increased uncertainty in $P(0)$.

If reaching the steady-state must be taken into account, or if it is of interest to know how the attenuation evolves along the fiber length, the measurement must require several cuts, which lead to a complete destruction of the fiber. This measurement is suitable only for technological purposes. In the field of transmission measurements, and also for experimental inspections on installed fibers, nondestructive techniques are required. Among them, the optical time domain reflectometry (OTDR) is nowadays the most popular. It is based on the detection, at the fiber input end, of the backscattered power decay.

9.1.1 The Cutback Technique

The classical cutback attenuation measurement is a destructive technique based on N measurements, which requires N-1 cuts of the fiber along its length. The results of the measurement of the output power are plotted *vs.* the distance z from the input section. Fig. 9.1 reports the behavior of a typical plot. A uniform spacing L/N of the cuts has been considered. The launching condition must be always the same for all these measurements. Therefore, the fiber, which is originally wound on a drum, must be accurately prepared, in order to avoid any stress at the input end, when the cuts are made. Before the measurement is performed, the turns of the fiber which correspond to the sections to be cut should be unwound, so that the fiber can be cleaved without moving the drum further. Particular care should also be devoted to prevent any drift of the optical source, especially when a laser diode whose emitted power is influenced by the temperature, is employed.

From the data reported in Fig. 9.1, one can observe that the optical power decay is not uniform. In fact, it is usually stronger at shorter distances. This can be ascribed to two different causes:

i) a considerable fraction of the optical power is not carried by guided rays, and is present, mainly in the cladding, as radiating power, owing to a poor launching condition

ii) the steady state condition of the optical power distribution among the guided rays has not been reached yet

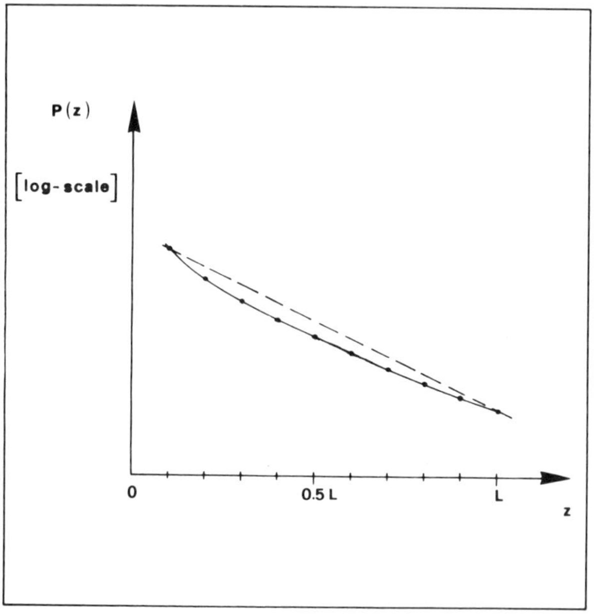

Fig. 9.1 - Typical behavior of the optical power measured along the fiber length by a destructive cutback technique.

The former cause always gives rise to an evident excess power decay at very short distances. The latter cause entails a variation in the power decay which can be either positive or negative (as discussed in 2.3.1), and is usually of small amplitude and relatively long extension, except when the fiber is strongly perturbed.

The correct evaluation of the attenuation per km must be made ignoring the first two points on the left, i.e., considering the optical power decay only when it reaches a uniform exponential behavior. The slope of the straight line which interpolates such points gives the attenuation per km. It can be considered as the steady-state one only when the first two points correspond to a distance of some hundreds of meters. In Fig. 9.1, the straight line which connects the two extreme points of the plot has been dashed, showing that a remarkable error may occur when only these two points are considered.

Nevertheless, a two point measurement is widely used, because only a few meters of fiber are lost. In this case, a correct excitation of the fiber is very important, to avoid at least the presence of any power in the cladding, particularly at the section where the cut has to be made. A correct excitation can be obtained using the following tools:

i) a mode-scrambler

ii) a cladding mode-stripper

iii) a particular light source

A mode scrambler is an apparatus which induces an external strain on the lateral surface of the fiber, forcing mode coupling over a very short fiber length. This is very useful to excite also those modes, near the guidance condition, which are not excited by an optical source having a too small beam divergence or spatial extension. However, reaching the steady-state is theoretically impossible using only a mode scrambler, because of the statistical nature of this definition. Nevertheless, recalling 2.3.1, the steady-state optical power distribution does not depend on the amplitude of the fiber perturbation, but only on its statistical properties, which influence the analytical form of operator $\hat{D}$. So, it would seem possible, in principle, to launch a power distribution very similar to the one which characterizes the steady state, obtaining a great reduction in the distance required for the reaching of the steady state itself. If the statistical properties of the perturbation affecting the fiber are known, the steady-state optical power distribution can practically be simulated also by the use of a suitable beam-launcher. Yet, since the above statistical properties are usually not known with the precision required, beam launchers are preferably used to excite all the guided rays, rather than to simulate the steady state distribution of optical power. Section 9.2.1 is devoted to the description of some types of mode scramblers and beam launchers.

Almost all the optical power which exceeds the guidance condition, after some turning points from the launching section, has flowed completely into the cladding. So, about 1 m after the first input end, all the radiating power which could affect the attenuation measurement is in the cladding, except for leaky rays, which are considered at the moment to be together with strictly guided rays. Radiating power, which is lost for the propagation, may remain in the cladding for hundreds of meters, owing to the index difference between the cladding and the outer medium. A cladding mode stripper is an apparatus which drains the optical power guided by the cladding. It is based on an index matching oil, in which a given length of fiber is immersed, provided that the primary coating has been removed. Mode stripping is improved by increasing the length of immersion. About 20 cm may be enough, especially if the index difference between oil and cladding is small.

The extension and the divergence of the beam of the optical source can be optimized in order to avoid the presence of radiating power. For instance, using a gas laser, which is characterized by a highly collimated beam, it is convenient to employ a microscope objective with the same *NA* of the fiber, in order to have a uniform excitation of all the rays. Similar considerations hold also for sources consisting of a wide spectrum lamp followed by a monochromator, for spectral measurement purposes. A laser diode is usually coupled to the fiber directly or through a pigtail. Sometimes, the launched beam may be too narrow inside the fiber, and mode scrambling is necessary to obtain a uniform excitation. Only when a LED is used, especially if it is top emitting, is the whole fiber light acceptance filled.

The cutback technique does not require a high level of launched power, so the use of LED sources appears to be the most suitable, except when a high spectral resolution is needed. Note that for graded index fibers, in which rays are curved by the index gradient along their trajectories, the angular divergence of the source beam is not so critical a parameter. In this case, however, the spatial extension of the optical source may limit the number of launched rays.

Fig. 9.2 reports the experimental setup for an attenuation measurement with the two point cutback technique. After the launching section, we have ordinately: a mode-scrambler, a cladding mode-stripper, and the cut section. A typical distance z_o between the launching section and the cut section is about four meters. Great care must be devoted to the measurements of $P(L)$ and $P(z_o)$. The two fiber cross sections must be orthogonal to the fiber axis and clean, in order to allow the transfer of the maximum fraction of the guided power to the photodetector. All the optical power going out of the fiber must be incident over the photodetector active area. In this way one can assume that the two measurements are made in the same detection conditions. The shortest time interval which is possible should separate the two measurements, in order to avoid possible drifts in the launched power.

The output voltage of the photodetector is proportional to the incident optical power, as shown in 3.2. So, letting $V(L)$ and $V(z_o)$ the output voltages which respectively correspond to $P(L)$ and $P(z_o)$, the attenuation per km is obtained as

$$A = \frac{1}{L - z_o} \; 10 \log_{10} \left[\frac{V(z_o)}{V(L)} \right] \left[\frac{dB}{km} \right] \tag{9.1}$$

where L and z_o are expressed in km. In this expression, the overall responsivity $\mathscr{R}_v$ of the photodetector vanishes, provided that it is independent of the detected optical power. In this sense, a linear response over the whole dynamic range foreseen for the measurement is requested of the photodetector. This is usually based on a PIN photodiode. The optical power to be detected, even when a spectral measurement is performed, is normally high enough that a noise induced uncertainty turns out to be negligible, especially when a light chopper and a lock-in amplifier are employed to enhance the sensitivity (see Appendix 2). Therefore, a simple direct detection on a suitable load resistance R_L, without any amplification in the photodetector, is preferable. Using a lock-in amplifier also allows one to operate at any level of ambient light.

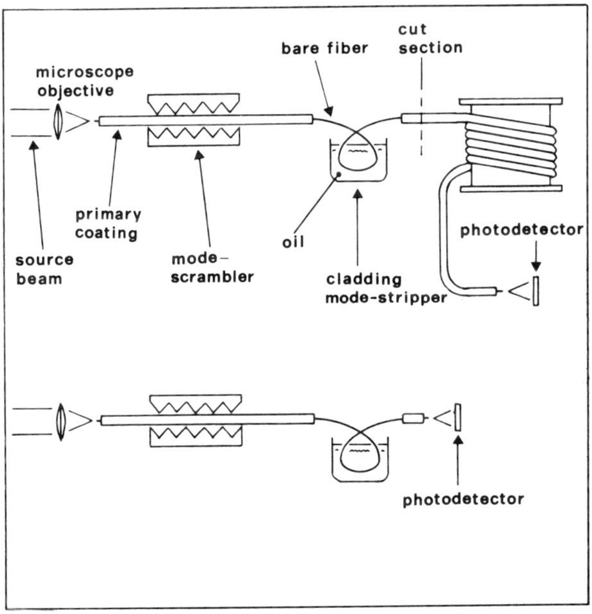

Fig. 9.2 - Experimental setup of the two-point cutback technique.

Although cutback attenuation measurements are probably the simplest techniques for fiber characterization, great care must be devoted to obtain reliable and repeatable results. In section 9.2 some details on this important measurement technique will be discussed. In particular, we will analyze the accuracy of the method, the dynamic range which is typically requested of the photodetector and of the voltmeter, and finally some particular applications of such a measurement technique. Among them: spectral loss measurements, to investigate the absorption properties of the fiber glass, and measurements performed varying the launching condition, to give some insight into the nature of the fiber perturbations.

9.1.2 The Backscattering Technique (OTDR)

In this section, to present an outline of the backscattering technique, some simplifying assumptions will be made. In Chapter 9.3 we will return to this technique, removing such assumptions for evaluating the accuracy of the method. An optical pulse of peak power P_o launched into a fiber travels with an approximate velocity c/n_1, where c is the free space velocity of light and n_1 is the core refractive index, considered for simplicity as a constant. After a time interval t, the peak power, owing to the fiber attenuation, results

$$P(t) = P_o \exp\left(-\gamma \frac{c}{n_1} t\right)$$

Over a unit distance, a fraction D of the peak power is scattered backwards. With the formalism introduced in 2.1.1, referring to a step-index fiber and to an isotropic scattering, it is $D = \gamma_s F$, so this is a very small fraction. At the input section of the fiber, the power density per unit distance,

$$p(t) = DP_o \exp\left(-2\gamma \frac{c}{n_1} t\right) \tag{9.2}$$

can be detected, giving all the information for evaluating the optical power decay along the fiber axial coordinate z. In fact, $p(t)$ can also be written as

$$p[t(z)] = DP_o \exp(-2\gamma z) \tag{9.3}$$

So, knowing factor c/n_1, it is possible to graduate the time scale replacing t with z, and to evaluate the intrinsic attenuation coefficient γ.

This schematic description gets clear why the backscattering technique is also called OTDR (Optical Time Domain Reflectometry).

Such a technique is characterized by some unquestionable advantages over the cutback technique:

i) it is non-destructive

ii) the access to only one fiber end is needed

iii) any kind of local additional loss, superimposed on the fiber attenuation, e.g., that caused by a fault or by a joint between two fiber pieces, can be located and appreciated in its amplitude and spatial extension

iv) it allows one to follow the evolution of the optical guided power along the whole fiber length, leading to the possibility of appreciating any variation in the slope of the optical power decay

Nevertheless, there are some problems in performing OTDR measurements which have to be stressed:

i) this technique requires a high intensity launched peak power, essentially as a consequence of the very small value of coefficient D

ii) there are some uncertainties due to possible fluctuations in the value of D along z, which may be caused by fluctuations either in the density of scattering centers (i.e., in γ_s), or in the numerical aperture (i.e., in F)

iii) the ray distribution of the backscattered power is usually different from the one which locally characterizes the power of the pulse travelling in the forward direction, so the effects of the launching condition cannot be appreciated

Considering especially the above advantages, this technique appears as particularly suitable for tests whose purposes are in the transmission area. So, OTDR can be efficiently employed during a cable manufacturing process in order to detect possible faults caused by strains on the fiber, as well as during the installation of a transmission system for the inspection of the joints. On the other hand, the uncertainties which affect the result of the measurement, due to possible fluctuations in the fiber parameters and to a difficult control of the launching condition, suggest avoiding the use of this technique for technological purposes.

As for the spatial resolution in the detection of local additional losses or faults, it depends first of all on the width Δt of the pulse. Denoting with Δz this spatial resolution, as a first approximation, we have

$$\Delta z = \frac{1}{2} \frac{c}{n_1} \Delta t \tag{9.4}$$

Typically, $\Delta z = 1$ m can be obtained with $\Delta t = 10$ ns. The factor 1/2 originates from the following considerations. Two fiber sections separated by a distance Δz give rise to two reflections of the same pulse. They travel backwards a distance $2\,\Delta z$, as shown schematically in Fig. 9.3.

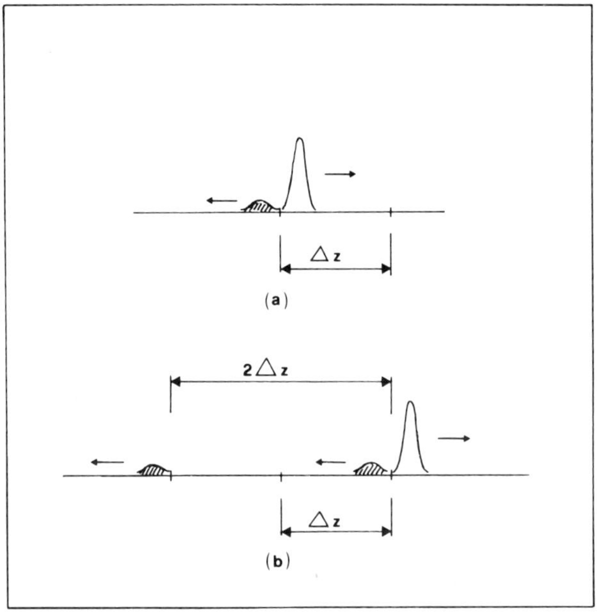

Fig 9.3 - Reflections of the same pulse in two fiber sections: (a) at a given time t, (b) at $t + (n_1/c)\,\Delta z$.

Nevertheless, the required spatial resolution influences the choice of the intensity of the peak power. In fact, the signal detected by the electrical instrumentation is not rigorously the power density per unit distance (9.2), but it is a power envelope that can be seen as the summation of discrete contributions of the type $p(t)\,\Delta z$. In other words, the detected power envelope results in

$$R(k\,\Delta t) = D\,P_o\,\Delta z \exp\left(-2k\,\gamma\,\frac{c}{n_1}\Delta t\right) \qquad k = 1, 2, \ldots \tag{9.5}$$

Therefore, fixing a minimum level R_m for the instantaneous value of R, imposed by requirements on the signal to noise ratio, the smaller the Δz the higher the intensity of the peak power P_o. Letting, for example, γ_s=0.2 km^{-1}, F = 5 · 10^{-3}, we have D = 10^{-3} km^{-1}. Then, if Δz = 1 m, it results in $P_o \geq 10^6\ L_A^2\ R_m$, in which L_A represents the fiber attenuation over which the measurement is extended. L_A, when expressed in dB, is usually called one-way loss, and measures the dynamic range of the instrument. Letting L be the whole fiber length which has to be inspected, we have

$$L_A = 10 \log_{10} [\exp(\gamma L)] \cong 4.34\, \gamma L \quad [dB]$$

This suggests that a laser diode, capable of emitting a considerable peak power must be employed as an optical source.

Great care must be devoted to the separation of the backscattered signal from the reflection on the fiber input end. In fact, what must be measured is several orders of magnitude smaller than the launched pulse, whose reflection on the fiber input end could saturate the receiver. Different solutions have been proposed to solve this problem. They will be discussed in 9.3.

9.1.3 Technological Measurements

In this section, we will give an outline of some technological measurement techniques, intended for:

i) absorption loss evaluation

ii) scattering loss evaluation

The former ones are performed either on the bulk glass or on fibers, the latter ones on unclad fibers, but sometimes also on cladded fibers. Absorption loss measurements are based on the evaluation of the temperature rise due to absorption. Scattering loss measurements exploit integrating or differential detectors which surround the fiber. An integrating detector measures the whole scattered power, and is devoted to separating the contributions γ_s and γ_a to the intrinsic attenuation coefficients γ. A differential detector can be used for the study of the angular distribution of the scattered power, in order to give an insight into the causes of scattering.

For absorption loss measurements, a high level of launched power is required in order to make the temperature rise appreciable. However, a power level of 1 W should not be exceeded, to avoid nonlinear effects in the light propagation, like Brillouin or Raman scattering.[1] This limitation does not hold for bulk glass. Since the outer surface of a bulk glass sample is much greater than that of a fiber having the same length, in order to have an equal temperature rise, a higher level of power is necessary for bulk glass samples.

The experimental apparatus for evaluating absorption loss consists of a calorimeter 20 – 40 cm long, which is used both for bulk glass samples and for fibers with little modifications. Some problems of accuracy arise for the bulk glass sample, which is completely immersed in the calorimeter, because of the possible presence of contaminants on the sample end faces, [2],[3] one of which is crossed by the source beam. In order to limit such effects, the two end faces should be immersed in an index matching oil, so that the sample can be considered as infinitely long. [4]

When a fiber is tested, it crosses the calorimeter, and various measurements can be made at different distances from the optical source, giving useful information about possible fluctuations of γ_a along the fiber length. [5],[6] In order to avoid systematic errors, the thermocouple which detects the temperature rise is suspended between the capillary tube containing the fiber and another capillary tube used as a thermal reference. The sensitivity of the method depends on the thermocouple responsivity, which, in the best case, is about $4 \cdot 10^{-5}$ V/°C, and on the possibility to measure very small voltages. Using a light-chopper and a lock-in amplifier (see Appendix 2), it is possible to estimate voltages as small as 1 nV. The resulting sensitivity for the absorption attenuation per km, A_a, turns out to be about $25/P(z)$ dB/km, where the optical guided power $P(z)$ is expressed in mW. [6]

A spectral loss measurement is then possible, using a wide spectrum lamp and a monochromator, only when the fiber exhibits remarkable peaks of absorption. In order to evaluate low values of absorption losses, a dye laser, or a set of collimated beam lasers operating at different wavelengths, must be employed. A typical value of $P(z)$ is in this case 100 mW, and the sensitivity becomes 0.25 dB/km.

A measurement technique which has been derived from the calorimetric method allows one to obtain separately γ_a and γ_s. It is based on the different velocities which characterize the transfer of energy from a bare fiber to the surroundings. The energy diffusion due to scattering has a very high transfer velocity. On the other hand, the heat due to absorption exhibits a transfer velocity which depends on the air pressure, because convection may be predominant, provided that an insulating material holds the fiber. This is inserted in a cylindrical metallic cavity, in which the air pressure can be controlled. The metallic surface of the cavity absorbs the scattered energy, leading to a rise of its temperature. A similar process, but much slower, occurs for the transfer of the heat due to fiber absorption. Launching a rectangular optical pulse of known energy into the fiber, the two fractions of this energy which are lost, respectively, owing to scattering and absorption can be determined. [(7)] The sensitivity which can be reached, using a gas laser as optical source, is about 1 dB/km.

The use of an electric resistance, as thermal sensor for calorimetric measurements, instead of a thermocouple, has been proposed too. [(8)] The main advantage of this technique is in the possibility of measuring variations of the resistance value, as for a bolometer, by the use of a Wheatstone bridge, with ac voltages. This technique has been derived from a previously proposed method [(9),(10)] which was based on the metallization of the fiber lateral surface, over a length of some centimeters. This thin layer of metal absorps the scattered light and also the heat produced by the fiber absorption, giving a measure of the total loss. Scattering and absorption cannot be separated, but a very high sensitivity characterizes this method. For a guided power of 1 mW, the sensitivity turns out to be about 0.1 dB/km.

An integrating scattering detector consists of a cavity in which the fiber can run. Several centimeters of fiber are tested. All the optical power scattered by the fiber must be detected. Two main arrangements have been proposed for this kind of detector:

i) a spherical cavity, with a highly reflecting internal surface, to convey all the scattered power onto a small area photodetector [(11)]

ii) a cavity whose walls are made up by photodetecting surfaces

For the latter type, two geometries have been used: a cube made up by six solar cells, [(12)] and a sandwich detector, where the fiber is inserted between two parallel solar cells. [(13)] In both these methods, the solar cells operate in the photovoltaic mode.

A uniform response for any direction of the scattered power must be assured. For an integrating sphere, this is possible if the optical power is incident upon the photodetector after at least one reflection. This condition entails a limitation in the sensitivity of the method.

For an integrating cavity made up by solar cells, a response calibration for every cell is needed. Since the complexity of calibration increases with the number of cells, the sandwich detector appears preferable. If the whole power scattered by the fiber P_s is detected, we can know

$$P_s(z) = \gamma_s P(z) \Delta z \tag{9.6}$$

where $P(z)$ is the optical guided power at the section of measurement, and Δz is the fiber length under test. Therefore, the knowledge of $P(z)$ gives γ_s. This type of measurement can be performed either on unclad fibers or on bare cladded fibers. In the latter case, it is necessary to assume that the cladding does not introduce any absorption loss on the power which crosses transversally its section. Owing to the small thickness of the cladding, this is certainly a reasonable assumption. However, a correction must be taken into account, both for unclad and for cladded fibers, in order to evaluate the total scattered power $P_s(z)$ from the measurement of the power $P_s'(z)$ which flows out of the fiber lateral surface.

Letting n_1 and n_2 be respectively the refractive indices of core and cladding for a step-index fiber, and considering isotropic scattering, the fraction of scattered power which flows from the core into the cladding is

$$F_c = \frac{\int_{\theta_M}^{\pi/2} \sin\theta \, d\theta}{\int_0^{\pi/2} \sin\theta \, d\theta} = \cos\theta_M = \frac{n_2}{n_1}$$

Assuming the refractive index of the air $n_a = 1$, the fraction of scattered power which flows from the cladding to the air is

$$F_a = \frac{1}{n_2}$$

Therefore we have

$$P_s'(z) = F_a F_c P_s(z) = \frac{1}{n_1} P_s(z)$$

This correction holds for unclad fibers as well. If the exact expression (2.3) of the angular dependence of the scattered power for Rayleigh scattering is considered, it is possible to demonstrate [14] that the following correction holds

$$P_0'(z) = \frac{3n_1^2 + 1}{4n_1^2} P_s(z)$$

A differential scattering detector consists of a small area photodetector which can scan the radiation field of the fiber, on a plane containing the fiber axis. [15] An index matching oil with the same refractive index of the core must be used, both for unclad and for bare cladded fibers, in order to maintain outside the fiber the same angular distribution of scattered power.

9.1.4 Measurements of Total Loss by Lateral Scattering Detection

Equation (9.6), assuming γ_s independent of z, can be used for a total loss measurement. It is performed detecting the power which is scattered in different fiber sections. Since $P_s(z)$ and $P(z)$ are proportional, considering two sections z_1 and z_2, we have

$$\frac{P_s(z_1)}{P_s(z_2)} = \frac{P(z_1)}{P(z_2)}$$

Dividing this ratio, expressed in dB, by the distance z_2-z_1, the total attenuation per km, A, of the fiber is obtained. [16] Scattering photodetectors having NEP on the order of 1 pW/$\sqrt{\text{Hz}}$ can be easily obtained, therefore, by the use of a light chopper and a lock-in amplifier (see Appendix 2), the sensitivity in the detection of the guided power $P(z)$ can be very high. For example, even for $\gamma_s = 0.05$ km^{-1}, which is very low value, assuming $\Delta z = 1$ cm, $P(z)$ as small as 0.02 mW can be appreciated. So, a spectral loss measurement, using a wide spectrum lamp and a monochromator, is possible.

The accuracy in the calculation of A depends, as well as for the two-point cutback measurement, on the distance z_2-z_1, and on the overall attenuation coefficient γ itself. The greater the quantity γ $(z_2$-$z_1)$ the better the accuracy. A differential measurement, performed simultaneously on the two fiber sections, with a pair of equal scattering detectors, may improve the accuracy remarkably. [17]

Nevertheless, some uncertainty about this method remains in the assumption that γ_s is uniform along the fiber length. For example, if $\gamma(z_2$-$z_1)$ is about 0.5, fluctuations in γ_s as small as 10 % may alter the measurement completely. Fluctuations in γ_s having a short correlation length can be corrected taking an average of each measurement over a set of fiber sections close to z_1 and z_2. [18] For instance, if 100 independent measurements of $P_s(z_i)$, $i = 1,2$, are performed in the neighborhood of z_i, within an axial distance of 1 m (typically $\Delta z = 1$ cm), and if one considers the average values $<P_s(z_1)>$ and $<P_s(z_2)>$, most of the γ_s fluctuation effects turn out to be cancelled.

9.1.5 The Insertion Loss Measurement

When a measurement of total attenuation must be performed in the field on an installed cable, neither the two point cutback technique nor the lateral scattering technique can be employed. The only alternative to the back-scattering technique is then an insertion loss measurement. Here, great care must be taken to obtain repeatable launching conditions to evaluate with high accuracy the launched power $P(0)$. There are two main launching techniques:

i) using another fiber of the same type as a beam-launcher

ii) the small-spot small-angle technique

With the former method, the power $P(0)$ is launched into the test fiber through a short piece of fiber of the same type. A good connector is used to assure a high repeatability. Typical uncertainties of about 0.2 dB on the connector insertion loss affect the measurement. With the latter method, a small spot of light having a small divergence is focused on the fiber input end. This is based on the assumption that all the focused power becomes guided power. Since only rays of the innermost core region are excited, the measured attenuation is often smaller than that at the steady state, as one can derive from the considerations reported in 2.3.3. The use of a mode scrambler just after the input section of the fiber can reduce this systematic error.

9.1.6 Joint Loss Measurements

The power loss of a joint, and in particular of a connector, has been investigated in 2.4.5 and 2.4.6. In principle, two kinds of measurements can be performed:

i) the measurement of the power loss L_o, as defined in 2.4.5, when the first fiber is uniformly excited

ii) the measurement of the power loss $L_o(z_1,z_2)$, as defined in 2.4.6, when the first fiber is at the steady state before the joint.

In order to perform a measurement of the latter kind, a very long distance z_1 is usually required. The use of a mode scrambler can reduce z_1. Note that this kind of measurement is of some interest especially when z_2 is so great it restores the steady-state. In this case, we obtain the effect of the joints on a sequence of cascaded long-length fibers, which is the typical case of an installed optical link.

The measurement under a uniform excitation is useful to compare different jointing techniques. In this sense, only a relative evaluation may be requested. On the other hand, a measurement under the steady-state excitation is useful for optical transmission system designs, but it depends strongly on the particular perturbation which affects the fibers.

A measurement of L_o can be performed with good accuracy by a destructive method very similar to the cutback technique for fiber loss measurements. With the formalism introduced in 2.4, we have

$$L_o = \frac{P^{(1)}}{P^{(2)}}$$

where $P^{(1)}$ and $P^{(2)}$ are the total guided powers respectively before and after the joint. $P^{(1)}$ is measured before the execution of the splice or connector. $P^{(2)}$ is measured after, cutting the second fiber some meters from the joint section. A cladding mode stripper is recommended for the latter optical power measurement, and even for the former, when the distance from the optical source is not large.

L_o also can be evaluated through the backscattering technique. This is particularly suited to test a sequence of cascaded joints, after their installation, along an optical link, requiring access to only one end of the link. Nevertheless, some discrepancies in the two backscattered measurements, which can be performed in either direction of the fiber, suggest to take the arithmetic mean of the two values. Also, the theory gives different values of L_o in the two directions, but only when the two fibers have noticeable differences in their parameters, which is not so frequent. Since these discrepancies occur sometimes for very low values of L_o, it is possible to measure even $L_o < 1$, which could not have any physical meaning. This is probably due to some uncertainties about the optical power distribution of the backscattered signal, to which we will return in 9.3.2.

A measurement of $L_o(z_1, z_2)$ requires a very high accuracy. In fact, the uncertainties on the fiber attenuation during the reaching of the steady state may be much greater than the power loss of the joint. In order to evaluate in the laboratory the behavior of a certain jointing technique, one can cut the fiber and then insert the splice or connector to be tested, comparing the received optical power before and after this operation. But in the field, it is very difficult to obtain even a rough evaluation of $L_o(z_1,z_2)$ over an installed link.

In the following, we will show how $L_o(z_1,z_2)$ can be evaluated through the knowledge of the steady-state attenuation coefficient of the two fibers, only for a connector, provided that the two fibers have equal parameters. Recalling 2.4.6, it is

$$L_o(z_1,z_2) = \frac{P^{(1)}(z_1+z_2)}{P^{(2)}(z_1,z_2)}$$

where $P^{(1)}(z_1+z_2)$ is the optical power guided by a single fiber at the section z_1+z_2, and $P^{(2)}(z_1,z_2)$ is the optical power at the same section in the presence of the connector. Considering a set of logarithmic quantities, it is

$$L_o(z_1,z_2)\Big|_{dB} = L_o\Big|_{dB} + L^{(2)}(z_1,z_2)\Big|_{dB} - L^{(1)}(z_2)\Big|_{dB}$$

where L_o is the intrinsic power loss of the connector when the first fiber is characterized by the steady-state optical power distribution, $L^{(2)}(z_1,z_2)$ is the attenuation of the second fiber between z_1 and z_1+z_2, and finally $L^{(1)}(z_2)$ is the attenuation at the steady state over the distance z_2.

From the knowledge of the steady-state attenuation coefficient, one obtains $L^{(1)}(z_2)$. The attenuation $L_o|_{db} + L^{(2)}(z_1,z_2)|_{db}$ can be expressed as the ratio between the optical power $P^{(1)}(z_1)$ at the output of the first fiber and the optical power $P^{(2)}(z_1,z_2)$ at the output of the second fiber in the presence of the connector. $P^{(1)}(z_1)$ can be measured demounting the connector. Since usually

$$L^{(2)}(z_1,z_2)\Big|_{dB} < L^{(1)}(z_2)\Big|_{dB}$$

$L_o(z_1,z_2)$ turns out to be smaller than L_o. The difference

$$\Delta L\Big|_{dB} = L^{(1)}(z_2)\Big|_{dB} - L^{(2)}(z_1 z_2)\Big|_{dB}$$

represents the loss reduction due to a new reaching of the steady-state, a consequence of the selective excitation which occurs at the joint.

For a splice, the uncertainty on $P^{(1)}(z_1)$ entails an uncertainty on the whole $L_o(z_1,z_2)$. A backscattering investigation over the distance z_1+z_2, which could be the only possible test for splice characterization, may yield unreliable results in the evaluation of $L_o(z_1,z_2)$. This is mainly due to difficulty in the evaluation of $L^{(2)}(z_1,z_2)$.

9.1.7 On-Line Tests during Fiber Fabrication

During the process of fiber pulling, a continuous monitoring of its attenuation can be performed.[19] Part of the light produced in the preform white-hot zone is guided by the fiber. A photodetector, which is mounted on the fiber take-up drum, collects the optical power guided at the only fiber end which is accessible. An interference filter is set in front of the photodetector, to select the wavelength region of interest for the measurement. A chart recorder gives a plot of the received power along the fiber length.

This technique is based on the assumption that the launched optical power is constant during the pulling process. This assumption appears reasonable, at least as a first approximation, owing to the controls in the temperature of the preform white-hot zone and in the speed of the drum. Repeated tests, performed in comparison with a two point cutback measurement, made after the pulling process, have shown a substantial agreement, with typical differences smaller than $\pm$ 1 dB/km. The overall dynamic range of this technique is about 30 dB, which is large enough for single pieces of fiber, even when a very long piece length is considered.

For what concerns the possibility of appreciating eventual local additional losses, on the other hand, some doubts are justified because of the launching condition. In fact, the first consequence of a fault during the pulling process is a variation in the geometrical parameters of the fiber. This can be revealed by the present monitoring technique, but the local variation of attenuation which could be measured is not reliable. So, the only experimental test for a quantitative estimate of local additional losses remains backscattering, which is, however, an off-line procedure.

A similar technique for monitoring fiber attenuation during extrusion coating has been proposed too.[20] The optical source and the photodetector are here set along the axes of the play-off reel and of the take-up reel, respectively. The two fiber ends must exactly positioned along these two axes, so that rotation should not have any effect on the launching and detecting efficiency. Clearly, this is true only as a first approximation, especially for what concerns the launching condition. The resulting measurement, therefore, can be used only as a check for revealing eventual stresses produced by the application of the primary coating, and even for controlling some process parameters, but not as a reliable test of the fiber attenuation.

9.2 DISCUSSION ON CUTBACK

In the present paragraph, the description of cutback techniques will be carried out with more detail. First of all, different types of beam launchers and mode scramblers are shown. Next, spectral loss measurements and tests devoted to the evaluation of the effects of the fiber perturbation are described. Unless otherwise stated, a two point measurement is always considered.

9.2.1 Beam Launchers and Mode Scramblers

Beam launchers and mode scramblers are used to shorten the distance required for reaching the steady-state. A beam launcher is an apparatus that transforms the radiance distribution of the optical power emitted by the source, and conveys the new radiance distribution into the fiber under test. A mode scrambler induces mode coupling by an external strain directly on the test fiber. In both cases, the optical power, which is available from the source, is reduced. Nevertheless, without beam launchers or mode scramblers, especially when the optical source is characterized by a collimate beam, the fiber must be very long to reach the steady-state, so that the optical power turns out to be reduced as well.

i) a short fiber with a rough output end face, to make the radiance distribution launched into the test fiber practically independent of the optical source radiance distribution [21]

ii) a sequence of equal cascaded fibers, separated by joints with small displacements, to force mode coupling before the input of the test fiber [22]

iii) a sequence of three different cascaded fibers (ordinarily step index, graded index, step index), so that a random transformation of the radiance takes place [23]

In Fig. 9.4 these beam launchers are schematically shown. The first requires a very short dummy fiber (about half a meter), but its features depend on the roughness of the output end face. The second employs some long pieces of fiber (typically 100 m long), however it is believed that the steady-state optical power distribution is launched into the test fiber, provided that this has the same parameters of the other ones. The third type of beam launcher is probably a good trade-off, each cascaded fiber being about 1 m long.

Mode scramblers are based on bending or microbending induced in the fiber over a length of several centimeters. Usually the fiber is forced between two irregular surfaces by a lateral pressure. These surfaces can be rough,[24] or made of a sequence of transverse knives, whose height and spatial period are adjustable.[25] In a modified version, the fiber is forced over a sequence of rollers.[26] A quite different method has been proposed in (27). It consists of a logitudinal metallic tube containing the fiber. This is maintained in its position by four twisted metallic wires. The tube shrinks, owing to the heat produced by the Joule effect, and the metallic wires induce a lateral pressure on the fiber.

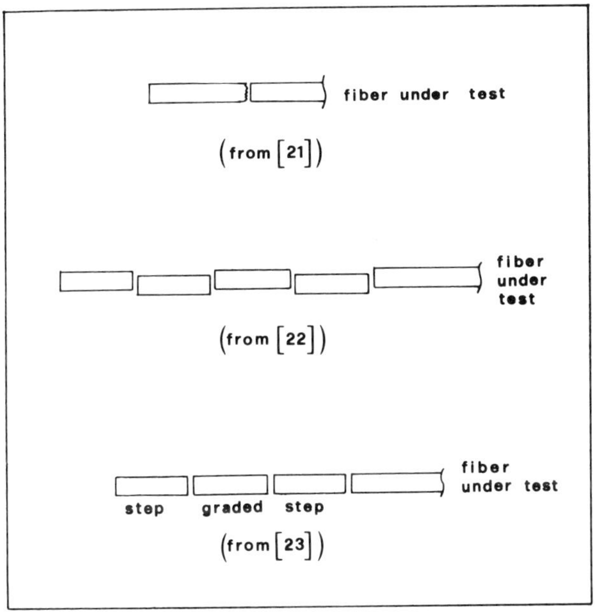

Fig. 9.4 - Main kinds of beam launchers.

These mode scramblers, shown in Fig. 9.5, should be as irregular as possible in order to simulate the light propagation in a randomly perturbed fiber. Therefore, periodic structures must be avoided, especially when their spatial period is comparable with the ray oscillation period, which is typically several millimeters. Long mode scramblers give a better mode mixing, but cause a greater loss.

Beam launchers and mode scramblers are used also for baseband response measurements. In CW operation, or at low modulation frequencies, as it is for light choppers, the transmission properties of the fiber are independent of the mode delay distribution. So, for attenuation measurements, the steady-state optical power distribution can be approximated only in amplitude. On the other hand a baseband response measurement, which is performed at higher modulation frequencies, can be strongly affected by the mode delay distribution. Therefore, for such a measurement, the steady-state optical power distribution must be approximated both in amplitude and in argument. Thus, beam launchers and mode scramblers, employed in attenuation measurements, are sometimes not suitable for baseband response measurements.

To test the features of a beam launcher or a mode scrambler, for attenuation measurements, it is possible to scan the fiber output pattern a few meters from the launching section. This test, when a graded index fiber is considered, must be performed both on the near and on the far field pattern, in order to compare the results obtained with the amplitude mode power distribution predicted by the theory at the steady state. An indirect test, often more accurate, consists of repeating the attenuation measurement by varying the launching condition. However, even when the measured attenuation is completely independent of the launching condition, we can state only that a very repeatable measurement technique has been obtained, but not necessarily that it gives the steady-state attenuation measurement.

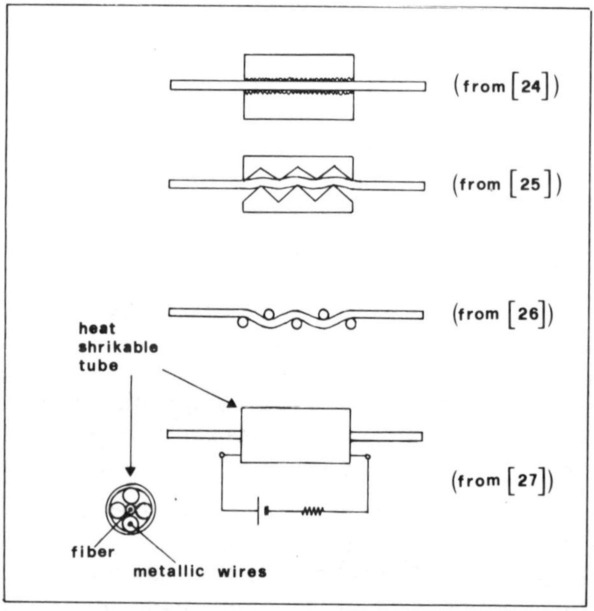

Fig. 9.5 - Some types of mode scramblers.

Nevertheless, in order to compare the characteristics of different types of fibers, it is sufficient to perform attenuation measurements whose repeatability is assured by the independence of the launching condition, and this can be done by choosing a suitable beam launcher and/or mode scrambler. When the design of a transmission system requires an extrapolation of attenuation measurements over longer distances, the knowledge of the only steady-state attenuation coefficient is of small interest. In this case, it should be very interesting to separate the contributions of the intrinsic loss and of the excess loss due to the fiber perturbation. This is possible through more complex measurement techniques, which do not require beam-launchers or mode-scramblers, and will be discussed in 9.2.4.

9.2.2 Accuracy of Cutback Measurements

The power loss L_A between the two measurement sections $z = z_o$ and $z = L$ of a two point cutback technique, from (9.1), can be written as

$$L_A = A(L - z_o) = 10 \log_{10} \left[\frac{V(z_o)}{V(L)} \right] \quad [dB] \tag{9.7}$$

where A is the attenuation per km of the fiber, and $V(L)$ and $V(z_o)$ represent the two photodetector output voltages. Differentiating this equation, we have

$$\frac{\Delta L_A}{L_A} = \frac{4.34}{L_A} \left[\frac{\Delta V(z_o)}{V(z_o)} + \frac{\Delta V(L)}{V(L)} \right] \tag{9.8}$$

which represents the maximum uncertainty of the evaluation of L_A as a function of the uncertainties in the measurements of $V(z_o)$ and $V(L)$. It can be seen that this uncertainty decreases for larger values of L_A. This means that for a given A, a better accuracy is achieved by increasing the distance $L - z_o$.

Even for spectral loss measurements, which are characterized by very low levels of launched power, the detected signal is well above the noise of the photodetector. So the uncertainties on $V(z_o)$ and $V(L)$ are mainly due to the following possible causes:

i) drifts of the optical source

ii) poor end face preparation

iii) reading uncertainties on the display of the voltmeter

All the systematic errors in the evaluation of L_A, e.g., the presence of radiating power at $z = z_o$, are not considered here. Finally, for what concerns the nonuniform responsivity on the active area of the photodetector, though fluctuations of the order of some per cent may be present, their effects can be easily reduced through a proper optical setup. [28]

If the optical source is suitably stabilized, and the fiber ends are cleaved with good repeatability (see Appendix 3), the corresponding uncertainties are as small as the reading uncertainty, so that we can assume the resulting $\Delta V(z_o)/V(z_o)$ and $\Delta V(L)/V(L)$ are on the order of 0.1 %. In Fig. 9.6 the maximum uncertainty $\Delta L_A/L_A$ is reported *vs.* $(L-z_o)$, with different values of A as a parameter. For instance, when $A = 1$ dB/km, in order to have $\Delta L_A/L_A \leq 1$ %, the fiber length must be $L > 0.9$ km.

The uncertainty on the evaluation of the distance $L - z_o$ must be added to $\Delta L_A/L_A$ to obtain the maximum uncertainty on A. As will be shown in 8.2, a measurement of the fiber length may be very accurate, therefore we have

$$\frac{\Delta A}{A} \cong \frac{\Delta L_A}{L_A}$$

The resulting sensitivity of the instrument which performs the measurement of A turns out to be

$$\Delta A \cong \frac{\Delta L_A}{L_A} A = \frac{4.34}{L - z_o} \left[\frac{\Delta V(z_o)}{V(z_o)} + \frac{\Delta V(L)}{V(L)} \right] \left[\frac{dB}{km} \right]$$

Letting $\Delta V(z_o)/V(z_o) = \Delta V(L)/V(L) = 10^{-3}$, we have

$$\Delta A \cong \frac{8.7}{L - z_o} \cdot 10^{-3} \left[\frac{dB}{km} \right]$$

Measurements on low loss fibers require a sensitivity better than 0.1 dB/km, therefore it must be that $L \geq 0.1$ km.

As regards the maximum dynamic range of a two point cutback technique, it can be expressed directly as the maximum permitted value of L_A. Assuming that the photodetector can appreciate optical powers as low as 1 pW (easily possible using a light chopper and a lock-in amplifier), this value can correspond to $\Delta V(L)/V(L) = 100\%$. In fact, the uncertainty $\Delta L_A/L_A$ results acceptably small, even for an extremely large $\Delta V(L)/V(L)$, owing to the very high value of L_A. So the maximum dynamic range L_{AM} can be expressed as a function of the optical launched power $P(0)$ by

$$L_{AM} = 90 + P(0)\Big|_{dBm} \quad [dB]$$

A typical value of $P(0)$ is 1 mW, therefore we have L_{AM} = 90 dB. However, when the photodiode is followed by an amplifier (usually a transimpedance preamplifier), the linear dynamic range of the amplifier coincides with that of the whole measurement, if it is smaller than the above L_{AM}.

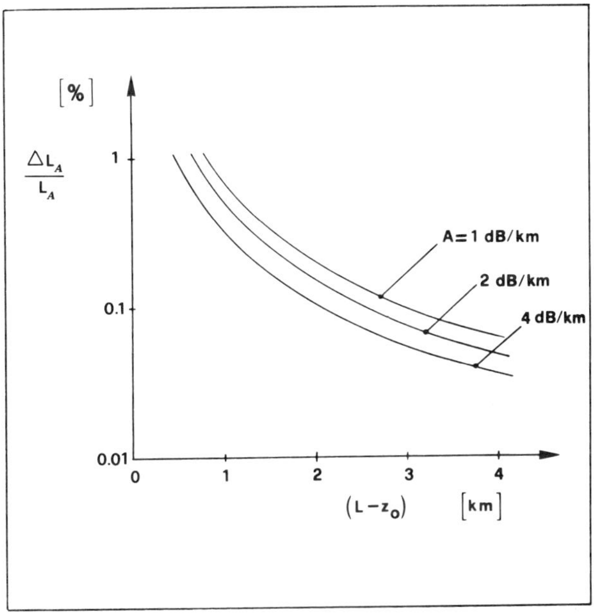

Fig. 9.6 - Maximum uncertainty in a two-point cutback measurement.

Considering a general N point cutback technique, the attenuation per km of the fiber under test is obtained by fitting the experimental data (z_i, V_i), i = 1, ... N, by a straight line, in logarithmic scale, through a least-square method. So we have

$$A = 10 \frac{\sum_{i=1}^{N} \log_{10}\left(\frac{V_i}{V_o}\right) - N \sum_{i=1}^{N} z_i \log_{10}\left(\frac{V_i}{V_o}\right) \Big/ \sum_{i=1}^{N} z_i}{\sum_{i=1}^{N} z_i - N \sum_{i=1}^{N} z_1^2 \Big/ \sum_{i=1}^{N} z_i} \left[\frac{dB}{km}\right]$$

where V_o is a reference voltage (for instance, 1 mV), which has no influence. Uniform weights have been assumed, considering that the uncertainties in the single measurements are practically independent of z. So, the error on A assumes a random nature, and the overall *rms* uncertainty decreases approximately as $1/\sqrt{N}$. A better accuracy is achieved increasing the number of measurement points, as expected.

A technique based on a modified cutback method has been suggested in order to test installed fibers. It requires two sources and two detectors. One source and one detector are placed at each fiber end.[20] This technique is based on two cuts and four measurements. The first measurement is performed at the fiber end 2 with detector 2, under the excitation of source 1, giving

$$V_{12} = Q_1 \eta_1 10^{-L_A/10} d_2 \mathcal{R}_{v2} \tag{9.9}$$

where Q_1 is the optical power emitted by source 1, η_1 the fraction of this power that is launched into the fiber, d_2 the fraction of the output power that is collected by detector 2, and $\mathcal{R}_{v2}$ the responsivity of detector 2.

After a cut of the fiber close to end 1, by using source 1 and detector 1, with similar formalism and second measurement gives

$$V_{11} = Q_1 \eta_1 d_1 \mathcal{R}_{v1} \tag{9.10}$$

Similarly, exciting from the other end, with source 2, we obtain

$$V_{21} = Q_2 \eta_2 10^{-L_A/10} d_1 \mathcal{R}_{v1} \tag{9.11}$$

and, after another cut of the fiber close to the end 2, it is

$$V_{22} = Q_2 \eta_2 d_2 \mathcal{R}_{v2} \tag{9.12}$$

The four equations from (9.9) to (9.12) yield

$$L_A = 10 \log_{10} \left[\left(\frac{V_{12} V_{21}}{V_{11} V_{22}} \right)^{1/2} \right] [\, dB \,]$$

in which all the unknown constants Q_i, η_i, d_i, $\mathcal{R}_{vi}$, $i = 1,2$, vanish. This method represents an interesting alternative to backscattering, but, especially when repeated tests are required during the life of a transmission system, the waste of fiber may be considerable. Moreover, an uncertainty exists about the two launching conditions, which, as known, influence the attenuation value, even when two equal sources are employed.

However, the present method appears accurate when a very long repeater spacing is tested, especially if joints are present along the link. In this case, in fact, the effects of the launching condition are quite negligible, and the value of L_A which is measured turns out to be close to that imposed by the steady-state condition.

9.2.3 Spectral Loss Measurements

A spectral loss measurement is performed using a wide spectrum lamp and a monochromator as optical source. As discussed in Appendix 1, two main kinds of wide spectrum lamp are used:

i) a tungsten halogen lamp, characterized by a quite uniform spectrum in the interval $0.6\ \mu m < \lambda < 2\ \mu m$

ii) an arc lamp, whose spectrum exhibits a sequence of peaks in the same interval

The amount of optical power which can be launched into the fiber depends on the following properties:

i) radiance of the lamp

ii) type of monochromator and wavelength resolution required

iii) core diameter and numerical aperture of the fiber

Considering commercial monochromators, adjusted for a wavelength resolution of 5 nm, and a fiber having one core diameter $2a$ = 50 μm and NA = 0.2, we have typically 0.1 mW of launched power with a tungsten halogen lamp, and 1 mW with an arc lamp. The difference is due to the higher radiance of arc lamps. A two point cutback measurement is performed as follows. For each wavelength λ of interest, the guided power is measured at $z = L$, giving $V(L,\lambda)$. Next the fiber is cut at $z = z_o$, and, selecting the same wavelengths, the values $V(z_o,\lambda)$ are measured. So it is

$$A(\lambda) = \frac{1}{L - z_o}\ 10 \log_{10}\left[\frac{V(z_o, \lambda)}{V(L, \lambda)}\right] \left[\frac{dB}{km}\right]$$

Wavelength selection must be very repeatable, since even small differences may entail considerable errors, especially in correspondence of a peak of absorption. This is particularly true when an arc lamp is used, whose peaks of emission add to the fiber peaks of absorption.

A monitor of the launching condition can be realized as shown in Fig. 9.7, where the optical power reflected by the input fiber end face provides, through a video camera, a test on the focusing process. The intensity of the beam is measured by the reflection on the blades of the light chopper, which is detected as a reference to avoid power fluctuations.[29]

However, an uncertainty on $V(z_o, \lambda)$ and $V(L, \lambda)$ remains. Considering (9.8), we can assume $\Delta V(z_o)/V(z_o) \cong \Delta V(L)/V(L) = 2\,\%$, and the resulting sensitivity in the measurement of A turns out to be

$$\Delta A \cong \frac{0.17}{L - z_o} \left[\frac{dB}{km}\right]$$

Since the maximum uncertainties on $V(z_o, \lambda)$ and $V(L, \lambda)$ have been considered as mainly due to the repeatability of the wavelength selection, they do not depend on λ, and hence ΔA is independent of λ too.

This assumption is justified except for those wavelengths which correspond to the fiber absorption peaks, where the value $V(L, \lambda)$ may be affected also by the noise of the photodetector. In this sense, the maximum dynamic range of the measurement is strongly influenced by the variations of attenuation which occur in the fiber varying the wavelength. Also the distance $L - z_o$ must be properly chosen. When it is too short, the accuracy in the determination of the minimum losses may be poor. On the other hand, for a very long distance, the maximum dynamic range may not be enough for appreciating the peaks of absorption. However, when ΔA is independent of λ, which is the most common case, the uncertainty $\Delta A/A$ decrease with increasing A. A typical result of a spectral loss measurement with constant error bars is reported in Fig. 9.8.

In order to perform a spectral loss measurement, a wide spectrum photodiode should be required. As shown in 3.2, Si photodiodes can be employed for $\lambda < 1$ μm, whereas Ge photodiodes have a more extended response, from about 0.8 μm up to 1.7 μm. Also compound semiconductors, like GaInAs can be used at the longer wavelengths. The spectral responsivity of the photodiode must be considered, together with the spectral emission of the lamp and the spectral response of the monochromator, in order to avoid errors due to the repeatability of the wavelength selection. The reference photodetector in Fig. 9.7 allows one to reduce these possible errors. When a very wide region of spectrum has to be tested, different photodetectors can be employed, like pyroelectric detectors, whose responsivity is practically uniform over a very large wavelength interval.

In (30), an interlaboratory comparison of spectral loss measurements has been reported. The wavelength region under test was from 0.7 μm to 1.8 μm. Using different monochromators, launching apparata and photodetectors, the values obtained agree within a maximum deviation of $\pm$ 5 %, for measurements performed on graded index fibers about 2 km long. From the analysis of the experimental results, the following observations have been made. The heights of the OH absorption peaks change for different launching conditions. Precisely, for a uniform excitation of all the rays, we have greater heights. At the lower wavelengths, where the attenuation per km $A(\lambda)$ practically coincides with $A_s(\lambda)$, which represents its part due to scattering, we can write

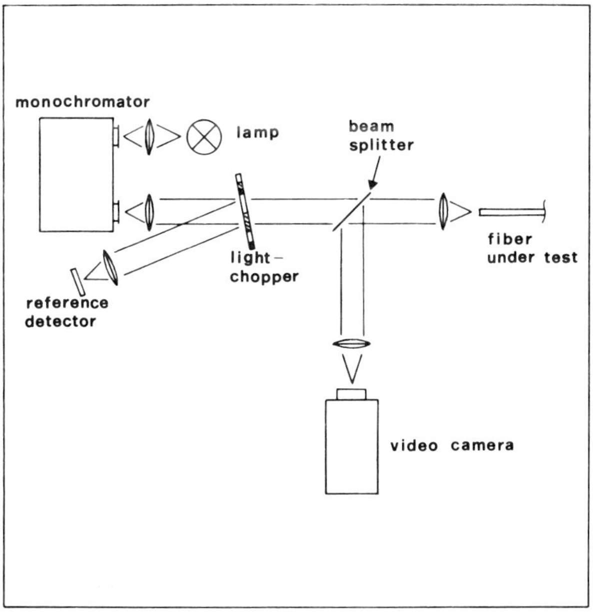

Fig. 9.7 - Launching apparatus with a monitor of the launching condition.

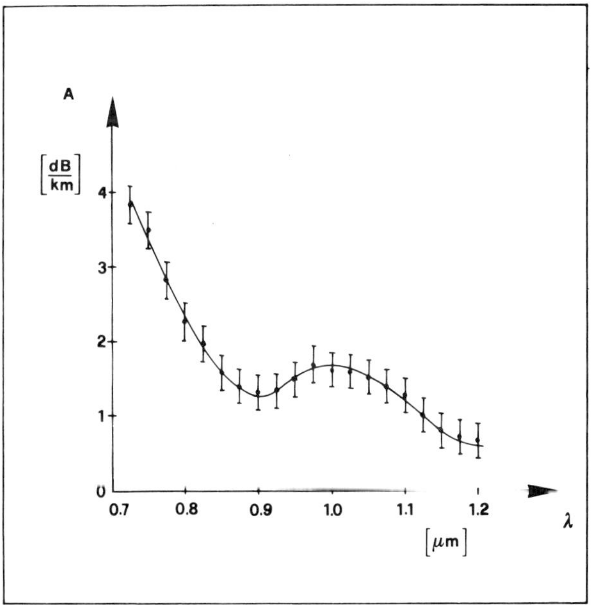

Fig. 9.8 - Typical plot of a spectral loss measurment performed by a two-point cutback technique.

$$A_s(\lambda) = c_o + \frac{c_1}{\lambda^4}$$

but the constants c_o and c_1 depend on the launching condition.

A possible explanation of the above experimental observations is that the OH-ion concentration, as well as the scattering center concentration, varies along the radial coordinate in the core region. In particular, the OH-ion concentration should decrease from the cladding towards the fiber axis. This appears reasonable; in fact, during preform preparation and fiber pulling, a diffusion of OH ions can effectively be present from the support tube towards the preform center.

9.2.4 Measurements Performed by Varying the Launching Condition

A two point cutback measurement can be repeated several times under different launching conditions, in order to evaluate the excess loss due to the fiber perturbation, when this is present. [31] This technique is based on the knowledge of the optical power flow equation which characterizes the fiber. Under this condition, it can give the intrinsic attenuation coefficient γ and the amplitude of the fiber perturbation, allowing an extrapolation of measurements performed on single pieces of fiber over any distance. In 2.2, we have shown that, independently of the model assumed for describing mode coupling effects, a differential equation of the type (2.45) can be considered, where $\hat{D}$ is a linear second order differential operator with respect to a general variable u denoting modes or rays.

Imposing the boundary condition $W(u_M,z) = 0$, where u_M is the maximum permitted value of u, the following general solution for the optical power distribution is obtained

$$W(u,z) = \sum_n c_n Q_n(u) \exp(-\lambda_{on} z)$$

$$\lambda_{on} = \gamma + K f(n)$$

in which K represents the amplitude of the fiber perturbation (in Chapter 2 it has been called A), $Q_n(u)$ is the nth statistical configuration, and c_n are the coefficients determined by the launching condition. Finally, $f(n)$ is an increasing function of n whose form depends on $\hat{D}$. The fiber loss over a distance z can be written as

$$L_A(z) = \frac{\int_0^{u_M} W(u,0)\,du}{\int_0^{u_M} W(u,z)\,du}$$

$$= \exp(\gamma z) \quad \frac{\sum_n c_n \int_0^{u_M} Q_n(u)\, du}{\sum_n c_n \exp[-K f(n) z] \int_0^{u_M} Q_n(u)\, du}$$

For a given operator $\hat{D}$, and fixed distance z, the quantity

$$S(z) = L_A(z) \exp(-\gamma z) = \frac{\sum_n c_n \int_0^{u_M} Q_n(u)\, du}{\sum_n c_n \exp[-K f(n) z] \int_0^{u_M} Q_n(u)\, du}$$

turns out to be a function only of the amplitude of the perturbation K and of the launching condition, expressed through coefficients c_n. Assuming a parameter h, characterizing the launching condition, a family of curves $S(z)$ can be plotted, taking h as an independent variable and K as a parameter. Comparing these theoretical curves with the experimental curves $L_A(z)$ obtained by the two point cutback method, under a different h, an indirect evaluation of K is possible, choosing the plot with the most similar shape. Next, γ can be determined through the absolute values of the measured $L_A(z)$ and of the theoretical $S(z)$ previously chosen.

The simple case of a step index fiber can be easily investigated, assuming, accordingly with 2.2, $\hat{D} = K(\partial^2/\partial \vartheta^2)$ where θ represents the ray propagation angle. The attenuation coefficient of the nth statistical configuration, from (2.54), is

$$\lambda_{on} = \gamma + K \left(\frac{n \pi}{2 \theta_M} \right)^2 \quad (n \text{ odd})$$

where θ_M is the maximum permitted value of θ. For a step index fiber, the launching condition can be varied, modifying the divergence of the source beam, for instance by a set of different microscope objectives. In this way, the parameter h coincides with that introduced in 2.3.3. Fig. 9.9 reports, on the left-hand side, the theoretical predictions of $S(z)$, and, on the right hand side, an example of measured $L_A(z)$. A distance z = 1 km and $\theta_M = \pi/20$ have been assumed, and finally h has been expressed as a fraction of θ_M. From the comparison between experimental data and theoretical predictions, it is possible to obtain $K = 0.2 \cdot 10^{-2}$ km^{-1} and $\gamma = 0.75$ km^{-1}.

The uncertainties which can be estimated to affect such indirect evaluations are 20 % on K and 5 % on γ.[31] Note that it is $2.2 \leq L_A \leq 2.9$, for h ranging from 0 (excitation of the axial ray only) to θ_M (uniform excitation of all the rays). Such measured values would correspond to an average attenuation coefficient from 0.75 to 1.1 km^{-1}. In this interval of values one should expect to find the actual attenuation coefficient of the fiber under test, without applying the present method. On the contrary, from the above estimates of K and γ, the steady-state attenuation coefficient can be evaluated as

$$\lambda_{o1} = \gamma + K \left(\frac{\pi}{2\,\theta_M} \right)^2 \cong 0.95 \quad [\,km^{-1}\,]$$

with an overall uncertainty smaller than 10 %. However, the main feature of this technique is not in an accurate evaluation of the steady-state attenuation coefficient, but in the possibility of separating intrinsic loss and effects of the perturbation. This separation appears very interesting not only for the design of transmission systems, but also for optimizing the cable manufacturing process, in the sense of 2.3.5.

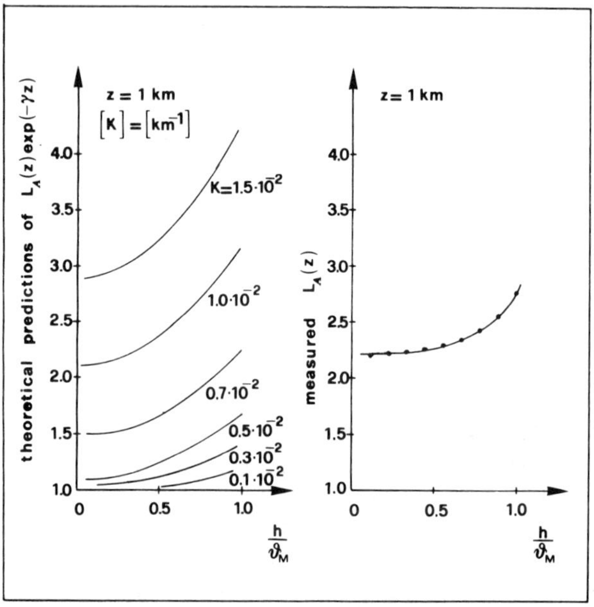

Fig. 9.9 - Theoretical predictions of $S(z)$ and experimental measurements of $L_A(z)$ varying the launching condition [from (31)].

The role of operator $\hat{D}$ in the model of mode coupling is important, but not dramatic. In fact, even for a Bessel-type operator, like that introduced in 2.2.5, the steady-state attenuation coefficients coincides approximately with the average attenuation coefficient measured for $h = \frac{3}{4}\,\theta_M$. This value of the launching parameter leads to an optical power distribution which approaches very rapidly the steady-state, for both the operators discussed in Chapter 2.

9.3 DISCUSSION ON BACKSCATTERING

In the present paragraph the possibility of avoiding the saturation of the receiver due to the reflected power is first discussed, and then different solutions to this problem are presented. Then, starting from the study of the backscattered optical power distribution, the meaning of the measurement results are investigated. Finally, the sensitivity limit and the dynamic range of this attenuation measurement are derived from an analysis of all the sources of noise. *

9.3.1 Launching Techniques

The launching technique based on a simple beam splitter, schematically shown in Fig. 9.10, is not suitable. In fact, the amount of power reflected on the fiber input end is about 10^2 times smaller than the launched power, while the backscattered power is only 10^6 times smaller than that power. Therefore the high level of reflected power saturates the photodetector, which becomes practically blind for the backscattered signal detection. Many different solutions have been proposed for this problem:

i) the use of an index matching cell
ii) injection of light through a tapered fiber section
iii) an electronic gating of the photodetector
iv) the use of a polarized incident light
v) the use of an acousto-optical light deflector

An index matching cell, replacing the simple beam splitter, is used in the experimental setup shown in Fig. 9.11. This cell is filled by an index matching oil, which strongly reduces the reflection on the fiber end face. [32] In this oil a traditional beam splitter is immersed. Remark that the difference between the oil index and the fiber core index must be smaller than 10^{-2}, in order to lower the reflected power level down to the backscattered power level.

The injection of light into the fiber core through a tapered fiber section was the erliest proposed technique. [33] The primary coating must be removed from the lateral surface of the fiber, which is then tapered to the point where some amount of optical power can pass, under a suitable angle, from the air to the fiber core. The scheme of this solution is illustrated in Fig. 9.12. A nonuniform excitation of

*See notes pg. 498

the guided rays characterizes this technique, because only those rays close to the guidance condition are excited. Thus, the attenuation measurement turns out to be usually overestimated.

An electronic gating of the photodetector [34] is possible only when a photomultiplier is used, and in this case a high voltage pulser must be employed. This appears feasible for laboratory measurements, but it is not practical for field measurements. A polarized incident light maintains its state of polarization during the reflection on the fiber end face, whereas the backscattered power is randomly polarized. Therefore the use of a crossed polarizer, between the beam splitter and the photodetector, reduces by far the influence of the reflected light. [35]

Alternatively, a Bragg deflector, controlled by an external acoustic wave, can be set between the beam splitter and the fiber input end.[36] The fiber pulse is injected into the fiber through the Bragg deflector in the absence of any acoustic wave. Next, the acoustic wave is applied for a short time, just to deflect the optical power reflected by the fiber input end face.

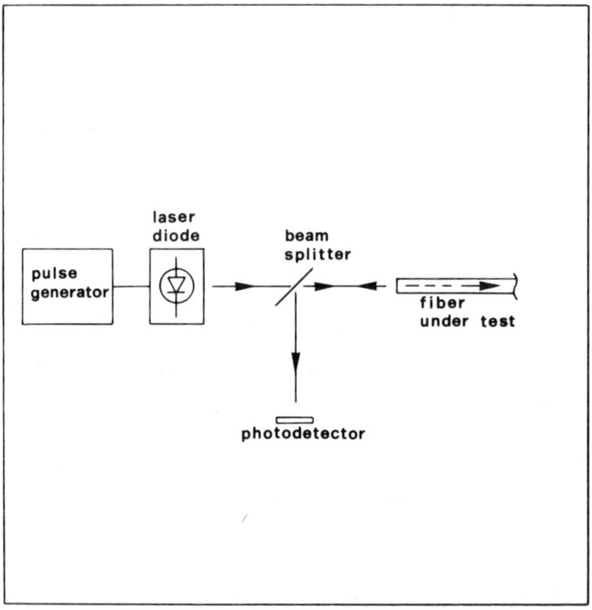

Fig. 9.10 - Launching technique based on a simple beam splitter.

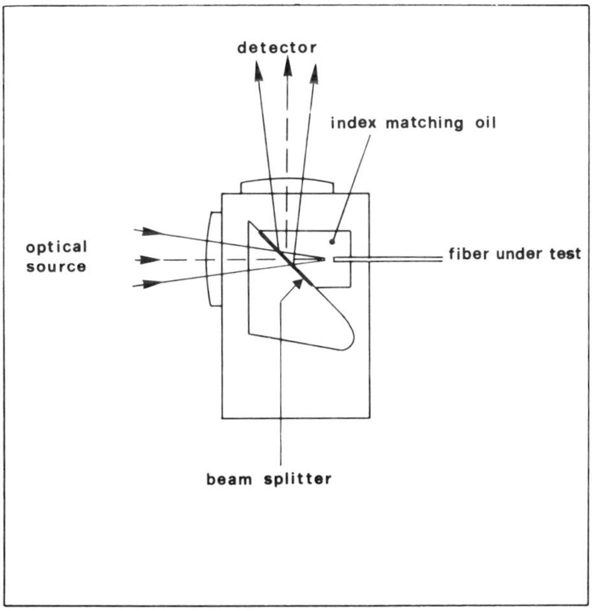

Fig. 9.11 - Index-matching cell for launching pulses in backscattering measurements [from (32)].

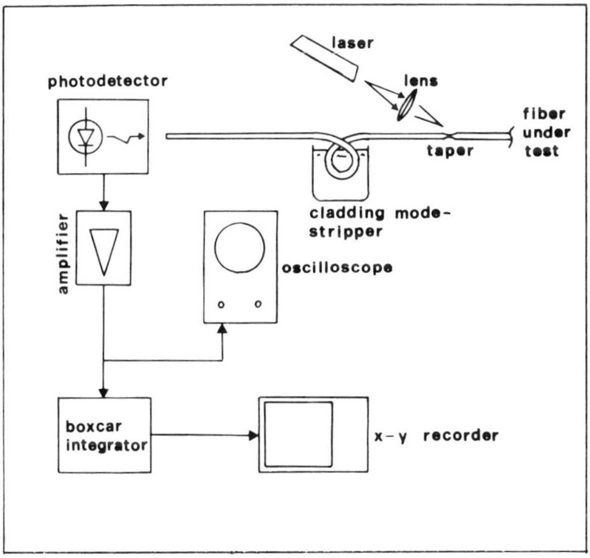

Fig. 9.12 - Scheme of the backscattering measurement using a light injection through a tapered fiber section.

With the electronic gating and the acousto-optical light deflector, the time interval in which the signal is taken off must be chosen as a trade-off between two opposite requirements. A short interval makes it possible to appreciate the power decay even in the first few meters of the fiber. A long interval simplifies the electronics of the setup. A typical choice is 1 μs, so the fiber length from the optical source which cannot be tested is about 100 m.

9.3.2 Ray Distribution of the Backscattered Power

In order to study the ray distribution of the backscattered power it is useful to denote the rays with a variable x ranging between 0 and 1. It can be related to the second constant of the motion E as follows

$$x = \frac{n_o^2 - E^2}{n_o^2 - n^2(a)}$$

where n_o is the on-axis refractive index and $n(a)$ is the cladding refractive index. For a general graded index profile of the type

$$n(r) = n_o \sqrt{1 - 2\Delta g(r)}$$

we have

$$x = \frac{1}{2\Delta}\left(1 - \frac{E^2}{n_o^2}\right)$$

For instance, when α profiles are considered, it is $x = \epsilon/\Delta$.

The optical power distribution in domain (E,h) depends on the radiance distribution $B(r, \phi, \theta, \theta_\phi)$ through the integral equation (1.49), i.e.,

$$W(E,h) = 4\int_0^2 d\phi \int_{r_1(E,h)}^{r_2(E,h)} \frac{B(r, \phi, \theta(r), \theta_\phi(r))}{n^2(r)\left(\frac{dr}{dz}\right)}\, dr \tag{9.13}$$

when B is independent of θ_ϕ, $W(E,h)$ turns out to be independent of h, so that, integrating over h, we have an optical power distribution $W(E)$ dependent only on E. Introducing the variable x, we are left with a radiance distribution which is a function of x only, and results in [37]

$$B(x) = \frac{2}{\pi i^2(x)} \int_0^\pi d\phi \int_0^{i(x)} B(r', \phi, \theta)\, r'\, dr' \tag{9.14}$$

where $r' = r/a$, and $i(x)$ is the inverse function of $g(r')$. For instance, when α profiles are considered, it is $i(x) = x^{1/\alpha}$.

Under the above assumption, the backscattered radiance distribution $B_b(x)$ from reference (38) can be written as

$$B_b(x) = \frac{3}{4}\gamma_s \Delta\left[\int_x^1 B(\zeta)\,d\zeta + \frac{1}{i^2(x)}\int_0^x B(\zeta)\,i^2(\zeta)\,d\zeta\right] < v_g > \Delta t \tag{9.15}$$

where γ_s is the attenuation coefficient due to Rayleigh scattering, $< v_g >$ is the average group velocity and Δt is the pulsewidth. This expression shows that the backscattered radiance distribution is not uniform. $B_b(x)$ is always a decreasing function of x, but it does not vanish for $x = 1$. So it is different also from the steady-state radiance distribution, characterized by the condition $B(1) = 0$.

In the absence of mode coupling, the forward and the backward radiance distributions remain unchanged. An optical pulse whose peak power is P_o, under this assumption, reaches the section z with a peak power

$$P(z) = P_o \exp(- < \gamma_1 > z)$$

where $< \gamma_1 >$ represents an average attenuation coefficient along the distance z, in the forward direction, imposed by the launching condition characterized by $B(x)$. The backscattered pulse has the following peak power

$$P_b(z) = DP(z)\exp(-< \gamma_2 > z) = DP_o \exp[-(< \gamma_1 > + < \gamma_2 >) z] \tag{9.16}$$

where D is a constant, for simplicity independent of z, and $< \gamma_2 >$ is an average attenuation coefficient along the distance z, in the backward direction, imposed by the backscattered radiance distribution $B_b(x)$. Expression (9.16) shows that the measured attenuation coefficient turns out to be the average of $< \gamma_1 >$ and $< \gamma_2 >$. It is possible to demonstrate[38] that it corresponds to the following equivalent radiance distribution

$$B_e(x) = K\left[\frac{B(x)}{P} + \frac{B_b(x)}{P_b}\right]$$

where K is a proportionality factor, P and P_b represent the total guided powers related to $B(x)$ and $B_b(x)$, respectively.

In Fig. 9.13 three radiance distributions *B(x)* are reported in continuous lines. The correspondent $B_b(x)$ and $B_e(x)$ are shown in dashed and in dotted lines respectively. This example has been carried out for a parabolic index fiber and for the following launching conditions in the forward direction:

i) a uniform *B(x)*

ii) a radiance distribution *B(x)* proportional to $1/\sqrt{x}\,[J_1(z_1(0)\sqrt{x})]$, which in the presence of mode coupling should represent the steady-state condition of the radiance [38]

iii) a radiance distribution $B_x(x)$ which remains unchanged in the forward and in the backward directions.

The present analysis, however, is only roughly approximate, because of the fluctuations in the fiber parameters which are always present in real fibers. In the following, we first consider that such fluctuations have a long spatial period, so that mode coupling does not produce appreciable effects. Next, we will study the evolution of the backscattered power even in the presence of mode coupling.

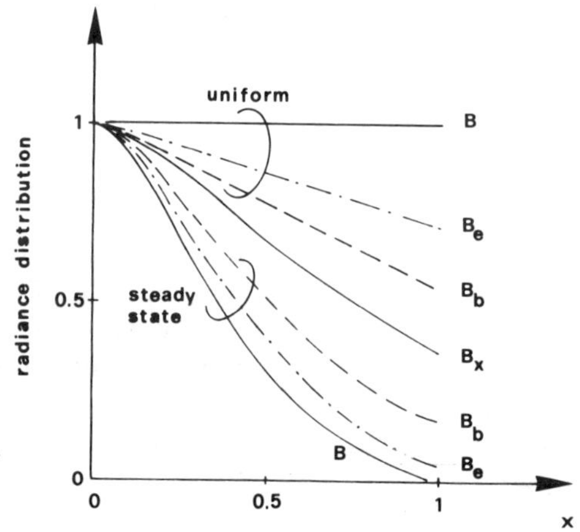

Fig. 9.13 - Radiance distributions under different lauching conditions [from (38)].

9.3.3 Effects of Fluctuations in the Fiber Parameters

Under a uniform excitation, i.e., for for $B(x) = B_0$, letting $P(z)$ and $P_b(z)$ the total optical powers in the forward and backward directions respectively, it is possible to demonstrate [39] that it is

$$P_b(z) = d(z)\,P(z)$$

$$d(z) = \frac{3}{8}\gamma_s \left(\frac{\displaystyle\int_0^a \frac{n^2(r) - n^2(a)}{n^2(r)}\, r\,dr}{\displaystyle\int_0^a [n^2(r) - n^2(a)]\, r\,dr} \right) < v_g > \Delta t$$

where d depends on the fiber section z through the local values of the fiber parameters, which are:

i) the scattering attenuation coefficient γ_s

ii) the numerical aperture NA, through parameter Δ

iii) the shape of the index profile, represented by function $g(r)$

Considering for example an α profile fiber, we have

$$d(z) \cong \frac{3}{4}\gamma_s\,\Delta\frac{\alpha}{\alpha + 1} < v_g > \Delta t \tag{9.17}$$

Since local fluctuations in α are usually small, one can conclude that the main causes of fluctuations in the amount of backscattered power are: the fluctuations in the density of the scattering centers, which influence the local value of γ_s, and the fluctuations in Δ. Remark that in this approximate analysis any core diameter fluctuation does not directly affect d. Nevertheless, considering that γ_s is produced not only by Rayleigh scattering, core diameter fluctuations influence d through γ_s itself. It must be pointed out again that in the present approach parameter fluctuations are assumed as characterized by a long spatial period, in order to neglect any mode coupling effect.

Local fluctuations in Δ, having a random nature, are usually the most used. Especially when their correlation length is short, they can be easily recognized in the trace of the backscattered power, so their effect can be eliminated. On the contrary, Δ fluctuations having a determinstic nature, like a monotonic increase or decrease along z, can alter the measurement results. Precisely, if $\Delta(z)$ increases or decreases, the attenuation coefficient turns out to be underestimated or overestimated, respectively.

It is possible to separate the effects of the actual power decay, which is the object of the measurement, from the effects of Δ fluctuations, comparing the two subsequent measurements which can be performed launching the pulse from both the fiber ends.[39] In this way, we obtain

$$P_{b1}(z) = F_1(z)\,P_1(z)$$
$$P_{b2}(z) = F_2(z)\,P_2(z)$$

it being possible to write

$$F_1(z) = F_o[1 + \delta(z)]$$
$$F_2(z) = F_o[1 + \delta(z)]$$

in which δ(z) accounts for the Δ fluctuations, assumed to be equal in the two directions. Note that this assumption could not be justified for γ_s fluctuations. Considering the same launched power P_o at the two fiber ends, one has

$$P_1(z) = P_o \exp(-2\gamma z)$$
$$P_2(z) = P_o \exp[-2\gamma(L - z)]$$

where γ is the attenuation coefficient to be measured and L the total fiber length. So we are left with

$$P_{b1} = F_o[1 + \delta(z)]\,P_o \exp(-2\gamma z)$$
$$P_{b2} = F_o[1 + \delta(z)]\,P_o \exp[-2\gamma(L - z)]$$

Therefore, the actual power decay can be evaluated through the parameter

$$D(z) = \sqrt{\frac{P_{b1}(z)}{P_{b2}(z)}} = \exp(\gamma L)\exp(-2\gamma z)$$

whereas the effects of the Δ fluctuations are taken into account by the parameter

$$I(z) = \sqrt{P_{b1}(z)\,P_{b2}(z)} = \exp(-\gamma L)\,F_o P_o[1 + \delta(z)]$$

In Fig. 9.14 an example of the above evaluations is reported.

In order to take into account mode coupling effects, an optical power flow equation, in which the radiance distribution *B(x)* is present, must be derived. The radial distribution of optical power produced by *B(x)* results in

$$W(r) = 4\pi^2 r \int_{n(a)}^{n(r)} B\left[\frac{1}{2\Delta}\left(1 - \frac{E^2}{n_o^2}\right)\right] E\, dE$$

$$= 4\pi^2 \Delta n_o^2 r \int_{g(r)}^{1} B(x)\, dx$$

The total guided power can be obtained integrating over r, as

$$P = 2\pi^2 \Delta n_o^2 \int_0^1 B(x)\, i^2(x)\, dx$$

From this expression it is possible to conclude that the optical power distribution in the domain of the variable x is proportional to $B(x)\, i^2(x)$. Therefore one can write

$$\frac{\partial}{\partial z}(Bi^2) = -\gamma (Bi^2) - \frac{\partial}{\partial x}(M_1 Bi^2) + \frac{1}{2}\frac{\partial^2}{\partial x^2}(M_2 Bi^2) \qquad (9.18)$$

This is a general Fokker-Planck type equation defined by the moments $M_1(x)$ and $M_2(x)$. It holds both in the forward direction, and in the backward direction replacing B with B_b. This analysis [(40)] leads to the determination of the steady-state attenuation coefficient, from the experimental evaluation of the backscattered power decay, knowing the launching condition and the fiber perturbation, which impose $M_1(x)$ and $M_2(x)$.

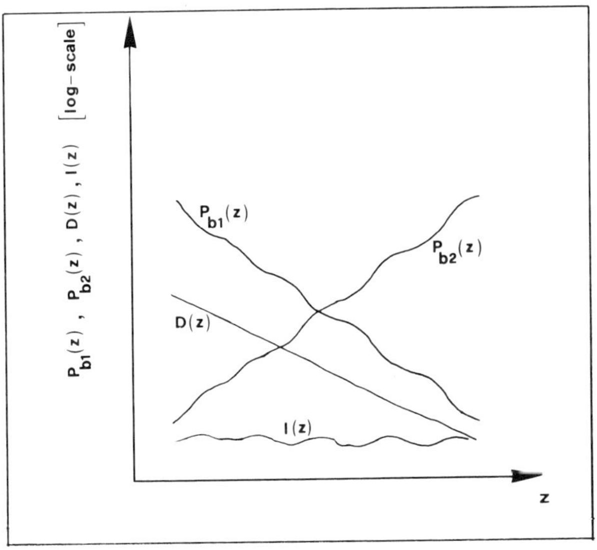

Fig. 9.14 - Typical behaviors of the backscattered powers $P_{b1}(z)$, $P_{b2}(z)$, and of the parameters $D(z)$, $I(z)$.

Nevertheless, the assumptions taken on the launching condition and on the fiber perturbation limit the validity of this method. In particular, the analytical forms of $M_1(x)$ and $M_2(x)$ are very critical in order to obtain the value of the steady-state attenuation coefficient. The study is useful however, in order to stress how mode coupling effects can influence an attenuation measurement performed by the backscattering technique, especially when joint losses must be evaluated. In this case disagreements as large as 50 % between measured and actual losses may occur. [41]

9.3.4 Measurements of Local Additional Losses

Core diameter fluctuations entail local additional losses which have been studied in 2.2.3. In an attenuation measurement, they must not be taken as error inducing fluctuations, but their effects must be evaluated as excess losses added to the intrinsic optical power decay. In (42) a technique for deriving information about core diameter fluctuations, from a backscattering investigation, has been proposed. Even the role of the launching condition can be taken into account. [43] This technique, which is based on the two measurements obtained launching from the two fiber ends, is meaningful only if there are no fluctuations in *NA*. Otherwise, the two effects interfere and it is not possible to separate them.

In (44) it has been demonstrated that core diameter fluctuations affect chiefly the back traveling signal, since its optical power is mainly carried by rays close to the guidance condition. On the contrary, the signal traveling in the forward direction, obtained from a laser source, is characterized by an optical power distribution usually concentrated in the inner core region. Since the measured attenuation is the average of the attenuations in the forward and in the backward directions, the evaluation of the effects of core diameter fluctuations may be underestimated. Nevertheless, using a mode scrambler, also in the forward direction we have a large amount of optical power close to the guidance condition, leading to a correct estimate of this excess loss.

The backscattering technique is particularly suitable for testing of cabled fibers, with the purpose of fault location. A fault is usually produced by a lateral pressure on the primary coating, which induces microbending or even fractures. Also temperature variations may lead to a fault, due to the axial stress between fiber and primary coating, owing to their different thermal expansion coefficients. Two kinds of faults can be displayed on the backscattering trace:

i) local perturbations of the light propagation, which appear as peaks in the backscattering trace

ii) local variations in the attenuation coefficient, which appear as slope variations in the backscattering trace

In order to locate with a high spatial resolution these two kinds of faults, the pulsewidth Δt must be reduced. As shown in 9.1.2, when both fault location and overall attenuation have to be obtained at the same time from a single backscattering test, Δt must be chosen as a trade-off between opposite requirements.

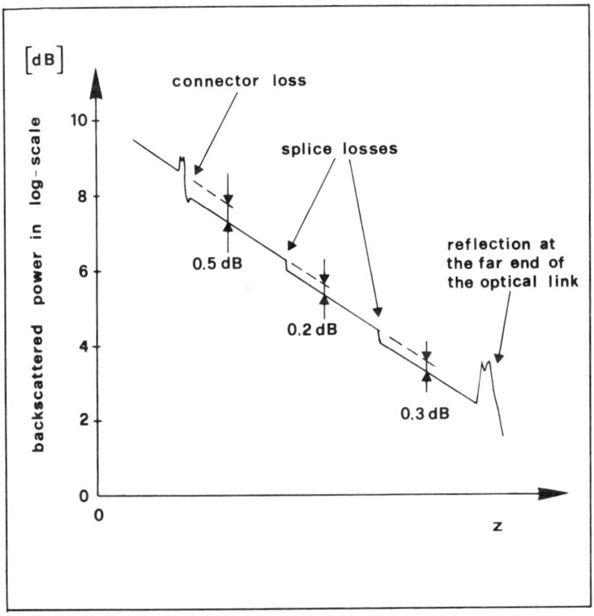

Fig. 9.15 - Example of a backscattered trace in which different contributions to the overall loss of the optical link are present.

When an optical link, consisting of some installed optical cables separated by joints, has to be tested, backscattering can give a large amount of information. In Fig. 9.15, an example of this type of backscattering trace is reported. In particular, a splice loss appears as a small lowering in the curve, whereas a connector loss is characterized usually by a greater lowering and by a peak due to a residual Fresnel reflection. The presence of a fault due to stress in the fiber produces an effect similar to that of a connector. The reflection at the far end of the optical link, where usually no matching oil is employed, would allow an approximate evaluation of the whole length of the link, if it were not known. In Fig. 9.15, a method for an approximate evaluation of the power loss L_o of a joint,[45] which has been defined in 9.1.6, is also shown.

For an accurate experimental test, a large dynamic range should be required by the measurement. In fact, it may occur that, during the measurement of an optical link, the power loss of the farthest joints cannot be evaluated at all if there are dynamic range limitations. The dynamic range of a backscattering investigation will be discussed in the following sections.

9.3.5 Analysis of All Noise Sources in Backscattering Measurements

There are three fundamental sources of noise which affect a backscattering trace. They are: [46]

i) noise due to the optical source

ii) shot noise of the photodetector

iii) thermal noise of the receiver circuit

These noises add a random error to all the systematic uncertainties discussed in the previous sections.

Since the optical source is pulse modulated with high peak powers, the random nature of the filament emission, described in 3.1.5, is the main cause of noise due to the optical source. When a signal averaging, over many backscattering traces, is performed, this leads to remarkable drifts in the level of the optical launched power. The above mentioned averaging, usually performed by a boxcar integrator (see Appendix 2), is needed for reducing the effects of the other sources of noise. The noise due to the optical source can be taken into account through a noise current $< i_{os}^2 >$.

An APD detector is usually preferred to a PIN detector, because it has better performances at the higher distances, when the signal-to-noise ratio (SNR) is lower. Thus shot noise must be taken into account. From (3.5), the shot noise current can be written as

$$< i_{sh}^2 > = 2q < M > F(<M>) < i_s > B_N$$

where q is the electron charge, $<M>$ is the mean avalanche gain, $<i_s>$ is the average useful signal current over all the repeated pulse processed by the boxcar integrator, $F(<M>)$ represents the excess noise factor for the avalanche detection, and finally B_N denotes the noise equivalent bandwidth of the receiver circuit.

The thermal noise is mainly due to the preamplifier circuit. Considering a FET front-end included in a transimpendance preamplifier, we have a circuit noise current, which, from the discussion reported in 4.1, can be written in the form

$$< i_c^2 > = a\,B_N + b\,B_N^3$$

where a and b are positive constants depending on the circuit parameters. In conclusion, all the sources of noise which affect a backscattering measurement can be taken into account through the equivalent circuit reported in Fig. 9.16.

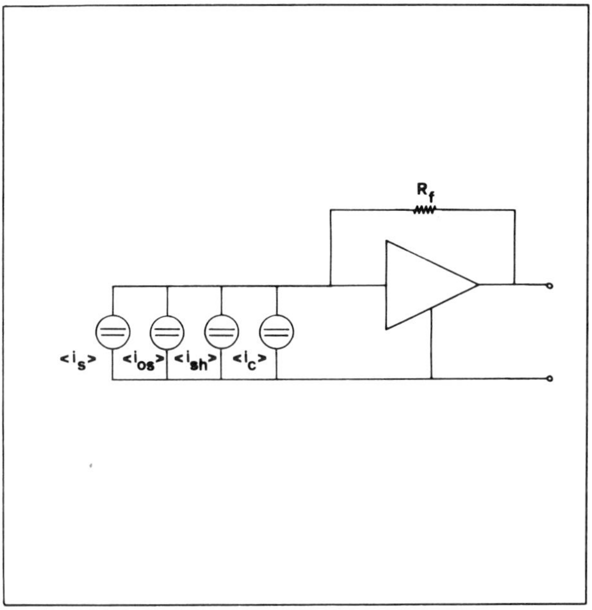

Fig. 9.16 - Noise equivalent circuit of the receiver in backscattering measurements [from (46)].

As for the choice of the bandwidth of the receiver, which imposes the value of B_N, one has to satisfy two opposite requirements. A wide bandwidth is needed in order to reproduce the backscattering trace without distortion, and even a bandwidth on the order of the inverse of the pulsewidth can be required if the purpose of the measurement is fault location. On the contrary, the narrower the bandwidth, the smaller the noise superimposed to the useful signal.

The signal-to-noise ratio (SNR), without averaging, can be written as

$$SNR = \frac{< i_s >^2}{< i_{os}^2 > + < i_{sh}^2 > + < i_c^2 >} \tag{9.19}$$

This quantity may be often too small for the precision requested, mainly owing to $< i_{sh}^2 >$ and $< i_c^2 >$. Therefore, signal averaging performed by boxcar integrator is needed. Assuming as negligible the additional noise introduced by this instrument, from (9.19) the new signal-to-noise ratio, SNR', obtained after the averaging over N samples, results in

$$SNR' = \frac{< i_s >^2}{< i_{os}^2 > + \frac{1}{N}\left[< i_{sh}^2 > + < i_c^2 > \right]}$$

If N is large, $< i_{os}^2 >$, which is independent of N, can become the most critical cause of uncertainty in the measurement. On the contrary, when $< i_{os}^2 >$ is negligible, through the signal averaging the uncertainty on the measurement of the backscattered power in a given fiber section reduces to $1/\sqrt{N}$.

An interesting technique [47] has been proposed in order to avoid the presence of $< i_{os}^2 >$. It consists of taking two samples of the same backscattered signal, which correspond to a pair of fiber sections z_1 and z_2. The average attenuation per km between these two sections can be written as

$$A = \frac{10}{z_2 - z_1} \log_{10} \left[\frac{V(z_1)}{V(z_2)} \right]$$

where $V(z_1)$ and $V(z_2)$ represent the two voltage samples. Then, a signal averaging can be performed directly on A, leading to a reduction, as $1/\sqrt{N}$, of the overall uncertainty. In this case the boxcar integrator can be replaced by a digital averaging processor, which is characterized by a lower cost and a smaller size, making this technique even better suited for field operation. This two point backscattering technique cannot be used for fault location, but only to enhance the accuracy of a measurement of total loss.

Nevertheless, a time delay generator, two sample-and-hold circuits, and two analog-digital converters are required in order to process the two voltage samples $V(z_1)$ and $V(z_2)$ numerically. Moreover, in order to avoid gain errors in the sampling and conversion circuits, the two sample gates must operate alternately on z_1 and z_2, over a sequence of pulses. Finally, a time resolution, typically of about 100 ps, [47] is required for the aperture duration of the sample-and-hold circuit. In conclusion, the cost reduction with respect to a traditional backscattering technique appears modest. The main features of this modified technique remain the small size of the instrumentation and the higher accuracy. In the next section we will show how also a wider dynamic range can be obtained by the two-point backscattering technique.

9.3.6 Dynamic Range Limitations

From (9.19), considering typical values of the parameters involved, and assuming in particular a launched pulse of 100 ns width and 1 W peak power, one can obtain some limit curves of the SNR along the fiber length z.[46] For a typical attenuation per km $A = 3$ dB/km, we have the curves reported in Fig. 9.17, where each cause of noise is separately considered. Two situations are examined:

i) $<M> = 1$, that is the case of a PIN detector, characterized by a strong limitation due to circuit noise

ii) $<M> = 80$, in which the limitation, except for very low values of SNR, is imposed by shot noise

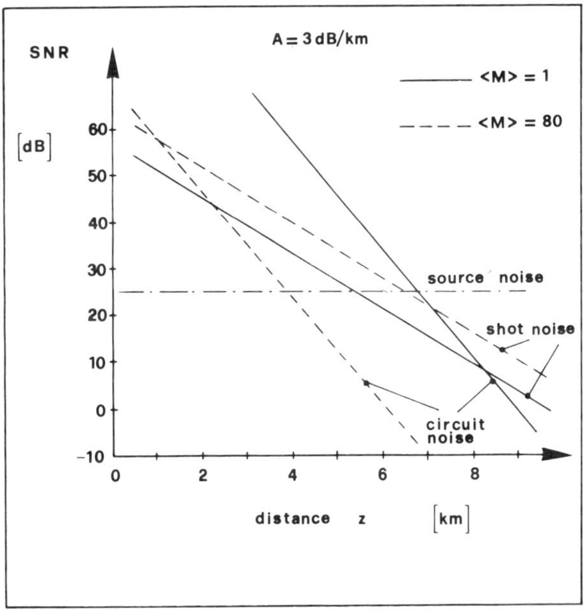

Fig. 9.17 - Asymptotic behaviors of the SNR in a backscattering measurement [from (46)].

The noise of the optical source, which is independent of $<M>$ and of the distance z, limits mainly the higher values of SNR. The maximum uncertainty in the evaluation of the overall power loss L'_A, between two points z_1 and z_2 on the backscattering trace, so that $V(z_1)$ and $V(z_2)$ are the two measured values, turns out to be

$$\frac{\Delta L'_A}{L'_A} \cong \frac{4.34}{L'_A}\left[\frac{\Delta V(z_1)}{V(z_1)} + \frac{\Delta V(z_2)}{V(z_2)}\right] \tag{9.20}$$

as it is for the cutback two point technique. Here, we must let

$$\Delta V(z_1) = \Delta V(z_2) = \Delta V$$

differently from a cutback measurement, where we have considered $\Delta V(z_1)/V(z_1) = \Delta V(z_2)/V(z_2)$, owing to simple reading uncertainties. In fact, for backscattering measurements, the level of detected power is so small that the noise superimposed turns out to be not negligible.

Neglecting the uncertainty on (z_2-z_1), the resulting sensitivity of the instrument which performs the measurement of A becomes

$$\Delta A = \frac{\Delta L'_A}{L'_A} A \cong \frac{1}{2}\frac{4.34}{z_2 - z_1}\Delta V\left[\frac{1}{V(z_1)} + \frac{1}{V(z_2)}\right]\left[\frac{dB}{km}\right] \tag{9.21}$$

where 1/2 is due to the double distance covered. In other words, it derives from the relation $L'_A = 2\,L_A$, where $L_A = A(z_2 - z_1)$ represents the one-way loss. In expression (9.21), ΔV is usually reduced, in the part due to circuit and shot noises, by signal averaging.

For the traditional backscattering technique, this is performed by a boxcar integrator. For the twopoint backscattering technique discussed in the previous section, it is more convenient to use a digital processor. Employing the later technique, moreover, no drifts in the launched power affect the measurement, therefore the reduction of ΔV is complete. On the other hand, if the quantization error introduced by the analog-digital converter is predominant, we have to consider

$$\Delta V = \frac{V(z_1)}{2^{n+1}}$$

where we have set $V(z_1)$ at the maximum available level of quantization, n being the bits for each sample. This error cannot be reduced by the signal averaging. Therefore, n must be high enough to make it practically negligible.

Now we will describe briefly how ΔV can be reduced by the signal averaging. This evaluation is limited to those ranges of $(z_2 - z_1)$ in which circuit and shot noise dominate, independently of the technique (traditional or two point) employed. Assuming that circuit and shot noises are gaussian with variance σ_v^2, the averaging process leads to a new variance σ_v^2/N. Then considering ΔV as the error which is exceeded with a probability of 5%, we have

$$\Delta V \cong \frac{1.96}{\sqrt{N}} \sigma_v$$

The sensitivity in the one-way loss, assuming $V(z_1)$ and $V(z_2)$ as independent measurements, turns out to be

$$\Delta L_A \cong \frac{1}{2} 4.34 \frac{1.96}{\sqrt{N}} \sigma_v \left[\frac{1}{V^2(z_1)} + \frac{1}{V^2(z_2)} \right]^{1/2} \quad [dB] \tag{9.22}$$

The expressions of $V(z_1)$ and $V(z_2)$ as functions of the source and fibers parameters [46] can be derived from

$$V(z) = 0.25 \, R_f \mathscr{R}_o < M > \gamma_s < v_g > \Delta t \, P_o \exp(-2\gamma z)$$

where $\mathscr{R}_o$ is the intrinsic responsivity of the photodiode, R_f is the feedback resistance of the transimpedence preamplifier, and the remaining symbols are known. The multiplication by 0.25 is due to a double beam spliting for separating launched and backscattered signals. Substituting into (9.22), it is

$$\Delta L_A \cong \frac{1}{2} 4.34 \frac{1.96}{\sqrt{N}} \sigma_v \left[\frac{\{1 + \exp[4\gamma(z_2 - z_1)]\}^{1/2}}{0.25 \, R_f \mathscr{R}_o < M > \gamma_s < v_g > \Delta t \, P_o} \right] \exp(\gamma z_1) \quad [dB] \tag{9.23}$$

This expression allows one to obtain the maximum dynamic range, i.e., the maximum one-way loss L_{AM}, over which a backscattering measurement can extend. For a given ΔL_A, it is

$$L_{AM} = 10 \log_{10}\{\exp[\gamma(z_2 - z_1)\} \cong 4.34 \, \gamma(z_2 - z_1) \, [dB]$$

which can be calculated inverting (9.23). The following approximate expression holds

$$L_{AM} \cong 1.1 \ln \left\{ 1 + \frac{N \Delta L_A^2}{300 \, \sigma_v^2} [R_f \mathscr{R}_o <M> \gamma_s <v_g> \Delta t \, P_o \exp(-\gamma z_1)]^2 \right\} \quad [dB] \tag{9.24}$$

Clearly, the smaller the value of z_1, the larger the maximum dynamic range. So, z_1 is chosen as a trade-off between the requirements of reaching the steady-state and of attaining a maximum dynamic range. Sometimes, in order to discuss the features of a certain backscattering instrument, it is useful to define an overall maximum dynamic range of the instrument, which turns out to be L_{AM} with $z_1 = 0$, and is obviously independent of γ.

Equation (9.24) is based on the assumption that the variance σ_v^2 is independent of $V(z_1)$ and $V(z_2)$. Yet, as shown in Fig. 9.17, SNR(z) exhibits different slopes depending on the predominant cause of noise. For the two point technique, which is characterized by only circuit and shot noises, especially when circuit noise turns out to be predominant over the range (z_2 -z_1) considered, (9.24) can be assumed as correct. On the contrary, for the traditional technique, a maximum SNR of about 27 dB, [46] due to source noise, is imposed at short distances. In this case, the evaluation of L_{AM} cannot be performed starting from (9.22), but directly from (9.20), in which $\Delta V(z_1)/V(z_1)$ must be assumed as $1.96/\sqrt{\text{SNR}}$, while only $\Delta V(z_2)$ can be reduced by increasing N.

In conclusion, letting $\Delta L_A = 0.1$ dB and considering typical values of the transmitter and receiver parameters, for the traditional technique, an overall maximum dynamic range of about 20 dB can be obtained. For the two-point technique, under the same conditions, the limit of about 40 dB can be achieved. [48] The above dynamic range limitations have been derived assuming a very large value of N (typically 10,000). These values of L_{AM} increases with increasing ΔL_A. Finally, other definitions of dynamic range are sometimes taken into account, for instance, that related to the possibility of appreciating the end reflection at the output section of the optical link under test. [34] Conforming to the last definition, the above values of L_{AM} turns out to be remarkably enlarged.

9.3.7 Accuracy of Backscattering Measurements

Rewriting (9.20), with $\Delta V_1 = \Delta V_2 = \Delta V$, in the form

$$\frac{\Delta L_A'}{L_A'} \cong \frac{4.34}{10\log_{10}[V(z_1)/V(z_2)]} \frac{\Delta V}{V(z_1)} \left[1 + \frac{V(z_1)}{V(z_2)}\right] \tag{9.25}$$

for a given maximum uncertainty $\Delta L_A'/L_A'$, it is possible to plot the maximum uncertainty on the first sample, $\Delta V/V(z_1)$, as a function of the ratio between the two samples $V(z_1)/V(z_2)$. A function of this type has a maximum for intermediate values of $V(z_1)/V(z_2)$, which leads to a relaxation in the accuracy requirement for the two sample measurements, because $\Delta L'_A/L'_A$, has been fixed. This maximum can be explained as follows. When $V(z_1)/V(z_2)$ is small, L'_A is small too, therefore, in order to maintain a given $\Delta L'_A/L'_A$, the accuracy in the measurements of the two samples must be high. When $V(z_1)/V(z_2)$ is large, the accuracy in the measurement of the second sample limits the accuracy of the overall measurement.

Taking into account the attenuation per km, A, of the fiber under test, such an optimum value of the ratio $V(z_1)/V(z_2)$ leads to an optimum value of the distance (z_2-z_1) over which the measurement should be performed. With the traditional backscattering technique, this entails only the choice of an optimum pair of points on the backscattering trace, for calculating the overall loss L'_A between them. On the contrary, when a two point backscattering technique is employed, the time delay between the two sampled voltages must be suitably optimized.[47] Considering a real situation, ΔV depends on the level of the received power, through source and shot noises, thus it cannot be considered as independent of z_1 and z_2. So, (9.25) turns out to be only roughly approximate, nevertheless an optimum value of the distance (z_2-z_1) exists. For instance, assuming $A = 3$ dB/km, with the usual typical values of the parameters involved, and in particular 100 ns pulsewidth and 1 W peak power for the launched signal, the optimum value of (z_2-z_1) is about 1 km.[47] This evaluation is practically independent of the prefixed value of $\Delta L'_A/L'_A$, which influences only the resulting accuracy requirement in the measurement of the two voltages samples.

As for the overall accuracy of a backscattering attenuation measurement, besides the uncertainty due to the noise superimposed to the useful signal, all those causes of uncertainty related to fluctuations in the fiber parameter, described in 9.3.3, must be considered. Even the uncertainty in the forward and backward optical power distributions must be taken into account if a mode scrambler is not employed. Though repeated backscattering measurements performed by the same apparatus exhibit a good repeatability, the agreement among measurements performed by different backscattering setups, may be sometimes not good. Finally, a comparison with the result of a classical cutback technique may exhibit even a remarkable difference.

In this sense, the result of a backscattering attenuation measurement can be reported with an estimated maximum error as small as

0.1 dB, only when the purpose is a relative comparison. On the contrary, when the absolute value of the fiber attenuation has to be measured, an estimated maximum error of about 1 dB must be always considered. Finally, an attenuation measurement performed by backscattering can be extrapolated over longer distances only as a first approximation.

For the evaluation of local additional losses (and in particular fault location), the accuracy of the backscattering technique is high enough. Nevertheless, when a high spatial resolution is required in fault location, the energy of the pulse must be reduced and the noise bandwidth B_N of the receiver must be enlarged. Therefore the SNR is doubly worsened when the spatial resolution is increased.

Before conclusion of the present section, we will describe briefly an alternative technique, which has been recently proposed in order to enhance the SNR, and hence the accuracy of the measurement. It is based on the use of a pseudo-random sequence of pulses, forming a word, instead of the single pulse of the traditional technique.[(49)] The optical source is continuously modulated. A typical pulse duration is on the order of 1 μs. A delay unit feeds a correlator, set in the receiver, which is able to detect the backscattered signal even at very low values of the SNR. This correlator is followed by a low-pass filter, whose cutoff frequency is determined by the word length. The longer this length, the better the improvement in the SNR. Typical word lengths are of some thousands of pulses. Varying the delay time, one can test different fiber sections. The spatial resolution of the measurement can be assumed as a fraction of the pulse duration.

The optical power emitted by the laser sources in continuous wave is about 10^3 times smaller than the peak power which characterizes the pulse operation. So, the process gain due to the correlation must be at least 10^3, in order to have comparable performance with the traditional technique. Practical tests have shown that the accuracy of an attenuation measurement performed by this modified backscattering technique, and even its dynamic range, can be excellent.[(49)] In particular, the operation in continuous wave leads to very stable launching conditions, avoiding the noise due to the optical source. Moreover, the experimental apparatus is simpler than the traditional one, which requires a boxcar integrator.

9.3.8 Outline of Optical Frequency Domain Reflectometry (OFDR)

Recently a new backscattering technique has been proposed. It consists of exciting the fiber with a swept frequency sinusoidally modulated signal.[50] The baseband response, in amplitude and argument, which is measured reproduces the Fourier transform of the backscattering trace under a Dirac pulse function as a launched pulse. Neglecting the reflections at the input and output fiber end faces, from (9.2) one can easily obtain that this baseband response results in

$$\bar{R}(f) = \frac{1}{1 + i\left(\frac{\pi f}{\gamma < v_g >}\right)}$$

where i is the imaginary unit and f the modulation frequency. Taking the modulus of $\bar{R}(f)$, one obtains

$$|\bar{R}(f)|^2 = \frac{1}{1 + \left(\frac{\pi f}{\gamma < v_g >}\right)^2}$$

Therefore, knowing the average group velocity $< v_g >$, from the measurement of the amplitude response only, it is possible to derive the attenuation coefficient γ. When the reflection at the output fiber end is not negligible, this amplitude response is not simply decreasing, but exhibits an oscillating behavior superimposed onto this decay. It is possible to demonstrate that the period of such an oscillation is related to the fiber length,[50] leading to the possibility of the estimate of this unknown too.

Since the modulation frequencies of interest for this test typically range from 0 to some hundreds of kHz, a lock-in amplifier can be used, so that there is a noticeable enhancement in the SNR with respect to the classical OTDR is achieved. On the other hand, OFDR can be employed only for the determination of the average attenuation coefficient and not for fault location or joint characterization. In this sense, especially for field measurements, OTDR remains a more complete attenuation measurement techniques.

9.4 DIFFERENTIAL MODE ATTENUATION MEASUREMENTS

Differential mode attenuation (DMA) measurements can be performed, as introduced in 5.1.3, either with an input or an output selection. Cutback techniques are always employed in practice, because differential backscattering, though proved experimentally,[35] leads to a backscattered optical power distribution which cannot be controlled by variations in the launching condition. The exitation is performed successively for input selection, and simultaneously for output selection. When the excitation can be considered to be uniform in the domain of guided rays, the cut is not necessary, and this technique becomes completely nondestructive.

In this paragraph we report the main results of DMA measurements, and a brief discussion on the accuracy which characterizes the different techniques. It is very important to recall that the mode dependent part of the intrinsic attenuation coefficient can be experimentally evaluated through this type of measurement, only if mode coupling effects are negligible.

9.4.1 DMA Measurements of Step Index Fibers

In (51), through an input far field selection, the θ dependence of the intrinsic attenuation coefficient γ has been experimentally evaluated. The optical source was a He-Ne laser, the fiber was 228 m long and a cut was performed at 2 m from its input section. From the data collected the behavior of $\gamma(\theta)$ reported in Fig. 9.18 has been obtained. It is characterized by a substantial uniformity for low values of θ, whereas, considerably before the maximum permitted value θ_M is reached, γ increases very rapidly. The fiber was uncabled and ıring the measurement it was loosely wound on a drum, with a uiameter of about 30 cm. The presence of mode coupling, however, has been excluded on the basis of the following test.

For three different excitation conditions ($\theta=0^\circ$, $\theta=4^\circ$, $\theta=7^\circ$), the output far field pattern of the fiber has been scanned both at $z = 228$ m and at $z = 2$ m. The agreement between the relative optical power distributions obtained confirms that mode coupling has negligible effects on the measurement. From this test it is possible to derive also the angular resolution of the measurement. In fact, the response to a selective excitation is a bell-shaped curve in θ, whose width decreases with increasing θ, from about 2° at $\theta = 0^\circ$, to about 1° at $\theta = 7^\circ$

A DMA measurement on step index fibers, performed through an output selection on both near and far fields, has been reported in (52). A LED and a PIN photodiode were used respectively as optical source and photodetector. The optical arrangement is that reported in Fig. 5.1. This investigation, however, has been performed together with a DMD measurement, in order to calculate the baseband response of the fiber from data of differential mode measurements. We will return on this type of indirect test in 10.4.

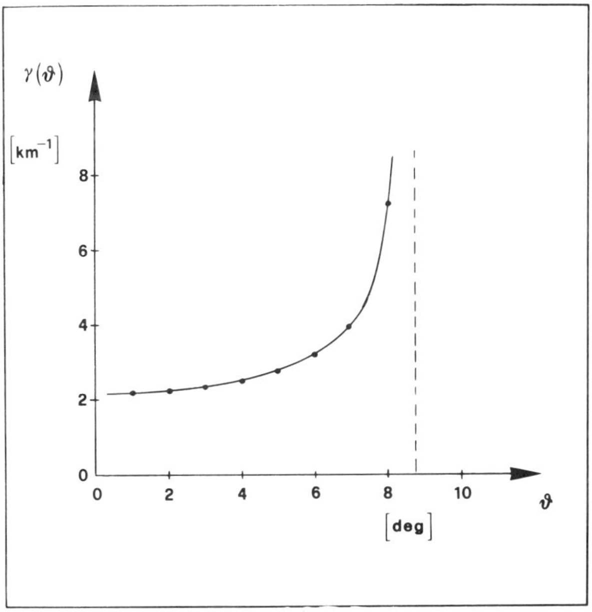

Fig. 9.18 - Intrinsic attenuation coefficient γ *vs.* the propagation angle θ, obtained from a DMA measurement on a step index fiber [from(51)].

9.4.2 DMA Measurements of Graded Index Fibers

In step index fibers, to select families of rays having the same second constant of the motion E, only a far field scanning must be made. On the contrary, in graded index fibers, both near and far field scanning procedures are required for a complete investigation. In (53) a partial output selection has been proposed, based on a para-axial excitation, so that only quasi-meridional rays are present in the fiber, and on a far field scanning at the fiber output end.

A complete output selection is presented in (54), using the optical arrangement reported in Fig. 5.1. This apparatus allows the selection of the two mode variables β and ν. A LED and a PIN photodiode were used. In Fig. 9.19 the results of such a measurement for two graded-index fibers, A and B, are reported. The normalized mode group number x_g introduced in 2.2.5 has been used as variables denoting the modes. Only modes having ν=0 have been considered in the figure, but modes with $\nu \neq 0$ (which correspond to skew rays) have also shown similar behaviors. In Table 9.1 the main characteristics of the two fibers are reported. It is rather surprising that both fibers exhibit an increase in the attenuation coefficient for low order modes, with respect to intermediate modes. A possible interpretation is that this excess loss is due to impurities located near the central region of the core (absorption loss), or to irregularities of very small spatial extension, but not negligible amplitude, in the inner part of the index profile (scattering loss).

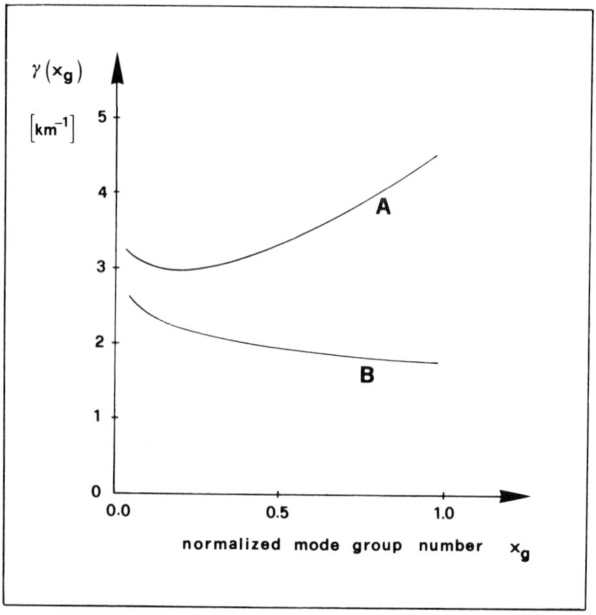

Fig. 9.19 - DMA measurements on two graded index fibers for modes having $\nu \cong 0$ [from (54)].

TABLE 9.1 – Main characteristics of fibers *A* and *B*.

Fiber	Length	Profile Parameter α	*NA*	Core Radius *a*
A	75 m	2.0	0.17	39 μm
B	1000 m	2.2	0.20	37 μm

In order to better understand the mechanisms leading to such results, in (55) a combined technique, able to give scattering and total DMA measurements, has been proposed. Krypton and He-Ne lasers were employed to perform an input selection. The focused spot diameter was about 10 μm, therefore the spatial resolution was not very high. An integrating sphere, like that described in 9.1.3, has been used for the measurement of the lateral scattering. By subsequent measurements under different launching conditions, at the same fiber section, it is possible to evaluate the relative mode dependent scattering loss. Repeating such measurements at another fiber section, it is possible to obtain also the absolute mode dependent total loss. From the comparison between these data, even the mode dependent absorption loss can be estimated, though the accuracy of the last indirect test turns out to be rather poor.

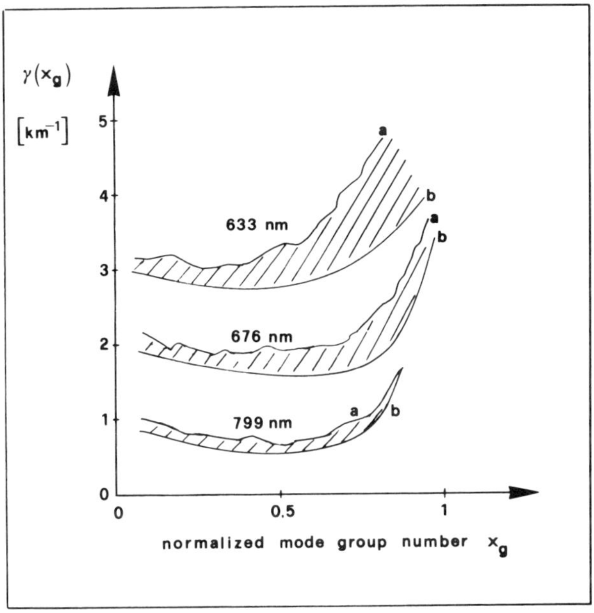

Fig. 9.20 - Differential mode measurements of the total loss (curves a), and scattering loss (curves b), and evaluation of absorption loss (shaded areas) [from (55)].

In Fig. 9.20 an example of such a measurement is reported, for three different wavelengths. Also in this case, a higher attenuation coefficient has been obtained for low order modes, and this effect appears to be mainly due to scattering loss. Other measurements on different fibers have shown smaller excess loss for the low order modes, [55] but again there is a remarkable scattering loss. In conclusion, while for step-index fibers $\gamma(\theta)$ is always an increasing function, for graded-index fiber $\gamma(x_g)$ is often a function characterized by a minimum, which corresponds to intermediate modes. In this case, the excess loss which occurs for low order modes is mainly due to Rayleigh scattering, caused by microscopic fluctuations in the refractive index distribution.

In order to reach an acceptable value of the SNR, since the detection of lateral scattering is required here a collimated beam laser is necessary as the optical source. On the other hand, when the total guided power has to be measured, at the fiber output end, a laser diode and even a LED can be employed. Nevertheless, for these types of optical sources, a uniform excitation and an output selection appears more suitable.

9.4.3 Discussion of the Accuracy of DMA Measurements

Let us consider an output selection, leading to N independent measurements on as many groups of modes or rays. When a uniform excitation must be realized, a LED source is preferable. Moreover, in order to simulate a uniform radiance distribution in air, a top emitting LED must be employed. From a device of this type, the maximum level of injected optical power into a typical multimode fiber ($a = 25\ \mu$m, NA=0.2) is about 10 μW. With such a low level of launched power, the angular and spatial resolutions of the scanning procedure must be chosen as a trade-off. In fact, a high resolution leads to a large number of measurement points, a low resolution allows one to obtain a better accuracy for each measurement point.

The maximum uncertainty in the evaluation of the power loss $L_A(u)$, where u is a general variable denoting the modes or rays, can be written as

$$\frac{\Delta L_A(u)}{L_A(u)} = \frac{4.34}{L_A(u)} \left[\frac{\Delta V(z_o, u)}{V(z_o, u)} + \frac{\Delta V(L, u)}{V(L, u)} \right]$$

where $V(z_o,u)$ and $V(L,u)$ are the measurement results at $z = z_o$ and $z = L$, selecting the same group of modes denoted by u. Since N is usually a rather high number (typically on the order of 10^2), and for backscattering measurements as well, we must let

$$\Delta V(z_o, u) = \Delta V(L, u) = \Delta V$$

which means that the noise of the photodetector is not negligible, especially at $z = L$. The sensitivity in the measurement of the attenuation per km, $A(u)$, which characterizes the group of modes denoted by u, is then given by

$$\Delta A(u) = \frac{\Delta L_A(u)}{L_A(u)} A(u)$$

$$\cong \frac{4.34}{L - z_o} \Delta V \left[\frac{1}{V(z_o, u)} + \frac{1}{V(L, u)} \right] \left[\frac{dB}{km} \right] \qquad (9.26)$$

In ΔV we must consider also an uncertainty in the reproducibility of the u selection.

The use of a lock-in amplifier is needed, because of the low level of optical power to be detected. Considering typical values of the quantities involved in practical measurements, from (9.26) it is possible to obtain a sensitivity of about 0.5 dB/km, whose dependence on u is not remarkable.

Using an edge-emitting LED or a laser diode, the maximum level of launched power increases by a factor 10 or 10^2. Nevertheless, $\Delta A(u)$ exhibits a strong dependence on u, owning to the very weak excitation of the modes just near the guidance condition. In this case, therefore, only for the groups of modes that belong to the core inner region in graded-index fibers, a very accurate DMA measurement can be performed. When the selection is made at the fiber input, a collimated beam laser is preferable, to obtain easily a small spot on the fiber input and face.

In this case, the sensitivity in the measurement of $A(u)$ is practically imposed only by the repeatability of the selective excitation, and could be noticeably better than in the previous case. Yet, the use of laser diodes, which are employed also in the transmission systems, should lead to a more interesting investigation, but focusing their beams appears more difficult.

An example of measurement that requires a very high sensitivity in the optical power detection is reported in (56), where the modal power distribution of a backscattered light has been measured on a step index fiber. This is not rigorously a DMA measurement, but it is, however, useful for investigating causes of attenuation. An APD photodetector was used in order to enhance the SNR, th optical source was a pulse modulated laser diode, with the pulsewidth $\Delta t = 75$ ns and repetition rate 1 kHz. Two launching conditions have been employed: one in the neighborhood of $\theta = 0°$, the other in the neighborhood of $\theta = 8°$. The selection has been performed on the far field pattern of the backscattered light. The apparatus used separating launched and backscattered powers is based on a polarized incident light (see 9.3.1.).

Three points on the backscattering trace have been chosen, by sampling the waveform at different temporal locations, to perform as many modal power distribution measurements. The first point belongs to the power reflected on the fiber input end (a), the second point corresponds to a given fiber section between input and output (b), the third point belongs to the power reflected at the fiber output end (c). In Fig. 9.21 the results of such measurements, under the two launching conditions described above, are reported. Curves a are altered by spurious reflections, e.g., on lenses and prisms of the launching apparatus. Curves c reproduce rather accurately the launching conditions, showing that mode coupling is practically negligible. Finally, curves b appear to be strongly influenced by the launching conditions.

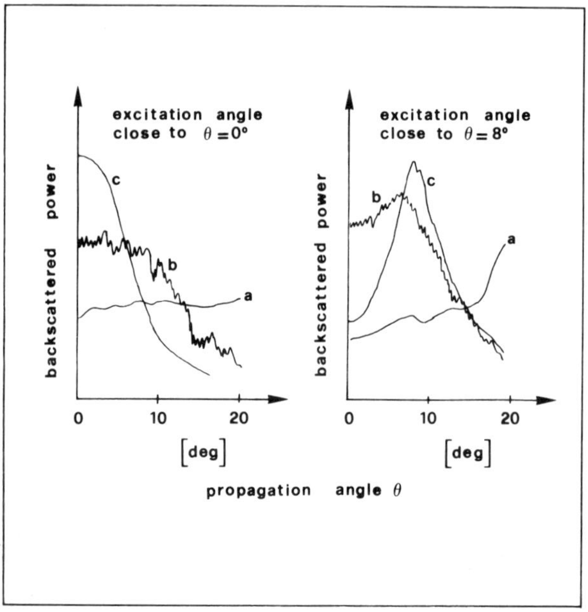

Fig. 9.21 - Modal power distributions obtained from the backscattered light in different points of the backscattering trace, for two launching conditions [from (56)].

In (38), a similar measurement has been extended to graded index fibers, confirming substantially the theoretical predictions about the ray distribution of the backscattered power discussed in 9.3.2. However, the selection of the single mode variable x, which has been exploited, appears somehow restrictive. On the other hand, a double selection probably is not possible, owing to the useful signal too small to be detected.

Before concluding the present paragraph, we will stress that an accurate evaluation of the modal power distribution not always leads to an accurate DMA measurement. In fact, the main cause of uncertainty in the attribution of a given experimental result to the mode dependent part of the attenuation per km, $A(u)$, is usually due to the presence of mode coupling. In fact, even for uncabled fibers wound on a drum with a large diameter, a little amount of mode coupling is often present. In particular, it leads to power exchanges between groups of modes whose separation Δu is well above the resolution limit of the mode selection, which characterizes the measurement setup. In this sense, what is measured is not only the effect of $A(u)$, but also that of an optical power flow among the modes.

REFERENCES

1. Smith, R.G. "Optical Power Handling Capacity of Low Loss Optical Fibers as Determined by Simulated Raman and Brillouin Scattering," *Appl. Opt.*,11, pp. 2489-2494 (1972).
2. Pinnow, D.A., and Rich, T.C. "Development of a Calorimetric Method for Making Precision Optical Absorption Measurements," *Appl. Opt.*, 12, pp. 984-992 (1973).
3. Rich, T.C., and Pinnow, D.A. "Optical Absorption in Fused Silica and Fused Quartz at 1.06 μ," *Appl. Opt.*, 12, pp. 2234-2234 (1973).
4. White, K.I., and Midwinter, J.E. "An Improved Technique for the Measurement of Low Optical Absorption Losses in Bulk Glass," *Opto-Electron.*, 5, pp. 323-334 (1973).
5. Pinnow, D.A., and Rich, T.C. "Measurement of the Absorption Coefficient in Fiber Optical Waveguides Using a Calorimetric Technique," *Appl. Opt.*, 14, pp. 1258-1259 (1975).
6. White, K.I. "A Calorimetric Method for the Measurement of Low Optical Absorption Losses in Optical Communication Fibers," *Opt. & Quant. Electron.*, 8, pp. 73-75 (1976).
7. Zaganiaris, A. "Simultaneous Measurements of Absorption and Scattering Losses in Bulk Glass and Optical Fibers by a Microcalorimetric Method," *Appl. Phys. Lett.*, 25, pp. 345-347 (1974).
8. Stone, F.T., Eichenbaum, B.R., and Lovelace, C.R. "Loss Measurememts in Optical Fiber Ribbons Using Calorimetry," *Appl. Opt.*, 16, pp. 2484-2486 (1978).

9. Cohen, R.L. "Loss Measurements in Optical Fibers. 1: Sensitivity Limit of Bolometric Techniques," *Appl. Opt.*, 13, pp. 2518-2521 (1974).
10. Cohen, R.L., West, K.W., Lazay, P.D., and Simpson, J. "Loss Measurements in Optical Fibers. 2: Bolometric Measuring Instrumentation," *Appl. Opt.*, 13, pp. 2522-2524 (1974).
11. Broquet, I. "La diffusion dans les fibres optiques multimodes," *Ann. Telec.*, 29, pp. 195-208 (1974).
12. Tynes, A.R. "Integrating Cube Scattering Detector," *Appl. Opt.*, 9, pp. 2706-2710 (1970).
13. Dakin, J.P. "A Simplified Photometer for Rapid Measurements of Total Scattering and Attenuation of Fiber Optical Waveguides," *Opt. Comm.*, 12, pp. 83-88 (1974).
14. Stone, J. "Measurement of Rayleigh Scattering in Liquids Using Optical Fibers," *Appl. Opt.*, 12, pp. 1824-1827 (1973).
15. Kaiser, P., and Astle, H.W. "Measurements of Spectral Total and Scattering Losses in Unclad Optical Fibers," *J. Opt. Soc. Am.*, 64, pp. 469-474 (1974).
16. Tynes, A.R., Pearson, A.D., and Bisbee, D.L. "Loss Mechanisms and Measurements in Clad Glass Fibers and Bulk Glass," *J. Opt. Soc. Am.*, 61, pp. 143-153 (1971).
17. Cancellieri, G., and Zoboli, M. "Enhanced-Sensitivity Nondestructive Measurement of Optical Fiber Loss," *Alta Freq.*, 47, pp. 424-427 (1978).
18. Bianciardi, E., Cancellieri, G., and Zoboli, M. "Statistical Treatment of Scattering Measurements for Investigating Optical Fiber Attenaution," *Alta Freq.*, 49, pp. 251-253 (1980).
19. Cannell, G.J. "Continuous Measurement of Optical Fiber Attenuation during Manufacture," *Electron. Lett.*, 13, pp. 125-126 (1977).
20. Midwinter, J.E. *Optical Fibers for Transmission*, New York: John Wiley & Sons (1979).
21. Ikeda, M., Sugimura, A., and Ikegami, T. "Multimode Optical Fibers: Steady-State Mode Exciter," *Appl. Opt.*, 15, pp. 2116-2120 (1976).
22. Geckeler, S. "Pulse Broadening in Optical Fibers with Mode Mixing," *Appl. Opt.*, 18, pp. 2192-2198 (1979).

23. Love, W.F. "Novel Mode Scrambler for Use in Optical Fiber Bandwidth Measurements," *Conf. Opt. Fiber Comm.*, Washington, pp. 118-120 (1979).

24. Eve, M., *et al.* "Launching Independent Measurements of Multimode Fibers," 2nd E.C.O.C., Paris, pp. 143-146 (1976).

25. Zeidler, G. "Effects of Mechanically Induced Periodic Bends on the Transfer Characteristics of Glass Fibers," 2nd E.C.O.C., Paris, pp. 105-109 (1976).

26. Tokuda, M.,Seikai, S., Yoshida, K., and Uchida, N. "Measurement of Baseband Frequency Response of Multimode Fibers by Using a New Type of Mode Scrambler," *Electron. Lett.*, 13, pp. 146-147 (1977).

27. Ikeda, M., Murakami, Y., and Kitayama, K. "Mode Scrambler for Optical Fibers," *Appl. Opt.*, 16, pp. 1045-1049 (1977).

28. Sharma, A.B., Hubach, E.J.R., Lahtenoja, M., and Halme, S.J. "Technique for Minimizing Effect of Detector Inhomogeneity on Fiber Attenuation Measurements," *Electron. Lett.*, 18, pp. 56-57 (1982).

29. Keck, D.B., and Tynes, A.R., "Spectral Response of Low Loss Optical Waveguides," *Appl. Opt.*, 11, pp. 1502-1506 (1972).

30. Heitmann, W., Day, C.R., Zwick, U., and Mengel, F. "Broadband Spectral Attenuation Measurements on Optical Fibers: An Interlaboratory Comparison by Members of COST 208," *Opt. & Quant. Electron.*, 13, pp. 47-54 (1981).

31. Cancellieri, G. "Attenuation Measurements on Single Pieces of Fibers and Extrapolation Over Longer Distances," *Electron. LEtt.*, 18, pp. 1062-1063 (1982).

32. Costa, B., and Sordo B. "Experimental Study of Optical Fiber Attenuation by a Modified Backscattering Technique," 3rd E.C.O.C., Munich, pp. 69-71 (1977).

33. Barnoski, M.K., and Jensen, S.M. "Fiber Waveguides: A Novel Technique for Investigating Attenuation Characteristics," *Appl. Opt.*, 15, pp. 2112-2115 (1976).

34. Personick,S.D."Photon Probe: An Optical Fiber Time Domain Reflectometer," *Bell Syst. Tech. J.*, 56, pp. 355-366 (1977).

35. Jensen, S.M. "Observation of Differential Mode Attenuation in Graded-Index Fiber Waveguides Using OTDR," Opt. Fiber Comm., Washington 1979, pp. 120-122.

36. Nakazawa, M., Tanifuji, T., Tokuda, M., and Uchida N.,"Photon Probe Fault Locator for Single Mode Optical Fiber Using an Acousto-Optical Light Deflector," *IEEE J. Quant. Electron.*, QE-17,pp. 1264-1269 (1981).
37. Di Vita, P., and Rossi, U. "Realistic Evaluation of Coupling Loss between Different Optical Fibers," *J. Opt. Comm.*, 1, pp. 26-32 (1981).
38. Costa, B. Di Vita, P., Morra, P., Rossi, U., and Sordo, B. "Backscattering Mode Power Distribution in Optical Fibers," 7th E.C.O.C., Copenhagen, pp. 29-32 (1981).
39. Di Vita, P., and Rossi, U., "The Backscattering Technique: Its Field of Applicability in Fiber Diagnostics and Attenuation Measurements," *Opt. & Quant. Electron.*, 11, pp. 17-22 (1980).
40. Di Vita, P. "Effects of Mode Conversion on Backscattering Measurements of Multimode Optical Fibers," *Electron. Lett.*, 18, pp. 637-638 (1982).
41. Di Vita, P. "Backscattering Measurements of Optical Fibers: Analysis of Mode Coupling and Joints," CSELT Rapp. Tec., 10, pp. 149-150 (1982).
42. Eriksrud, M., Mickelson, A.R., and Andersen-Gott, T. "Backscattering Signatures from Graded Index Fibers with Diameter Variations," *Electron. Lett.*, 17, pp. 200-201 (1981).
43. Mickelson, A.R., and Eriksrud, M. "Role of Modal Distribution in Determining Power Backscattered from Fibers with Diameter Perturbation," *Electron. Lett.*, 17, pp. 658-659 (1981).
44. Conduit, A.J., Payne, D.N., Hartog, A.H., and Gold M.P. "Optical Fiber Diameter Variations and Their Effect on Backscattering Loss Measurements," *Electron. Lett.*, 17, pp. 308-310 (1981).
45. Costa, B., Esposta, F., D'Orio, C., and Morra, P. "Splice Loss Evaluation by Means of Backscattering Technique," *Electron. Lett.*, 15, pp. 550-551 (1979).
46. Hullett, J.L., and Jeffery, R.D. "Noise in Optical Fiber Backscatter Measurements," *Opt. & Quant. Electron.*, pp. 117-124 (1981).
47. Conduit, A.J., Hullett, J.L., Hartog, A.H., and Payne, D.N. "An Optimized Technique for Backscatter Attenuation Measurements in Optical Fibers," *Opt. & Quant. Electron.*, 12, pp. 169-178 (1980).

48. Hullett, J.L., and Jeffery, R.D. "Long Range Optical Fiber Backscattering Loss Signatures Using Two-Point Processing," *Opt. & Quant. Electron.*, 14, pp. 41-49 (1982).

49. Okada, K., Hashimoto, K., Shibata, T., and Nagaki, Y. "Optical Cable Fault Location Using Correlation Technique," *Electron. Lett.*, 16, pp. 629-630 (1980).

50. Kapron, F.P., Kneller, D.G., and Garel-Jones, P.M. "Aspects of Optical Frequency Domain Reflectometry," I.O.O.C. 81, San Francisco, pp. 106-107 (1981).

51. Keck, D.B. "Spatial and Temporal Transfer Measurements on a Low Loss Optical Waveguide," *Appl. Opt.*, 13, pp. 1882-1888 (1974).

52. Cancellieri, G. "Time-Dispersion Measurement in Optical Fiber by Near and Far Field Scanning Technique," *Electron. Lett.*, 14, pp. 465-467 (1978).

53. Eve, M. "Novel Technique for Measurement of Differential Mode Attenuation in Graded Index Fibers," *Electron. Lett.*, 13, pp. 744-745 (1977).

54. Olshansky, R., Oaks, S.M., and Keck, D.B. "Measurement of Differential Mode Attenuation in Graded Index Fiber Optical Waveguides," Top. Meet. Opt. Fiber Transm., Williamsburg, TuE5 (1977).

55. Olshansky, R., and Oaks, S.M. "Differential Mode Attenuation Measurements in Graded Index Fibers," *Appl. Opt.*, 17, pp. 1830-1835 (1978).

56. Payne, R. "Modal Distribution of Backscattered Light in a Step Index Multimode Fiber," *Electron. Lett.*, 17, pp. 568-570 (1981).

CHAPTER 10

TIME DISPERSION MEASUREMENTS

In Chapters 1 and 2 we have shown that several causes contribute to the determination of the baseband response of a multimode optical fiber. It is possible to implement some measurement techniques to single out the effects of the different causes. This is not restricted only to technological purposes, as it is for attenuation measurements. For instance, the choice of the optical source to be employed in a transmission system is greatly helped by accurate material dispersion measurements, especially when the system operates in the second window. Alternatively, it is possible to perform measurements on the overall effect, without any interest in the separation of the different causes of time dispersion.

In the following, we will consider first those measurements which are based on the total guided power. They appear suitable for appreciating overall effects, or sometimes material dispersion. Section 10.4 describes differential mode measurements of time dispersion, which can give information essentially on intermodal time dispersion, but also on mode coupling effects.

When the fiber perturbation is so small it induces only slight effects in mode coupling, as it is for uncabled fibers, a baseband response measurement performed over a reference distance can be used for an extrapolation of the results over longer distances, following a $1/z$ law. This is not true for cabled fibers, since the perturbation is usually considerable. Rigorously, the measurement should be performed under the steady-state condition. In this case, the results could be extrapolated following a $1/\sqrt{z}$ law. Yet, a steady-state optical power distribution should be simulated in amplitude and argument. This is very difficult to obtain in practice.

However, the presence of imperfect joints and the compensation effects of fiber chains, described in 1.3.10, makes this requirement less interesting than in the case of attenuation measurements where, nevertheless, only the amplitude of the steady-state optical power distribution must be simulated. Therefore, the problem of extrapolation of these measurement results remain open, and will be a topic of discussion for years. In the past, the dominant attitude of designers to this problem resulted in an over-design of the systems for what concerns the bandwidth limitation of the repeater spacing.

Instead of determining the bandwidth of a fiber chain, it could be more interesting to test if the baseband response remains unchanged in shape, except for the numerical value of the cutoff modulation frequency, after a minimum distance. This would allow one to exploit completely the features of the transmission medium, by using a quasi-standard equalizing circuit having only few parameters to be adjusted at the time of the installation.

10.1 GENERAL CONSIDERATIONS

Here, after a brief discussion on the problem of the extrapolation of the experimental results, we will describe the two kinds of time dispersion measurement usually performed:

i) measurement in the time domain, based on the evaluation of the pulse broadening along the fiber length

ii) measurement in the frequency domain, based on the modulation of the optical source with a sinusoidal signal, whose frequency is continuously varied

Both these techniques can be employed to evaluate either the overall effect, or single causes of time dispersion, depending on the arrangement used. *

10.1.1 Extrapolation of the Experimental Results of Time Dispersion Measurements

For attenuation measurements, only the amplitude of the steady-state optical power distribution must be simulated. This can be done, although as a first approximation, using beam launchers or mode scramblers, as discussed in 9.2.1. For baseband response measurements, on the contrary, both the amplitude and the argument of the steady-state optical power distribution should be reached. As shown in 2.3, the argument of this distribution can be expressed by a relative delay which, in principle, depends on the modulation frequency. Moreover, as drawn from Fig. 2.9, it depends also on the amplitude of the fiber perturbation. Neither beam launchers nor mode scramblers can simulate such a steady-state condition.

As discussed in 2.4.7, the effects of a joint on the baseband response of an optical link may be remarkable, moreover compensation ef-

*See notes pg. 498

fects of fiber chains may occur. So, the baseband response prediction based on the steady-state conditions could be unrealistic.[1] The computation of the baseband response of a link made up with a sequence of equal fibers is easy only in two limit situations: when no mode conversion or a complete mode mixing takes place in each joint.[2] In the former case, the overall behavior is that of a single fiber, of the same length as the whole link. In the latter case, the overall baseband response can be obtained as the product of the baseband responses of the single fibers. If the fibers are unperturbed, i.e., there is no mode coupling, the two results coincide. On the contrary, when a fiber perturbation is present, in the former case the reaching of the steady-state is not influenced by the joints, in the latter case, a new reaching of the steady-state must be considered for each fiber of the sequence.

Let us consider a sequence of n fibers, separated by $(n\text{-}1)$ joints where no mode conversion takes place. The ith fiber is characterized by a bandwidth decay having general form

$$B_i(z) = \frac{B_{oi}}{z^q} \tag{10.1}$$

in which constant q, ranging from 0.5 to 1.0, is imposed by the perturbation, whereas constant B_{oi} depends both on the perturbation and on the dominant cause of time dispersion which characterizes the fiber. For example, if intermodal time dispersion dominates, it depends on the profile departure from the optimum. If intramodal time dispersion is predominant, it depends on the source spectral width.

Under the above assumptions, the overall bandwidth can be obtained from those of the single fibers, as

$$B_{tot}^{-1/q} = \sum_{i=1}^{n} B_i^{-1/q} \tag{10.2}$$

An extrapolation law of this type is extremely rough and can lead to remarkable errors. In fact, it is impossible to define a single value of q, varying the distance. In other words, equation (10.1) holds for distances on the order of each fiber length, but cannot be extrapolated with the same value of q over the whole link. Nevertheless, equation (10.2) can be efficiently used just for determining the evolution of parameter q. The bandwidth is measured after the first two, three, etc., fibers. Correspondingly, from (10.2), and knowing the single B_i, one obtains a series of values $q_2, q_3 \ldots$. This series of values describe the evolution of q versus z.[3]

This technique allows one to determine the distance required for reaching the steady state, which is strictly related to the amplitude of the fiber perturbation, when this distance is very large with respect to the fiber piece length, imposed usually by requirements of the pulling process. This means that even very small fiber perturbations can be studied through such a method. It is, however, of critical importance that all the joints are practically perfect.

When the fiber perturbation is strong, so that the steady state is reached in a single piece of fiber, a series of bandwidth measurements can be performed, at different distances from the optical source, through fiber cutting.[4] This is indeed a destructive method, unsuitable for industrial environments.

A situation characterized by joints in which a complete mode mixing takes place is unrealistic. Nevertheless, the use of a traditional mode scrambler, like those described in 9.2.1, placed just after each joint, allows one to approximate this condition. Following this way, in (2) a good agreement has been found between the theoretical prediction and the experimental evaluation of the baseband response of four cascaded fibers.

About the use of mode scramblers in time dispersion measurements, it is necessary to stress the following features. They assure, under suitable conditions, repeatable experimental results. Therefore, they can be efficiently employed when the characteristics of different fibers have to be compared. On the contrary, great care must be paid when results are to be extrapolated over longer distances. It is not very important to evaluate experimentally the steady-state time dispersion behavior, in fact a fiber in a link is never at the steady-state rigorously, or it approaches this state only in its last meters. Differential mode delay measurements, as will be described in Chapter 11, can give some information about mode coupling. It is probably more interesting to derive the asymptotic properties of time dispersion from these indirect data.

10.1.2 Measurements in the Time Domain

The experimental setup of a measurement of time dispersion in the time domain is schematically shown in Fig. 10.1. A pulse generator drives the optical source (usually a laser diode), and triggers the sampling oscilloscope, through an adjustable delay line. Beam splitting of the emitted light provides, through an appropriate optical apparatus, the test signal launched into the fiber, and a reference signal. The two output signals are revealed by two indentical photodetectors, and sent to a double-trace sampling oscilloscope for comparison.

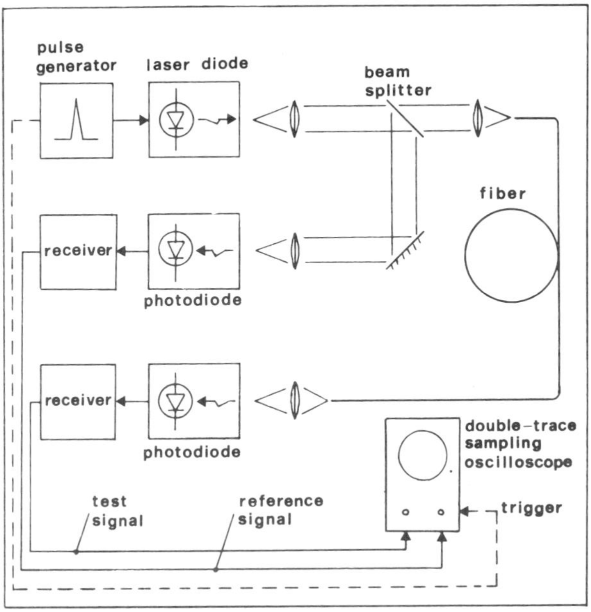

Fig. 10.1 - Experimental setup of a time dispersion measurement in the time domain.

In order to perform an accurate measurement, the launched pulse must be short with respect to the expected impulse response of the fiber under test. For today's low dispersion fibers, this may require considerable effort, in fact pulse durations as short as 100 ps may be required. [5] Some modifications to this traditional technique have been proposed, in order to relax the above requirement, and to achieve accuracy as well. Two modified techniques will be briefly described here:

i) the shuttle pulse measurement

ii) the circulating pulse measurement

A shuttle pulse measurement [6] is based on the use of end reflectors, which force the pulse to shuttle back and forth along the fiber. Both end reflectors exhibit a slight transmittivity to allow the injection and the extraction of the pulse. A circulating pulse measurement [7] is made aligning the two fiber ends, so that the pulse can make round trips along the fiber. Injection and extraction of the pulse are obtained using an acousto-optical Bragg deflector, like that employed for backscattering measurements and described in 9.3.1.

In both cases, one can observe that the optical power distribution is modified by the presence of the end reflectors and of the Bragg deflector. Nevertheless, when different fibers are joined to form an optical link, at each joint the optical power distribution suffers modifications. In this sense, the above techniques allow a good simulation of the situation in which a complete mode mixing occurs at each joint. In spite of this fact, using a shuttle pulse technique, without any destructive test, in (8) a sublinear law has been experimentally shown for the *rms* pulse broadening at very large distances.

Both traditional and modified techniques, in order to obtain high peak power without damaging the laser diode, employ a very low repetition rate typically about 1 kHz. The peak power launched into the fiber by a laser diode can be of significant wattage. Higher power levels are reached employing other optical sources, e.g., Nd:YAG lasers [(9)] whose use is, however, limited to laboratory applications. The sampling oscilloscope operates an average over many received pulses, and this leads to a noticeable enhancement of the signal to noise ratio SNR, provided that the shape of the pulse remains unchanged. The random nature of the filament emission, described in Chapter 3, which characterizes laser diodes in pulse modulation with high peak power, represents the most important cause of uncertainty of such measurements. In fact, this characteristic represents an error of stability of the optical source, and cannot be reduced by the averaging. This kind of problem occurs also for backscattering measurements, but here, in addition, not only the total pulse energy, but also its shape, strongly affect the measurement.

Only when the pulse duration is much shorter than the width of fiber impulse response, then such drifts in the launched pulses have negligible effects, because they influence spectral components which are well above the fiber useful bandwidth. Note that also the bandwidth of the two receivers must be much more extended than that of the fiber. Finally, the sampling oscilloscope must be characterized by a suitable rising time. In order to perform measurements on very good graded index fibers, typical bandwidths, required for transmitters and receivers, are of the order of several GHz, and the rising time of the sampling oscilloscope must consequently be smaller than 100 ps.

The photodetector is usually based on a PIN photodiode, [5] but avalanche photodiodes are also sometimes employed,[6,7] especially when long distances are tested, and therefore the signal is greatly attenuated. Anyway, in order to achieve a very wide receiver bandwidth, the photodiode active area must be very small. Therefore, sometimes suitable optics must be used in order to focus the optical power onto this active area. [5]

A measurement in the time domain is often employed only to give an estimate of the pulse broadening. In this application, the following relation is assumed

$$\sigma_o^2 = \sigma_i^2 + \sigma_f^2 \tag{10.3}$$

σ_i, σ_o,σ_f being respectively, the *rms* pulsewidth at the fiber input, at the fiber output, and of the fiber impulse response.

To demonstrate the validity of this relationship, we consider a normalized output pulse $P_{out}(t)$, for which it is

$$M_o = \int_{-\infty}^{+\infty} P_{out}(t)\, dt = 1$$

So, from (1.87) and (1.88), we have

$$\sigma_o^2 = \int_{-\infty}^{+\infty} t^2 P_{out}(t)\, dt - [\int_{-\infty}^{+\infty} tP_{out}(t)\, dt]^2 \tag{10.4}$$

Recalling the definition of the fiber impulse response $h_f(t)$, and letting $P_{in}(t)$ be the input pulse, one obtains

$$P_{out}(t) = \int_{-\infty}^{+\infty} P_{in}(t - \tau)\, h_f(\tau)\, d\tau \tag{10.5}$$

and substitution into (10.4) yields

$$\begin{aligned} \sigma_o^2 = &\int_{-\infty}^{+\infty} t^2\, dt \int_{-\infty}^{+\infty} P_{in}(t - \tau)\, h_f(\tau)\, d\tau \\ &- [\int_{-\infty}^{+\infty} t\, dt \int_{-\infty}^{+\infty} P_{in}(t - \tau)\, h_f(\tau)\, d\tau]^2 \end{aligned} \tag{10.6}$$

We assume also

$$\int_{-\infty}^{+\infty} P_{in}(t)\, dt = 1$$

which is justified because a scale factor is implicitly considered comparing input and output pulses. So, the two integrals in (10.6) can be factorized, leading to the demonstration of (10.1), provided that

$$\sigma_f^2 = \int_{-\infty}^{+\infty} t^2 h_f(t)\, dt - [\int_{-\infty}^{+\infty} t\, h_f(t)\, dt\,]^2$$

which is the definition of fiber *rms* pulsewidth, from its impulse response.

A more complete measurement of time dispersion consists of the determination of $h_f(t)$, from (10.5), using the sampled values of $P_{in}(t)$ and $P_{out}(t)$. This may introduce remarkable computational errors, which, however, could be greatly reduced when $\sigma_i^2 \ll \sigma_f^2$. Finally, the computation of the fiber baseband response, from the knowledge of $h_f(t)$ is usually performed through a numerical Fast Fourier Transform (FFT).[10] In this case, the intrinsic uncertainty due to the SNR, which is added to the errors due to the computation, gives rise to an overall uncertainty which normally increases with the modulation frequency considered.

10.1.3 Measurements in the Frequency Domain

A time dispersion measurement in the frequency domain directly gives the fiber baseband response, although usually only in amplitude. This is obtained comparing input and output signals, whose modulation frequency is continuously varied. Two setups will be considered here, depending on the instrument used for the electric output detection:

i) setup based on the use of a spectrum analyzer [11] (Fig. 10.2)

ii) setup based on the use of a vector voltmeter [12] (Fig. 10.3)

The vector voltmeter requires a reference signal directly from the wave generator. Using a vector voltmeter, it is possible to also obtain some information on the argument of the fiber baseband response, if the experiment is suitably arranged.[13,14] Since the noise equivalent bandwidth of a vector voltmeter is usually larger than that of a spectrum analyzer, a lower accuracy is typically reached employing the former instrument.

Semiconductor devices (LEDs or laser diodes), whose light can be directly intensity modulated by a current signal superimposed to a bias current, are suitable optical sources. Sometimes the light produced by a collimated beam laser or by a lamp is externally modulat-

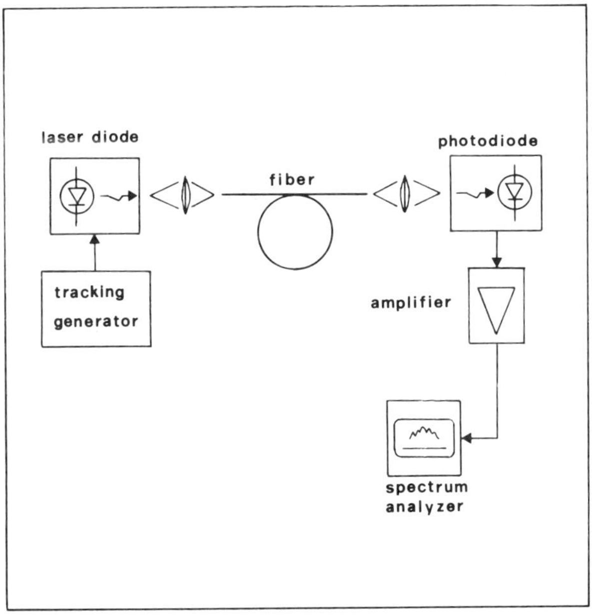

Fig. 10.2 - Experimental setup of a frequency domain measurement using a spectrum analyzer.

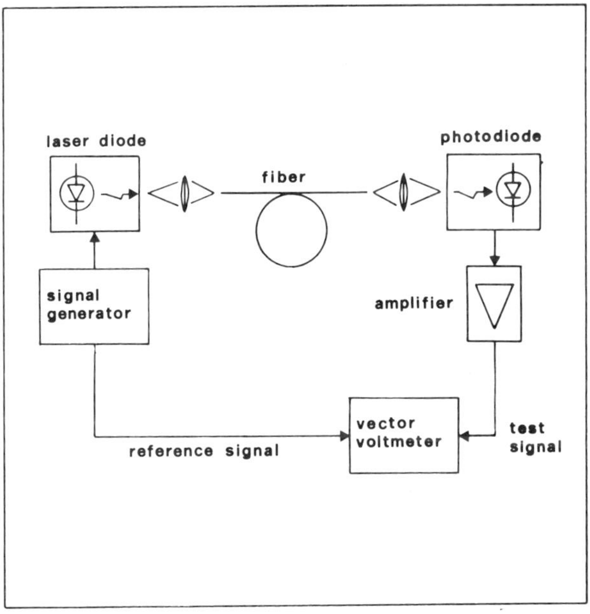

Fig. 10.3 - Experimental setup of a frequency domain measurement using a vector voltmeter.

ed through an electro-optical modulator, e.g., a Pockel's cell [15] (see Appendix 1). Alternative methods exploit the presence of an intrinsic modulation of the launched power, such as the beat frequencies of a collimated beam laser in free running operation, [16] or the intensity fluctuation noise in a laser diode. [17] These methods allow bandwidth measurements up to 2 GHz or even more. In 6.1.4, we have described a method, based on a pair of laser diodes, to obtain beat frequencies up to some GHz (see Fig. 6.7). This method, which has been proposed for the measurement of photodiode baseband responses, can be efficiently employed also for the measurement of fiber baseband responses in the frequency domain.

PIN, [5] but especially avalanche photodiodes, [11] are employed as photodetectors, as well as for measurements in the time domain, but high speed photomultipliers are sometimes used. [15] In principle, transmitter and receiver baseband responses are not required to be flat in the whole range of modulation frequencies interested by the measurement. In fact a comparison is performed at each modulation frequency, between the signals obtained before and after the fiber. Nevertheless, two main reasons suggest using transmitter and receiver bandwidths close to the expected fiber bandwidth:

i) sensitivity problems in the light detection

ii) accuracy problems in the baseband response measurement

Sensitivity problems in the light detection may arise when the transmitter and/or receiver, whose baseband responses have an intrinsic low-pass behavior, operate beyond their cutoff modulation frequency. Accuracy problems in the baseband response measurement may occur when the electric compensation, made to obtain a wide flat bandwidth, causes considerable ripples. Clearly, if the compensation is made within a tolerance of less than 1 dB and the possible ripples are smoothed, accuracy problems have negligible effects. However, it is convenient to have a transmitter-receiver bandwidth that is not too wide, for reducing noise.

In principle, the measurement of the argument of the fiber baseband response could also be obtained through a vector voltmeter from the comparison between the phases of the launched and received waveforms at each modulation frequency. Yet, even when the frequency interval between two test modulation frequencies is very small, a phase difference much greater than 2π may occur, owing to the absolute value of the propagation delay along the fiber. For instance, let us consider the modulation frequencies $f_l = 1000$ MHz and

$f_2 = 1010$ MHz. Letting z represent the fiber length, this phase difference can be written as

$$\Delta\phi = 2\pi(f_2 - f_1)\frac{< n_1 >}{c} z$$

where $<n_1>$ represents the average value of the core refractive index, and c is the free space velocity of light. Assuming $<n_1> = 1.5$ and $z = 1$ km, we have $\Delta\phi \cong 50\cdot 2\pi$.

On the other hand, if the 3 dB bandwidth of the fiber is about 1 GHz, the relative mode delay distribution is bound inside a time interval of some ns, and this means that a phase difference smaller than $10^{-2}\cdot 2\pi$ must be appreciated. Methods for avoiding this limitation have been proposed in (13) and in (14). The first method is based on the following idea. The emitted light is divided by a beam splitter into two beams: one excites the fiber, and the other is detected to give a reference signal. Test and reference signals are sent to two identical mixers, fed by the same tunable oscillators. The outputs are then compared by a vector voltmeter. The second method is based on the use of a monomode fiber as a delay line for the reference signal. We will return to these methods in 10.3.

10.1.4 Measurements of Intermodal Time Dispersion

The measurements in the time or in the frequency domain can be employed sometimes to evaluate the effects of single causes of time dispersion. A measurement of intermodal time dispersion can be performed at a fixed wavelength, exciting the fiber with a quasi-monochromatic optical source. In reference (17) the power of different longitudinal modes of a laser diode, selected by a monochromator, has been used to excite the fiber under test. Alternatively, a mode-locked collimated beam laser [(9)] can be employed.

Considering a measurement in the time domain, we recall (1.111), under the condition $\sigma_s^2/\lambda_o^2 \to 0$, where λ_o and σ_s are respectively the central wavelength and the *rms* spectral width of the optical source. So the *rms* pulsewidth of the fiber

$$\sigma_f^2(z) \cong \sigma_{inter}^2(z) \cong z^2[< \tau^2(\lambda_o) > - < \tau(\lambda_o) >^2]$$

can be easily obtained from the measurement. The other variables have been defined in 1.4.4.

A different but complementary analysis must be developed considering a measurement in the frequency domain. The phasor $\overline{H}(\omega, z)$, introduced in 1.3, which gives amplitude and argument of the fiber baseband response, can be written as

$$\overline{H}(\omega, z) = \int_0^\infty \sum_m K_m S(\lambda) \exp[-i\,\omega\,\tau_m(\lambda)\,z]\,d\lambda$$

where K_m is a constant accounting for the excitation of the mth mode, $S(\lambda)$ is the spectral distribution of the optical source, suitably normalized, and $\tau_m(\lambda)$ is the group delay per unit distance of the mth mode at the wavelength λ. Using the expansion

$$\tau_m(\lambda) \cong \tau_m(\lambda_o) + \tau_m'(\lambda_o)(\lambda - \lambda_o)$$

where the prime denotes a differentiation with respect to λ, and considering an average value over the modal distribution, $<\tau'_m(\lambda_o)>$, instead of $\tau'_m(\lambda_o)$, [17] we are left with

$$\overline{H}(\omega, z) \cong \overline{H}_{inter}(\omega, z)\,\overline{H}_{intra}(\omega, z) \tag{10.7}$$

where

$$\begin{aligned} \overline{H}_{inter}(\omega, z) &= \sum_m K_m \exp[-i\omega\,\tau_m(\lambda_o)\,z] \\ \overline{H}_{intra}(\omega, z) &= \int_0^\infty S(\lambda) \exp[-i\,\omega <\tau_m'(\lambda_o)> (\lambda - \lambda_o)\,z]\,d\lambda \end{aligned} \tag{10.8}$$

Now, if the optical source has a very narrow spectral width, $S(\lambda)$ can be considered as a Dirac pulse function, so that $H_{intra}(\omega, z) = 1$, and a measurement in the frequency domain gives

$$\overline{H}(\omega, z) \cong \overline{H}_{inter}(\omega, z)$$

Clearly, in this case much more information is obtained than from the evaluation of $\sigma^2_{inter}(z)$. However, a measurement of $H_{inter}(\omega, z)$ turns out to be difficult, especially when the fiber exhibits a very low intermodal time dispersion. In this case, in fact, σ_s^2 / λ_o^2 must be as small as 10^{-10}. Only an optical source whose spectrum is extremely narrow can satisfy this condition.

More accurate measurements of intermodal time dispersion may be performed with a differential mode technique. [18] In this case the requirement to have a quasi-monochromatic optical source is less critical. In fact, the differential mode delay distribution, even when averaged over the source wavelength spectrum, gives a large amount of information on intermodal time dispersion. This technique will be detailed in 10.4.

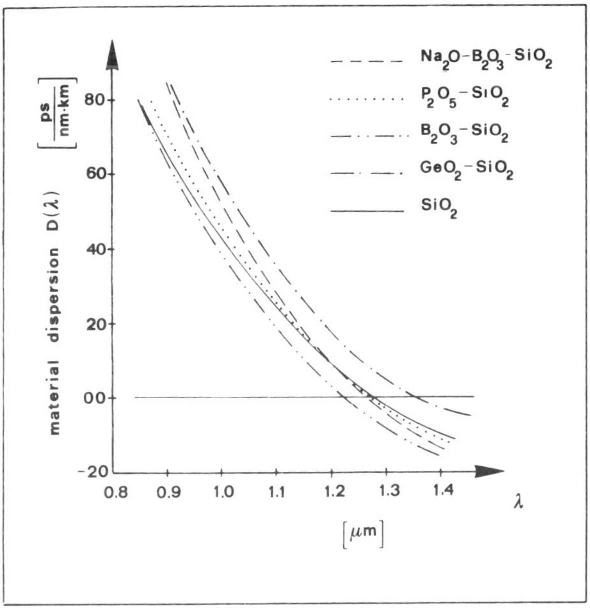

Fig. 10.4 - Experimental evaluation of material dispersion for different glasses [from (20)].

The presence of mode coupling may heavily affect a measurement of intermodal time dispersion. In fact mode coupling reduces this kind of time dispersion, as shown in 2.3, leading to a steady-state distribution of relative delays independent of z. This means that intermodal dispersion measurements, mainly devoted to technological investigations, aiming at the optimization of index profile design, should always be made on uncabled fibers.

10.1.5 Material Dispersion Measurements

One could arrange a measurement of intramodal time dispersion exciting a single family of modes or rays by means of a wide spectrum source. In this way the measurement would depend on the particular source employed and would suffer mode coupling in a very critical way. It is then more convenient to measure material dispersion, which in 1.4 has been defined as

$$D(\lambda) = -\frac{\lambda}{c}\frac{d^2 n_1}{d\lambda^2}$$

First of all, the knowledge of $D(\lambda)$ over a wide wavelength region allows one to determine the wavelength λ_m at which $D(\lambda)$ vanishes, for the particular glass considered. Moreover, the evaluation of intramodal time dispersion for a given transmission link is possible from the knowledge of $D(\lambda)$, provided that the spectral distribution $S(\lambda)$ of the optical source employed is known as well. In fact, considering the central wavelength of emission λ_o, and neglecting waveguide dispersion, we have

$$\tau_m'(\lambda_o) = D(\lambda_o) \qquad \left[\frac{ps}{nm \cdot km}\right]$$

An experimental evaluation of $D(\lambda)$ can be obtained in different ways:

i) through the measurement of the variations of the refractive index values, varying the wavelength

ii) through the measurement of the variations in the transit times of the optical power along the fiber, varying the wavelength

iii) from a double baseband response measurement at different wavelengths

A measurement of the absolute value of $n(\lambda)$ directly gives $n\, dn/d\lambda$, and $\lambda/c\,(d^2n/d\lambda^2)$. The first one is usually plotted as a function of n^2, [19] in order to predict effects of profile dispersion for quasi-monochromatic sources. The second one is normally plotted *versus* λ, when the wavelength of zero material dispersion λ_m, or the intramodal time dispersion related to a given optical source must be evaluated. [20] This measurement is performed on a bulk sample of lightguide glass. [21] In Fig. 10.4 the behaviors of $D(\lambda)$ for some lightguide glasses are reported.

The experimental results obtained by this technique are usually fitted by a three term Sellmeier's law, of the type (1.103). The main advantages of this direct method with respect to the indirect tests are:

i) the measurement is unaffected by any waveguide effect

ii) the chemical composition of the sample can be determined with high accuracy

In particular, the latter feature allows one to relate an experimentally obtained material dispersion to a precise glass composition.

Transit time variations at different wavelengths can be measured either through a pulse modulation or through a sinusoidal modulation of the optical source. Using a pulse modulation, the relative delay distribution of peak power is considered. Various techniques have been proposed, in order to obtain uniform launching conditions for varying wavelengths. If this is not done, other effects might influence $d\tau/d\lambda$. Moreover, a very narrow spectral width of the source and a very short pulse duration are required for an accurate evaluation of $D(\lambda)$.

In (22) a tunable mode locked krypton ion laser, at 520, 568 and 647 nm, has been employed. In (23) four laser diodes have been used, emitting from 810 to 900 nm, with a spectral width of about 2.5 nm. A dye laser, whose light is externally modulated by a Pockel's cell (see Appendix 1), has been proposed in (24) for the measurement of the wavelength of zero material dispersion. A source with a high peak power over an extended spectrum can be obtained using a Raman cell, which generates new wavelength components in addition to that of an incident beam, usually from a laser. Wavelength selection is obtained then by interference filters. [25] Also the test fiber can generate Raman spectral lines. [26] In this case, the wavelength selection is performed at the fiber output.

The use of a sinusoidal modulation was suggested in (27), and experiments have been described in (28), where the source was a LED filtered by a monochromator. The main advantage of this technique is in the use of a vector voltmeter, instead of a sampling oscilloscope, which allows a better time resolution (on the order of few ps), even when the optical power launched into the fiber is not extremely high. In (28) the wavelength range between 0.7 and 1.4 μm was covered by the use of four LEDs. A modification of the above technique was proposed in (29). The optical source is a gas laser, sinusoidally modulated by an external Pockel's cell (see Appendix 1). A sampling oscilloscope is used to detect phase differences through Lissajous' figures. However, the time resolution is not better than that achievable by a vector voltmeter.

Finally, we will briefly describe a technique based on a double measurement of baseband response at different wavelengths. [30] From (10.7) and (10.8), considering a gaussian normalized spectral distribution of the optical power launched by the source

$$S(\lambda) = \frac{A}{2\sqrt{\pi}} \exp\left[-A(\lambda - \lambda_o)^2\right]$$

where A is a suitable constant, one obtains

$$\overline{H}(\omega, z) = \exp\left[-\frac{(\omega < \tau_m'(\lambda_o) > z)^2}{4A}\right] \sum_m K_m \exp\left[-i\omega\tau_m(\lambda_o) z\right]$$

Filtering the optical power at the fiber output through an interference filter, centered at $\lambda = \lambda_o$, whose spectral response is $T(\lambda)$, we have

$$\overline{H}_1(\omega, z) = \int_0^{\infty} T(\lambda) \exp[-i\,\omega < \tau_m'(\lambda_o) > (\lambda - \lambda_o) z]\, d\lambda$$

$$\sum_m K_m \exp[-i\omega\tau_m(\lambda_o) z]$$

so that the ratio between $\overline{H}(\omega, z)$ and $\overline{H}_1(w, z)$ yields

$$\frac{\overline{H}(\omega, z)}{\overline{H}_1(\omega, z)} = \frac{\exp\left[-\dfrac{(\omega < \tau_m'(\lambda_o) > z)^2}{4A}\right]}{\int_0^{\infty} \mathrm{T}(\lambda) \exp[-i\,\omega < \tau_m'(\lambda_o)(\lambda - \lambda_o) z]\, d\lambda} \qquad (10.9)$$

in fact the modal distribution in the two cases remains unchanged. Both $\overline{H}(\omega, z)$ and $\overline{H}_1(\omega, z)$ can be experimentally measured. Thus, from (10.9), $< \tau_m'(\lambda_o) >$ is obtained, which, neglecting waveguide dispersion, coincides with $D(\lambda_o)$. In particular, when $T(\lambda)$ can also be considered gaussian, it being

$$T(\lambda) = \frac{A_T}{2\sqrt{\pi}} \exp[-A_T(\lambda - \lambda_o)^2]$$

where A_T is a constant, the following relation holds

$$< \tau_m'(\lambda_o) > \cong \frac{1}{2\pi B z} \sqrt{0.345 \frac{4A}{1 - A/A_T}}$$

in which B is the 3 dB bandwidth of the transfer function $\overline{H}(\omega, z)/\overline{H}_1(\omega, z)$.

Clearly, the greater the difference between A and A_T, the more accurate the method. So, a wide spectrum LED source and a narrow interference filter are recommended. In this way, the measurement of $< \tau_m'(\lambda_o) >$ turns out to be practically independent of the ratio $A/A_T \ll 1$. However, the method may be very accurate only when intermodal time dispersion is small, i.e., when intramodal time dispersion is the main effect which limits the fiber baseband response. One advantage of this technique is its independence from mode coupling effects, which influence only the intermodal factor, and this factor, vanishes in the computation.

10.1.6 Experimental Evaluation of Profile Dispersion

Profile dispersion effects can be evaluated through a series of intermodal time dispersion measurements, like those described in 10.1.4, at different wavelengths.[15],[17] Yet, it is better to perform a direct evaluation of profile dispersion. The following four techniques may be employed for this purpose

i) a direct measurement of variation of the refractive index with λ

ii) transverse interferometry measurement of index profile at different λ

iii) measurement of variation in the numerical aperture NA with λ

iv) differential mode delay measurements

The first technique has already been described in the previous section. The values of $n\, dn/d\lambda$ obtained from the measurement are plotted *versus* n^2, [19] in order to see if profile dispersion is or is not linear, as discussed in 1.4.7. With a transverse interferometry technique, as described in 7.1.5, each fringe represents a curve of equal phase. Any point (x,y) of the interferogram is characterized by a fringe displacement $S(x,y)$, which is proportional to the profile function $f(r,\lambda)$ defined in (1.112). The elementary modification ΔS due to a small variation $\Delta\lambda$ of the wavelength then is proportional to $\partial f/\partial\lambda$.[31] In this case, a wide spectrum optical source, e.g., a Xe arc lamp followed by interference filters or a monochromator, must be used.

Measurements of profile dispersion performed through the experimental evaluation of the λ dependence of NA [32] are based on the assumption that profile dispersion is linear. In this case, the parameter

$$q(\lambda) = \frac{n_1(0)}{N_1(0)}\,\frac{\lambda}{\Delta}\,\frac{d\Delta}{d\lambda}$$

defined in 1.4.5, can be easily determined from the measurement of $NA(\lambda)$.

Consider for simplicity a step index fiber, since it is

$$NA^2 \cong 2\,n_1^2\,\Delta \cong 2\,n_2^2\,\Delta$$

it is possible to write

$$q(\lambda) = \frac{n_1}{N_1}\,\lambda\left[\frac{1}{NA^2}\,\frac{d}{d\lambda}\,(NA^2) - \frac{1}{n_1^2}\,\frac{dn_1^2}{d\lambda}\right]$$

$$q(\lambda) = \frac{n_1}{N_1}\,\lambda\left[\frac{1}{NA^2}\,\frac{d}{d\lambda}\,(NA^2) - \frac{1}{n_2^2}\,\frac{dn_2^2}{d\lambda}\right]$$

The first equation is used when the core is made of pure silica, so that $dn_1^2/d\lambda$ is known. The second equation is used when the cladding is made of pure silica. Note that with traditional techniques from separate measurements of $dn_1^2/d\lambda$ and $dn_2^2/d\lambda$, one could calculate $q(\lambda)$ as

$$q(\lambda) = \frac{1}{n_1 N_1} \frac{\lambda}{NA} \left(n_2^2 \frac{dn_1^2}{d\lambda} - n_1^2 \frac{dn_2^2}{d\lambda} \right)$$

Yet, since $n_2^2(dn_1^2/d\lambda)$ and $n_1^2(dn_2^2/d\lambda)$ differ by a very small quantity, evaluation turns out to be less accurate. On the contrary, in the above equations we have a difference between two dissimilar terms. This method can also be applied to graded index fibers, though great care must be taken to select different points on the core transverse section, and to avoid the effects of leaky modes.

Finally, differential mode delay measurements, performed at different wavelengths, [18],[33],[34] allow one to evaluate the effects of profile dispersion, even when it is not linear. This kind of test, provided that mode coupling is negligible, makes it possible to find the wavelength of minimum intermodal time dispersion, i.e., the optimum wavelength for a given index profile. We will return to these techniques in 10.4. For now, we will restrict ourselves to showing how $\partial f/\partial\lambda$ can be derived from a differential mode delay measurement.

The time delay per unit distance of a ray which emerges from the output fiber end face, at the radical coordinate r_o, parallel to the fiber axis, is

$$\tau(r_o) = \frac{1}{c} \frac{1}{n(r_o)} \frac{\displaystyle\int_0^{r_o} \frac{n(r)\left[n(r) - \frac{\partial n}{\partial \lambda}\right]}{\sqrt{n^2(r) - n^2(r_o)}} dr}{\displaystyle\int_0^{r_o} \frac{dr}{\sqrt{n^2(r) - n^2(r_o)}}}$$

in which equation (1.105) has been used, letting $E = n(r_o)$ and $h = 0$. Now, assuming

$$n^2(r, \lambda) = n_o^2 [1 - f(r, \lambda)]$$

it is

$$\tau(r_o) = \frac{1}{c}\frac{1}{n(r_o)}\frac{\displaystyle\int_0^{r_o}\frac{n_o^2[1-f(r,\lambda)-\frac{\lambda}{2}\frac{\partial f}{\partial\lambda}]}{\sqrt{f(r_o,\lambda)-f(r,\lambda)}}dr}{\displaystyle\int_0^{r_o}\frac{dr}{\sqrt{f(r_o,\lambda)-f(r,\lambda)}}}$$

It is possible to invert this function, and, for monotonic profiles, we have simply [35]

$$n_o^2[1-f(r,\lambda)-\frac{\lambda}{2}\frac{\partial f}{\partial\lambda}]$$

$$= -\frac{1}{\pi c}\frac{d}{dr}\int_0^r \frac{n(r_o)\,\tau(r_o)}{\sqrt{n^2(r_o)-n^2(r)}}\left.\frac{dn^2}{dr}\right|_{r=r_o} dr_o$$

$$\cdot\int_0^{r_o}\frac{d\zeta}{\sqrt{n^2(\zeta)-n^2(r_o)}}$$

$\tau(r_o)$ can be evaluated from a differential mode delay measurement, and $f(r,\lambda)$ is assumed to be known from a measurement of index profile, so that $\partial f/\partial\lambda$ is obtained. However, the accuracy of this method is limited because of possible errors in the evaluation of r_o, $\tau(r_o)$ and $f(r,\lambda)$.

10.2 DISCUSSION OF THE MEASUREMENTS IN THE TIME DOMAIN

Different techniques have been proposed in order to produce and measure very short light pulses. Laser diodes, which are directly modulated, provide a pulse duration which is rarely shorter than 100 ps. Shorter pulse durations can be obtained using collimated beam lasers in mode locking operations. For example, in (9) a mode locked Nd:glass laser, by frequency doubling, has been employed for generating ultra-short light pulses (duration of about 7ps), at $\lambda = 0.53$ μm. A sampling technique, based on an ultra-fast optical Kerr shutter, was employed for input and output pulse detection.

Nevertheless, frequently the purpose of a time domain measurement is to test the fiber behavior before and after the installation of an optical link. In this case, it is more interesting to use the same sources and detectors which are employed in the transmission system. Therefore, laser diodes, PIN or APD photodetectors are preferred. A further advantage in the use of such components is their small size and in the rather simple electronics required. So, in the following, we will restrict ourselves to the analysis of all the causes of uncertainty which affect measurements in the time domain, using laser diodes as optical sources, and PIN or APD photodectors.

10.2.1 Main Causes of Uncertainty in Time Domain Measurements

The main causes of uncertainty in time dispersion measurements, performed in the time domain are:

i) noise due to the optical source
ii) shot noise of the photodiode
iii) thermal noise of the receiver
iv) noise due to the sampling oscilloscope

When a signal processing is necessary to deduce the fiber impulse response or baseband response, the following two causes of uncertainty must be added:

i) quantization errors due to the analog-digital conversion
ii) computational errors in the numerical deconvolution or in the numerical FFT

All these causes of uncertainty affect a short-term repeatability of such measurements, i.e., the repeatability of results obtained within a few hours, without removing the optical source. A long term repeatability, which could be tested over some days, having removed and restored the optical source, is also affected by uncertainties in the launching condition, which can be reduced only by the use of a suitable mode-scrambler, as discussed in 10.1.1.

The noise due to the optical source is produced by the random nature of the filament emission which characterizes a laser diode in pulse modulation with high peak powers. Since the sampling oscilloscope processess many pulses, this leads to remarkable errors due to drifts in amplitude and shape of the launched pulses. This error is very similar to that present in backscattering attenuation measurements performed through an electronic average, but here it is of much greater importance because, not only the amplitude, but also the shape of the launched pulse is involved in the measurement.

An additional cause of uncertainty is due to a jitter of the optical pulses with respect to the repetition rate of the pulse generator. This jitter is a consequence of the random nature of the processes which produce the optical pulse itself. This is usually obtained by driving the laser diode with an electric pulse from the discharge of a delay line, through a mechanical mercury reed relay [(5)] or through an avalanche transistor. [(36)] The last cause of uncertainty can be greatly

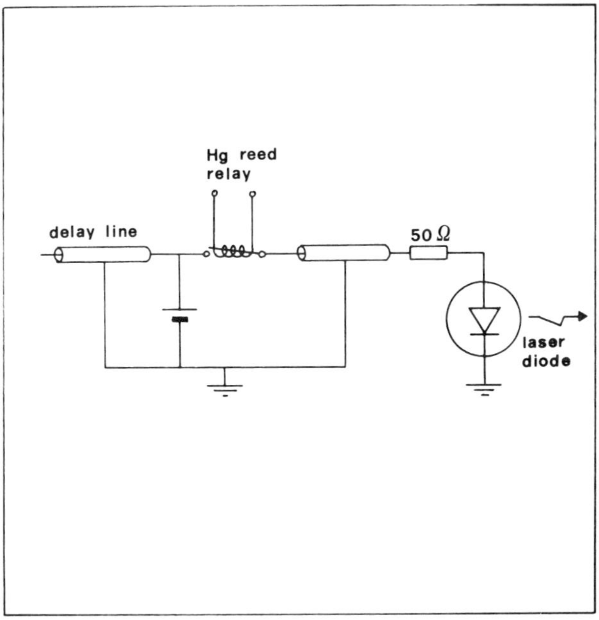

Fig. 10.5 - Laser driver based on a mechanical mercury reed relay [from (5)].

reduced if the trigger signal for the sampling oscilloscope is obtained from a photodetection of the launched optical pulse. The reference optical signal shown in Fig. 10.1 can be used for this purpose.

In Fig. 10.5 the electric arrangement of the laser driver based on a mechanical mercury reed relay is reported. [(5)] Fig. 10.6 shows a typical electric scheme of the laser driver based on an avalanche transistor. The polarity of the electric signal waveform is depicted in some points of the circuit.

As for shot and thermal noises, in order to attain the maximum bandwidth possible, the photodiode is usually connected directly to a load resistance R_L, without any amplification. So, when an avalanche photodiode is employed, as it is for backscattering measurements, the following shot noise current must be considered

$$< i_{sh}^2 > = 2q < M > \quad F(<M>) < i_s > B_N$$

where q is the electron charge, $<M>$ the mean avalanche gain, $<i_s>$ the average useful signal current, $F(<M>)$ represents the excess noise factor of the avalanche detection, and finally B_N denotes the noise equivalent bandwidth of the receiver. Both for avalanche and for PIN photodiodes, the circuit noise current reduces to

$$< i_c^2 > = \frac{4\,K\,T}{R_L}\,B_N$$

where K is Boltzmann's constant, and T is the absolute temperature. Finally B_N is given by

$$B_N = \frac{1}{2\,\pi\,R_L\,C_t}$$

where C_t represents the overall capacity, due to the photodiode junction and to the input of the instrument probe.

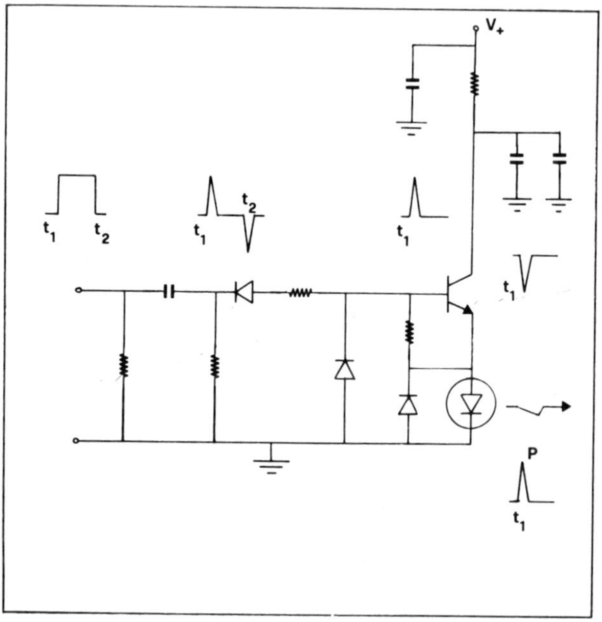

Fig. 10.6 - Laser driver based on an avalanche transistor.

The signal to noise ratio SNR without averaging, as it is for back-scattering attenuation measurements, can be written as

$$SNR = \frac{< i_s >^2}{< i_{os}^2 > + < i_{sh}^2 > + < i_c^2 >} \tag{10.10}$$

where $< i_{os}^2 >$ represents the noise due to the optical source. When N pulses are processed by the sampling oscilloscope, the new signal to noise ratio is

$$SNR' = \frac{<i_s>^2}{<i_{os}^2> + \frac{1}{N}\,(<i_{sh}^2> + <i_c^2>) + <i_o^2>} \tag{10.11}$$

in which $<i_o^2>$ denotes the noise due to the sampling oscilloscope. For example, assuming $R_L = 10\ \Omega$, $C_t = 3$ pF, then $B_N \cong 5$ GHz, and $<i_c^2> \cong 8 \cdot 10^{-14}\ A^2$, which corresponds to an equivalent incident optical power of about 0.6 μW. Considering only small values of $<i_s>$, $<i_{sh}^2>$ is usually negligible. Finally $<i_o^2>$ depends on the class of the instrument, but may be of the same magnitude as $<i_c^2>$. The time required for displaying the measured pulse is proportional to N, and can be imposed by a suitable regulation of the horizontal sweep of the oscilloscope. In principle, this is equivalent to introducing a low-pass filter at the output of the instrument, whose bandwidth is N times smaller than B_N.

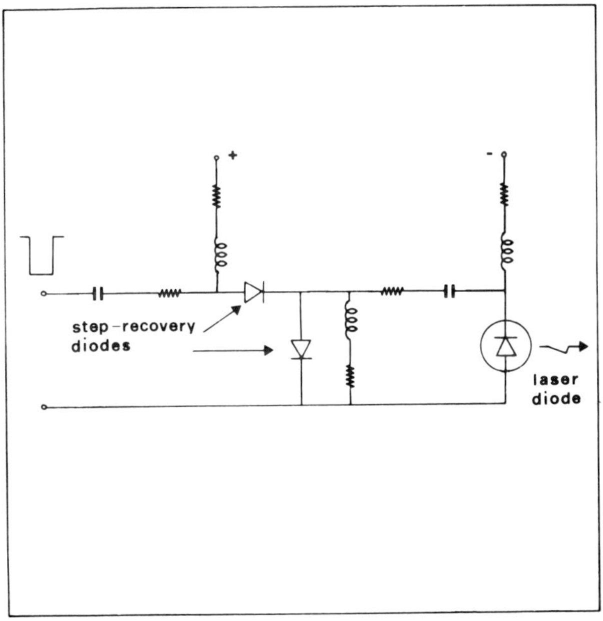

Fig. 10.7 - Scheme of a laser driver using step-recovery diodes [from (37)].

The term $<i_{os}^2>$ can not be reduced for increasing N. Nevertheless, effects of variations of the launched optical pulse can be avoided in part, using a laser diode suitable for CW operation, modulated by step-recovery diodes.[37] In this case, the peak power level is not as high as in traditional pulse modulation, since the laser diode always operates over the threshold, but filament emission has much lower effects. Typical pulse repetition frequency of this operation is 10 MHz. In Fig. 10.7 a scheme for the laser driver, based on the use of two step-recovery diodes, is reported.

As for quantization errors due to the analog-digital conversion, for the numerical computation, they are related to the bit number n of a sample through the well known expression

$$\Delta V = \frac{V(t)}{2^{n+1}}$$

where $V(t)$ is the analog sample and ΔV the maximum error due to the quantization. Since a remarkable number of samples must be stored to perform numerical processing, the choice of n should be trade-off between accuracy and the memory requirement.

Finally, in order to take into account the computational errors, we have to consider that, for instance, a procedure of FFT is based on the following equations [10]

$$R(K\omega_o) = \sum_{m=0}^{M-1} p(mT)\cos mK\omega_o, \quad K = 1, 2, \ldots$$

$$I(K\omega_o) = \sum_{m=0}^{M-1} p(mT)\sin mK\omega_o, \quad K = 1, 2, \ldots$$

where R and I represent respectively its real and imaginary parts, $\omega_o = 2\pi/MT$ is the modulation circular frequency, $p(t)$ is the pulse waveform to be transformed, M is the number of samples taken at intervals T. Denoting i and o, respectively, the input and output quantities, the baseband response can be obtained in amplitude and argument as

$$|\overline{H}(K\omega_o, z)| = \sqrt{\frac{R_o^2(K\omega_o) + I_o^2(K\omega_o)}{R_i^2(K\omega_o) + I_i^2(K\omega_o)}}$$

$$\arg[\overline{H}(K\omega_o, z)] = \tan^{-1}\left[\frac{I_o(K\omega_o)}{R_o(K\omega_o)}\right] - \tan^{-1}\left[\frac{I_i(K\omega_o)}{R_i(K\omega_o)}\right]$$

Now, $p_i(mT)$ and $p_o(mT)$ are affected by noise and quantization errors. Therefore, letting $\Delta p_i(mT)$ and $\Delta p_o(mT)$ the total maximum errors on $p_i(mT)$ and $p_o(mT)$, the overall accuracy in the measurement of $\overline{H}(w,z)$ can be obtained by differentiating the above relations. Nevertheless $\Delta p_i(mT)$ and $\Delta p_o(mT)$ are not constant values, in fact they depend on the instantaneous level of the signal at the sampling time. When the main causes of uncertainty are photodiode and circuit noises, $\Delta p_o(mT)/p_o(mT)$ will be much greater than $\Delta p_i(mT)/p_i(mT)$ due to fiber attenuation. In other words, the measurement of the output pulse is more critical. However, an exhaustive analysis of the overall accuracy of the measurement, and therefore of its sensitivity, turns out to be rather complicated, so that it may be more convenient to develop a statistical analysis over a set of repeated measurements. A practical technique suitable for investigating the repeatability of the experiments through its procedure, and even to reduce the overall uncertainty, will be described in 10.2.3.

10.2.2 Use of a Lock-In Amplifier to Improve the SNR

The spectral noise power density, which is superimposed to a measured pulse waveform, is not uniform. It is possible to show that it exhibits minima which correspond to integer multiples of the trigger frequency, i.e., of the pulse repetition frequency. This suggests a technique to improve the signal to noise ratio SNR, based on the experimental set up schematically shown in Fig. 10.8. [38] A trigger frequency which is twice the pulse repetition frequency leads to an amplitude modulation of the signal at the output of the sampling oscilloscope. The spectrum of the new signal consists of harmonics whose frequencies are integer multiples of the trigger frequency.

Then, connecting the output of the sampling oscilloscope to the input of a lock-in amplifier, whose reference signal is obtained directly from the repetition rate generator, allows one to reconstruct the pulse waveform, reducing the effects of noise. Precisely, letting T_c the time constant of the lock-in amplifier (see Appendix 2), we obtain $B_N = 1/\pi T_c$, and moreover this noise equivalent bandwidth must be considered around frequencies which are integer multiples of the trigger frequency, where the spectral noise power density is minimum.

The horizontal sweep of the oscilloscope must be sufficiently slow with respect to the inverse of the time constant T_c. Typical values of T_c are of the order of 1 s, imposing that the reproduction of the pulse by the oscilloscope requires a time of at least 30 s. No limitations are imposed, on the contrary, to the pulse repetition frequency. To summarize, the technique described in the present section can provide a strong reduction in $<i_{sh}^2>$, $<i_c^2>$, and even in $<i_o^2>$, so that the main cause of uncertainty remains the noise due to the optical source, $<i_{os}^2>$.

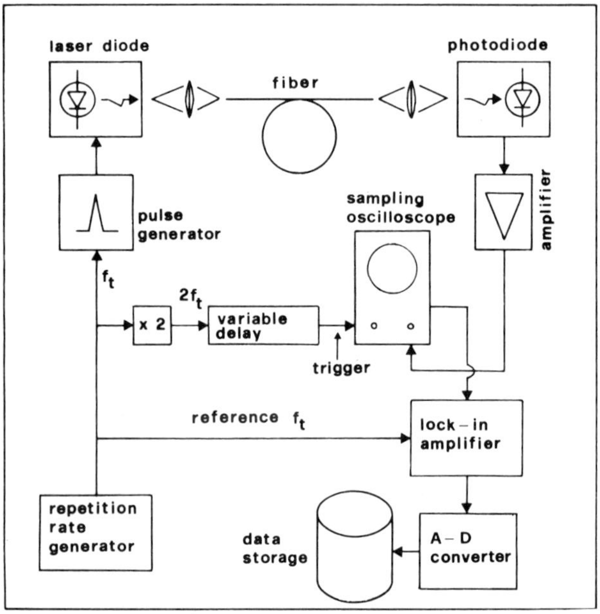

Fig. 10.8 - Experimental setup of a time domain measurement using a lock-in amplifier to improve the SNR [from (38)].

10.2.3 Use of a Pick-Up Procedure to Reduce the Overall Uncertainty

Input and output pulses are sampled at a frequency which is usually greater than that required by Shannon's theorem. [39] This makes it possible to perform a pick-up procedure on these redundant data, and to calculate different pulse waveforms or different Fourier transforms from the same pulse sequence. For instance, if the upper limit of modulation frequencies to be tested is 5 GHz, Shannon's theorem imposes the choice of a sampling time of at least 100 ps. Nevertheless, the sampling oscilloscope, for each pulse sequence, takes a sample of waveform every 10 ps or more frequently. Thus, it is possible to perform ten computations or more, which can be considered independent of each other. By applying this procedure both on the input pulse and on the output pulse, 100 impulse responses or baseband responses of the fiber are easily obtained.

If all the errors which affect the measurement can be considered as random variables, we are with 100 independent results, suitable for statistical treatment. [40] The dispersion of the measurement gives an estimate of all the uncertainties listed in 10.2.1. This dispersion can be reduced by a suitable averaging procedure. Since an exact analysis of all the causes of uncertainty which affect such a measurement is too difficult, this procedure appears the only one able to provide insight into the reliability of the whole measurement.

Considering a fiber baseband response calculated in such a way, and comparing all the resulting amplitudes and arguments for each modulation frequency, the uncertainty can be evaluated as a function of the modulation frequency f. An analysis of this type [41] shows that the uncertainty increases with increasing f, as expected, because of less power which characterizes the high frequency spectral components of the input signal (finite extension of the transmitter bandwidth), and especially of the output signal (fiber filtering, together with finite extension of the receiver and oscilloscope bandwidths).

The uncertainty turns out to be particularly severe for the argument of the baseband response, which could be obtained, instead, from the data of the amplitude, less affected by uncertainty, via the Hilbert transform. [42] However, it is not yet completely clear when the fiber baseband response can be considered as a minimum phase transfer function, which is the only case in which amplitude and argument are related through a Hilbert transform. In 10.3.1 we will briefly discuss this point. Moreover, a numerical Hilbert transform may induce a worsening of the overall accuracy, introducing further computational errors. In fact, it has been shown, [41] using a pick up procedure, that the uncertainty of the argument of the baseband response obtained via the Hilbert transform may be greater than that obtained directly. Precisely, since in the Hilbert transform all the frequencies of the amplitude response are considered to derive the argument at any single frequency, the uncertainty of the argument turns out to be large, even at low modulation frequencies.

A practical way for an approximate evaluation of computational errors in a convolution or in a transformation consists in performing convolution and deconvolution, or transformation and antitransformation, and then comparing initial and resulting data. Also the choice of the optimum upper limit f_{Mo} of the modulation frequency interval considered for the numerical Fourier transform is possible by comparing different anti-transforms, each of them obtained stopping the computation at different frequency upper limits f_M. For $f_M < f_{Mo}$, the shape of the pulse is too smooth, owing to the absence of the higher spectral components. For $f_M > f_{Mo}$, the shape appears to be too indented because of the relevant noise which affects the higher spectral components.

10.2.4 Evaluation of the Effective Input Pulse

Using the experimental setup reported in Fig. 10.1, the launched pulse is split before entering the fiber. Because of spatial and angular relative delays of emission, combined with the spatial and angular filtering operated by the optical apparatus and by the fiber itself, the pulse which is monitored may be considerably different from that which effectively enters the fiber. Moreover, different spectral components of the optical power may be characterized by different angles of emission, as described in 6.2.5. Then, if some of these angles exceed the *NA* of the fiber, the alignment of the optical source with the fiber axis at the input section, through the optical apparatus, represents a possible cause of repeatability failures for the measurement.

In order to avoid these problems, which are causes of uncertainties in the knowledge of the effective input pulse waveform, the experimental setup can be modified as follows. The reference signal is not obtained by splitting the optical power emitted by the laser diode, which is directly coupled to the test fiber, but it is taken at a different time. First, the output pulse is measured and recorded, and afterwards the fiber is cut a few meters from the input end, and the pulse is measured again at the new fiber end. This technique is very milar to a two point cutback method for attenuation measurements.

Nevertheless, in this case, since the measurement is performed at two different times, drifts of the launching condition may affect the result. These drifts are mainly caused by thermal fluctuations, which lead to variations of the pulse shape, and even of the spectral distribution of the emitted power. Finally, fiber end preparation must be particularly accurate, especially when the two point technique is used, to ensure a good repeatability of the detection conditions.

10.3 DISCUSSION OF THE MEASUREMENTS IN THE FREQUENCY DOMAIN

The traditional techniques for measuring the baseband response of a fiber in the frequency domain are based on a direct measurement of the amplitude response, and on the calculation of the phase response via the Hilbert transform. This procedure is correct only when the fiber can be considered as a linear minimum-phase two-port network. Such an assumption will be discussed here, together with the main causes of uncertainty which affect the overall result.

Also some techniques proposed for a direct evaluation of the phase response will be briefly presented.

10.3.1 The Fiber Transfer Function

A multimode optical fiber can be seen as a two-port network. The input is the optical guided power $P(t, 0)$, and the output, calling z the fiber length, is the optical guided power $P(t,z)$. As discussed in Chapter 2, the linearity is not rigorously verified for the optical powers, but only for the electromagnetic fields. Yet, assuming an optical source with a spectral width on the order of 1nm, the fluctuations of the optical carrier frequency can be averaged out, and the fiber turns out to be linear also in the optical powers.

So it is possible to define a transfer function, as

$$\overline{H}(\omega, z) = \frac{\overline{P}(\omega, z)}{\overline{P}(\omega, 0)} \tag{10.12}$$

where $\overline{P}(\omega, z)$ and $\overline{P}(\omega, 0)$ respectively are the Fourier transforms of $P(t,z)$ and $P(t,0)$. Equation (10.12), except for an attenuation factor, gives the baseband response of the fiber. Considering M modes in propagation, and neglecting for simplicity, mode coupling and any mode dependent attenuation, we have

$$\overline{P}(\omega, z) = \exp(-\gamma z) \sum_{m=1}^{M} p_m \exp(-i\,\omega\,\tau_m z)$$

$$\overline{P}(\omega, 0) = \sum_{m=1}^{M} p_m$$

where γ is the intrinsic attenuation coefficient, p_m represents the excitation of the mth mode and τ_m its group delay per unit distance. Substitution into (10.12), neglecting the uniform attenuation exp $(-\gamma z)$, yields

$$\overline{H}(\omega, z) = \frac{\sum_{m=1}^{M} p_m \exp(-i\,\omega\,\tau_m z)}{\sum_{m=1}^{M} p_m}$$

In order to have a minimum-phase two-port network, the transfer function, written in Laplace from as $\overline{H}(s, z)$, where $s = \sigma + i\omega$, must have no zeros in the closed right half plane. [42] It has been shown [43] that, ordering $\tau_1, \tau_2, \ldots \tau_M$ so that $\tau_1 < \tau_2 < \ldots \tau_M$, this condition is always satisfied when

$$p_1 > \sum_{m=2}^{M} p_m$$

This means that the fastest mode must be predominant. Also the minimum-phase condition may be satisfied in other cases, depending on the particular distribution of the group delays $\tau_1, \tau_2, \ldots \tau_M$. However, a transfer function without minimum-phase behavior is possible from the theoretical point of view, especially when higher order modes carry a considerable amount of optical power. Mode coupling effects, which help to gather the optical guided power within the lower order modes, contribute to making the assumption of a minimum-phase behavior more reasonable. [44]

To verify the minimum-phase behavior of the fiber, it is possible to measure amplitude and argument of the fiber baseband response, by a time domain technique, and to check the phase response obtained via a Hilbert transform. Unfortunately, the uncertainties in the two phase evaluations are so high, that it is difficult to have an experimental evidence of the existence of fibers without minimum phase transfer function. Moreover, sometimes a good agreement between the two phase evaluations has been found. [15]

In conclusion, the phase response obtained using the Hilbert transform is often acceptable, although from the theoretical point of view, some doubts remain about the general validity of this procedure. In 10.3.4 we will discuss an algorithm for the numerical computation of Hilbert transforms, and the effects of the noise superimposed on the signal using this algorithm.

It is interesting to note that the definition of fiber baseband response

$$\overline{H}(\omega, z) = \frac{\overline{P}(\omega, z)}{P(0, z)}$$

already given in this book, coincides with (10.12), except for an attenuation factor. In fact, in $P(\omega, z)$ the product ωz is always present. Also when material dispersion limits the fiber baseband response, we have a product of this type, so that such an expression is true again. This property holds at the steady state too, and even during the reaching of the steady state, since these situations are characterized by baseband responses in which there are products of the types ωz and $\omega^2 z$, as shown in (2.70).

In other words, from the mathematical point of view it is completely equivalent, whether $z = 0$ or $\omega = 0$. However it is quite different considering the experimental point of view. In fact, the former condition entails the execution of two measurements, at the fiber input end, and at the fiber output end, and the subtraction of an attenuation factor, which is independent of the modulation frequency. On the contrary, the latter condition can be satisfied by only one measurement, at the fiber output end.

Yet, what is measured in this case is the transfer function of the system consisting of transmitter-fiber-receiver. So, in order to assume the $\overline{H}(\omega, z)$ obtained as the actual fiber baseband response, the bandwidth limitation of transmitter and receiver must be negligible. This means that the baseband response of the transmitter-receiver system must be practically flat over all the modulation frequency interval which is expected to be covered by the baseband response of the fiber.

10.3.2 Main Causes of Uncertainty in Frequency Domain Measurements

In this section we will only discuss the uncertainty related to the amplitude response measurement. The main causes of this uncertainty are:

i) shot noise of the photodiode

ii) thermal noise of the receiver

iii) noise due to the measuring equipment

For shot noise $< i_{sh}^2 >$ and thermal noise $< i_c^2 >$, the considerations reported in 10.2.1 regarding time domain measurements can be repeated as well. The signal to noise ratio SNR, before the vector voltmeter or the spectrum analyzer, is then

$$SNR = \frac{< i_s >^2}{< i_{sh}^2 > + < i_c^2 >}$$

where $< i_s >^2$ represents the useful signal.

Since the preamplifier, if it exists, or the instrument, is characterized by a low impedance front end, to obtain a sufficiently large bandwidth, we can assume that $< i_c^2 >$ is proportional to the noise equivalent bandwidth of the receiver B_N, i.e., that the first term dominates in (4.2). Since both vector voltmeters and spectrum analyzers have a tunable receiver circuit, the effective noise equivalent bandwidth B_N' is usually much smaller than B_N. So, the signal to noise ratio, considering also the measuring instrument, can be written as

$$SNR' = \frac{< i_s >^2}{(< i_{sh}^2 > + < i_c^2 >)B_N'/B_N + < i_o^2 >}$$

where $< i_o^2 >$ represents the noise current due to the instrument.

Typical minimum values of B_N' are 10 Hz for a spectrum analyzer and 1 kHz for a vector voltmeter. If the receiver is based on an amplifier circuit, $< i_o^2 >$ turns out to be negligible, therefore it is possible to write

$$SNR' \cong \frac{B_N'}{B_N} SNR$$

In this case, the use of a spectrum analyzer appears preferable. On the contrary, if the receiver does not have an amplifier circuit, i.e., the photodiode output is directly connected to the instrument, $< i_o^2 >$ may have a critical role.

The determination of SNR′ gives the maximum dynamic range of the measurement, using a certain transmitter-receiver system, and with a fixed power loss due to the test fiber. Letting P_t be the optical power launched into the fiber at its input end, and P_r be the optical power detected at its output end, the current generated by the photodiode can be written as

$$I = \mathscr{R} P_r = \mathscr{R} \frac{P_t}{L_A}$$

where L_A represents the power loss of the fiber and $\mathscr{R}$ is the responsivity of the photodiode. The electric power produced by I on a load resistance R_L is

$$P_e = R_L I^2 = R_L \mathscr{R}^2 \frac{P_t^2}{L_A^2}$$

So, if one fixes a margin for appreciating the amplitude response decay *versus* the modulation frequency of about 20 dB in optical power, we have

$$L_A = \frac{\sqrt{R_L}}{10^2 \sqrt{P_{em}}} \mathscr{R} P_t$$

where P_{em} represents the minimum level of detectable signal. We will give an example of evaluation of L_A, when, in order to obtain a receiver bandwidth as large as possible, no amplifiers are used, so that $< i_o^2 >$ turns out to be the dominant cause which limits the maximum dynamic range. For R_L = 50 Ω, a vector voltmeter is typically characterized by $P_{em} = 10^{-13}$ W, whereas a spectrum analyzer can reach a minimum level as low as $P_{em} = 10^{-16}$ W.* Considering for simplicity a PIN photodiode, one can assume $\mathscr{R}$ = 0.5 A/W, and a

*See notes pg. 498

launched power $P_t = 1$ mW, which is typical for laser diode sources. So, we have $L_A \cong 20$ dB for a vector voltmeter, and $L_A \cong 35$ dB for a spectrum analyzer.

Therefore, an inspection over a fiber length of 5-10 km appears feasible, at least for low loss fibers. Nevertheless, as will be shown in 10.3.6, the accuracy over long distances turns out to be poor, especially when a two-point measurement is required to avoid the effects of the finite bandwidths of transmitter and receiver. In conclusion, for practical cases, inspections over distances not greater than a few km are possible.

10.3.3 A Method for Improving the Maximum Dynamic Range

A cause which further limits the maximum dynamic range L_A of a frequency domain measurement is due to the limited bandwidth of the transmitter. In fact, it is very difficult to obtain with usual techniques a flat spectrum for the launched signal above 1 GHz. If the fiber bandwidth to be tested exceeds this limit, it is necessary to compare two measurement results, with and without the fiber between transmitter and receiver. In this case, the higher modulation frequencies suffer a weaker excitation, and the value of L_A must be reduced.

A method for avoiding this limitation is based on the use of a pair of single mode laser diodes, whose electromagnetic fields are superimposed, leading to a beat frequency which can be tuned from 0 to about 5 GHz. [(45)] This technique has already been described in 6.1.4, for photodiode baseband response measurements. Here, assuming the baseband response of the transmitter to be flat over the whole modulation frequency interval of interest, the fiber baseband response can be obtained by subtracting the photodetector contribution from the total response (fiber plus photodetector).

Since a photodetector having a bandwidth larger than a few GHz can be realized without much difficulty, the subtraction is frequently avoided, leading to an improvement of the overall accuracy. In Fig. 10.9 an example of a baseband response measurement performed with such a technique is shown. The maximum error, determined by observing the rapid fluctuations of the data obtained, has been reported within two continuous lines. It is practically constant up to 5 GHz, and does not exceed 0.5 dB.

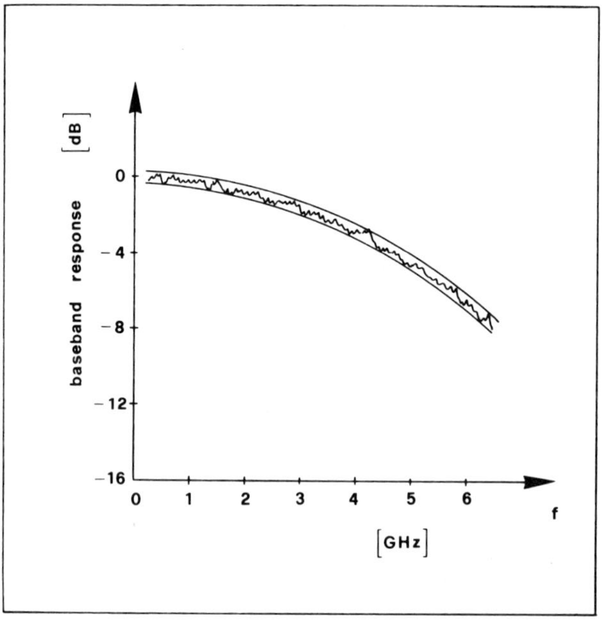

Fig. 10.9 - Baseband response measurement on a low-dispersion fiber [from (45)].

10.3.4 Evaluation of the Phase Response via a Hilbert Transform

The phase response arg [$\overline{H}(\omega, z)$] can be obtained from the amplitude response [$\overline{H}(\omega, z)$] via a Hilbert transform, through the following expression [(42)]

$$\arg [\overline{H} (\omega, z)] = \frac{2\,\omega}{\pi} \int_0^\infty \frac{\ln | \overline{H} (x, z) |}{x^2 - \omega^2} \, dx \qquad (10.13)$$

It has been shown that, when the noise affecting the measurement is not negligible, an integration performed by Simpson's formula (parabolic integration) smooths out part of the noise, whereas a trapezoidal integration emphasizes noise effects. [(41)] Also the choice of the upper limit f_M of the modulation frequencies, which must be taken in practice for the numerical integration of (10.13), influences the overall accuracy. This is a direct consequence of the worse SNR′ which characterizes higher modulation frequencies.

In Fig. 10.10 an example of the estimated percent of uncertainty related to the phase evaluation is reported as a function of the modulation frequency, considering different upper limits f_M, both for the trapezoidal integration rule (continuous lines) and for the parabolic integration rule (dashed lines). When a trapezoidal rule is employed, the uncertainty turns out to be practically uniform, and the effects due to different choices of f_M are remarkable. On the other hand when a parabolic rule is used, the uncertainty increases with increasing f, and the choice of f_M influences appreciably only the lower modulation frequencies.

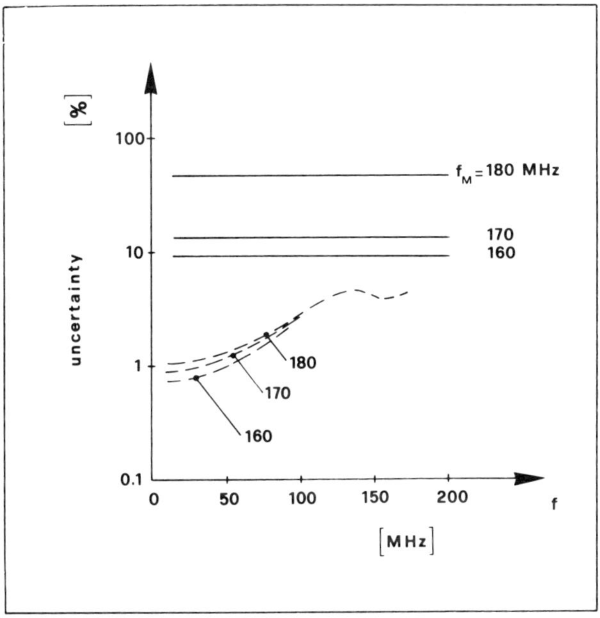

Fig. 10.10 - Example of the estimated percent uncertainty in the phase evaluation via a numerical Hilbert transform [from (41)].

10.3.5 Direct Measurement of the Phase Response

In this section, we will briefly describe two methods for a direct evaluation of the phase response. The first one is based on the use of an IF modulation applied by electric mixers. The other one employs a monomode fiber to delay a reference signal.

The experimental setup on which the first method is based is shown in Fig. 10.11.[13] A sinusoidal RF generator feeds two mixers. The first one of them gives the modulation frequency

$$f = f_o + f_m$$

where f_o is the RF frequency and f_m an IF frequency. The other one of them gives the frequency

$$f_m = f - f_o$$

Frequency f is sent to the laser driver, whereas frequency f_m at the output of the second mixer is detected by a vector voltmeter. This frequency is also used as a reference, taking it directly from the IF generator.

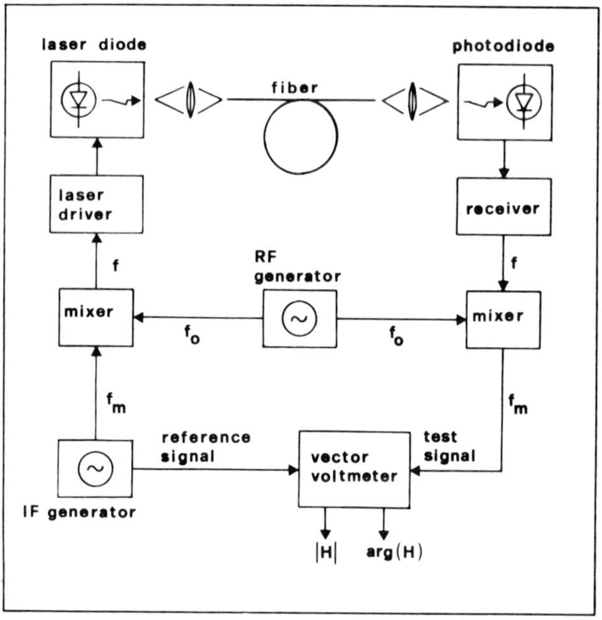

Fig. 10.11 - Experimental setup for a direct measurement of the fiber baseband response in amplitude and argument, using mixers [from (13)].

In this situation, it is possible to write the baseband response of the whole transmission system (including transmitter, fiber, and receiver) as

$$\overline{H}_t(\omega, z) = |\overline{H}_t(\omega, z)| \exp(-i\,\omega\,t_o) \exp[-i\,\theta(\omega)] \qquad (10.14)$$

where $\omega = 2\pi f$, t_o represents the absolute value of the time delay, and $\theta(\omega)$ is the phase response to be measured.

The phase measurement performed by the vector voltmeter [13] gives

$$\Delta \phi = \frac{\theta(\omega + \omega_m) - \theta(\omega - \omega_m)}{2}$$

where $\omega_m = 2\pi f_m$. If $f_m \ll f$, the quantity $\Delta\phi/\omega_m$ can be considered as the derivative $d\theta/d\omega$, which allows us to obtain $\theta(\omega)$, except for an unimportant additive constant. The modulation frequency f is varied to cover the whole baseband response of the fiber, while f_m is left constant. A typical value for the IF frequency is $f_m = 10$ MHz, and the maximum value of f may vary from hundreds of MHz to some GHz, depending on the vector voltmeter employed.

In such a measurement, as well as for material dispersion measurements or differential mode delay measurements, the very high sensitivity in the phase evaluation obtained by the use of a vector voltmeter is of fundamental importance. This sensitivity can be on the order of 0.1 degrees, provided that the SNR is high enough. Also a good frequency stability of the RF generator is very important for the accuracy of the phase evaluation. This accuracy deteriorates with increasing f, leading to an upper limit for the frequency interval which can be tested for phase evaluation. However, when the SNR is high at low frequencies, an acceptable accuracy is usually obtained well above the 3 dB bandwidth of $|\overline{H}_t(\omega, z)|$. Finally, in order to obtain the baseband response of the fiber, as usual, two measurements must be performed, with and without the fiber itself.

The experimental setup on which the second method is based is shown in Fig. 10.12. [14] A monomode fiber, having the same length and the same value of the on-axis index as the test fiber, is used for cancelling the factor $\exp(-i\omega t_0)$ in expression (10.14). In this way, the vector voltmeter directly gives $\theta(\omega)$ as the phase difference between the test and the reference signals. All the considerations of the accuracy of phase evaluation made for the previously described method, can also be repeated. Here, in addition, we have to observe that the fiber lengths and the on-axis indices must be equal with a very high precision. In fact, an error as small as one part per 10^3 may lead to phase differences on the order of several degrees between two adjacent test modulation frequencies. In (14) a very good agreement has been obtained between phase response evaluations performed with such a technique and with a time domain technique.

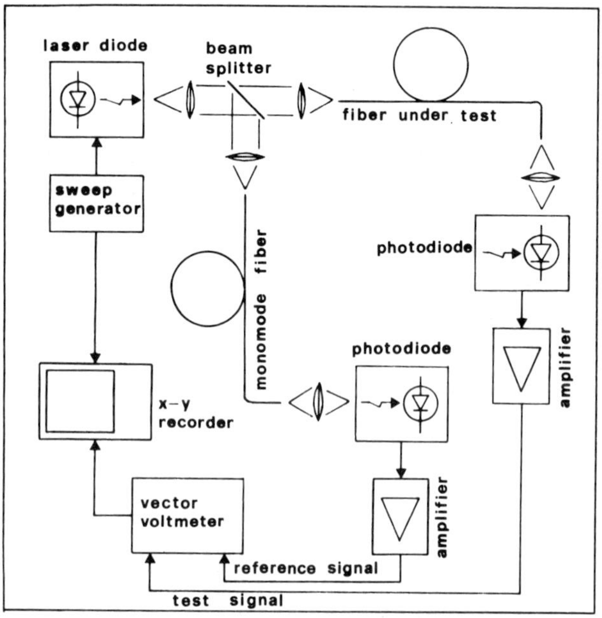

Fig. 10.12 - Experimental setup for a direct measurement of the fiber baseband response in amplitude and argument, using a monomode fiber as a reference [from (14)].

10.3.6 Overall Accuracy of Frequency Domain Measurements

A measurement in the frequency domain, when all the modulation frequencies are well below the 3 dB bandwidth of the transmitter-receiver system, can be considered practically unaffected by systematic errors. In this case a single measurement is enough. On the other hand when there are modulation frequencies also above the 3 dB bandwidth of the transmitter-receiver system, a double measurement is necessary, with and without the fiber. Moreover, in order to avoid the effects of particular launching conditions, a two point measurement based on a cut of the fiber a few meters from the optical source is often advisable. A similar case holds for time domain measurements when the launched pulsewidth is not much smaller than the width of the fiber impulse response.

In particular, the presence of appreciable spatial relative delays of emission (see 6.2) in the optical source could lead to remarkable systematic errors. This also affects time domain measurements, although the evaluation of the consequent uncertainty is more immediate in frequency domain measurements. Letting Δt_o (f, f') the difference between the overall temporal width of emission at two tested frequencies f, f', it is typically $\Delta t_o = 100$ ps for LED sources, and $\Delta t_o = 10$ ps for laser sources. So the maximum phase uncertainty, $2\pi f \Delta t_o$, and the consequent amplitude uncertainty, especially when a laser diode is employed, become more than negligible only when f is rather high. At such modulation frequencies, the uncertainty due to the finite bandwidth of the transmitter-receiver system is usually dominant. In conclusion, it is often necessary to perform a double measurement, but the cut of the fiber is only sometimes needed.

However, calling $\overline{H}_f(\omega, z)$ the fiber baseband response and $\overline{H}_c(\omega, z)$ the baseband response of the transmitter-receiver system, the whole baseband response which is measured the first time is

$$\overline{H}_m(\omega, z) = \overline{H}_c(\omega, z)\,\overline{H}_f(\omega, z)$$

By removing or cutting the fiber, one obtains $\overline{H}_c$ (ω, z), and it is possible to evaluate $\overline{H}_f$ (ω, z) through the following relationships

$$|\overline{H}_f(\omega, z)| = \frac{|\overline{H}_m(\omega, z)|}{|\overline{H}_c(\omega, z)|} \qquad (10.15)$$

$$\arg[\overline{H}_f(\omega, z)] = \arg[\overline{H}_m(\omega, z)] - \arg[\overline{H}_c(\omega, z)]$$

Now, as for time domain measurements, two methods can be followed to obtain $\overline{H}_m$ (ω, z) and $\overline{H}_c$ (ω, z):

i) a reference signal is derived through a beam splitter before the input end of the fiber, and the two measurements are performed at the same time, provided that the two photodetectors exhibit equal performances

ii) after the first measurement performed over the whole fiber length, a short piece of fiber is cut and the measurement is repeated without removing the launching condition

The first method allows one to avoid possible drifts in the level of the launched power. The second method assures that $\overline{H}_c$ (ω, z) is measured with the same spectral and spatial components of the optical power which were present also in the measurement of $\overline{H}_m$ (ω, z). Since no appreciable drifts in the level of the launched power are usually present when the optical source is sinusoidally modulated differently from the case of pulse modulation, even the second method results are very repeatable. From the above considerations, frequency domain measurements appear to be more reliable than time domain measurements.

Finally, the overall maximum uncertainty which affects an amplitude response measurement can be evaluated by differentiating the first part of (10.15), that is

$$\frac{\Delta\,|\,\overline{H}_f(\,\omega, z\,)\,|}{\overline{H}_f(\,\omega, z\,)} = \frac{\Delta\,|\,\overline{H}_m(\,\omega, z\,)\,|}{\overline{H}_m(\,\omega, z\,)} + \frac{\Delta\,|\,\overline{H}_c(\,\omega, z\,)\,|}{|\,\overline{H}_c(\,\omega, z\,)\,|}$$

The overall maximum uncertainty which affects a phase response measurement, performed directly using one of the methods described in the previous section, may be considerably higher, since the evaluation is obtained from a difference. Precisely, differentiating the second part of (10.15), we have

$$\frac{\Delta\,\arg[\,\overline{H}_f(\,\omega, z\,)\,]}{\arg \overline{H}_f(\,\omega, z\,)}$$

$$= \frac{\arg[\,\overline{H}_m(\,\omega, z\,)\,]}{\arg[\,\overline{H}_m(\,\omega, z\,)\,] - \arg[\,H_c(\,\omega, z\,)\,]}\;\frac{\Delta\,\arg[\,\overline{H}_m(\,\omega, z\,)\,]}{\arg[\,\overline{H}_m(\,\omega, z\,)\,]}$$

$$+ \frac{\arg[\,\overline{H}_c(\,\omega, z\,)\,]}{\arg[\,\overline{H}_c(\,\omega, z\,)\,] - \arg[\,\overline{H}_c(\,\omega, z\,)\,]}\;\frac{\Delta\,\arg[\,\overline{H}_c(\,\omega, z\,)\,]}{\arg[\,\overline{H}_c(\,\omega, z\,)\,]}$$

10.4 DIFFERENTIAL MODE DELAY MEASUREMENTS

We will describe here the main results of differential mode delay (DMD) measurements, introduced in 5.1.4. These measurements can be performed by means of a pulse or a sinusoidal modulation, and the mode selection may be made at the fiber input or at the fiber output, as well as for DMA measurements. In this paragraph, mode coupling will be considered negligible. The description of indirect tests on mode coupling effects, by DMA and DMD measurements will be made in Chapter 11.

10.4.1 Purposes of a DMD Measurement

Recalling 5.1.4, the main objects of DMD measurements can be listed as:

i) an indirect evaluation of the fiber baseband response

ii) the determination of the index profile at a given wavelength

iii) the investigation of profile dispersion effects

The baseband response of a fiber, due to intermodal time dispersion [46,47] can be calculated from the following expression

$$H_{inter}(\omega, z) = \sum_n p_n \exp(-i\,\omega\,\tau_n\,z) \qquad (10.16)$$

where z is the fiber length, ω is the circular modulation frequency, p_n are coefficients accounting for the power distribution among the modes, and τ_n is the group delays per unit distance of the nth mode. Now, considering (10.16) for n groups of modes, selected by a suitable scanning technique, τ_n can be obtained experimentally by a DMD measurement, whereas p_n could be determined theoretically from a measurement of the fiber index profile and assuming a uniform launching condition, which is the excitation usually taken for baseband response predictions. An experimental evaluation of p_n is also possible, measuring the amplitude of the optical power distribution at the fiber output, so that the actual baseband response (characterized by the launching condition imposed during the experiment) is measured. When intramodal time dispersion has negligible effects, as known, $\overline{H}_{inter}$ tends to the overall baseband response of the fiber.

This measurement can be performed either using pulse or sinusoidally modulated signals, and gives an indirect evaluation of both the amplitude and the phase response. Considering a sinusoidally modulated signal, a single modulation frequency f_o can be employed for evaluating τ_n. This frequency f_o is usually much smaller than the upper limit of the frequency interval, tested in frequency domain measurements of the fiber baseband response, leading to a much simpler circuital arrangement. In fact, assuming f_o = 27.7 MHz, and considering a phase resolution of 0.1 degrees, one obtains a time resolution as small as 10 ps in the DMD measurement, which is enough to test fibers whose bandwidth is well above 1 GHz.

In Fig. 10.13 we report an example of $\overline{H}_{inter}$ determined by a DMD measurement for a graded index fiber, with a sinusoidally modulated signal at f_o = 27.7 MHz. [47] Dashed lines refer to results computed considering meridional rays only. The summation has been made on the basis of the optical power distribution at the fiber output. In this case, the weights p_n are determined by multiplying the experimental data of the optical power distribution by a suitable quantity, which accounts for geometrical and guidance properties. An approximate evaluation of the index profile is enough to provide p_n, since the determination of the guidance properties is not critical for near-parabolic fibers.

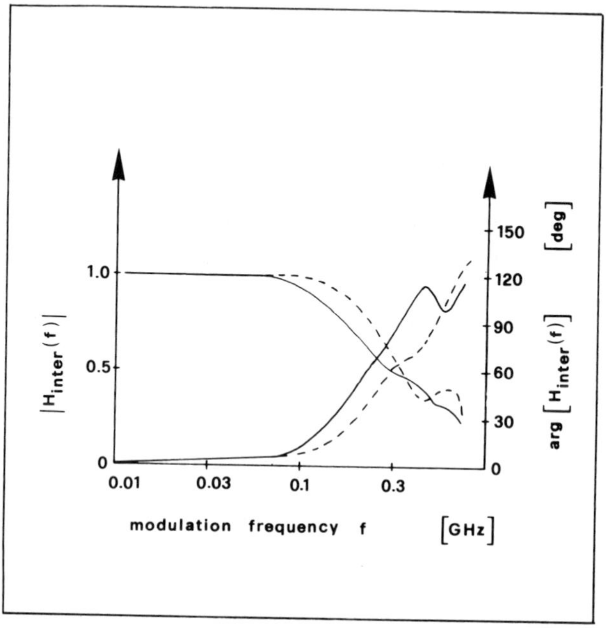

Fig. 10.13 - Evaluation of H_{inter} by a DMD measurement and an optical power distribution measurement [from (47)].

In order to give an example of calculation of the factor, taking into account the geometrical properties, we can procede as follows. The phasor which represents the total optical power at the fiber output is

$$\overline{P}(\omega) = \int_0^{2\pi} d\psi \int_0^{2\pi} d\phi \int_0^{a} r dr \int_0^{\frac{\pi}{2}} \overline{B}(r, \psi, \phi, \theta_a ; \omega) \sin\theta_a \, d\theta_a \tag{10.17}$$

where r, ψ, ϕ, θ_a are the usual ray coordinates on a fiber end face (θ_a denotes the propagation angle in air). $\overline{B}$, whose dimension is a radiance, is assumed so that

$$| \overline{B}(r, \psi, \phi, \theta_a ; \omega) | = | \overline{B}(r, \psi, \phi, \theta_a ; \omega_o) | \tag{10.18}$$

$$\arg [\overline{B}(r, \psi, \phi, \theta_a ; \omega)] = \frac{\omega}{\omega_o} \arg [\overline{B}(r, \psi, \phi, \theta_a ; \omega_o)]$$

where $\omega_o = 2\pi f_o$, and $B(r, \psi, \phi, \theta_a; \omega_o)$ is the result of the measurement. Since it is

$$\overline{H}_{inter}(\omega, z) = \frac{\overline{P}(\omega)}{P(0)}$$

the effects of the finite dimension of the photodetector are not significant, even though it is present in the weights p_n. It influences only the total number n of different mode groups. A similar role is played by the finite dimension of the spot on the fiber input end face when the selection is performed at the fiber input.

The method based on equation (10.17) is completely independent of the index profile. In other words, it gives the p_n only through geometrical properties. Yet, it requires a large number of measurement points. It is then more convenient to also consider the guidance properties of the fiber. This leads to a double integration in the domain of guided modes, instead of a quadruple integration over geometrical coordinates.

The determination of the index profile from a DMD measurement is possible in principle for any index profile, provided that a complete near and far field scanning is performed. Nevertheless, it is sometimes more convenient to determine the value of α which best fits the actual profile, using an α profile law. This is possible, considering only meridional rays, and performing only a near field scanning, with $\theta_a \cong 0$.[48] Under this condition, for an α profile, a relation exists between the radial coordinate r and the mode variable ϵ. Fig. 5.2 shows the optics which can be employed for this purpose, considering an input selection. In Fig. 10.14 the results of three DMD measurements of this type are reported. The horizontal axis has also been graduated linearly in the variable ϵ according to (1.97) which holds for near parabolic profiles.

Recalling the behavior of the group delay per unit distance $\tau(\epsilon)$ for an α profile fiber (see 1.3), from Fig. 10.14 it is clear that only fiber A can effectively be considered as an α profile fiber, and the resulting α is 2.1. On the contrary, fibers B and C exhibit a relative group delay distribution with strong fluctuations, while, over a such an extended time interval, an α profile fiber should be characterized by almost linear behavior of $\tau(\epsilon)$. A perturbed α profile could be considered, [48],[49] but this is a questionable approach, since, if a perturbed α profile is present, τ is not function of a single mode variable. In conclusion, some doubts remain on the accuracy of this method, when the fiber index profile departs from an α profile shape.

The investigation of profile dispersion effects by DMD measure ments consists of an evaluation of the wavelength dependence of intermodal time dispersion. This is possible by performing DMD measurements at different wavelengths. [18],[33] In Figs. 10.15 and

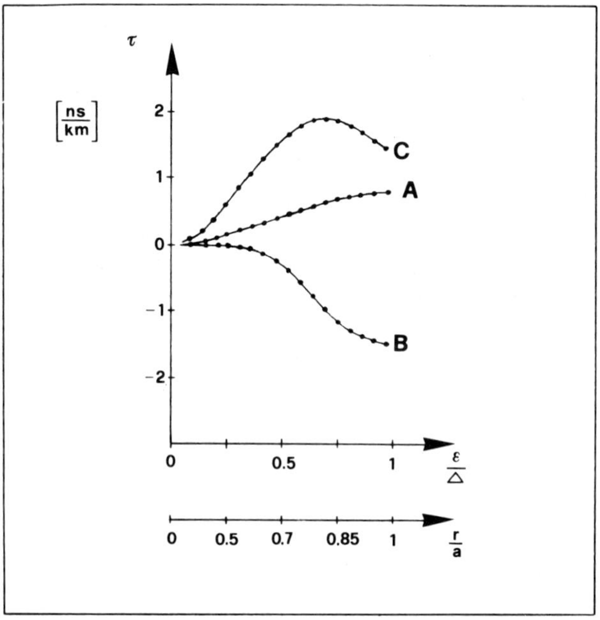

Fig. 10.14 - DMD measurements on three graded index fibers [from (48)].

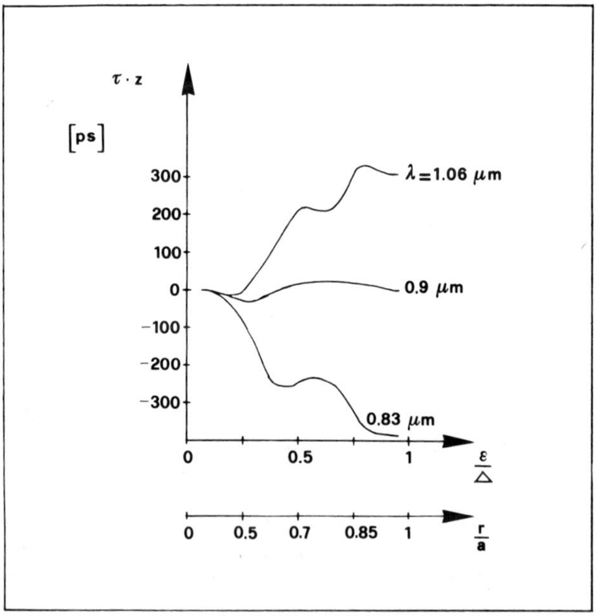

Fig. 10.15 - DMD measurements on the same fiber at different wavelengths by input selection [from (18)].

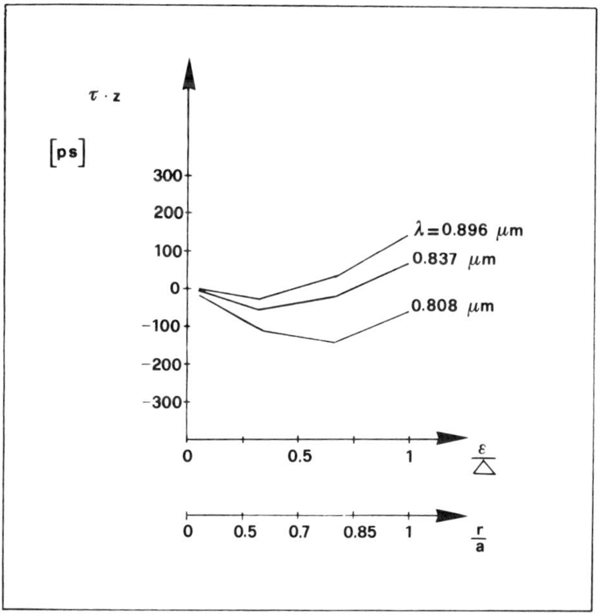

Fig. 10.16 - DMD measurements on the same fiber at different wavelengths, by output selection [from (33)].

10.16 some experimental results are given, respectively for an input [18] and for an output selection. [33] In both cases an α profile has been assumed and only the radial coordinate r has been selected. A different pulse modulated laser diode was used for each wavelength considered.

10.4.2 DMD Measurements on Step Index Fibers

Only the evaluation of the fiber baseband response is of interest for DMD measurements of step index fibers. The first description of a DMD measurement has been given in (50). The index profile was not a rigorous step index profile, smoothed near the core-cladding interface, as shown by measurements. An input selection of the propagation angle θ was performed, and a pulse modulated laser diode was employed. Fig. 10.17 reports the received pulses for different input angles with θ as a parameter. It is possible to observe that the groups of modes characterized by the higher values of θ exhibit a smaller peak power (due to the mode dependent attenuation coefficient reported in Fig. 9.18), and also a greater intrinsic time dispersion, probably due to mode coupling.

Measured relative group delays, obtained as the distribution of the peaks of these pulses, are reported in Fig. 10.18 in a continuous line. For comparison, also the theoretical predictions of the relative group delay distribution for $\alpha = \infty$, and $\alpha = 6$, over the same distance $z = 228$ m, are given. From this comparison, one can conclude that mode coupling is present.

A similar conclusion may be drawn also from the experimental results obtained in (46), where a DMD measurement on a step index fiber has been performed using an output selection and a sinusoidally modulated LED source. Both the propagation angle θ and the eccentricity ρ/a of the rays were selected by means of a scanning apparatus of the type reported in Fig. 5.1. The results of this DMD measurement are shown in Fig. 10.19. The dashed line refers to the theoretical prediction of relative group delay distribution for a step index fiber of the same length, $z = 130$ m. In particular, rays having smaller eccentricity exhibit greater mode coupling effects.

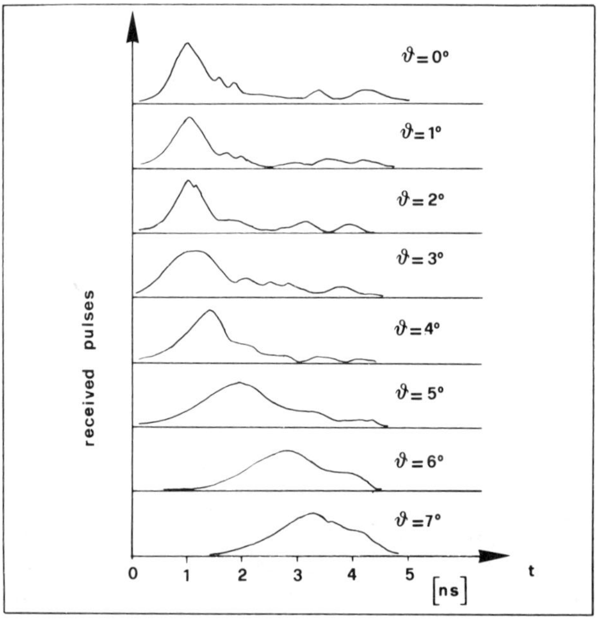

Fig. 10.17 - Received pulses with the input angle θ as a parameter [from (50)].

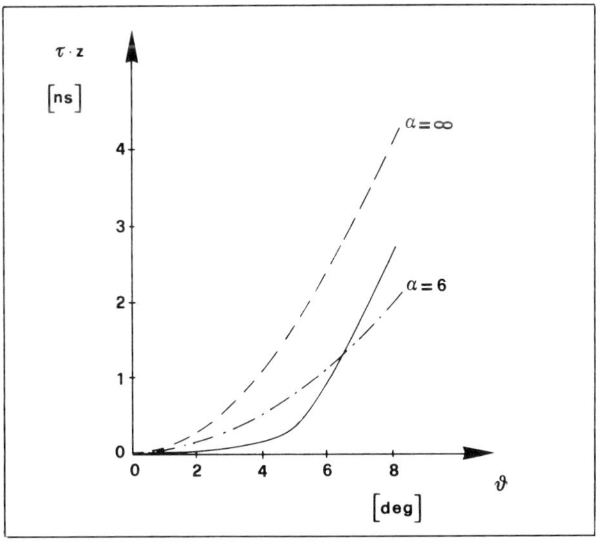

Fig. 10.18 - Relative group delay distribution obtained from the data of Fig. 10.17, and comparison with theoretical predictions [from (50)].

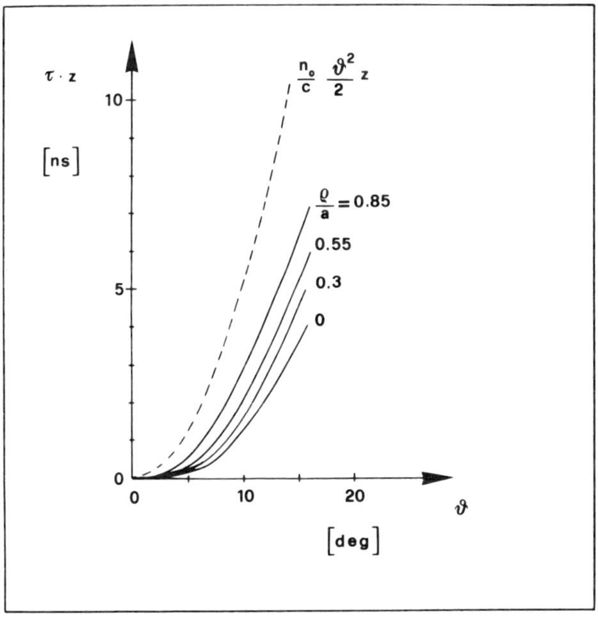

Fig. 10.19 - DMD measurement on a step index fiber considering also different values of the ray eccentricity [from (46)].

10.4.3 DMD Measurements on Graded Index Fibers

For measurements on graded index fibers a noticeable time resolution is required. Moreover, intramodal time dispersion is more than negligible in most cases. Sinusoidal modulation allows one to obtain a better time resolution and even a higher signal-to-noise ratio. Nevertheless, pulse modualtion also provides some information about intramodal time dispersion. Input selection is usually more suited for the evaluation of mode coupling effects, although an output selection leads to a better mode resolution and is of simpler implementation.

For example, in order to obtain information about the index profile, an input selection was performed in (48), using a pulse modulated signal. The estimated time resolution of about 100 ps leads, over a typical distance $z = 1$ km, to an uncertainty of a few per cent in the determination of the profile parameter α. In (18), using again an input selection and a pulse modulated signal to test profile dispersion, the estimated time resolution was reduced to 20 ps, owing to a more sophisticated instrumentation. A nonlinear profile dispersion was assumed, but the index profile shape was considered approximately to be of the α-type, so that α turns out to be a function of the wavelength λ. Because of the enhanced time resolution of the set-up, α was determined with an uncertainty smaller than 1 per cent.

In (51) an input selection and a sinusoidal modulation of the signal were employed to evaluate intermodal time dispersion in very low dispersion fibers. The modulation frequency f_o was 277 MHz, leading to a time resolution due to the vector voltmeter of about 1 ps. Also, considering the frequency stability of the signal generator, the overall estimated time resolution was about 5 ps.

An output selection is described in (33), where the measurement has been performed with a sinusoidally modulated signal. The signal to noise ratio has been enhanced by means of a simplified mode selection, based on the optical arrangement shown in Fig. 5.3. An output selection of both angle and radial coordinates has been shown to be possible in (47) as a consequence of the better SNR value assured by the sinusoidal modulation and by the use of a vector voltmeter.

10.4.4 Discussion of DMD Measurement Accuracy

The overall uncertainty of DMD measurements is determined by the noise superimposed on the useful signal, and by the spatial relative delays of an emission characterizing the optical source. The latter cause of uncertainty is reduced when an input selection is performed. However, a two point measurement is able to overcome this problem, also for an output selection, although it is characterized by an increase of the uncertainty due to the noise since a difference must be taken between two sets of results.

As for the signal to noise ratio SNR, the same observations given in 10.2.1 and in 10.3.2, respectively for pulse and sinusoidal modulations, can be repeated as well. The only difference is in the reduction of the useful signal power, represented by $< i_s >^2$, owing to the mode selection, which characterizes DMD measurements. In practice, it is rather difficult to test, distances greater than 1 km with a high accuracy.

A comparison between pulse and sinusoidal modulations yields the following results.[52] A sinusoidal modulation assures a higher SNR, and thus a reduction of the overall uncertainty. Moreover, it leads to a better time resolution. When $f_o > 300$ MHz, it is possible to reach time resolutions smaller than 5 ps. On the other hand, pulse modulation allows one to appreciate the combined effect of intramodal time dispersion and of mode coupling, as an intrinsic pulse broadening. However, a very high peak power must be launched into the fiber for an accurate evaluation of the shape of the received pulse, after the mode selection. Such a high level of optical power can be obtained for example by means of a collimated beam laser in mode locking. In this case the measureme it becomes rather sophisticated and expensive.

For DMD measurements, the fundamental cause of uncertainty in the experimental evaluation of mode delay distributions is due to mode coupling. In practice, the measured relative group delay distribution is due to both time dispersion and power exchange among the guided modes, as described by an equation of the type (2.49). In Chapter 11, we will return to DMD measurements devoted to investigating mode coupling effects specifically.

REFERENCES

1. Bjerkan, L., Stephansen, H., and Eriksrud, M. "Pulse Dispersion in Concatenated Fiber Optic Cables Spliced with Different Techniques," *J. Opt. Comm.*, 3, pp. 111-113 (1982).
2. Ikeda, M., and Kitayama, K. "Transfer Function of Long Spliced Graded Index Fibers with Mode Scramblers," *Appl. Opt.*, 17, pp. 63-67 (1978).

3. Wright, J.V., and Nelson, B.P. "Bandwidth Studies of Concatenated Multimode Fibre Links," Symposium on Opt. Fiber Measurements, Boulder, pp. 9-12 (1982).
4. Koyama, M., and Kobayashi, I. "Transmission Theory of Multimode Fibers Based upon a Perturbation Method," *Trans. IECE of Japan*, J60-B, pp. 395-402 (1977).
5. Gloge, D., Chinnock, E.L., and Lee, T.P. "Self-pulsing GaAs Laser for Fiber Dispersion Measurements," *IEEE J. Quant. Electron.*, QE-8, pp. 844-846 (1972).
6. Cohen, L.G. "Shuttle Pulse Measurement of Pulse Spreading in an Optical Fiber," *Appl. Opt.*, 14, pp. 1351-1356 (1975).
7. Tanifuji, T., and Ikeda, M. "Pulse Circulation Measurement of Transmission Characteristics in Long Optical Fibers," *Appl. Opt.*, 16, pp. 2175-2179 (1977).
8. Cohen, L.G., *et al.* "Transmission Properties of Low-Loss Near-Parabolic Index Fiber," *Appl. Phys. Lett.*, 26, pp. 472-474 (1975).
9. Gloge, D., Tynes, A.R., Duguay, N.A., and Hansen, J.W. "Picosecond Pulse Distortion in Optical Fibers," *IEEE J. Quant. Electron.*, QE-8, pp. 217-221 (1972).
10. Gold, B., and Rader, C.M. *Digital Processing and Signals*, New York: McGraw-Hill Book Co. (1969).
11. Personick, S.D., Hubbard, W.M. and Holden, W.S. "Measurements of the Baseband Frequency Response of a 1 km Fiber," *Appl. Opt.*, 13, pp. 266-268 (1974).
12. Jeunhomme, L., Cozannet, A., Boullie, R., and Hazan, J.P. "Measurements Methods of Optical Conductors Transmission Characteristics," 2nd E.C.O.C., Paris, pp. 123-134 (1976).
13. Nicolaisen, E., and Ramskow Hansen, J.J. "A Novel Technique for Investigating Dispersion Characteristics of Fiber Waveguides," 4th E.C.O.C., Geneva, pp. 171-178 (1978).
14. Okoshi, T., Change, J.C., Saito, S. "Measuring the Complex Frequency Response of Multimode Optical Fibers," *Appl. Opt.*, 20, pp. 1414-1417 (1981).
15. Cohen, L.G., Astle, H.W., and Kaminow, I.P. "Wavelength Dependance of Frequency-Response Measurements is Multimode Optical Fibers," *Bell Syst. Tech. J.*, 55, pp. 1509-1523 (1976).

16. Gloge, D., Chinnock, E.L., and Ring, D.H. "Direct Measurement of the (Baseband) Frequency Response of Multimode Fibers," *Appl. Opt.*, 11, pp. 1534-1538 (1972).

17. Machida, S., Imoto, N., and Ohmori, I. "Multimode Fiber Baseband Frequency Response Measurement with Single Frequency Extracted from Modulated InGaAsP Laser," *Electron. Lett.*, 15, pp. 607-609 (1979).

18. Jeunhomme, L., Pocholle, J.P., and Raffy, J. "Wavelength Dependence of Modal Dispersion in Graded Index Optical Fibres," *Electron. Lett.*, 14, pp. 364-366 (1978).

19. Arnaud, J.A., and Fleming, J.W. "Pulse Broadening in Multimode Optical Fibres with Large $\Delta n/n$: Numerical Results," *Electron. Lett.*, 12, pp. 167-169 (1976).

20. Fleming, J.W. "Material Dispersion in Lightguide Glasses," *Electron. Lett.*, 14, pp. 326-328 (1978).

21. Fleming, J.W., "Material and Mode Dispersion in GeO_2 -B_2O_3 -SiO_2 Glasses," *J. Am. Ceram. Soc.*, 59, pp. 503-507 (1976).

22. Gloge, D., and Chinnock, E.I. "Fiber-Dispersion Measurements Using a Mode-Locked Krypton Laser," *IEEE J. Quant. Electron.*, QE-8, pp. 852-854 (1972).

23. Miyashita, T., Horiguchi, M., and Kawana, A. "Wavelength Dispersion in a Single Mode Fibre," *Electron. Lett.*, 13, pp. 227-228 (1977).

24. Payne, D.N., and Hartog, A.H. "Determination of the Wavelength of Zero Material Dispersion in Optical Fibres by Pulse-delay Measurements," *Electronics Lett.*, 13, pp. 627-629 (1972).

25. Luther-Davis, B., Payne, D.N., and Gambling, W.A. "Evaluation of Material Dispersion in Low Loss Phosphosilicate Core Optical Fibres," *Opt. Comm.*, 13, pp. 84-89 (1975).

26. Lin, C., Cohen, L.G., French W.G., and Presby, H.M. "Measuring Dispersion in Single Mode Fibers in the 1.1-11.3 μm Spectral Region — A Pulse Synchronization Technique," 5th E.C.O.C., Amsterdam, 14.3 (1979).

27. Ozeki, T., and Watanabe, A. "Measurements of Wavelength Dependance of Group Delay in a Multimode Silica Fiber," *Appl. Phys. Lett.*, 22, pp. 382-383 (1976).

28. Costa, B., Mazzoni, D., Puleo, M., and Vezzoni, E. "Phase Shift Technique for the Measurement of Chromatic Dispersion in Optical Fibers using LEDs," *IEEE Trans. Mic. Th. & Tech.*, MTT-30, pp. 1497-1503 (1982).

29. Daikoku, K., and Sugimura, A. "Direct Measurement of Wavelength Dispersion in Optical Fibers: Difference Method," *Electron. Lett.*, 14, pp. 149-151 (1978).

30. Tanifuji, T., and Ikeda, M., "Simple Method for Measuring Material Dispersion in Optical Fibres," *Electron. Lett.*, 14, pp. 367-369 (1978).

31. Gloge, D., Kaminow, I.P., and Presby, H.M. "Profile Dispersion in Multimode Fibres: Measurement and Analysis," *Electron. Lett.*, 11, pp. 469-471 (1975).

32. Sladen, F.M.E., Payne, D.N., and Adams, M.J. "Measurement of Profile Dispersion in Optical Fibres: a Direct Technique," *Electron. Lett.*, 13, pp. 212-213 (1977).

33. Costa, B., Esposto, F., and Sordo, B. "Wavelength Dependence of Differential Group Delay in Grade Index Optical Fibers: Application to Fiber Links Characterization," *Opt. Fiber Comm.*, Washington, pp. 122-125 (1979).

34. Klein, K.F., Heinlein, W.E., and Witte, K.H. "Excitation Dependent Material Dispersion in Graded Index Fibers," *Electron. Lett.*, 18, pp. 100-102 (1982).

35. Coppa, G., Di Vita, P. "Determination of Index Dispersion in Optical Fibres from DMD Measurements," *CSELT Rapp. Tec.*, 8, pp. 175-176 (1980).

36. Garside, B.K., and Park, R.E. "Ultrashort Pulses from Semiconductor Diode Lasers," *Opt. & Laser Tech.*, 15, pp. 91-94 (1982).

37. Sicilia, M., Spano, P., Trioli, M. "Misura di impulsi ottici brevi," *Note Rec. Notizie*, 31, pp. 21-26 (1982).

38. Bassi, P., and Zoboli, M. "Increasing the Sensitivity of Time-domain Techniques for Measuring Optical Fiber Transfer Function," *Appl. Opt.*, 20, pp. 3-4 (1981).

39. Taub, H., and Schilling, D.L. *Principles of Communication Systems*, Tokyo: McGraw-Hill Book Co. (1971).

40. Bassi, P., and Zoboli, M. "Statistics of Optical Fiber Transfer-Function Measurements: A Computer Aided Approach," SPACECAD 79, Bologna, pp. 339-344 (1979).

41. Bassi, P., and Zoboli, M. "Experimental Noise Effects on Transfer Function-Phase Evaluation via Hilbert Transforms: Application to Optical Fibres," *Circ. Th. & Appl.*, 9, pp. 59-65 (1981).

42. Guillemin, E.A. *Theory of Linear Physical Systems*, New York: J. Wiley & Sons (1963).

43. Sandberg, I.W., Kaminow, I.P., Cohen, L.G., and Mammel, W.L. "On the Phase of the Modulation Transfer Function of a Multimode Optical Fiber Guide," *Bell Syst. Tech. J.*, 57, pp. 99-109 (1978).

44. Sandberg I.W. "A Note Concerning Optical Waveguide Modulation Transfer Functions," *Bell Syst. Tech. J.*, 57, pp. 3047-3056 (1978).

45. Piccari, L., and Spano, P. "New Method for Measuring Ultra-wide Frequency Response of Optical Detectors," *Electron. Lett.*, 18, pp. 116-118 (1982).

46. Cancellieri, G. "Time Dispersion Measurement in Optical Fibre by Near and Far Field Scanning Technique," *Electron. Lett.*, 14, pp. 465-467 (1978).

47. Cancellieri, G., Mezzetti, M., Someda, C.G., and Zoboli, M. "Simplified Procedure for Indirect Tests of Optical Fibre Transfer Functions," *Electron. Lett.*, 15, pp. 234-236 (1979).

48. Olshansky, R., and Oaks, S.M. "Differential Mode Delay Measurement," 4th E.C.O.C., Geneva, pp. 128-132 (1978).

49. Petermann, R. "Simple Relationship Between Differential Mode Delay in Optical Fibres and the Deviation from Optimum Profile," *Electron. Lett.*, 14, pp. 793-794 (1978).

50. Keck, D.B. "Spatial and Temporal Power Transfer Measurements on a Low-Loss Optical Waveguide," *Appl. Opt.*, 13, pp. 1882-1888 (1974).

51. Jeunhomme, L., and Lamouler, P. "Intermodal Dispersion Measurements and Interpretation in Graded Index Optical Fibres," *Opt. & Quant. Electron.*, 12, pp. 57-64 (1980).

52. Costa, B., Morra P., and Sordo, B. "Factors Affecting Accuracy and Interpretation of Differential Mode Delay Measurements," 7th E.C.O.C., Copenhagen, 5.5 (1981).

CHAPTER 11

INDIRECT TESTS ON MODE COUPLING EFFECTS

Scanning techniques, as introduced in 5.1.5, may be efficiently employed for indirect tests on mode coupling effects, which, depending on the quantities to be measured, can be divided into two classes:

i) tests based on amplitude mode power distributions

ii) tests based on mode power distributions, in amplitude and delay

Tests belonging to the former class are simpler, but give only approximate evaluations. Moreover, they have been employed only for step index fibers. The other ones can give a large amount of information, especially when the index profile is accurately known.

Either input or output selection can be performed, considering a single or a pair of mode variables, therefore assuming a one-dimension or a two-dimension model, respectively, for the description of mode coupling. Especially over distance in which the steady-state condition has not been reached yet, rays having different values of the constant of the motion h may show different behaviors.[1] Then the use of a pair of mode variables appears justified. The use of an optical power flow equation, of the type introduced in 2.2, is not always convenient. Sometimes a set of single transition equations is preferable.

11.1 TESTS BASED ON AMPLITUDE MODE POWER DISTRIBUTIONS

We report here two examples of tests on mode coupling effects, performed on step index fibers, by the measurement of the amplitude mode power distribution. The first one is based on an input selection and also on an output selection. The other test is founded only on an output selection, provided that the steady state has been reached. Both these techniques take into account an optical power flow equation, in which the propagation angle θ is the only independent mode variable. The round symmetry of the fiber is considered. In other words, an equation of the type discussed in 2.2.7 is assumed, but considering a Bessel-type operator, instead of a harmonic one.

This equation can be written as

$$\frac{\partial W}{\partial z} = - \gamma(\theta)\, W + \frac{(\Delta\theta)^2}{\theta} \frac{\partial}{\partial\theta} \left[C(\theta)\, \theta\, \frac{\partial W}{\partial\theta} \right] \tag{11.1}$$

where $W(\theta, z)$ is the optical power distribution, $\gamma(\theta)$ the intrinsic attenuation coefficient, $\Delta\theta$ the angular separation between adjacent modes, and $C(\theta)$ a coupling coefficient. Regarding the overall accuracy of these methods, recall that they are indirect tests, based on the assumption that equation (11.1) is satisfied. All the considerations about differential mode attenuation measurements, reported in 9.4.3 may be repeated here, for what concerns the intrinsic accuracy of the measurement of the optical power distribution.

11.1.1 Measurements of Amplitude Mode Power Distributions under a Selective Excitation

When a step index fiber is excited by a collimated laser beam, under a propagation angle θ, the expected optical power distribution is an annular ring in the far field pattern at the fiber output. The width of this annular ring, compared with that of the input beam, can give some information about mode coupling effects, [2,3] as shown in Fig. 11.1. Letting θ_o equal the propagation angle which has been selected at the fiber input, the parameters which can be measured on the output far field pattern are the mean value θ', and the rms width $\Delta\theta'$ of the annular ring. The input beam divergence, which is about 0.3 degrees for typical He-Ne lasers operating with the TEM_{oo} mode, can be neglected with respect to $\Delta\theta'$. Reducing θ_o, the output annular ring tends to degenerate into a disk.

For a quantitative evaluation of mode coupling from experimental data, equation (11.1) has been considered in (3), assuming

$$\gamma(\theta) = b\,\theta^2$$

where the intrinsic attenuation coefficient common to all the rays has been ignored, since the measurement is performed on a relative mode power distribution only. Moreover, for the sake of simplicity, the coupling coefficient has been assumed to be independent of θ, letting

$$(\Delta\theta)^2\, C\,(\theta) = D$$

Under the above assumptions, D can be determined from the measurement, independently of the knowledge of constant b. In fact, it can be shown that, for distances z shorter than that required for the reaching of the steady state, the input angle θ_{ox}, at which the output ring degenerates into a disk, [3] turns out to be

$$\log \theta_{ox} = \frac{1}{2} \log z + \log (2\sqrt{D})$$

Therefore, to evaluate D it is sufficient to know z and θ_{ox}. The latter one is available from repeated measurements of the output far field patterns under selective excitation.

The accuracy of the present method may be strongly affected by an uncertainty on the actual value of the angle θ_{ox}, caused by a poor end fiber preparation. A second measurement of the output far field pattern, made at a fiber section very close to the input end, cutting the fiber itself after the first measurement, can greatly reduce this uncertainty on θ_{ox}.

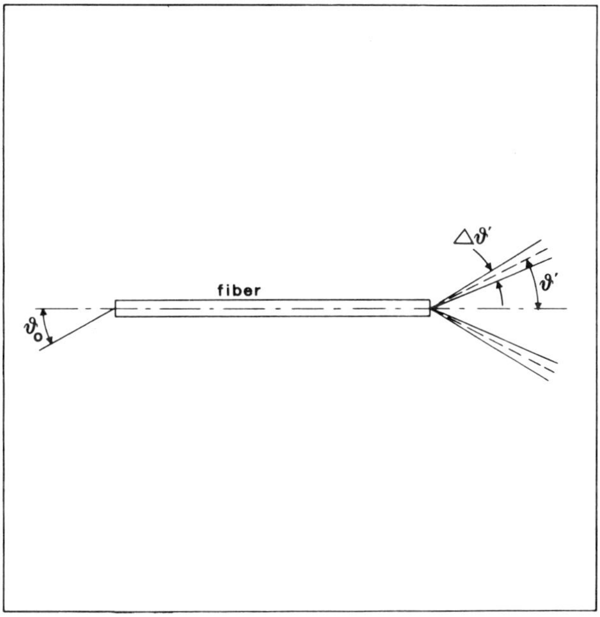

Fig. 11.1 - Measurement based on the evaluation of the amplitude mode power distribution under a selective excitation.

11.1.2 Measurement of Amplitude Mode Power Distributions at the Steady State

Equation (11.1), considering γ as independent of θ, for simplicity can be rewritten in the following form

$$\frac{\partial^2 W}{\partial \theta^2} + \left[\frac{1}{\theta} + \frac{1}{C(\theta)}\frac{dC}{d\theta}\right]\frac{\partial W}{\partial \theta} - \frac{1}{(\Delta\theta)^2 C(\theta)}\left[\gamma W + \frac{\partial W}{\partial z}\right] = 0 \tag{11.2}$$

The steady state optical power distribution is

$$W(\theta,z) = Q_1(\theta)\exp(-\lambda_1 z)$$

where symbols Q_1 and λ_1 have been introduced in 2.2.7. Substitution into (11.2) yields the coupling coefficient, [4] which results

$$C(\theta) = -\frac{\lambda_1 - \gamma}{(\Delta\theta)^2}\,\frac{\int_0^\infty Q_1(\zeta)\,\zeta d\zeta}{\theta\,\dfrac{dQ_1}{d\theta}}$$

From the measurement of the steady state optical power distribution $Q_1(\theta)$, $C(\theta)$ is determined, except for a proportionality factor, depending on the steady-state attenuation coefficient λ_1. In (4), the reaching of the steady state was tested by repeated measurements of the output far field pattern, under different launching conditions. The optical source was a He-Ne laser, whose beam was slightly focused. Varying the input angle of incidence from 0 to 12 degrees, the output far field pattern did not change appreciably. The behavior of $C(\theta)$ obtained from this indirect test is in good agreement with Marcuse's theory of mode coupling, briefly presented in 2.2.2. From this kind of test, it should be possible to obtain even an estimate of the correlation length and of the amplitude of the fiber perturbation.

Nevertheless, the uncertainty on the assumption of a negligible mode dependent part of the intrinsic attenuation coefficient, and on the effective reaching of the steady state, may limit the overall accuracy of this indirect test. In particular, an output far field pattern, practically independent of the launching condition, could be obtained before the reaching of the steady state, especially when the modes just near the guidance condition are strongly attenuated.

However, the test described here gives an approximate evaluation of the mode dependent part of the coupling coefficient, which is often assumed to be negligible in other kinds of tests on mode coupling effects.

11.2 TESTS BASED ON AMPLITUDE AND DELAY MODE POWER DISTRIBUTIONS

We will describe here two kinds of tests on mode coupling effects, based on a time dependent model, which takes into account modulated signals. In the first method, a pulse modulated signal is measured by a sampling oscilloscope. In the other one, a sinusoidally modulated signal is measured by a vector voltmeter. In both cases, an output selection is made.

Since these measurements are based on the evaluation of the time dispersion of a single mode group, as a consequence of mode coupling, the main cause of uncertainty is due to the possible presence of material dispersion. Therefore, the use of a quasi-monochromatic optical source is needed. The considerations reported in 10.4.4 about the intrinsic accuracy of DMD measurements are applicable here too.

11.2.1 Theoretical Model for the Measurement of the Mode-Dependent Attenuation and the Coupling Coefficient

A time dependent set of coupled power equations can be written, adding a time dispersion term to (2.9), as

$$\frac{\partial W_i}{\partial z} + \tau_i \frac{\partial W_i}{\partial t} = -\gamma_i W_i + \sum_j h_{ij} (W_j - W_i) \tag{11.3}$$

where W_i is the optical power carried by the ith group of modes, τ_i and γ_i are, respectively, the group delay per unit distance and the attenuation coefficient of this group of modes, and $h_{ij} = h_{ji}$ are the coupling coefficients. Since this model is intended for differential mode measurements, we have taken here mode groups instead of single modes, owing to some limitations in the practical mode selection. However, in principle, equation (11.3) should hold only for groups of quasi-degenerate modes.

Letting

$$E_i(z) = \int_{-\infty}^{+\infty} W_i(z, t)\, dt \qquad\qquad A_i(z) = \int_{-\infty}^{+\infty} W_i(z, t)\, dt$$

we have[5]

$$\frac{dE_i}{dz} = -\gamma_i E_i + \sum_j h_{ij}(E_j - E_i)$$

$$\frac{dA_i}{dz} - \tau_i E_i = -\gamma_i A_i + \sum_j h_{ij}(A_j - A_i) \tag{11.4}$$

The mean transit time, over a distance z, of the ith group of modes is given by

$$T_i(z) = \frac{A_i(z)}{E_i(z)}$$

So, eliminating γ_i from (11.4), we obtain

$$\frac{dT_i}{dz} = \tau_i + \sum_j h_{ij} \frac{E_j}{E_i}(T_j - T_i) \tag{11.5}$$

which describes how T_i evolves along z. At the steady state, one can assume that dT_i/dz tends to a constant value, independent of i (called τ_∞ in 2.2.8), so that equation (11.5) is simplified.

Neglecting mode coupling between nonadjacent modes, the coefficients h_{ij} can be determined from the set of equations (11.5) and from the experimental measurements of T_i and E_i. The latter one represents the energy carried by the ith mode group. The group delays per unit distance τ_i of the uperturbed fiber, obtained from an index profile measurement, must be known with a noticeable accuracy. Finally, when the steady state is reached, the measurement of dT_i/dz is not required. On the contrary, when it is not reached, an approximate evaluation of this quantity can be obtained as

$$\frac{dT_i}{dz}(z) \cong \frac{T_i(z) - T_i(z_o)}{z - z_o}$$

where z_o is an intermediate distance. $T_i(z_o)$ can be measured cutting the fiber at z_o.

All the parameters in the first of (11.4), except γ_i, can be determined from the measurement, so that even γ_i can be derived from this set of equations. The requirement of an evaluation of the energies of mode groups imposes the use of pulse modulated signals. To enhance the signal to noise ratio, a boxcar integrator is necessary (see Appendix 2).[6]

In conclusion, the technique presented here, compared with the indirect tests described in 11.1, is able to give a greater amount of information, essentially because it employs data from both amplitude and delay measurements.

This technique can be modified as follows.[6] Considering an optical power flow equation, instead of a set of coupled power equations, we can assume the model

$$\frac{\partial W}{\partial z} + \tau(\theta)\frac{\partial W}{\partial t} = -\gamma(\theta)W + \frac{(\Delta\theta)^2}{\theta}\frac{\partial}{\partial\theta}\left[C(\theta)\theta\frac{\partial W}{\partial\theta}\right] \tag{11.6}$$

where, restricting ourselves to step index fibers for simplicity, we have added a time dispersion term to (11.1). Letting

$$M_j = \int_{-\infty}^{+\infty} t^j\, W(\theta, z, t,)\, dt, \quad j = 0, 1, 2, \ldots$$

equation (11.6) yields

$$\begin{aligned} \frac{\partial M_0}{\partial z} &= -\gamma(\theta)M_0 + \frac{(\Delta\theta)^2}{\theta}\frac{\partial}{\partial\theta}\left[C(\theta)\theta\frac{\partial M_0}{\partial\theta}\right] \\ \frac{\partial M_j}{\partial z} &+ j\tau(\theta)M_{j-1} \\ &= -\gamma(\theta)M_j + \frac{(\Delta\theta)^2}{\theta}\frac{\partial}{\partial\theta}\left[C(\theta)\theta\frac{\partial M_j}{\partial\theta}\right], \quad j = 1, 2, \ldots \end{aligned} \tag{11.7}$$

From these relationships, it results that

$$C(\theta) = \frac{1}{(\Delta\theta)^2\theta M_0^2\frac{\partial}{\partial\theta}\left(\frac{M_j}{M_0}\right)}\int_0^\theta M_0^2\left[\frac{\partial}{\partial z}\left(\frac{M_j}{M_0}\right) - j\tau(\zeta)\frac{M_{j-1}}{M_0}\right]\zeta\, d\zeta$$

For the experimental determination of $C(\theta)$, it is sufficient to measure M_0 and M_1, using the theoretical prediction of $\tau(\theta)$ derived from an index profile measurement, and approximating $\partial/\partial\theta\,(M_1/M_0)$ as

$$\frac{\partial}{\partial\theta}\left(\frac{M_1}{M_0}\right) \cong \frac{1}{\Delta\theta}\left[\frac{M_1(\theta+\Delta\theta, z)}{M_0(\theta+\Delta\theta, z)} - \frac{M_1(\theta, z)}{M_0(\theta, z)}\right]$$

Then $\gamma(\theta)$ can be obtained from the first equation of (11.7).

11.2.2 Measurement of Mode Dependent Attenuation and Coupling Coefficient

The measurement technique proposed in the previous section has been employed for step index fibers.[6] The experimental setup is schematically shown in Fig. 11.2. A GaAs laser diode and a Si APD were used. The signal was pulsed modulated, with 400 ps pulse-width. The mode selection was performed at the fiber output end. A boxcar integrator enables us to apply this technique even over long distances (of the order of 10 km), where an effective steady-state condition is reached.

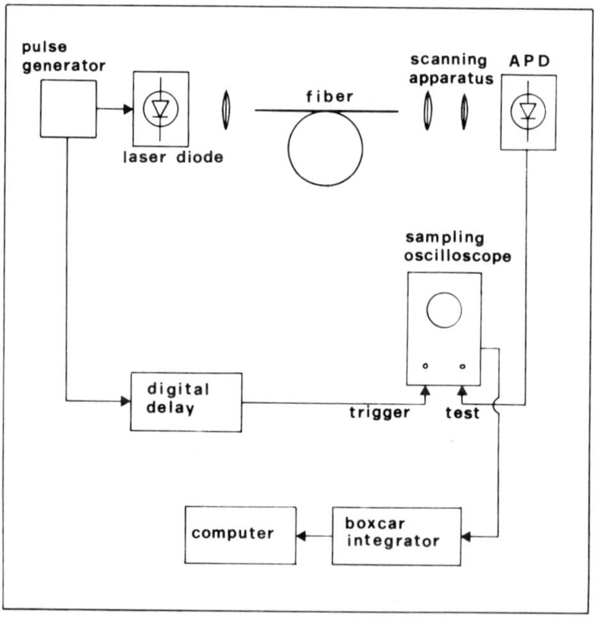

Fig. 11.2 - Experimental setup for the evaluation of mode-dependent attenuation and coupling coefficient (from [6]).

Two examples of the output selected pulse waveforms are reported in Fig. 11.3. The first refers to a 8.5 km-long fiber, the other one to a 0.5 km–long fiber of the same type. For both families of curves, which are normalized to have the same amplitude, the waveforms selected between $\theta = 0^\circ$ and $\theta = 4^\circ$ are practically superimposed. On the other hand, a very small spread in the mean transit times characterizes the first family of curves, whereas a large dispersion in the mean transit times, and even in the pulse shapes, characterizes the other one.

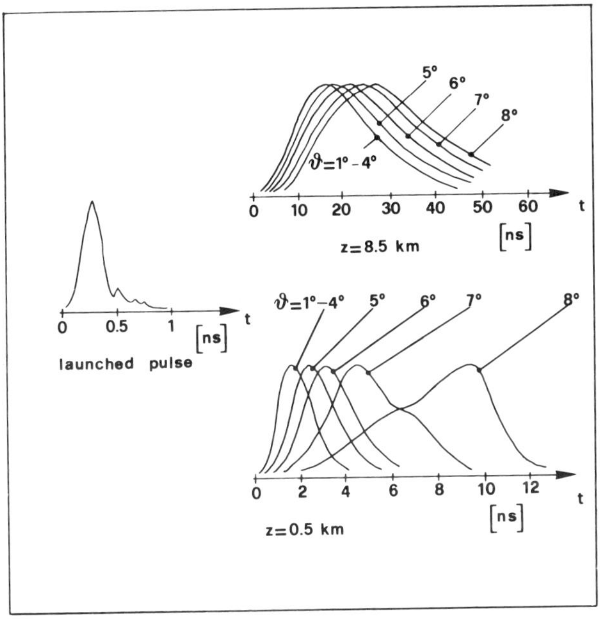

Fig. 11.3 - Output selected pulse waveforms, with the angle as a parameter (from [6]).

We can conclude that the equilibrium of the power exchange among the modes is reached for modes characterized by small values of θ close to the input fiber section. On the other hand, the modes having greater values of θ reach the equilibrium much faster. Moreover, for a long distance, all the curves are approximately gaussian, showing the validity of the theoretical model which describes the time dependent effects of mode coupling, discussed in 2.3.2.

From the experimental data obtained, assuming that mode coupling is imposed by microbending, a correlation length of the perturbation of about 2 mm and an amplitude of the order of 10^{-4} mm^{-1} could be assumed. However, such estimates must be considered with some care, owing to the uncertainties which affect the measurement of γ (θ) and $C(\theta)$. This technique is probably more accurate than those based simply on amplitude mode power distributions. Nevertheless it might lead to remarkable errors when the data obtained are used to extrapolate a given behavior to longer distances, or for a quantitative evaluation of the cause of mode coupling.

11.2.3 Theoretical Model for an Indirect Evaluation of the Fiber Baseband Response in the Presence of Mode Coupling

The indirect test presented here was initially proposed to estimate the fiber baseband response, using a differential mode delay measurement to extend its validity to the case of appreciable mode coupling. In 10.4.1, we have shown that DMD measurements, performed at a single modulation frequency f_o, can give all the information to calculate the fiber baseband response due to intermodal time dispersion, provided that mode coupling has negligible effects. In (10.16), neither material dispersion nor mode coupling effects are taken into account. In the presence of mode coupling, this equation should be modified as follows:[7] the phasor which represents the total optical power at the fiber output is assumed to be

$$\overline{P}(\omega, z) = \frac{1}{A} \exp(-i\,\omega\,\Delta\,t) \sum_n \sum_m K_{nm}\, q_m \exp(-i\,\omega\,\tau_m^g\, z) \tag{11.8}$$

where q_m takes into account the amplitude of the optical power distribution at the fiber input, τ_m^g is the relative delay distribution per unit distance calculated on the basis of the index profile, Δt is a possible shift between the measured relative delay distribution and the theoretical one, A is an attenuation factor, and K_{nm} is a transition coefficient between the mth and the nth selected mode groups. In other words, K_{nm} represents the fraction of optical power which, along the distance z, passes from the mth group to the nth group. Generally, $K_{nm} \neq K_{mn}$. For the evaluation of the baseband response, based on the expression

$$\overline{H}(\omega, z) = \frac{\overline{P}(\omega, z)}{P(0, z)}$$

the knowledge of A and Δt might appear to be necessary. However, they are unknowns of the model, as are the coefficients K_{nm}, and therefore must be determined.

Equation (11.8) can be seen as a summation of n contributions having the form

$$\overline{p}_n(\omega) = \frac{1}{A} \exp(-i\,\omega\,\Delta t) \sum_m K_{nm}\,\overline{q}^g_m(\omega)$$

$$\overline{q}_m^{\,g}(\omega) = q_m \exp(-i\,\omega\,\tau^g_m\,z) \tag{11.9}$$

The measurement is performed at the modulation circular frequency $\omega_o = 2\pi f_o$. Therefore, replacing ω with ω_o in (11.9), $P_n(\omega_o)$ is known as a result of the measurement. q_m can be determined by another scanning procedure, performed by cutting the fiber near its input section. If spatial relative delays of emission can be assumed negligible, the q_m are real quantities, and only a measurement of the amplitude mode power distribution is required near the fiber input section. If it is possible to neglect the optical power transferred to radiating modes, a conservation, [7] of the type

$$\sum_n K_{nm} = 1 \tag{11.10}$$

can be added.

Before discussing the determination from the experimental data of all the unknowns of the model, we will describe the asymptotic behavior of such a model of mode coupling. Asymptotically, we can assume that the measured relative delay distribution T_m is much smaller than the predicted one $\tau^g_m\,z$. Therefore, for the modulation frequencies of interest, the $\overline{p}_n(\omega)$ are simply real quantities. So, it can be written

$$\frac{1}{A} \exp(-i\,\omega\,\Delta t) \sum_m K_{nm}\,\overline{q}^g_m(\omega) = p_n(\omega)$$

$$\frac{1}{A} \exp(-i\,\omega\,\Delta t) \sum_m K_{lm}\,\overline{q}^g_m(\omega) = p_l(\omega)$$

Dividing these two relations, we have

$$\sum_m K_{nm}\,\overline{q}^g_m(\omega) = \frac{p_n(\omega)}{p_l(\omega)} \sum_m K_{lm}\,\overline{q}^g_m(\omega)$$

This equation is characterized by the particular solution

$$\frac{K_{nm}}{K_{lm}} = \frac{p_n(\omega)}{p_l(\omega)} \tag{11.11}$$

which can be considered a steady-state solution, because it is independent of the launching condition $\overline{q}^g_m(\omega)$. Substituting (11.11) into (11.10), we are left with

$$K_{lm} = \frac{P_l(\omega)}{P_t(\omega)} \quad , \quad P_t(\omega) = \sum_n p_n(\omega) \tag{11.12}$$

This asymptotic transition coefficient is independent of m, i.e., the lth mode group takes an equal amount of power from each mode group. Such an asymptotic situation is in perfect agreement with the classical definition of steady-state, which is characterized by an optical distribution which reproduces itself.

To have solutions independent of the choice of the number N of the mode groups, we must balance equations and unknowns. Since (11.9) represents N complex equations, considering (11.9) and the conservation condition (11.10), we have $3N$ scalar equations, with $N^2 + 2$ scalar unknowns (the coefficients K_{nm}, A and Δt). The N transition coefficiets $K_{n1}, K_{n2}, \ldots, K_{nN}$ can be considered as belonging to a gaussian function of the type

$$K_{nm} = B_n \exp - \left[\frac{(\mu_n - m \Delta u)^2}{2 \sigma_n^2} \right]$$

where B_n, μ_n and σ_n are free parameters, and finally Δu represents the difference of the mode variable u between two adjacent mode groups.[7]

This assumption means that the process of power transfer begins from adjacent mode groups, and then propagates to all other mode groups. In other words, at a short distance, the nth group takes its power mainly from the $(n$-1)th and the $(n$+1)th groups, whereas, for higher distances, the presence of μ_n as an unknown allows one to take into account also the existence of a preferential direction in the power flow, e.g., towards the high or the low values of u. It is however a resonable assumption that the group of innermost modes, denoted by $n = 1$, takes most of its power from the group $m = 1$. So $\mu_1 = 0$. In conclusion, the unknowns reduce to $3N$+1. Then Δt is considered a free parameter. Solving the whole set of equations, Δt may vary within a rather small interval, leading to an uncertainty in the determination of K_{nm} and A, which is practically negligible.[7]

When a strong mode coupling affects the fiber, the description of the power exchange through one summation only, could appear somehow inexact. So, a model based on several summations can be assumed. Considering the situation shown in Fig. 11.4, where the coupling condition is reported along the distance z from the fiber input section, the dashed line represents the actual unknown evolution of the coupling condition, which, as discussed in 2.2, is a continuous process. Letting L equal the whole fiber length, the one step solution described here is based on the theoretical prediction from point 0 to point B_l, and to one summation from B_l to A_l. Note that the known initial and final conditions are represented by the point 0 and by the point A_l (or A_o). In this sense, a zero step solution can be considered (from 0 to A_o) as the method for determining the fiber baseband response, neglecting mode coupling, through the only summation (10.16). Finally, to obtain a better correspondence between the model and the actual process, a two step solution can be developed, which, fixing an intermediate point A_2, follows the path $OB_2A_2B_3A_l$.

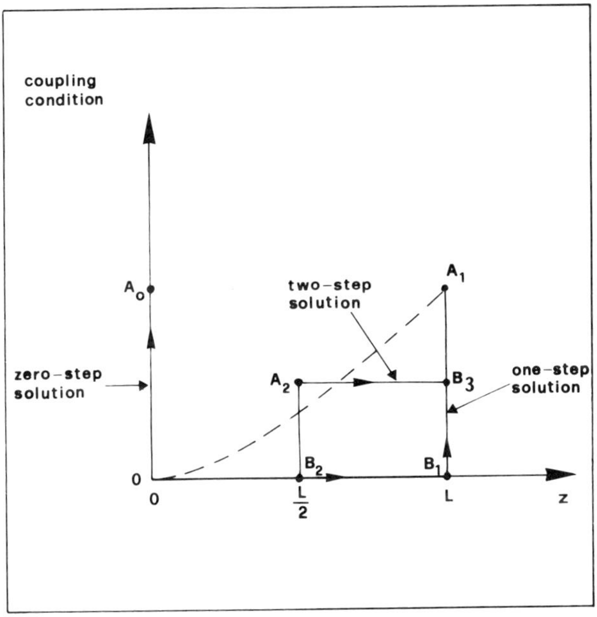

Fig. 11.4 - Evolution of the coupling condition along z and models based on different numbers of summations (from [7]).

It consists of an intermediate and a final summation. In Fig. 11.4, A_2 has been chosen in correspondence with $z = L/2$, and is characterized by half of the final amount of mode coupling. The comparison between the measured T_m and the predicted $\tau_m^g z$ gives an estimated amount of mode coupling present. From experimental data,[7] a one step solution can be considered well approximated provided that T_m is greater than about 1/5 $\tau_m^g z$.

11.2.4 Measurement of the Fiber Baseband Response through an Indirect Test, in the Presence of Mode Coupling

The method presented in the previous section has been applied with two different summation techniques:

i) performing the summation on the basis of geometrical properties only[7]

ii) taking both geometrical and guidance properties into account,[8] as suggested in 10.4.1

In the latter case, domain (r_l, θ_l) has been assumed, leading to a remarkable simplification in the scanning procedure, as pointed out in 5.1.6.

In (7) a step index fiber whose relative delay distribution, obtained by a DMD measurement, is reported in Fig. 10.19, was analyzed. Four families of rays have been considered, depending on the values of the eccentricity ρ/a, and five groups of rays ($n = 0,1 \ldots 4$), denoted by the mode variable $u = \theta$, have been assumed for each family. It has been supposed that negligible power transitions occur between rays with different eccentricities, in agreement with the approximate models of mode coupling discussed in 2.2. In Fig. 11.5 some of the results obtained are reported. They refer to $\rho/a = 0$ and $\rho/a = 0.85$. Three gaussian curves of K_{nm} are shown, for n=0,2,4. The vertical dashed lines denote the value $n\Delta\theta$, which should coincide with μ_n, if no preferential direction in the power flow is present. Finally, the two dashed curves of relative delays represents the maximum uncertainty in the position of the predicted values of $\tau_m^g z$, i.e., on Δt.

The computation of the fiber baseband response through a zero step solution gives about 98 MHz and 400 MHz, respectively, for the 3 dB and 10 dB bandwidth. The computation through one step solution gives about 85 MHz and 300 MHz, respectively, showing that the model based on the assumption that mode coupling has negligible effects is a better approximation at low modulation frequencies.

As for the behavior of the transfer coefficients reported in Fig. 11.5, we can observe that there are no significant differences in the amplitude of the gaussian curves. At higher θ, wider curves characterize $\rho/a = 0$ with respect to $\rho/a = 0.85$, meaning that meridional rays exchange a greater amount of power. This could be understood in a qualitative way also from a simple observation of the measured relative delay distribution (Fig. 10.19).

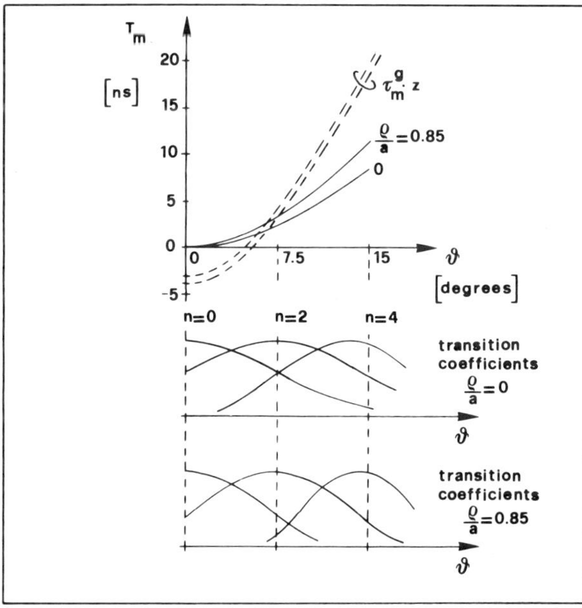

Fig. 11.5 - Relative delay distributions measured and predicted (dashed lines), and behaviors of some transition coefficients in a step index fiber (from [7]).

In (8) a graded index fiber has been tested. Also in this case, it has been shown that rays having a smaller r_l suffer a greater mode coupling. In Fig. 11.6 a comparison is reported between measured and predicted curves of the relative delay distributions. The dashed lines refer to helical rays. In the computation of the group delays per unit distance τ_m^g, the effect of material dispersion for quasi-monochromatic sources, i.e., the quantity $\lambda/n(dn/d\lambda)$, has also been considered.

For the measurement on step index fibers described in (7), a LED source was employed assuming, however, that intermodal time dispersion is dominant. But for the measurement on graded index fibers reported in (8), a laser diode was required. In both cases, a sinusoidally modulated signal with f_o = 5.55 MHz has been used. As for the output mode selection, it has been performed by means of a double scanning appararus, like that shown in Fig. 5.1.

In conclusion, we can state that these indirect tests on mode coupling effects, though rather simple from the experimental point of view, are able to give a large amount of information about the evolution of the optical power distribution along the fiber length, especially for what concerns the evaluation of the fiber baseband response. Yet, since the transfer coefficients K_{nm} here used are not related to any type of coupling coefficient present in the classical optical power flow equations, it is not possible to use these experimental results to predict steady-state excess loss or baseband response. In other words, an extrapolation over longer distances appears difficult. Moreover, some restrictive assumptions, e.g., the conservation condition (11.10) or the absence of mode coupling between modes having different eccentricities, contribute to making these techniques suitable only when a weak mode coupling is present.

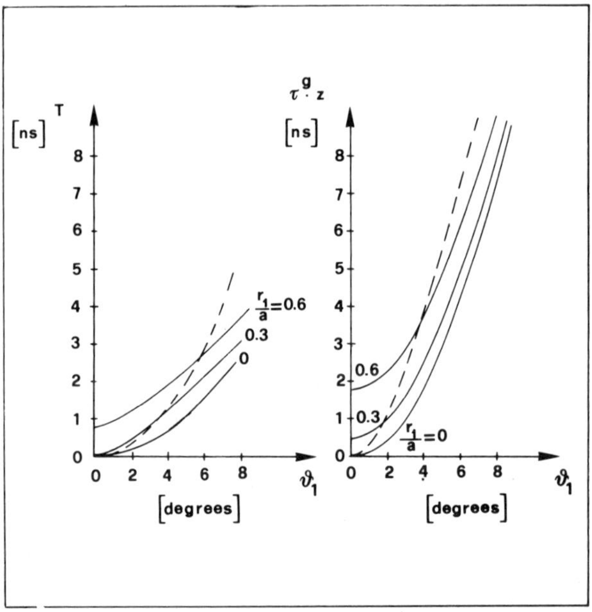

Fig. 11.6 - Measured (T) and predicted ($\tau^{g} \cdot z$) curves of relative delay distribution in a graded index fiber (from [8]).

REFERENCES

1. Cancellieri, G. "Time Dispersion Measurement in Optical Fibre by Near and Far Field Scanning Technique," *Electron. Lett.*, 14, pp. 465-467 (1978).
2. Jeunhomme, L., and Pocholle, J.P. "Experimental Determination of the Radiation Pattern of Optical Fibres," *Opt. Comm.*, 12, pp. 89-92 (1974).
3. Gambling, W.A., Payne, D.N., and Matsumura, H. "Mode Conversion Coefficients in Optical Fibers," *Appl. Opt.*, 14, pp. 1538-1542 (1975).
4. Jeunhomme, L., and Pocholle, J.P. "Angular dependence of the Mode-coupling Coefficient in a multimode optical fiber," *Electron. Lett.*, 11, pp. 425-426 (1975).
5. Kawakami, S., "New Principle to Measure Loss and Mode-conversion Parameters of Multimode Fibres," *Electron. Lett.*, 13, pp. 706-707 (1977).
6. Kitayama, K., and Ikeda, M. "Mode Coupling Coefficient Measurements in Optical Fibres," *Appl. Opt.*, 17, pp. 3979-3983 (1978).
7. Cancellieri, G., and Mezzetti, M. "Scanning Technique for Investigating Optical Fibre Dispersion," *Opt. & Quant. Electron.*, 11, pp. 305-317 (1979).
8. Cancellieri, G., Fantini, P., and Mezzetti, M. "Misura degli effetti dell'accoppiamento tra i modi in fibre ottiche a profilo d'indice graduale," 3.a Riun. Naz. Elettrom. Appl., Bari 1980, pp. 227-279.

CHAPTER 12
MEASUREMENTS OF MONOMODE FIBERS

In this chapter we discuss some measurement techniques which are particularly important for monomode fibers. These kinds of fibers, offer very large bandwidths and are well suited for fabrication with ultra-low attenuation. They are not yet widely used, but it is not unwise to forecast the employment of monomode rather than multimode fibers in long haul transmission systems, such as transoceanic links.

Real monomode fibers suffer from birefringence, which was not in the brief theoretical description of ideal monomode fibers reported in Chapter 1. This effect, which acts on light polarization, owing to elliptical deformation of the fiber core, is very sensitive to bends, temperature, and pressure. It is probably responsible for the main limitations of real monomode fibers, since it introduces polarization noise and bandwidth limitations.

The measurement tests on monomode fibers can be divided into the following two classes:*

i) index-profile and geometrical measurements

ii) transmission measurements

Some of the measurements of the former class are peculiar to monomode fibers, e.g., the cutoff wavelength measurement, the fundamental mode spot size measurement, and the determination of an equivalent step index profile. Transmission measurements are simple extensions of the corresponding procedures already described for multimode fibers, but requiring greater care, for the particular propagation which characterizes monomode fibers. An exhaustive treatise of monomode fibers is well beyond our purposes, and the aim of the present chapter is to provide some essential information useful to technicians who are not likely to deal specifically with monomode fibers in the future.

*See notes pg. 498

12.1 INDEX PROFILE AND GEOMETRICAL MEASUREMENTS ON MONOMODE FIBERS

The most precise way for characterizing monomode fibers consists of measuring their index profile *n(r)*. Since this is usually a very accurate and expensive test procedure, it is possible to determine, alternatively, an equivalent step index profile.[1] The cutoff wavelength of the second order mode guided by the fiber is measure in order to predict the modal nature of the light propagation in the test fiber. For fibers having a rigorous step index profile, this cut off wavelength can be indirectly obtained from a measurement of core diameter and numerical aperture. But since a rigorous step index profile is rather difficult to obtain, a direct measurement is recommended.

The near field pattern of the fundamental mode has a gaussian distribution. The knowledge of the spot size which characterizes this optical power distribution is very important for joint loss prediction. Also, the measurements of the core ellipticity and of the numerical aperture are very useful for this purpose. These measuremets, derived from those on multimode fibers, require an extremely high accuracy, due to the modal nature of the light propagation which must be carefully considered here.

12.1.1 Index Profiling in Monomode Fibers

In 7.1.8, we have stressed that only some of the several techniques proposed for multimode fiber profiling can be applied to monomode fibers too. In practice, all the methods based on the optical guided power measurement, e.g., the near field and the far field scanning procedures, suffer very strong limitations either in the accuracy of the index determination or in the spatial resolution. Alternatively, a modified near field scanning technique has been proposed.[2] As already discussed in 8.1.2, where this method has been suggested for inspection of geometrical parameters, it exploits the presence of modes guided by the cladding outer surface to obtain information on both core and cladding refractive index distributions. However, the results obtained by this technique show a good accuracy only when a high quality imaging optics is employed. Moreover, the fiber output end face must be accurately prepared, in the core and in the cladding regions.

A further modification of this technique has been recently proposed[3] to overcome the above limitations. In this method, a small divergence optical source is used, instead of a uniform launching condition. A diffuser having a fixed transverse diameter is inserted between a lamp and the fiber input end. By varying the distance L between the diffuser and this fiber input end, it is possible to obtain different small aperture launching conditions.

An example of near field power distributions obtained on a monomode fiber using this technique, through an output scanning procedure for different values of L, is reported in Fig. 12.1. When $L = 0$, we have an approximately uniform launching condition, which entails the presence of leaky rays. This presence can be avoided for $L > 1.5$ cm. Nevertheless, further increasing the distance L may reduce the accuracy in the refractive index determination. In this sense, the value $L = 3.5$ cm appears to be the optimum tradeoff for this monomode fiber. [3] The accuracy in the evaluation of $n(r)$ by this technique has been estimated to be better than one part per 10^4. On the other hand, the spatial resolution is not excellent.

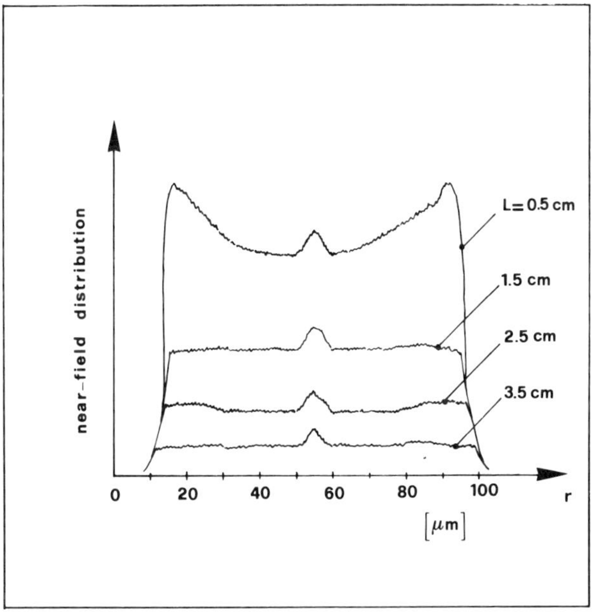

Fig. 12.1 - Measured near field power distributions on a monomode fiber [from (3)].

A correction must be made for a short fall in the launched guided modes. This correction is necessary, since, with a small divergence optical source, the envelope of the measured near field pattern is deformed by the presence of an undesired parabolic term in r/b, where b denotes the cladding outer radius. The only possible correction to be applied is that based on the assumption of a uniform value of the refractive index in the cladding region. Therefore, possible imperfections in the cladding refractive index distribution cannot be appreciated. In conclusion, this method appears simple and can be of great interest to an industrial environment, yet it does not offer the precision required for improving the technological processes of fiber fabrication. In this field, refracted near field and slice interferometry appear to be the preferred methods.

12.1.2 Cutoff Wavelength Measurement

The normalized frequency V, defined in 1.1.3, gives the optical frequency range for monomode operation. Letting λ_c equal the cutoff wavelength of the second order mode, it is possible to define a corresponding cutoff normalized frequency as

$$V_c = \frac{2\pi a}{\lambda_c} \sqrt{n_1^2 - n_2^2} \tag{12.1}$$

a being the core radius, and $\sqrt{n_1^2 - n_2^2}$ the numerical aperture NA of the fiber, considered for simplicity as characterized by a rigorous step index profile. The wavelength below which the second order mode can propagate is represented by λ_c, whereas at any wavelength λ the fundamental mode propagates. Therefore, when a fiber is used with a normalized frequency $V > V_c$, it is not a monomode one. Yet, the value of V must not be too low, since the influence of bends and microbends would cause heavy losses, and coupling from the optical source becomes easier with increasing V. In conclusion, the fiber should be designed for operation close to cutoff.

The near field optical power distribution offers an easy way to determine experimentally the cutoff wavelength λ_c. In fact, the fundamental mode exhibits an intensity distribution with a maximum along the axis, and round symmetry. When the wavelength is decreased beyond cutoff, it can be seen that a dip appears at the center of this intensity distribution, as shown in Fig. 12.2. This central dip is due to the superimposition of the second order mode optical power distribution, which is zero at the center.

The experimental setup is then based on a lamp, followed by a monochromator, and on an output scanning arrangement, to perform the near field investigation. The fiber length should be on the order of several centimeters. In fact, it has been shown [4] that values of the cutoff wavelength considerably shorter than the actual ones occur with increasing the length of the fiber under test because of losses induced by bends.

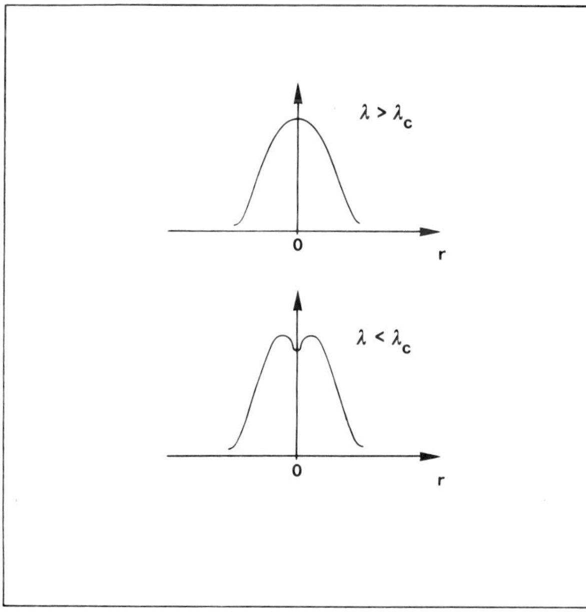

Fig. 12.2 - Mode intensity distributions at the output of a monomode fiber.

For ideal step index monomode fibers, as obtained in 1.1.3, always $V_c = 2.405$. In this case, with the knowledge of NA and a, it is possible to obtain λ_c from (12.1) without any further measurement. Alternatively, a direct measurement of λ_c, performed by the above technique, can give useful information about the product $a \cdot NA$. For graded index fibers, equation (12.1) holds provided that for n_1 we consider the maximum refractive index in the core region, but the value of V_c is a function of the profile distribution. For instance, considering an α profile, it is a function of parameter $\propto$. In particular, for near-parabolic profiles, $V_c \cong 3.5$. In this case, since it is not possible to determine the profile distribution with an accuracy good enough to obtain V_c theoretically, it is always necessary to measure λ_c directly.

Other techniques have been proposed for the experimental evaluation of λ_c. One of the earliest proposed methods was founded on the observation that the second order mode just above cutoff is only weakly guided, and a fiber bend causes a remarkable loss in the guided power due to radiation. When $\lambda > \lambda_c$, practically no loss is present, since the fundamental mode is well above cutoff and no other modes are guided. But when λ becomes smaller than λ_c, radiation loss occurs. The experimental setup has been arranged as follows: two spectral response measurements at the fiber output are performed, the one over a straight fiber and the other over a bent fiber of the same type and length.[5] The ratio between the two sets of measured values is used to determine λ_c. The accuracy of this method is questionable, since microbending may cause second order mode radiation even when the fiber is kept straight. *

An interesting technique, recently proposed, [6] is based on the coherence properties of the electomagnetic field at the fiber output end. This method is able to appreciate a power ratio as small as 0.05 between the second order mode and the fundamental mode. The experimental setup is shown in Fig. 12.3. A 2 m -long fiber is excited by an arc lamp, filtered by a monochromator, with spectral resolution which can be as fine as 15 Å, corresponding to a coherence time greater than about 0.1 ps.

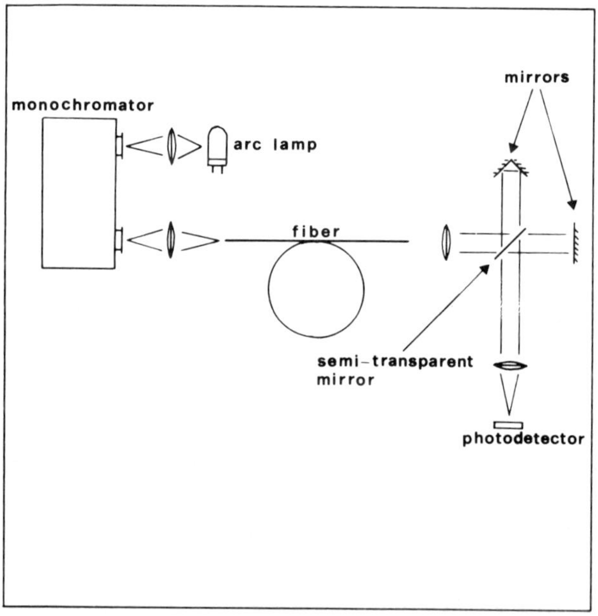

Fig. 12.3 - Experimental setup for the measurement of λ_c with the interferometric technique [from (6)].

*See notes pg. 498

Considering rectangular coordinates (x,y) over the fiber end face, the electric field component, polarized as y for a reference axis $y = 0$, is given by

$$\mathscr{E}_y(x,0,t) = \sum_n a_n f_n(x) \exp\left\{i\left[\omega t + \phi_n(t)\right]\right\}$$

where $f_n(x)$ is the spatial configuration of the nth guided mode, a_n is its amplitude along the y axis, $\phi_n(t)$ its phase, and ω the mean circular frequency of the optical source. Overlapping two electric field configurations, $\mathscr{E}_y(x,0,t)$ and $\mathscr{E}_y(-x,0,t+\tau)$, which is provided by the interferometer, we have

$$E_y(x,0,t,\tau) = \sum_n a_n f_n(x) \exp\left\{i\left[\omega t + \phi_n(t)\right]\right\}$$

$$+ \sum_n a_n f_n(-x) \exp\left\{i\left[\omega(t+\tau) + \phi_n(t+\tau)\right]\right\}$$

The near field intensity distribution at the fiber output, under a spatially-incoherent quasi-monochromatic excitation, except for a proportionality factor, can be written as

$$I(x,0,\tau) = \langle E_y(x,0,t,\tau)\, E_y^*(x,0,t,\tau) \rangle$$

where $\langle\,\rangle$ denotes a temporal average over a time interval longer than the one which characterizes the fluctuations of $\phi_n(t)$.

When τ is much greater than the coherence time, the modulus of the complex rate of coherence along the x axis is given by

$$|\mu(x,-x)| = \frac{I_{max}(x) - I_{min}(x)}{I_{max}(x) + I_{min}(x)} \tag{12.2}$$

where I_{max} and I_{min} are the maximum and minimum values of I during time τ, provided that the near field intensity distribution is symmetrical with respect to the y axis, i.e. $I(x) = I(-x)$. Thus, viewing the interference fringes by an image intensifier, it is possible to relate the power ratio between the two modes to the point on the fiber end face, where the fringes begin to become indistinguishable. Some examples of results are given in Fig. 12.4. At λ = 700 nm a single mode is present, since the visibility of fringes is perfect. At λ = 550 nm a second order mode is present; in fact, there is an imperfect fringe visibility. $|\mu(x,-x)|$, given by (12.2), coincides with the fringe visibility, so that the value of x for which $I_{max} = I_{min}$ allows one to determine the ratio $(a_2/a_1)^2$ between the powers of the two modes.[6]

On the right-hand side of Fig. 12.4, examples of the measured values of this ratio *versus* the wavelength, for two fiber samples, are reported. Fiber 1 exhibits a remarkable dispersion in the cutoff region, which has been attributed to axial disuniformity.

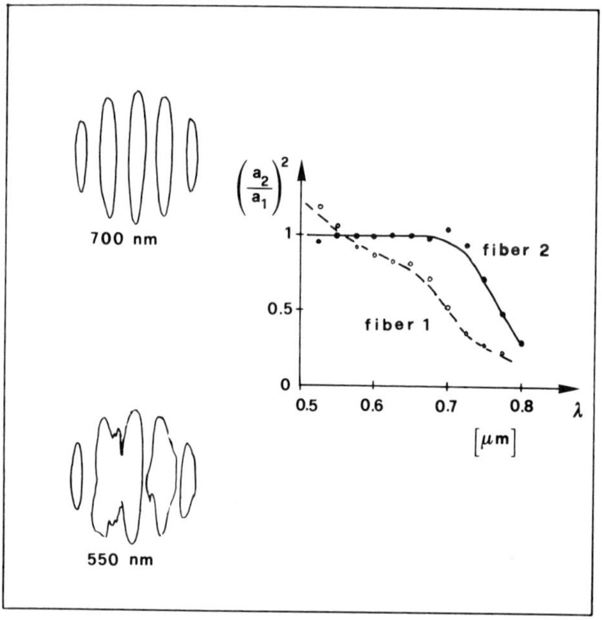

Fig. 12.4 - Interference fringes and ratio between the powers of the fundamental and the second-order modes, at various wavelengths [from (6)].

12.1.3 Fundamental Mode Spot Size Measurement

The spot size of the fundamental mode is conventionally defined as the radial coordinate at which the intensity of the light in the fiber transverse section is reduced by a factor $1/e^2$, with respect to its axial value. Two complementary ways can be followed to obtain experimentally the fundamental mode spot size r_s. The first one is based on a uniform illumination, parallel to the fiber axis, of the fiber input section, and on an output selection, performed by a photodetector having a very small pinhole in front of its active area. The other one consists of a selective illumination of the fiber input section and of the detection of the whole transmitted optical power. In both cases, near field selections are used, which must be characterized by very fine spatial resolution. The output selection, usually based on a magnification of the fiber output end face through a microscope objective, is preferable for practicality.

The measurement of a general spot size is strongly influenced by the wavelength. An experimental plot of spot size *versus* λ exhibits a sharp drop due to the cutoff of the second order mode.[7] Therefore, to measure the fundamental mode spot size, it is necessary to determine accurately if only the fundamental mode propagates, and, in particular, if the wavelength λ is high enough. When the index profile is not of the step type, the near field optical power distribution of the fundamental mode is not gaussian, so that the definition of fundamental mode spot size may become ambiguous.[(1)*] For rigorous step index profiles, it is possible to measure r_s by a far field scanning technique; and, in fact, for gaussian beams the half-width angle is uniquely related to the spot size. Yet for real monomode fibers this technique does not appear very accurate, either.

Another method for determining r_s, which is rather independent of the actual fiber profile, is based on coupling two pieces of the same monomode fiber.[1] The radial offset which characterizes the joint is varied, and is assumed equal to r_s when the coupling efficiency reaches the value $1/e$. It has been shown that even for index profiles which deviate very strongly from the step form, the variation of the coupling efficiency *versus* the offset remains gaussian.

12.1.4 Equivalent Step Index Profile Determination

It is convenient to represent the propagation behavior of a monomode fiber by a few key parameters. For ideal step index profiles, this is effectively possible. Yet, the actual shape of the index profile of monomode fibers may be considerably different from the ideal step index one. Fabrication processes usually introduce a central dip, and spikes at the core edge. For monomode operation, the radial intensity distribution, i.e., the mode profile, remains gaussian in shape when the index profile does not depart too much from an ideal step index profile, as confirmed by experiment. Thus, it is possible to determine an equivalent step index profile, characterized by the same mode profile. This equivalent profile can replace the actual one, and is used to predict the main transmission properties of fiber joints. Good results can be obtained even in the case of pronounced dips.

There are some procedures to obtain the only two parameters of the equivalent step index profile, i.e., the numerical aperture NA_e and the core radius a_e.[8] One of these procedures requires a measurement of the actual index profile $n(r)$,[9] but it appears to be too sophisticated and expensive to use, considering also that the representation

*See notes pg. 499

through an equivalent step index profile must simplify fiber testing. Another procedure is based on the derivation of NA_e and a_e by calculations from experimental curves of the spot size r_s *versus* the wavelength.[10] These two parameters can also be obtained from measurements of the far field pattern,[11] which consists of a pronounced central lobe and small side lobes, which are at least 40 dB down the fundamental one. This technique will be described in 12.1.6.*

12.1.5 Measurements of Core Cladding Concentricity and Core Ellipticity in Monomode Fibers

Microscope inspection of the monomode fiber core is often inadequate to detect small errors in core-cladding concentricity and slight core ellipticity, because of the very small dimensions involved. Yet, the former defect can lead to remarkable joint losses, and the latter defect can cause a pulsewidth spread of several ps/km, due to birefringence.

In (12), a procedure for measuring core-cladding concentricity has been proposed. It is based on a high precision connector, in which one of the two fibers can rotate on its axis. The two cladding outer surfaces are assumed references for the alignment. The relative behavior of the joint loss is recorded *versus* the rotation. Performing a 360° rotation, provided that the axis is maintained coincident with that of the other fiber, allows one to evaluate core-cladding concentricity errors within $\pm$ 1 μm. Nevertheless, some doubts remain about the very high precision and cost of the chuck which is employed for providing rotaton.

A test for the measurement of core ellipticity is described in (13). Because of birefringence, a circularly polarized light launched into the fiber changes polarization along the fiber length periodically, varying gradually from circular to linear and *vice versa*. The light scattered through the fiber outer wall is detected, and the azimuthal pattern obtained depends on the light polarization. Thus, by moving a photodetector in the aximuthal direction and along the fiber axis, an amplitude modulated signal can be obtained. The fiber ellipticity is related to the period of this signal. Differences as small as a few per cent between the maximum and minimum core diameters can be appreciated.

Alternatively, the core ellipticity can be evaluated through a scanning of the far field pattern at the fiber output end by a narrow slit

*See notes pg. 499

(Ronchi frame).[14] In the gaussian approximation the asymptotic divergence is related to the spot size, so by rotating the sample it is possible to measure the ellipticity of the fundamental mode spot size, closely related to that of the core's.

12.1.6 Measurements of NA in Monomode Fibers

The determination of the numerical aperture *NA* in monomode fibers through a simple far field scanning measurement is not suitable, as it was for multimode fibers, because of the diffraction effects from the small aperture represented by the core cross section. A very accurate but expensive technique is based on a measurement of the actual index profile. Alternative techniques allow one to determine the value NA_e, which has been defined for the equivalent step index profile.

Among the latter techniques, we will describe an indirect procedure proposed in,[11] already quoted, which gives the equivalent core radius a_e. The far field pattern of the fundamental mode consists of a central lobe and a series of side lobes. Letting θ_{ah} equal the angle in air which corresponds to one-half of the central maximum, an unambiguous relationship then exists between the normalized frequency V and the parameter

$$A = \frac{2\pi}{\lambda} a_e \sin \theta_{ah} \tag{12.3}$$

The angular position θ_{ax} of the first order side lobe maximum is related to θ_{ah} and V by an unambiguous relationship. This relationship can be written as

$$\frac{\sin \theta_{ax}}{\sin \theta_{ah}} = f(V) \tag{12.4}$$

where f represents a given function.

A simple measurement of θ_{ah} and θ_{ax} gives V through (12.4), which yields the equivalent core radius a_e by means of the relation between A and V, and expression (12.3). Finally, from the known relationship

$$V = \frac{2\pi}{\lambda} a_e NA_e$$

it is possible to determine NA_e. Repeating this test at different values of wavelength λ gives the dependence of NA_e and a_e on λ, provided that a single mode remains in propagation.

12.2 TRANSMISSION MEASUREMENTS OF MONOMODE FIBERS

Transmission measurements of monomode fibers consist of:

i) attenuation measurement

ii) dispersion measurement

Attenuation measurements, from the transmission point of view, as well as for multimode fibers, can be performed by cutback techniques and backscattering techniques. Time dispersion measurements are devoted to testing material dispersion and some effects of waveguide dispersion, as discussed in 1.4, or of birefringence.

Since very small attenuation coefficients can be obtained, especially at longer wavelengths, the accuracy of attenuation measurements must be excellent. Moreover, typical values of the bandwidth per km reached on monomode fibers, in the region of zero material dispersion (see 1.4), can be on the order of hundreds of GHz · km. Therefore, the traditional techniques for measuring baseband responses, i.e., the time domain and frequency domain techniques, can be applied only over very large distances. Other methods, whose time resolution is on the order of 1 ps or less, must be employed sometimes.

12.2.1 Attenuation Measurements in Monomode Fibers

The optical power launched into a monomode fiber is typically about 10 dB less than that launched, from the same source, into a multimode fiber. A cutback measurement, whose accuracy is not influenced appreciably by noise, does not suffer from this reduction, whereas it is very cumbersome for backscattering measurements. However, a cutback technique employed for appreciating losses of roughly 0.5 dB or less must be performed with the aid of a substitution method, using a high precision calibrated optical attenuator. The launching condition is not critical, since neither the mode dependent attenuation nor the mode coupling effects influence the propagation, provided that higher order modes are not effectively present. On the other hand, birefringence and microbending may introduce additional losses, which should be separated from the intrinsic losses. In particular, the result of a spectral loss measurement may exhibit noticeable peaks, due to the cutoff wavelengths of higher order modes as discussed in 12.1.2, leading sometimes to misinterpretations as impurities' absorption peaks.

An interlaboratory comparison of spectral attenuation measurements performed by the cutback technique has been reported in reference (15). In the low-loss windows around $\lambda = 1.0, 1.3$ and $1.5\ \mu m$, loss variations have been obtained within $\pm$ 8%. This discrepancy can be attributed not only to the possible presence of higher order modes, though weakly excited, but also to birefringence and microbending. In reference (16), a better agreement has been obtained during a similar comparison performed just at the wavelength $\lambda = 1.3\ \mu m$. The standard deviation obtained has been evaluated as about 0.03 dB/km, showing that the cutoff wavelength of the second order mode can play a dramatic role in attenuation measurements on monomode fibers when the wavelength is varied.

As for backscattering measurements on monomode fibers, they were first proposed in reference (17), where it was shown, surprisingly, that the fraction of scattered power which is guided backward by a monomode fiber is nearly the same as for multimode fibers with an equal index difference. The launching techniques which are traditionally used for multimode fibers (see 9.3.1), to avoid reflection effects on the fiber input end, must be employed here with some caution. In particular, the use of a polarized incident light does not appear to be suitable, at least in its usual form, since the light maintains its polarization state during the propagation along a monomode fiber. In reference (18), a polarization scrambler has been subsequently inserted in the first meters of the test fiber. The use of an index matching cell, or an acousto-optical light deflector, is probably more efficient.

To enhance the sensitivity of the measurement ,different detection schemes have been suggested. In reference (19), heterodyne detection has been arranged, performing the optical mixing of the backscattered power with only a fraction of the power emitted by the optical source, which has not been launched into the fiber input end. In this way, a test over a 30 km-long installed monomode fiber has been performed. In reference (20), an avalanche photodiode has been used as a very sensitive photon detector by biasing it above the breakdown voltage. Breakdown is initiated by a single carrier entering the space charge region. A Ge-APD detector, cooled by liquid nitrogen, has been employed, leading to the limit behavior of the photon counter. This technique allows experimental investigations characterized by maximum one-way losses of about 17 dB.

The possibility of employing very high launched powers, for example by means of Q-switched Nd:YAG laser, [21] should be considered with some care. The nonlinear generation of other optical frequencies by the Raman effect may occur at optical power levels which are not extremely high. In reference (22) such levels have been demonstrated to be about 1 W.

A backscattering investigation of monomode fiber attenuation and fault location must be accurately interpreted. Some uncertainties about the evolution of the optical power distribution characterizing the backscattered light, discussed in 9.3.3, which were present for multimode fibers are not present here, owing to the propagation of a single mode. On the other hand, in the evaluation of joint losses, a local presence of more than one mode must be considered possible. For example, in reference (22), two measurements on an installed link, 24 km long, have been reported: at $\lambda = 1.06\ \mu m$ and $\lambda = 1.32\ \mu m$. Different behaviors of the splice losses, and even of the overall attenuation, have been pointed out, showing that only at the longer wavelength is the fiber rigorously monomode.

From the ratio between the backscattered power and the input pulse energy, it is possible to obtain even the scattering attenuation coefficient γ_s,[(23)] which expresses, particularly for monomode fibers, the lower limit for the achievable attenuation. Although γ_s may be determined by other indirect methods, e.g., those described in 9.1.3 which can be extended also to monomode fibers, the above technique appears to be more accurate by far. The use of a *Q*-switched Nd:YAG laser is suggested to give a sufficiently high input power level. The value of γ_s thus obtained represents the mean of the scattering attenuation coefficient in the core and cladding, weighted by the optical power distribution of the fundamental mode. This is exactly what must be measured.

12.2.2 Time Dispersion Measurements in Monomode Fibers

Traditional time dispersion measurement techniques can also be applied to monomode fibers, provided that the ranges of values evaluated by the instruments are suitable for these very-low-dispersion fibers. For example, a measurement in the time domain must be able to appreciate a pulse broadening as small as a few hundred picoseconds, or, equivalently, a measurement in the frequency domain requires spectral investigations up to 10-20 GHz. To avoid the very high cost of a procedure so sophisticated, it is sometimes convenient to develop particular techniques which operate on different principles. These alternative methods are essentially based on the finite coherence time of a partially coherent optical source.

In reference (24), an interferometic method has been proposed, which is founded on a comparison between autocorrelation and cross correlation measurements. A laser diode is used as the optical source. In Fig. 12.5 the experimental arrangement is schematically shown. The upper setup gives the measurement of the autocorrelation of the field at the fiber input, the lower setup the measurement of the cross correlation between the input and output fields. The fiber baseband response can be obtained as the ratio between the Fourier transforms of the cross correlation and of the autocorrelation. On fiber samples whose length is on the order of 1 m, effects of material dispersion as low as 70 ps/nm · km can be revealed by this technique.

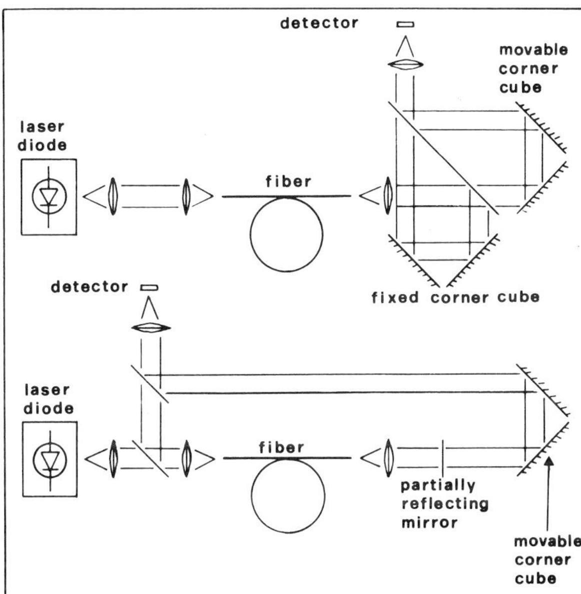

Fig. 12.5 - Experimental setup for measuring the baseband response of a monomode fiber through a comparison between autocorrelation and crosscorrelation of coherent light [from (25)].

A similar technique is founded on a Mach-Zehnder interferometer, [25] whose branches are the test fiber and a reference fiber. [26] In this way it is possible to resolve 0.3 ps as group delay difference over a 4 m-long test fiber. Using a launching fiber in which a quasi-continuous wavelength spectrum of optical power is excited by the Raman effect, under a pump of a Nd:YAG laser, and a monochromator, such a technique allows an accurate determination of the minimum material dispersion wavelength. Experimental tests on practical monomode fibers have shown a minimum dispersion on the order of 25 ps/nm·km.

This technique has been improved to avoid the requirement of a high power laser as optical source. In reference (27), the possibility of using a tungsten halogen lamp has been proved. A lock-in amplifier, followed by a x-y recorder, instead of a sampling oscilloscope, is used for detecting the cross correlation signal. This is obtained by a PIN photodiode, as the voltage

$$V(\tau) = \mathscr{R}_v \left[I_1 + I_2 + 2\sqrt{I_1 I_2}\, S_{12}(\tau) \cos\phi \right]$$

where τ is the relative delay of the two beams, I_1 and I_2 are their intensities, $S_{12}(\tau)$ is the mutual coherence function between them, ϕ is a phase difference, and $\mathscr{R}_v$ is the overall responsivity of the photodetector.

Since the interferometer is not perfectly stable, ϕ varies randomly between 0 and 2π, so that the envelope of the signal gives $S_{12}(\tau)$. In principle this meausrement could be performed by a dc voltmeter, but, owing to the small value of the useful signal, a light chopper and a lock-in amplifier (see Appendix 2) are preferable. On the x-y recorder, axis x represents the time τ and is graduated in ps. Axis y reports $V(\tau)$. Finally, τ can be varied by inserting a suitable optical delay line along one of the two interferometer branches.*

REFERENCES

1. Streckert, J., and Brinkmeyer, E. "Characteristic Parameters of Monomode Fibers," *Appl. Opt.*, 21, pp. 1910-1915 (1982).
2. Sabine, P. V. H., Donaghy, F. A., and Irving, D. H. "Fibre Refractive-Index Profile by Modified Near Field Scanning," *Electon. Lett.*, 16, pp. 882-883 (1980).
3. Irving, D.H. "Near Field Scanning Technique for Profiling Single Mode Fibres," *Electon. Lett.*, 19, pp. 190-191 (1983).
4. Murakami, Y., Kawana, A., and Isuchiya, H. "Cutoff Wavelength Measurement for Single Mode Optical Fibers," *Appl. Opt.*, 18, pp. 1101-1105 (1979).
5. Katsuyama, Y., Tokuda, M., Uchida, N., Nakahara, M. "New Method for Measuring V-Value of a Single Mode Optical Fibre," *Electon. Lett.*, 12, pp. 669-671 (1976).
6. Grosso, G., Spano, P., De Marchis, G. "Una nuova tecnica per

*See notes pg. 499

la misura della lunghezza d'onda di taglio in fibre singolo modo," 4.1 Ruin. Naz. Electon. Appl, Flircnzc, pp. 161-163 (1982).

7. Millar, C. A. "Direct Method of Determining Equivalent Step Index Profiles for Monomode Fibres," *Electron. Lett.*, 17, pp. 458-460 (1981).
8. Fox, M. "Calculation of Equivalent Step Index Parameters for Single Mode Fibers," Symposium on Opt. Fiber Measurements, Boulder, pp. 93-96, (1982).
9. Stewart, W. J. "Simplified Parameter-Based Analysis of Single Mode Optical Guides," *Electron. Lett.*, 16, pp. 380-382 (1980).
10. Millar, C. A. "Application of Automated Equivalent Step Index Profiling to a 31.6 Km Monomode Fiber System," *Opt. Fiber Comm.*, Pheonix, pp. 62-65, (1982).
11. Gambling, W. A., Payne, D. N. , and Matsumura, H. "Propagation Studies on Single Mode Phosphosilicate Fibres," 2nd E.C.O.C., Paris, pp. 95-100, (1976).
12. Ridgway, D. N., and Freeman, L. J. "A Simple Technique for High Accuracy Core-Cladding Concentricity Measurement of Single Mode Fibers," Symposium of Opt. Fiber Measurements, Boulder, pp. 139-142, (1982).
13. Heickhoff, W., and Krumpholtz, O. "Determination of the Ellipticity of Monomode Glass Fibres from Measurements of Scattered Light Intensity," *Electron. Lett.*, 12, pp. 405-406 (1976).
14. Pocholle, J. P., and Auge, J. "New Simple Method for Measuring the Mode Spot Size in Monomode Fibres," *Electron. Lett.*, 19, pp. 191-193, (1983).
15. Heitmann, W., Day, C. R., Zwick, U., and Mengel, F. "Broadband Spectral Attenuation Measurements on Optical Fibres: An Interlaboratory Comparison by Members of COST 208," *Opt. & Quant. Electron.*, 13, pp. 47-54 (1981).
16. Gardner, W. B. "Single Mode Fiber Loss Round Robin," Symposium of Opt. Fiber Measurements, Boulder, pp. 85-87, (1982).
17. Brinkmeyer, E. "Backscattering in Single Mode Fibers," *Electron. Lett.*, 16, pp. 329-330 (1980).

18. Aoyama, K., Nakazawa, K., and Itoh, T. "Optical Time Domain Reflectometry in a Single Mode Fiber," *IEEE J. Quant. Electron.,*QE-17, pp. 862-868 (1981).
19. Healey, P., and Malyon, D. J. "OTDR in Single Mode Fiber at 1.5 μm Using Heterodyne Detection," *Electron. Lett.,*18, pp. 862-863 (1982).
20. Healey, P."OTDR in Monomode Fibres at 1.5 μm Using Semiconductor Laser," *Electron. Lett.,*17, pp. 62-64 (1981).
21. Philen, D. L."Optical Time Domain Reflectometry on Single Mode Fibers Using a Q-switched Nd:YAG Laser," Symposium on Opt. Fiber Measurements, Boulder, pp. 97-100, (1980).
22. Philen, D. L., White, I. A., Kuhl, J. F., and Mettler, S. C. "Single Mode Fiber OTDR: Experiment and Theory," *IEEE Trans Mic. Th. & Tech.*, MIT-30, pp. 1487-1496 (1982).
23. Gold, M. P., and Hartog, A. H. "Measurement of Backscattered Factor in Single Mode Fibres," *Electron. Lett.*, 17, pp. 965-966 (1981).
24. Bomberger, W. D. and Burke, J. J. "Interferometric Measurement of Dispersion of a Single Mode Optical Fibre," *Electron. Lett.*, 17, pp. 495-496 (1981).
25. Baird, K. M. and Hanes, G. R. "Interferometers," *Applied Optics and Optical Engineering* (Vol. 4), R. Kingslake, editor, New York: Academic Press (1967).
26. Cohen, L. G. and Stone, J. "Interferometric Measurements of Minimum Dispersion in Short Lengths of Single Mode Fibres," *Electron. Lett.*, 18, pp. 564-566 (1982).
27. Stone, J. and Cohen, L. G. "Minimum Dispersion Spectra of Single Mode Fibres Measured with Subpicosecond Resolution by White Light Cross Correlation," *Electron. Lett.*, 18, pp. 716-718 (1982).

Appendix 1

Optical Instruments Frequently Used in Optical Fiber Measurements

Among the optical instruments employed for optical fiber characterization, the following play a very important role and require a brief description:

i) monochromators

ii) light modulators

iii) microscope objectives

iv) optical directional couplers

Monochromators are used for optical source characterization, attenuation, and time dispersion measurements. Light modulators allow one to launch a high power, modulated light beam into the fiber. The light beam is obtained from a collimated beam laser, which cannot be directly modulated. Microscope objectives are the fundamental optical components in launching, receiving, and scanning arrangements. Optical directional couplers are used for monitoring the intensity of an optical power signal, and for separating input and output signals when they are present at the same fiber end, e.g., in backscattering attenuation measurements.

1.a Monochromators

A light monochromator is a dispersing instrument, able to bring sequentially each wavelength of the light to the same output position, so that a narrow wavelength band is isolated from a wide source spectrum. A similar device is the spectrograph, which disperses the incident light into a spatial array, allowing the whole spectrum to be recorded simultaneously. The monochromator must isolate a single wavelength at a time. Monochromators are commonly used when light absorption or reflection are the characteristics to be measured. Sequential recording, however, requires a good time stability of the optical source.[(1)] The element on which both monochromators and spectrographs are based is called the light disperser. The light disperser may consist of a prism or a diffraction grating.

The material used for prisms should not only be transparent in the spectral region to be investigated, but also highly dispersing. For instance, although quartz is transparent, it would be useless in the red and near-infrared region because of its very low dispersion. In the infrared region suitable materials are: calcium fluoride, sodium chloride, potassium bromide, and cesium bromide. In the visible region, liquid prisms may be used. They are usually very highly dispersive, but require high temperature stability to prevent convection currents.

Diffraction gratings are optical surfaces with many parallel and equally spaced straight grooves. Gratings, instead of prisms, are normally preferred because of their higher dispersion characteristics. A typical material from which gratings are made is aluminum. The surface of a grating may be planar or curved. Plane gratings are preferred for monochromators, since they need no focus adjustments when the light spectrum is scanned. Planar gratings are also cheaper and are usually able to cover the square or round aperture of typical optical systems.(2)

The angle β at which the light of a given wavelength λ will appear, when the light beam to be analyzed is incident at an angle α, as shown in Fig. A1.1, is given by the grating equation

$$m\lambda = d\,(\sin\alpha \pm \sin\beta) \tag{A1.1}$$

where d is the groove spacing, and m represents the order of diffraction considered. For $m = 0$, $\beta = \alpha$, and the grating acts as a simple mirror. In this case all the wavelengths are superimposed. For $m \geq 1$, the value of β is a function of the wavelength, and the various wavelengths are diffracted at different angles.

The quality of a grating is expressed through its resolving power, which is defined as its ability to separate adjacent spectrum lines. It is given by $\lambda/\Delta\lambda$, where $\lambda + \Delta\lambda$ is the wavelength of a line which can just be distinguished from the line of wavelength λ. For a grating with N grooves at a spacing d,(2) the resolving power is

$$\frac{\lambda}{\Delta\lambda} = mN = \frac{Nd\,(\sin\alpha \pm \sin\beta)}{\lambda}$$

A high resolving power is achieved by using large gratings and angles. Yet, the resolving power is strongly dependent on the optical quality of the grating surface. The tolerance of the groove spacing is about one hundredth of the wavelength of the investigated spectrum.

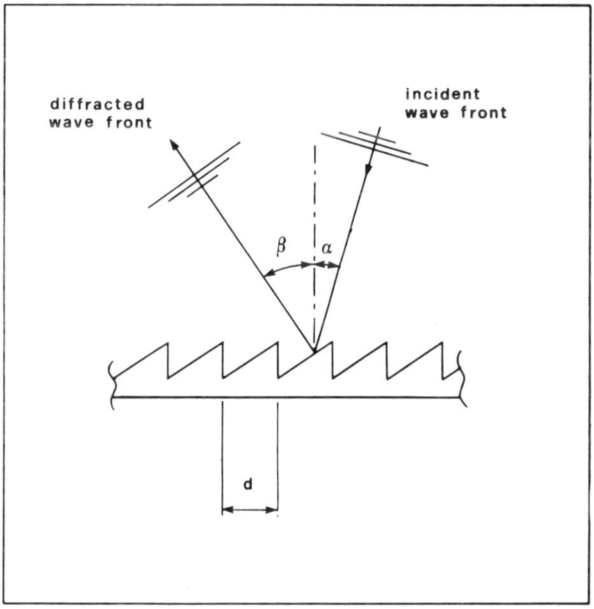

Fig. A1.1 - Diffraction grating.

Examples of prism and grating monochromator arrangements are given in Fig. A1.2 and Fig. A1.3. The selection of different wavelengths is performed by rotating the prism or the diffraction grating. The wavelength resolution depends on the band pass of the monochromator, which is imposed by the optics employed and by the width of the entrance and output slits. Commercially available monochromators usually have interchangeable or continuously variable slits. The width of the slits must be chosen as a trade-off between two opposite requirements: a high wavelength resolution or a high level of the transferred optical power. The latter requirement is relaxed when a high-radiance optical source is used. A suitable wavelength resolution for optical fiber attenuation measurements is about 5 nm.

When higher order harmonics are present filtering may be required. In this case an absorption filter is usually added, before or after the monochromator, to select a wavelength interval for which m in equation (A1.1) assumes a single value. For attenuation measurements in the first window, harmonics of the visible spectrum may be effectively present. They would lead to small errors, since they are more attenuated by the fiber than near-infrared wavelengths. When

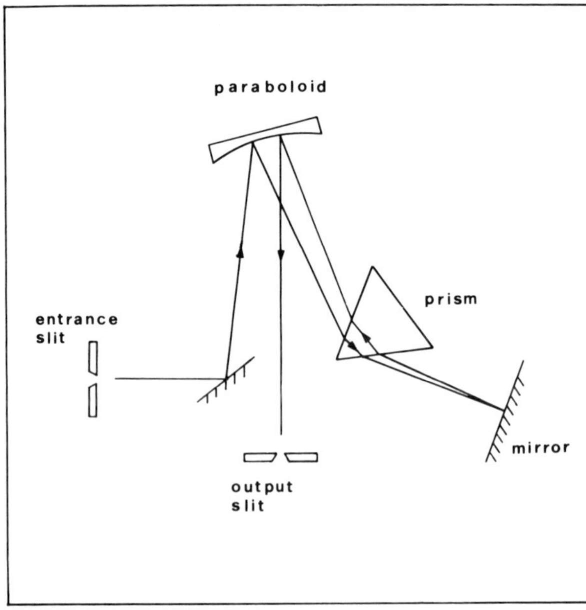

Fig. A1.2 - Prism monochromator.

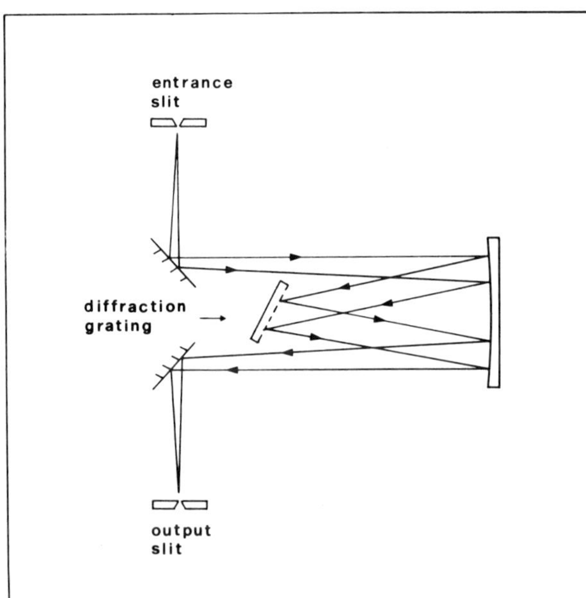

Fig. A1.3 - Grating monochromator.

the second window is investigated, a filter is always necessary to stop the first window harmonics.

A suitable source for monochromator illumination in spectral attenuation measurements is a high radiance tungsten halogen lamp, which exhibits an approximately flat spectrum of emission in the whole spectral region of interest. Higher levels of optical power are obtained from arc lamps, but they are characterized by sharp spectral peaks. A 250 W tungsten lamp is usually enough for spectral attenuation measurements when low loss fibers are investigated over distances of several kilometers.

An alternative method for spectral investigation, especially for the characterization of optical sources, is based on Fourier Transform Spectroscopy. Instead of selecting a small fraction of the optical power obtained from the source by a monochromator, the entire amount of optical power available is processed by a scanning interferometer, as shown in Fig. A1.4. For each wavelength a characteristic frequency dependent on the velocity of the movable mirror is generated by the photodetector. In fact, the intensity of a given spectral component is sinusoidally varied because of the mirror movement, and the amplitude of the generated signal is proportional to the intensity of this spectral component.

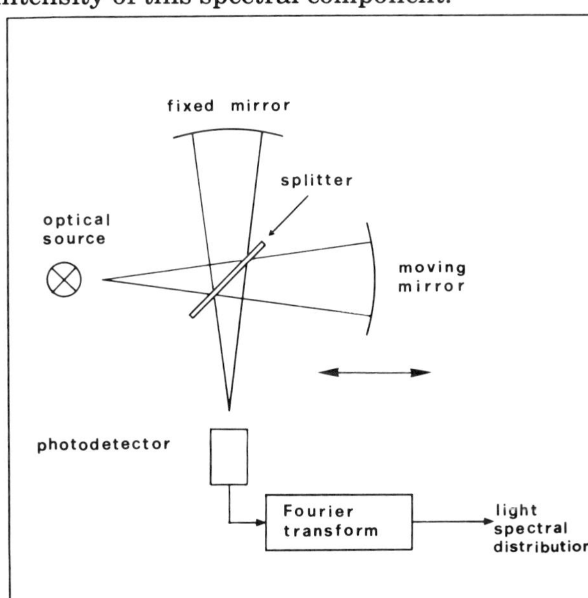

Fig. A1.4 - Scanning interferometer for spectral investigations.

A scanning interferometer acts as a wavelength-time converter. Fourier transforming the output of the photodetector makes it possible to obtain the original spectral distribution of the light power. Yet, there are a multiplicity of bands which can be originated by a scanning interferometer. It is necessary, therefore, to employ complementary apparata for separating spectral regions narrow enough to avoid overlapping of spectral components with different order numbers.[3] However, for measurements on laser diodes, the spectrum to be covered is so narrow that such apparata are not necessary.

Different types of interferometers can be used for spectral investigations. Among them, the one which exhibits the finest spectral resolution is the Fabry-Perot interferometer. It is based on two partially reflecting surfaces, separated by a suitable, varying distance. The spectral resolution of the interferometer $\Delta \nu_\lambda$ can be on the order of several MHz, therefore it appears to be particularly suited for resolving the spectral distributions of different modes in the emission spectrum of a laser diode.

1.b Light Modulators

Direct light modulation is very easy when a LED or a laser diode is used, but it is not possible with other light sources, like lamps or collimated beam lasers. In the range of modulation frequencies from hundreds of MHz to several GHz, it is sometimes convenient to employ an external modulation for LEDs and laser diodes because of their intrinsic modulation bandwidth limitations. To perform an external modulation, mechanical systems based on shutters or moving mirrors are efficient over a few kHz, because of their inertia.

Modulation of the light at higher frequencies can be obtained through the optical interactions with electric, magnetic, or acoustic waves, and an exhaustive list of suitable materials may be found in reference (4).

The most interesting devices for fiber optics applications are those based on the linear electro-optic effect, or Pockel's effect, and on the magneto-optic effect (Faraday's rotation). Acousto-optic devices are instead very promising for optical switching applications in communication systems.[5] External light modulators can be classified as:

i) bulk modulators, in which the incident light propagates freely

ii) waveguide modulators, in which the light beam is guided by a film structure

The modulation is achieved through a variation of the refractive index of a crystal, caused by an electric or magnetic field externally applied. This causes an amplitude, phase, frequency, or polarization modulation of the incident coherent light. Preferably, amplitude modulation is obtained in electro- and magneto-optic devices by interference of two cross polarized and phase modulated waves. It may also be obtained by grating deflection or semiconductor electroabsorption.[6]

Among the most studied electro-optic materials are $LiNbO_3$ and $LiTaO_3$, which are employed in the spectral region from 0.4 to 4 μm. Modulation bandwidths of up to 5 GHz have been achieved using a $LiTaO_3$ traveling wave modulator. Bulk modulators are best suited for optical fiber measurements, whereas waveguide modulators are especially good for integrated optics applications. Bulk modulators consume much more power than waveguide modulators; nevertheless, they can always be used for high power laser sources.

A simple scheme of a bulk electro-optic amplitude modulator is shown in Fig. A1.5. The amplitude modulation is obtained through interference of the two components E_x and E_y of the incident coherent wave with the modulating signal $V(t)$, which induces birefringence in the crystal. Letting $\Delta\phi(t)$ be the phase delay caused by the crystal when the external electric signal is applied, and if $\Delta\phi(t) \ll \pi/2$, the output optical power can be written as

$$P(t) = P_o \sin^2\left[\frac{\pi}{4} + \frac{\Delta\phi(t)}{2}\right] \cong \frac{1}{2} P_o \left[1 + \Delta\phi(t)\right]$$

where P_o is a constant, related to the input optical power.

For sinusoidal modulation with circular modulation frequency ω_o, the phase delay is

$$\Delta\phi(t) = m_a \sin\omega_o t$$

where m_a is the depth of modulation. It represents the main figure of merit of the device, and results in:[7]

$$m_a = 2\pi \frac{L}{\lambda} (n_x - n_y)$$

n_y and n_x being the refractive indices along x and y respectively, λ the wavelength of the light, and L the axial length of the crystal.

An example of a magneto-optic modulator is shown in Fig. A1.6. A linearly polarized wave which enters the crystal maintains its state

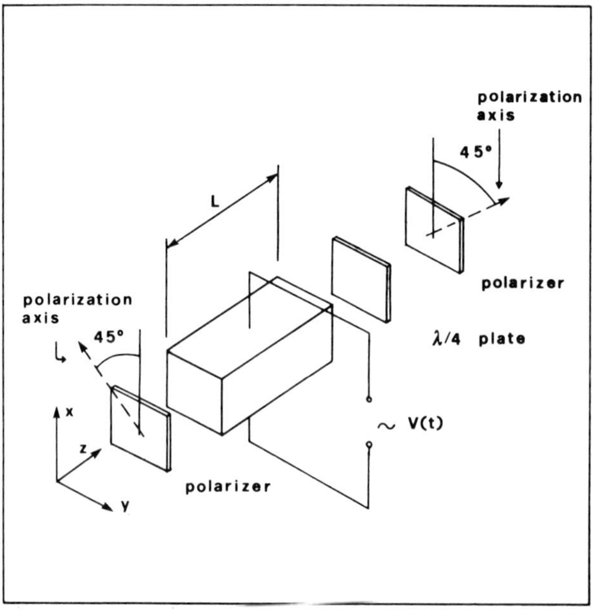

Fig. A1.5 - Electro-optic modulator.

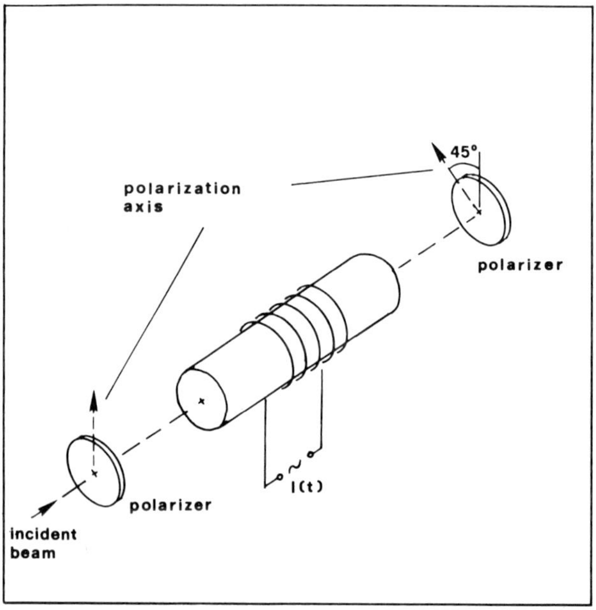

Fig. A1.6 - Magneto-optic modulator.

of polarization, but the direction of polarization rotates because of the applied signal. The use of a pair of polarizers is then necessary for achieving an amplitude modulation. Note that in electro-optic modulators the linear polarization is turned into an elliptical one.

1.c Microscope Objectives

First, a microscope objective is characterized by the length of the microscope tube for which it is fabricated. This length (usually 160 or 170 mm) gives the distance between the reference shoulder of the objective and the plane on which the image is formed (observation plane). Next, a particular objective has the following three main characteristic parameters:

i) magnification M

ii) numerical aperture NA_m

iii) focal length f

A plane object, set at a distance f from the objective input lens, produces an image on the observation plane whose linear dimensions are magnified M times.

There are several kinds of microscope objectives. However, the following general property characterizes all of them: the greater the magnification M, the larger the NA_m, and the shorter the focal length f. In Table A1.1, some typical values of M, NA_m, and f are given for commercially available objectives. These data are only indicative, and the characteristics of particular objectives should be checked according to manufacturer's specifications. An exhaustive treatise on microscope objectives is reported in (8).

Table A1.1 - Typical values for commercially available microscope objectives.

Magnification M	NA_m	Focal Length f[mm]
4 X	0.10	30
10 X	0.25	15
20 X	0.40	10
40 X	0.65	4
60 X	0.85	3

1.d Optical Directional Couplers

We will describe two optical directional couplers which can be used, in particular, as beam splitters for backscattering attenuation measurements. In the following, we assume that two waves traveling in opposite directions are present in the fiber, and at the fiber end these two waves must be separated. This problem becomes harder with increasing difference between the transmitted and received signal levels.*

The first example of a directional coupler uses a *Y* assembly obtained by fusing two optical fiber cores so that the core of the bidirectional fiber can overlap both these cores, as shown in Fig. A1.7. This device is difficult to fabricate, and needs accurate polishing of the input-output face, connected to the bidirectional fiber, to avoid any reflection which could blind the photodetector of the receiver. The use of an index matching oil is recommended. A directivity isolation from transmitter to receiver of about 40 dB can be achieved. It is, however, rather poor for backscattering tests.

Among the techniques which assure a higher directivity isolation, the one based on the use of the polarization difference, already mentioned in 9.3.1, appears to be the most attractive. Two orthogonal polarizers are set at the ports connected to the transmitter and

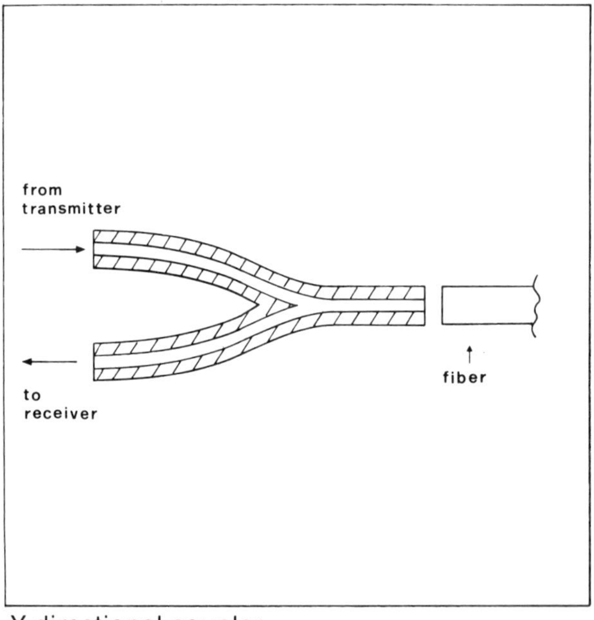

Fig. A1.7 - Y directional coupler.

*See notes pg. 498

receiver. The wave going out of the bidirectional fiber is assumed to be randomly polarized. This is certainly a reasonable assumption for multimode fibers. The device, usually obtained from a calcite crystal, is schematically shown in Fig. A1.8. A directivity isolation of up to 55 dB can be reached. The only drawback is in the insertion loss, due to polarization selection. Considering both the ends of the bidirectional fiber and the assumption that the light of the optical source is randomly polarized, we have an overall penalty of 6 dB with respect to unidirectional links.

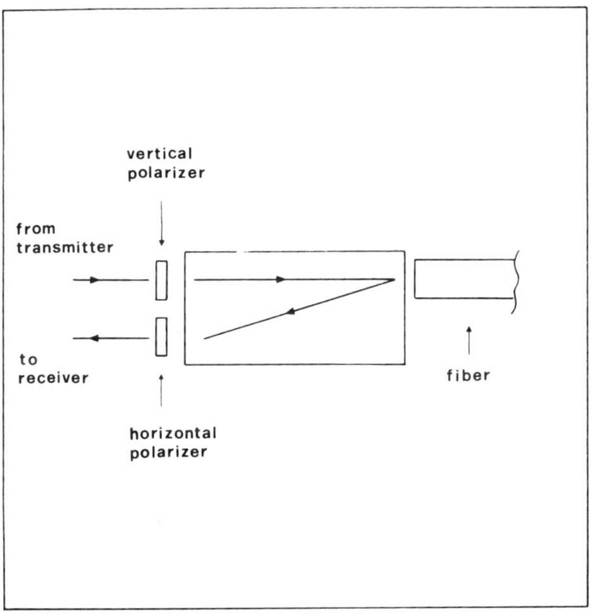

Fig. A1.8 - Directional coupler based on polarization difference.

REFERENCES

1. Meltzer, R.J. "Spectrographs and Monochromators," from *Applied Optics and Optical Engineering,* R. Kingslake, editor, Vol. V, New York: Academic Press (1969).
2. Richardson, D. "Diffraction Gratings," from *Applied Optics and Optical Engineering,* R. Kingslake, editor, Vol. V, New York: Academic Press (1969).
3. Baird, K.M., and Hanes, G.R. "Interferometers," from *Applied Optics and Optical Engineering,* R. Kingslake, editor, Vol. IV, New York: Academic Press (1969).

4. Kaminow, I.P., and Turner, E.H. "Electrooptic Light Modulators," *Appl. Opt.,* 5, pp. 1612-1628 (1966).
5. Graf, P. "Modulators and Deflectors," *Alta Freq.,* 41, pp. 726-743 (1972).
6. Campbell, J.C., De Winter, J.C., Pollack, M.A., Nahory, R.E. "Buried Heterojunction Electroabsorption Modulator," *Appl. Phys. Lett.,* 32, pp. 471-473 (1978).
7. Armenise, M.N., and De Sario, M. "Dispositivi elettroottici e magnetoottici," from *Optic Integrate e Microottica,* Ed. Scient. Telettra, Bologna, Italy: Patron(1982).
8. Benford, J.R. "Microscope Objectives," from *Applied Optics and Optical Engineering,* R. Kingslake, editor, Vol. III, New York: Academic Press (1969).

APPENDIX 2

Electric Instruments Frequently Used in Optical Fiber Measurements

We will describe here two electric instruments which have interesting applications in optical fiber measurements:

i) the lock-in amplifier

ii) the boxcar integrator

The former instrument is employed for source and detector characterization, for fiber attenuation measurements, and for index profile and geometrical measurements. Sometimes it is used with a mechanical light-chopper. The latter instrument is very important in backscattering attenuation measurements. Both these instruments are used exclusively to evaluate very small signals in the presence of a strong noise, and are based on the enhancement of the signal-to-noise ratio through averaging.

2.a Lock-in Amplifier and Light-Chopper

A lock-in amplifier is an ac voltmeter which exploits a reference signal to obtain a very narrow noise equivalent bandwidth around the modulation frequency of the useful signal. Its basic electric scheme is reported in Fig. A2.1. Considering both the test and reference signals as sinusoidal waveforms, we can write

$$v(t) = V\cos\omega_o t$$

$$v_o(t) = V_o\cos(\omega_o t + \phi)$$

where ω_o is the modulation circular frequency of the two signals, V and V_o are their amplitudes, and ϕ is a phase difference between them.

After the multiplier,

$$u(t) = \left(\frac{V_o}{2}\cos\phi\right) V + \frac{V_o}{2}\cos(2\omega_o t + \phi)\, V$$

After the low-pass filter, we are left with

$$V_{out} = \left(\frac{V_o}{2}\cos\phi\right) V$$

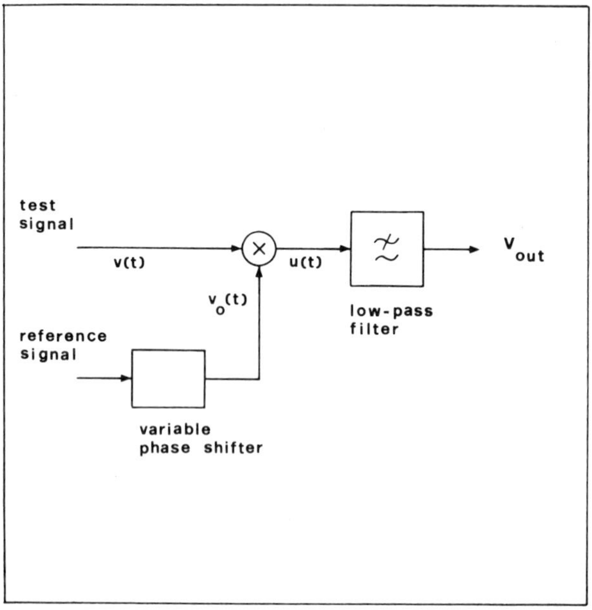

Fig. A2.1 - Basic scheme of a lock-in amplifier.

By monitoring V_{out} with a dc voltmeter it is possible to adjust the phase difference between the two signals, adding a phase shift $\Delta\phi$, so that V_{out} is at a maximum. The relationship between ϕ and $\Delta\phi$ is

$$\phi + \Delta\phi = 0$$

In this way it is possible to determine the phase of the test signal with respect to a given reference phase. Moreover, one can measure V very accurately through the equation

$$V_{out} = \frac{V_o}{2} V$$

in which $V_o/2$ is a known constant, since the reference signal is amplified up to a fixed level, using a suitable automatic gain control.[1]

The time constant T_c of the low-pass filter imposes the noise equivalent bandwidth which characterizes the measurement as

$$B_N = \frac{1}{\pi T_c}$$

This noise equivalent bandwidth is placed around the modulation frequency $f_o = \omega_o/2\pi$. For example, using $T_c = 0.3$ s, we have $B_N \cong 1$ Hz, so that the total optical power which corresponds to the noise of a photodetector, followed by the lock-in amplifier, coincides with its NEP. Time constants as large as 30 s can be used. The only drawback is in the long time required for a measurement. All the considerations reported above for sinusoidal waveforms can be repeated for other types of periodic waveforms. Only the proportionality constant in (A2.1) changes correspondingly. Commercially available instruments typically have a range of modulation frequencies which extends from a few Hz to several hundreds of kHz.

When an ac optical signal is not directly available, a mechanical light-chopper can be employed. A typical maximum modulation frequency which can be obtained in this way is on the order of several kHz. A light-chopper consists of a rotating disk out of which a series of blades have been cut. A light beam is chopped by these blades. The greater the number of blades, the higher the modulation frequency which can be obtained, thus fixing the angular velocity. The waveform is nearly rectangular when the light beam has a diameter much smaller than the separation between two adjacent blades. When it is not, a trapezoidal waveform occurs.

By the use of a low-noise preamplifier before the lock-in amplifier, ac voltages as small as 10 n V can be detected. In fact, roughly 10 n V is the noise level introduced by this instrument. Another advantage of using a lock-in amplifier for optical measurements is in the possibility of performing such measurements with the ambient light, since the useful signal can be revealed by frequency coincident with the reference signal. A lock-in amplifier, when employed for the measurement of phase differences, exhibits a phase resolution of approximately one degree.

2.b Boxcar Integrator

The boxcar integrator is an electronic signal averager which can be applied to periodic waveforms so they may be displayed on a monitor or a plotter, reducing noise. These waveforms can vary with time, provided that such a variation is slow with respect to the averaging time. A series of signal samples are taken from as many subsequent waveforms at the same waveform point as possible. The signal to noise ratio SNR, for a white noise background, can be enhanced by a factor N, coincident with the number of repeated waveforms during the averaging time. Gate widths for taking samples from 100 ps to several ms are now readily available. When more than one waveform point has to be tested, and in particular when the whole waveform must be examined, it is possible to make a scanning , either manually or automatically.

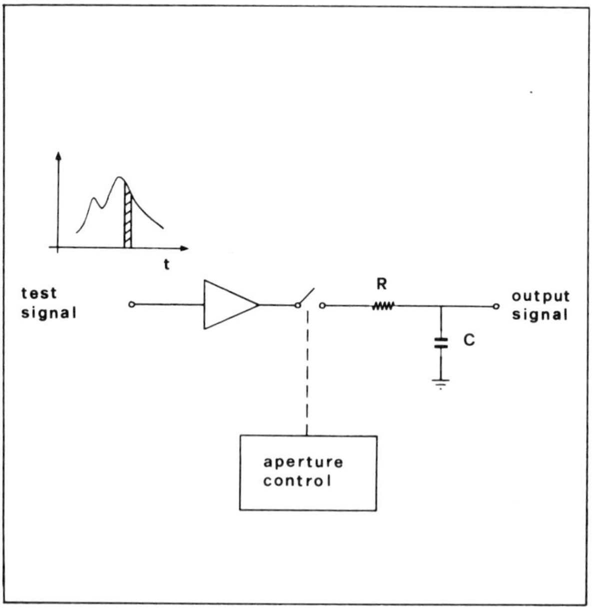

Fig. A2.2 - Basic scheme of a boxcar integrator.

The basic electric scheme of a boxcar integrator is shown in Fig. A2.2. Averaging is made by a simple RC circuit, whose low-pass time constant is T_{RC}. Letting T_o be the waveform time period and T_w the gate width, we can consider the following time interval as an observation time

$$\Delta\tau = \frac{T_o}{T_w} T_{RC}$$

The white noise superimposed onto the useful signal is similarly sampled, but its effect at the output of the RC circuit is reduced. The SNR improvement can be calculated as $\Delta\tau/T_o$. Since $N = \Delta\tau/T_o$ represents the number of waveforms repeated during an observation time, one obtains a SNR enhancement by a factor of N. Yet, the time required for a measurement of this type is not the whole $\Delta\tau$, but only the T_{RC}. Therefore, many waveform points can be tested during $\Delta\tau$; in fact, the whole waveform could be examined.

In this sense, for a fixed T_{RC}, the finer the time resolution for testing a given waveform, which is expressed through the ratio T_w/T_o, the higher the SNR improvement. For example, the typical repetition rate of a back-scattering measurement is 1 kHz, therefore it is $T_o = 1$

ms. To have $N = 10^3$ samples, we must let $\Delta\tau = 1$ s. Assuming $T_w = 0.1$ μs, which corresponds to a spatial resolution of about 10 m, we are left with $T_{RC} = 0.1$ ms. Note that here the whole backscattering trace does not have a duration T_o (which is imposed by considerations of laser heating). Its typical duration is 10 μs, which corresponds to a 1 km fiber length. A SNR improvement by a factor $N = 10^3$ reduces the uncertainty due to white noise by a factor of $\sqrt{N} \cong 32$. Yet, when the uncertainty is mainly due to fluctuations of the launched power, as discussed in 9.3.5, this reduction is not possible.

REFERENCES

1. De Vito, S. "High Sensitivity Optical Power Measurement by Lock-in Amplifier," *XXIII Congr. Int. per l'Elettronica,* Rome, Italy, pp. 395-403 (1976).

APPENDIX 3

Fiber End Preparation

For measurement purposes and for fiber jointing procedures in transmission systems, fiber end faces should be perpendicular to the axis and optically flat. We will describe a fiber cleaving technique and an inspection procedure for testing the quality of the fiber cut.

3.a Fiber Cleaving Technique

Conventional grinding techniques make it possible to obtain perfect results, but they are expensive and are usually time consuming. After a rough cut, the face can be polished using abrasives. Commercial equipment gives good quality and repeatability, and in particular allows a fair control of the end angle, whose tolerance with respect to the perpendicularity should be less than 1 degree. Some simple kits for manual grinding are commercially available, but their use is restricted to end preparation for low cost connectors, and are not likely to give reliable results, especially for the end angle.

A more appealing way to obtain high quality end preparations, like those required for measurements, is using a cleaving tool able to give a good cut in a few seconds. It is well known that glass fibers break with perpendicular and flat faces when they are suitably stressed and scored by a diamond scorer.[1] A fracture originates at the scoring point, and a plane crack propagates perpendicularly to the fiber axis with increasing velocity up to a critical value at which the kinetic energy is high enough to generate surface cracks. This critical velocity is reached when the plane crack has traveled the distance

$$d_c = \frac{K}{\sigma_B^2}$$

from the scoring point, where σ_B is the tensile strength at which the fiber breaks, and K is a constant which can be experimentally evaluated. For fused silica $K \cong 56$ $kg^2 \cdot mm^{-3}$. Silica fibers are characterized by a σ_B on the order of 25 kg $\cdot$ mm^{-2} when a diamond scorer is employed.

To achieve a mirrorlike surface over the whole section, d_c must be greater than the fiber diameter. Otherwise, beyond the distance d_c from the scoring point, mist and hackle zones appear, as shown in Fig. A3.1. Fibers whose diameter is greater than d_c can be cut with perpendicular and flat ends if they are bent over a curved surface. The fiber must be put on this curved surface on the side opposite the scoring point. In this way the region beyond d_c is somehow relieved of the stress required at the scoring point to initiate the break, so that the critical velocity is not reached.[1]

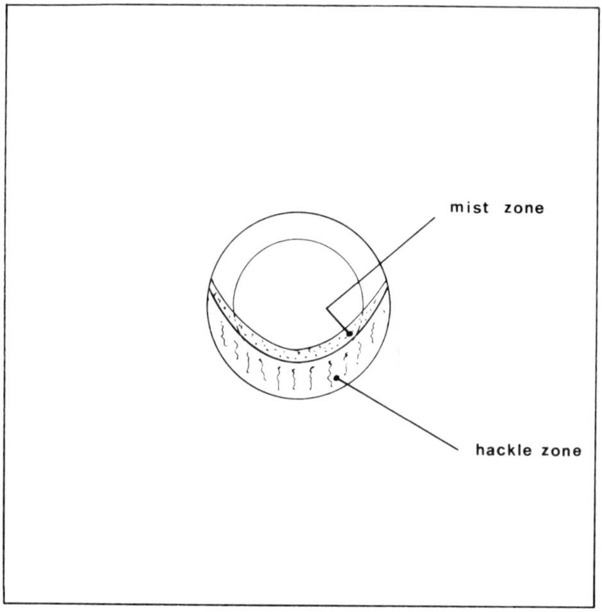

Fig. A3.1 - End face of a fractured fiber.

Letting D be the fiber diameter and E the elastic modulus of the glass, the curvature radius of the surface must be greater than

$$\rho_o = \frac{D}{\sigma_B} E$$

For fused silica $E = 7.2 \cdot 10^{-3}$ kg $\cdot$ mm^{-2}, therefore $\rho_o = 3 \cdot 10^2\ D$, approximately. It must be pointed out that the resulting cut is completely flat if $D\,\sigma^2 \leq K$, σ being the local tensile strength across the fiber section. When the curvature radius is exactly ρ_o, $D \leq 3.5 d_c$. The preceding results hold for bare fibers without any protective coating. If the coating is very thin, fair cuts might be obtained, but without repeatability. It is advisable to strip the coating off in this case.

A typical arrangement for fiber cleaving is shown in A3.2. Provided that a calibration is made for each type of fiber, adjusting the curvature radius ρ and the tensile strength through a force F, end angles with tolerances smaller than 1 degree are obtained with good repeatability with some commercial equipment. A practical realization of a cleaving machine is described in reference (2).*

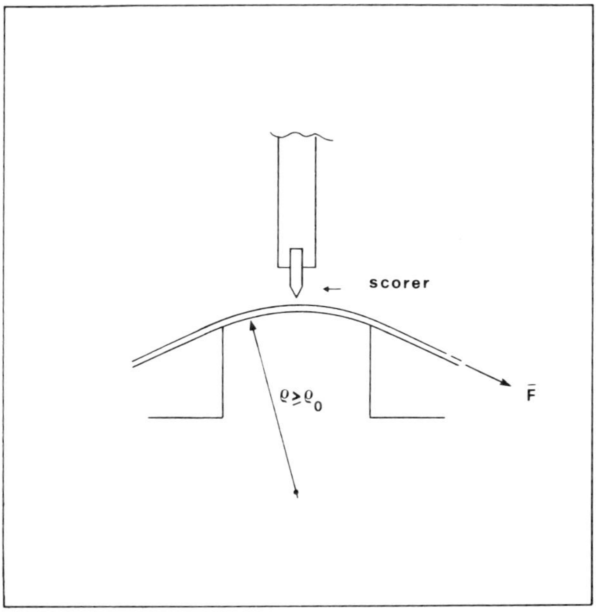

Fig. A3.2 - Fiber cleaving tool arrangement (from [1]).

3.b Fiber Break Angle Inspection Procedures

After cleaving, an accurate end face inspection is needed. Using a microscope, the far end of the fiber should be illuminated to distinguish core and cladding regions. The latter region should appear darker. Even if surface imperfections are present on the cladding region, they do not influence the light propagation, and the cut can be considered good. Accurate measurements of fiber geometry may

*See notes pg. 499

be obtained by coupling a video camera to the microscope, as described in Chapter 8.

In reference (3), a method based on the diffraction of laser light is proposed. It allows one to measure break angles to an accuracy of a fraction of a degree. The light of a He-Ne laser is focused so that the break surface is partially illuminated. The fiber rotates on its axis. For a perfect perpendicular end face, the emergent pattern is unaffected by rotation. When even small deviation from perpendicularity is present, the emergent pattern depends univocally on the break angle, which can be determined from the measurement of this pattern.

REFERENCES

1. Gloge, D., *et al.* "Optical Fiber End Preparation for Low Loss Splices," *Bell Syst. Tech. J.,* 52, pp. 1579-1588 (1973).
2. Tamburello, M. "Un Apparecchio Semplice, Robusto e di Basso Costo per il Taglio di Fibre Ottiche," *Alta Freq.,* 47, pp. 223-224 (1978).
3. Millar, C.A. "A Measurement Technique for Optical Fiber Break Angles," *Opt. & Quant. Electron.,* 13, pp. 125-131 (1981).

Footnotes

1) page 73, between line 26 and 27

Even for SiO_2-based monomode fibers, provided that they are characterized by a small core radius, it is possible to shift the zero of intramodal dispersion from about 1.3 μm to the third window of spectral attenuation. This is due to waveguide dispersion, which may have opposite sign with respect to material dispersion, leading to a sort of compensation. See for example: B.J. Ainslie et al. "Monomode fibre with ultra-low loss and minimum dispersion at 1.55 μm," *Electron. Lett.*, 18, pp. 842-844 (1982).

2) page 133, between line 11 and 12

In recent years this optimization procedure has lost part of its initial interest, because the main limitation to the extention of repeater spacing in long-haul optical transmission systems based on multimode fibers is usually attenuation. This is an indirect consequence of the diffusion of monomode fibers for high capacity transmission systems. Nevertheless the optimization procedure here described could be of some interest for the reader who would like to check his comprehension of the effects of microbending on attenuation and time dispersion, mathematically treated in the previous sections.

3) page 214, between line 16 and 17

Another advantage of coherent systems over traditional systems is in the possibility of a direct optical amplification of the signal along the link. For a description of the features of hypothetical links, characterized by some amplification sections within a regeneration span, see for example: T. Kimura, Y. Yamamoto, "Review. Progress of coherent optical fibre communication systems," *Opt. & Quant. Electron.*, 15, pp. 1-39 (1983).

4) page 215, between line 10 and 11

Recently a new laser configuration, particularly suitable for single mode operation, with very narrow spectral width, has been proposed. It consists of two laser cavities, coupled through a narrow gap. This type of device is called cleaved coupled-cavity laser, or briefly C^3 laser. See for example: C.H. Henry, R.F. Kazarinov, "Stabilization of single frequency operation of coupled-cavity laser," *IEEE J. Quant. Electron.*, QE-20, pp. 733-744 (1984).

5) page 351, between line 10 and 11

Many OTDR instruments are now commercially available. They can be employed also in the field. All the problems which have been solved for the instrument production and line-up turn out to be transparent for the users. Nevertheless some information about these problems and their solutions is of interest for a correct interpretation of results.

6) page 386, between line 23 and 24

Only few complete instruments for time dispersion measurements are presently commercially available. This is probably due to the following reasons. With any solution, we deal with very expensive instruments, containing blocks (like sampling oscilloscopes, or vector voltmeters, or spectrum analyzers) usually employed also for other measurements. Moreover, repeater spacing of multimode links, for which such instruments are suited, appears essentially attenuation-limited. A different situation can be foreseen for the monomode environment, which is characterized, however, by a different instrumentation.

7) page 416, in line 32, after $P_{em} = 10^{-16}$ W.

Recently, some very-low-noise vector voltmeters became commercially available. Their features are similar to those of a spectrum analyzer, making the above comparison still correct only for second-class instruments.

8) page 457, between line 16 and 17

In order to avoid these limitations, single-polarization monomode fibers have been recently proposed. In a fiber of this type, by an intentional increase of birefringence, it is possible to make leaky one of the two polarization states of the fundamental HE_{11} mode. See for example: A.W. Snyder, F. Rühl, "Single-mode single-polarization fibers made of birefringent material," *J. Opt. Soc. Am.*, 73, pp. 1165-1174 (1983).

9) page 462, between line 10 and 11

CCITT has recently recommended a measurement of this type over a fiber piece 2 m long. Such a short length assures that microbending has no appreciable effects, especially if the fiber is not in a cable. See for example: "Revised version of Recommendation G.652. Characteristics of a single-mode optical fibre cable," CCITT document COM. XV/TD 46-E (May 1984).

10) page 465, in line 9, after ambiguous.[1]

More precisely, a gaussian optical power distribution holds rigorously only for a parabolic unbounded index distribution. Nevertheless, a gaussian approximation is acceptable for step-index and other monotonic bounded profiles. On the contrary, it could be not very precise when the index profile exhibits an axial dip.

11) page 466, between line 8 and 9

Recently, the use of the equivalent step-index profile has been discouraged, since its experimental evaluation appears not very reliable. See for example: M. Fox, "Calculation of equivalent step-index parameters for single-mode optical fibres," *Opt. & Quant. Electron.*, 15, pp. 451-455 (1983).

12) page 472, between line 22 and 23

In order to measure the total dispersion of a monomode fiber, it is possible to use some techniques described here for intra-modal dispersion evaluation (10.1.5). In particular, automatized benches are now commercially available, based on the four-lasers technique, described in reference 23 of Chapter 10.

13) page 484, between line 8 and 9

Another use of such components is in optical LAN's (Local Area Networks), particularly for the bus topology. The diffusion of optical LAN's is expected to be considerable in the next few years, especially in the area of broadband computer information transfer.

14) page 495, between line 11 and 12

Many fiber cleaving arrangements are now commercially available. However, the direct experience of the operator is of maximum importance in order to improve the quality of the fiber cut, with any of these arrangements.